Engineering Design with SOLIDWORKS 2018
and Video Instruction

A Step-by-Step Project Based Approach
Utilizing 3D Solid Modeling

David C. Planchard
CSWP & SOLIDWORKS Accredited Educator

SDC
Publications

SDC Publications
P.O. Box 1334
Mission, KS 66222
913-262-2664
www.SDCpublications.com
Publisher: Stephen Schroff

Examination Copies
Books received as examination copies are for review purposes only and may not be made available for student use. Resale of examination copies is prohibited.

Electronic Files
Any electronic files associated with this book are licensed to the original user only. These files may not be transferred to any other party.

Trademarks
SOLIDWORKS®, eDrawings®, SOLIDWORKS Simulation®, SOLIDWORKS Flow Simulation, and SOLIDWORKS Sustainability are registered trademarks of Dassault Systèmes SOLIDWORKS Corporation in the United States and other countries; certain images of the models in this publication courtesy of Dassault Systèmes SOLIDWORKS Corporation.

Microsoft Windows®, Microsoft Office® and its family of products are registered trademarks of the Microsoft Corporation. Other software applications and parts described in this book are trademarks or registered trademarks of their respective owners.

The publisher and the author make no representations or warranties with respect to the accuracy or completeness of the contents of this work and specifically disclaim all warranties, including without limitation warranties of fitness for a particular purpose. No warranty may be created or extended by sales or promotional materials. Dimensions of parts are modified for illustration purposes. Every effort is made to provide an accurate text. The author and the manufacturers shall not be held liable for any parts, components, assemblies or drawings developed or designed with this book or any responsibility for inaccuracies that appear in the book. Web and company information was valid at the time of this printing.

The Y14 ASME Engineering Drawing and Related Documentation Publications utilized in this text are as follows: ASME Y14.1 1995, ASME Y14.2M-1992 (R1998), ASME Y14.3M-1994 (R1999), ASME Y14.41-2003, ASME Y14.5-1982, ASME Y14.5-1999, and ASME B4.2. Note: By permission of The American Society of Mechanical Engineers, Codes and Standards, New York, NY, USA. All rights reserved.

Additional information references the American Welding Society, AWS 2.4:1997 Standard Symbols for Welding, Braising, and Non-Destructive Examinations, Miami, Florida, USA.

ISBN-13: 978-1-63057-147-4
ISBN-10: 1-63057-147-4

Printed and bound in the United States of America.

Introduction

Engineering Design with SOLIDWORKS® 2018 and video instruction is written to assist students, designers, engineers and professionals. The book provides a solid foundation in SOLIDWORKS by utilizing projects with step-by-step instructions for the beginner to intermediate SOLIDWORKS user featuring machined, plastic and sheet metal components.

Desired outcomes and usage competencies are listed for each project. The book is divided into five sections with 11 projects.

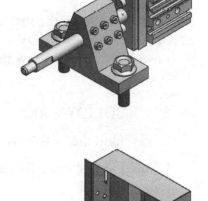

Project 1 - Project 6: Explore the SOLIDWORKS User Interface and CommandManager, Document and System properties, simple and complex parts and assemblies, proper design intent, design tables, configurations, multi-sheet, multi-view drawings, BOMs, and Revision tables using basic and advanced features. Additional techniques include the edit and reuse of features, parts, and assemblies through symmetry, patterns, configurations, SOLIDWORKS 3D ContentCentral and the SOLIDWORKS Toolbox.

Project 7: Understand Top-Down assembly modeling and Sheet Metal parts. Develop components In-Context with InPlace Mates, along with the ability to import parts using the Top-Down assembly method. Convert a solid part into a Sheet Metal part and insert and apply various Sheet Metal features.

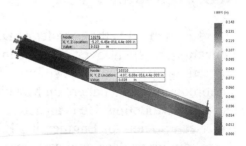

Project 8 - Project 9: Recognize SOLIDWORKS Simulation and Intelligent Modeling techniques. Understand a general overview of SOLIDWORKS Simulation and the type of questions that are on the SOLIDWORKS Simulation Associate - Finite Element Analysis (CSWSA-FEA) exam. Apply design intent and intelligent modeling techniques in a sketch, feature, part, plane, assembly and drawing.

Project 10: Comprehend the Differences of Additive vs. Subtractive Manufacturing. Understand 3D printer terminology along with a working knowledge of preparing, saving, and printing a 3D CAD model on a low cost ($500 - $3,000) printer.

Project 11: Review the Certified Associate - Mechanical Design (CSWA) program. Understand the curriculum and categories of the CSWA exam and the required model knowledge needed to successfully take the exam.

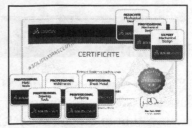

The author developed the industry scenarios by combining his own industry experience with the knowledge of engineers, department managers, vendors and manufacturers. These professionals are directly involved with SOLIDWORKS every day. Their responsibilities go far beyond the creation of just a 3D model.

Redeem the code on the inside cover of the book. Download the ENGDESIGN-W-SOLIDWORKS 2018 folder and videos to your local hard drive. View the provided videos to enhance the user experience.

- Bracket
- Chapter 2 Homework
- Chapter 3 Homework
- Chapter 4 Homework
- Chapter 5 Homework
- Chapter 6 Homework
- Chapter 7 Homework
- Chapter 9 Intelligent Modeling
- Chapter 11 CSWA Models
- CopiedModels
- CSWSA-FEA Models
- Extra Model Templates
- Graph paper
- LOGO
- Vendor Components

- Start a SOLIDWORKS session.

- Understand the SOLIDWORKS Interface.

- Create 2D Sketches, Sketch Planes and use various Sketch tools.

- Create 3D Features and apply Design Intent.

- Create an Assembly.

- Create fundamental Drawings Part 1 & Part 2.

The book is designed to complement the SOLIDWORKS Tutorials contained in SOLIDWORKS 2018.

About the Author

David Planchard is the founder of D&M Education LLC. Before starting D&M Education, he spent over 27 years in industry and academia holding various engineering, marketing, and teaching positions. He holds five U.S. patents. He has published and authored numerous papers on Machine Design, Product Design, Mechanics of Materials, and Solid Modeling. He is an active member of the SOLIDWORKS Users Group and the American Society of Engineering Education (ASEE). David holds a BSME, MSM with the following professional certifications: CCAI, CCNP, CSDA, CSWSA-FEA, CSWP, CSWP-DRWT and SOLIDWORKS Accredited Educator. David is a SOLIDWORKS Solution Partner, an Adjunct Faculty member and the SAE advisor at Worcester Polytechnic Institute in the Mechanical Engineering department. In 2012, David's senior Major Qualifying Project team (senior capstone) won first place in the Mechanical Engineering department at WPI. In 2014, 2015 and 2016, David's senior Major Qualifying Project teams won the Provost award in Mechanical Engineering for design excellence.

David Planchard is the author of the following books:

- **SOLIDWORKS® 2018 Reference Guide with video instruction**, 2017, 2016, 2015, 2014, 2013, 2012, 2011, 2010, and 2009

- **Engineering Design with SOLIDWORKS® 2018 and video instruction**, 2017, 2016, 2015, 2014, 2013, 2012, 2011, 2010, 2009, 2008, 2007, 2006, 2005, 2004, and 2003

- **Engineering Graphics with SOLIDWORKS® 2018 and video instruction**, 2017, 2016, 2015, 2014, 2013, 2012, and 2011

- **SOLIDWORKS® 2018 Quick Start with video instruction**

- **SOLIDWORKS® 2017 in 5 Hours with video instruction**, 2016, 2015, and 2014

- **SOLIDWORKS® 2018 Tutorial with video instruction**, 2017, 2016, 2015, 2014, 2013, 2012, 2011, 2010, 2009, 2008, 2007, 2006, 2005, 2004, and 2003

- **Drawing and Detailing with SOLIDWORKS® 2014**, 2012, 2010, 2009, 2008, 2007, 2006, 2005, 2004, 2003, and 2002

- **Official Certified SOLIDWORKS® Professional (CSWP) Certification Guide with video instruction, Version 4: 2015 - 2017**, Version 3: 2012 - 2014, Version 2: 2012 - 2013, Version 1: 2010 - 2010

- **Official Guide to Certified SOLIDWORKS® Associate Exams: CSWA, CSDA, CSWSA-FEA Version 3: 2015 - 2017**, Version 2: 2012 - 2015, Version 1: 2012 - 2013

- **Assembly Modeling with SOLIDWORKS® 2012**, 2010, 2008, 2006, 2005-2004, 2003, and 2001Plus

- **Applications in Sheet Metal Using Pro/SHEETMETAL & Pro/ENGINEER**

Acknowledgements

Writing this book was a substantial effort that would not have been possible without the help and support of my loving family and of my professional colleagues. I would like to thank Professor John M. Sullivan Jr., Professor Jack Hall and the community of scholars at Worcester Polytechnic Institute who have enhanced my life, my knowledge and helped to shape the approach and content to this text.

The author is greatly indebted to my colleagues from Dassault Systèmes SOLIDWORKS Corporation for their help and continuous support: Avelino Rochino and Mike Puckett.

Thanks also to Professor Richard L. Roberts of Wentworth Institute of Technology, Professor Dennis Hance of Wright State University, Professor Jason Durfess of Eastern Washington University and Professor Aaron Schellenberg of Brigham Young University - Idaho who provided vision and invaluable suggestions.

SOLIDWORKS certification has enhanced my skills and knowledge and that of my students. Thank you to Ian Matthew Jutras (CSWE) who is a technical contributor and the creator of the videos and Stephanie Planchard, technical procedure consultant.

Contact the Author

We realize that keeping software application books current is imperative to our customers. We value the hundreds of professors, students, designers, and engineers that have provided us input to enhance the book. Please contact me directly with any comments, questions or suggestions on this book or any of our other SOLIDWORKS books at dplanchard@msn.com or planchard@wpi.edu.

Note to Instructors

Please contact the publisher **www.SDCpublications.com** for classroom support materials (.ppt presentations, labs and more) and the Instructor's Guide with model solutions and tips that support the usage of this text in a classroom environment.

Trademarks, Disclaimer and Copyrighted Material

SOLIDWORKS®, eDrawings®, SOLIDWORKS Simulation®, SOLIDWORKS Flow Simulation, and SOLIDWORKS Sustainability are a registered trademark of Dassault Systèmes SOLIDWORKS Corporation in the United States and other countries; certain images of the models in this publication courtesy of Dassault Systèmes SOLIDWORKS Corporation.

Microsoft Windows®, Microsoft Office® and its family of products are registered trademarks of the Microsoft Corporation. Other software applications and parts described in this book are trademarks or registered trademarks of their respective owners.

The publisher and the author make no representations or warranties with respect to the accuracy or completeness of the contents of this work and specifically disclaim all warranties, including without limitation warranties of fitness for a particular purpose. No warranty may be created or extended by sales or promotional materials. Dimensions of parts are modified for illustration purposes. Every effort is made to provide an accurate text. The author and the manufacturers shall not be held liable for any parts, components, assemblies or drawings developed or designed with this book or any responsibility for inaccuracies that appear in the book. Web and company information was valid at the time of this printing.

The Y14 ASME Engineering Drawing and Related Documentation Publications utilized in this text are as follows: ASME Y14.1 1995, ASME Y14.2M-1992 (R1998), ASME Y14.3M-1994 (R1999), ASME Y14.41-2003, ASME Y14.5-1982, ASME Y14.5-1999, and ASME B4.2. Note: By permission of The American Society of Mechanical Engineers, Codes and Standards, New York, NY, USA. All rights reserved.

Additional information references the American Welding Society, AWS 2.4:1997 Standard Symbols for Welding, Braising, and Non-Destructive Examinations, Miami, Florida, USA.

References

- SOLIDWORKS Help Topics and What's New, SOLIDWORKS Corporation, 2018.

- Beers & Johnson, Vector Mechanics for Engineers, 6th ed. McGraw Hill, Boston, MA.

- Gradin, Hartley, Fundamentals of the Finite Element Method, Macmillan, NY 1986.

- Jensen & Helsel, Engineering Drawing and Design, Glencoe, 1990.

- Lockhart & Johnson, Engineering Design Communications, Addison Wesley, 1999.

- Olivo C., Payne, Olivo, T, Basic Blueprint Reading and Sketching, Delmar 1988.

- Walker, James, Machining Fundamentals, Goodheart Wilcox, 1999.

- 80/20 Product Manual, 80/20, Inc., Columbia City, IN, 2012.

- SMC Corporation of America, Product Manuals, Indiana, USA, 2012.

- Emerson-EPT Bearing Product Manuals and Gear Product Manuals, Emerson Power Transmission Corporation, Ithaca, NY, 2009.

- Emhart - A Black and Decker Company, On-line catalog, Hartford, CT, 2012.

During the initial SOLIDWORKS installation, you are requested to select either the ISO or ANSI drafting standard. ISO is typically a European drafting standard and uses First Angle Projection. The book is written using the ANSI (US) overall drafting standard and Third Angle Projection for drawings.

Screen shots in the book were made using SOLIDWORKS 2018 SP0 running Windows® 10.

Table of Contents

Instructor's information contains over 45 classroom presentations along with helpful hints, what's new, sample quizzes, avi files of assemblies, projects and all initial and final SOLIDWORKS models.

Redeem the code on the inside cover of the book. Download the ENGDESIGN-W-SOLIDWORKS 2018 folder and videos to your local hard drive. View the provided videos to enhance the user experience.

Videos

Name

SOLIDWORKS_Basics_Videos 3 of 3.zip
SOLIDWORKS_Basics_Videos 2 of 3.zip
SOLIDWORKS_Basics_Videos 1 of 3.zip

Bracket
Chapter 2 Homework
Chapter 3 Homework
Chapter 4 Homework
Chapter 5 Homework
Chapter 6 Homework
Chapter 7 Homework
Chapter 9 Intelligent Modeling
Chapter 11 CSWA Models
CopiedModels
CSWSA-FEA Models
Extra Model Templates
Graph paper
LOGO
Vendor Components

Overview of Projects

Project 1: Overview of SOLIDWORKS and the SOLIDWORKS User Interface

SOLIDWORKS is a design software application used to create 2D and 3D sketches, 3D parts, 3D assemblies and 2D drawings.

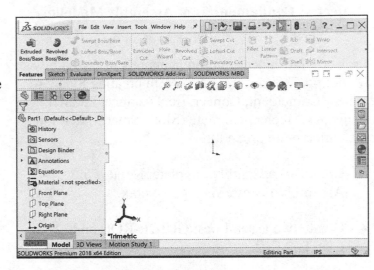

- Project 1 introduces the user to the SOLIDWORKS 2018 Welcome dialog box, User Interface (UI) and the CommandManager: Menu bar toolbar, Menu bar menu, Drop-down menus, Context toolbars, Consolidated drop-down toolbars, System feedback icons, Confirmation Corner, Heads-up View toolbar, Document Properties and more.

- Redeem the code on the inside cover of the book. View the provided videos and models to enhance the user experience.

- Start a new SOLIDWORKS session. Create a new part. Open an existing part and view the created features and sketches using the Rollback bar. Design the part using proper design intent.

Project 2: Fundamentals of Part Modeling

Project 2 begins by creating file folders and sub-folders to manage projects. Apply System Options and Document Properties. Develop a Custom Part Template.

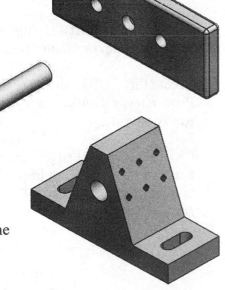

- Create three parts: PLATE, ROD, and GUIDE.

- Utilize the following features: Extruded Boss/Base, Instant3D, Extruded Cut, Fillet, Mirror, Chamfer, Hole Wizard and Linear Pattern. Apply materials and appearance.

- Learn the SOLIDWORKS interface, how to select the correct Sketch plane, fully define the sketch, edit sketches and features and copy and paste features.

Project 3: Fundamentals of Assembly Modeling (Bottom up)

Project 3 introduces the fundamentals of Assembly Modeling (Bottom-up) along with creating Standard mates (Coincident, Concentric, Distance, Parallel, Tangent), Mechanical mate (Slot), SmartMates and the Quick mate procedure.

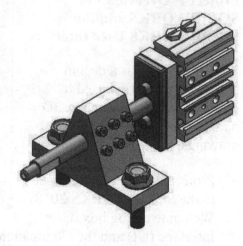

- Create an Assembly template. Review the Assembly FeatureManager syntax.

- Create two assemblies: GUIDE-ROD and CUSTOMER.

- Edit component dimensions and address tolerance and fit.

- Incorporate design changes into an assembly. Obtain additional SOLIDWORKS parts using 3D ContentCentral and the SOLIDWORKS Toolbox.

Project 4: Fundamentals of Drawing

Project 4 covers the development of a customized drawing template.

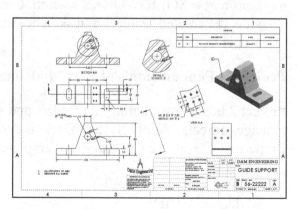

- Learn the two Sheet Format modes: 1.) Edit Sheet Format and 2.) Edit Sheet

- Develop and insert a Company logo from a bitmap or picture file.

- Create the GUIDE drawing with Custom Properties, various drawing views and a Revision table.

- Create the GUIDE-ROD drawing with Custom Properties, an Exploded Isometric view with a Bill of Materials and balloons.

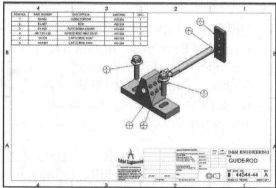

Project 5: Extrude and Revolve Feature

Project 5 focuses on the customer's design requirements. Create four key FLASHLIGHT components: BATTERY, BATTERYPLATE, LENS and BULB. Develop an ANSI - IPS Part Template.

- Create the BATTERY and BATTERYPLATE part with the Extruded Boss/Base feature and the Instant3D tool.

- Create the LENS and BULB with the Revolved Boss/Base feature. Utilize the following features: Extruded Boss/Base, Extruded Cut, Revolved Base, Revolved Cut, Dome, Shell, Fillet and Circular Pattern.

- Utilize the Mold tools to create the cavity plate for the BATTERYPLATE.

☼ Tangent edges are displayed for educational purposes in the book.

Project 6: Swept, Lofted and Additional Features

Project 6 develops four additional components to complete the FLASHLIGHT assembly: O-RING, SWITCH, LENSCAP and HOUSING.

- Utilize the following features: Swept Boss/Base, Lofted Boss/Base, Rib, Linear Pattern, Circular Pattern, Draft and Dome.

- Insert the components and sub-assemblies and create the Mates to finish the final FLASHLIGHT assembly.

Project 7: Top-Down assembly and Sheet Metal parts

Project 7 focuses on the Top-Down assembly modeling approach. Develop a Layout Sketch.

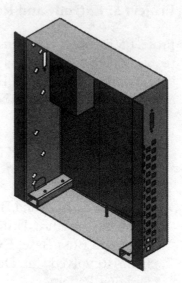

- Create components and modify them In-Context of the assembly.

- Create Sheet metal features. Utilize the following features: Rip, Insert Sheet metal Bends, Base Flange, Edge Flange, Miter Flange, Break Corners, Hem and more.

- Utilize the Die Cut Feature and Louver Form tool.

- Add IGES format part files from the Internet.

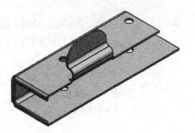

- Replace fasteners in the assembly and redefine mates.

- Utilize equations, Global Variables and a Design Table to create multiple configurations of the BOX assembly.

☼ The book is designed to expose the new SOLIDWORKS user to many tools, techniques and procedures. It may not always use the most direct tool or process.

Project 8: SOLIDWORKS Simulation

Project 8 provides a general overview of SOLIDWORKS Simulation and the type of questions that are on the SOLIDWORKS Simulation Associate - Finite Element Analysis (CSWSA-FEA) exam. On the completion of this project, you will be able to:

Recognize the power of SOLIDWORKS Simulation.

Utilize SOLIDWORKS Simulation to:

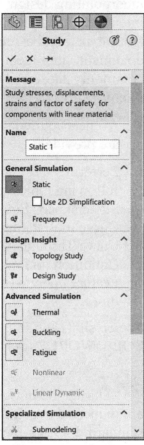

- Define a Static Analysis Study.

- Apply Material to a part model.

- Work with a Solid and Sheet Metal model.

- Define Solid, Shell and Beam elements.

- Define Standard and Advanced Fixtures and External loads.

- Define Local and Global coordinate systems.

- Understand the axial forces, sheer forces, bending moments and factor of safety.

- Define Connector properties such as Contact Sets, No Penetration and Bonded.

- Set and modify plots to display in the Results folder.

- Work with Multi-body parts as different solid bodies.

- Select different solvers as directed to optimize problems.

- Determine if the result is valid.

- Understand the type of problems and questions that are on the CSWSA-FEA exam.

- Ability to use SOLIDWORKS Simulation Help.

Perform short stand-alone step-by-step tutorials to practice and reinforce the subject matter and objectives.

Project 9: Intelligent Modeling Techniques

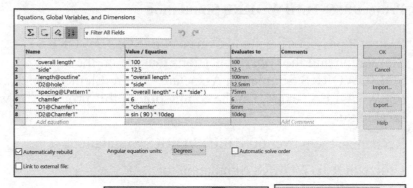

Project 9 introduces some of the available tools in SOLIDWORKS to perform intelligent modeling.

Intelligent modeling is incorporating design intent into the definition of the sketch, feature, part and assembly or drawing document. Intelligent modeling is most commonly addressed through design intent.

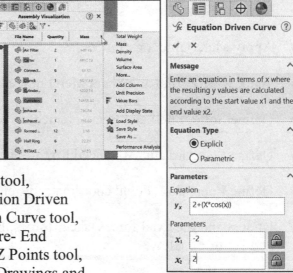

- All needed models for this project are provided.

- Perform short tutorials on the following topics: Fully Defined Sketch tool, SketchXpert, Equations, Explicit Equation Driven Curve tool, Parametric Equation Driven Curve tool, Curve Through XYZ Points tool, Feature- End Condition options, Curve Through XYZ Points tool, FeatureXpert, Symmetry, Assemblies, Drawings and more.

Screen shots and illustrations in the book display the SOLIDWORKS user default setup.

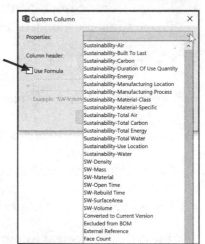

🔅 During the initial SOLIDWORKS installation, you are requested to select either the ISO or ANSI drafting standard. ISO is typically a European overall drafting standard and uses First Angle Projection. The book is written using the ANSI (US) overall drafting standard and Third Angle Projection for drawings.

Project 10: Additive Manufacturing - 3D Printing

Project 10 provides a basic understanding between the differences of Additive vs. Subtractive Manufacturing. Comprehend 3D printer terminology along with a working knowledge of preparing, saving, and printing a 3D CAD model on a low cost ($500 - $3,000) printer.

On the completion of this project, you will be able to:

- Discuss Additive vs Subtractive Manufacturing.

- Determine the differences between a Cartesian printer and a Delta printer.

- Create a STereoLithography (STL) file in SOLIDWORKS.

- 3D print directly from SOLIDWORKS using an Add-In.

- Save an STL file to G-code.

- Discuss printer hardware.

- Select the correct filament type:

- PLA (Polylactic acid), ABS (Acrylonitrile butadiene styrene) or Nylon.

- Prepare the G-code.

- Address model setup, print orientation, extruder temperature, and bed temperature.

- Comprehend the following 3D printer terminology:

- (STereoLithography) file - STL.

- Fused Filament Fabrication - FFF.

- Fused Deposition Model - FDM.

- Digital Light Process - DLP.

- Dissolvable Support System - DDS.

- Raft, Skirt and Brim.

- Support and Touching Buildplate.

- Slicer.

- G-code.

- Address fit tolerance for interlocking parts.

- Define general 3D Printing tips.

Project 11: Introduction to the Certified Associate - Mechanical Design (CSWA) Exam

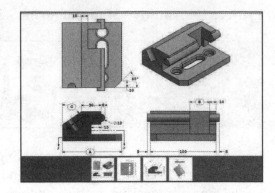

Project 11 provides a basic introduction into the curriculum and categories of the Certified Associate - Mechanical Design (CSWA) exam. Awareness to the exam procedure, process, and required model knowledge needed to take the CSWA exam. The five exam categories are:

- Drafting Competencies.

- Basic Part Creation and Modification.

- Intermediate Part Creation and Modification.

- Advanced Part Creation and Modification.

- Assembly Creation and Modification.

The CSWA certification indicates a foundation in and apprentice knowledge of 3D CAD design and engineering practices and principles. The main requirement for obtaining the CSWA certification is to take and pass the two part on-line proctored exams.

This first exam (Part 1) is 90 minutes; minimum passing score is 80, with 6 questions.

The second exam (Part 2) is 90 minutes; minimum passing score is 80 with 8 questions.

What is SOLIDWORKS®?

SOLIDWORKS® is a mechanical design automation software package used to build parts, assemblies and drawings that takes advantage of the familiar Microsoft® Windows graphical user interface.

SOLIDWORKS is an easy to learn design and analysis tool (SOLIDWORKS Simulation, SOLIDWORKS Motion, SOLIDWORKS Flow Simulation, etc.), which makes it possible for designers to quickly sketch 2D and 3D concepts, create 3D parts and assemblies and detail 2D drawings.

In SOLIDWORKS, you create 2D and 3D sketches, 3D parts, 3D assemblies and 2D drawings. The part, assembly and drawing documents are related. Additional information on SOLIDWORKS and its family of products can be obtained at their URL, www.SOLIDWORKS.com.

PART   ASSEMBLY

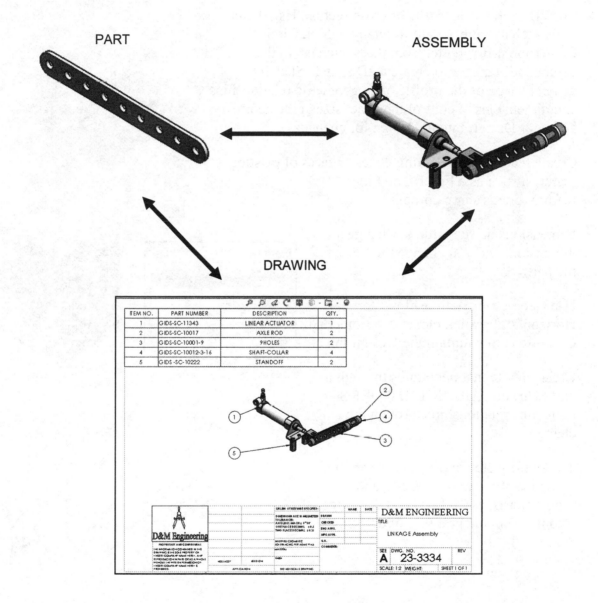

DRAWING

ITEM NO.	PART NUMBER	DESCRIPTION	QTY.
1	GIDS-SC-11343	LINEAR ACTUATOR	1
2	GIDS-SC-10017	AXLE ROD	2
3	GIDS-SC-10001-9	9 HOLES	2
4	GIDS-SC-10012-3-16	SHAFT-COLLAR	4
5	GIDS-SC-10222	STANDOFF	2

D&M ENGINEERING
TITLE:
LINKAGE Assembly
DWG. NO. 23-3334
SCALE: 1:2   SHEET 1 OF 1

Features are the building blocks of parts. Use feature tools such as Extruded Boss/Base, Extruded Cut, Fillet, etc. from the Features tab in the CommandManager to create 3D parts.

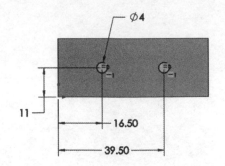

Extruded features begin with a 2D sketch created on a Sketch plane.

The 2D sketch is a profile or cross section. Use sketch tools such as Line, Center Rectangle, Slot, Circle Centerline, Mirror, etc. from the Sketch tab in the CommandManager to create a 2D sketch. Sketch the general shape of the profile. Add geometric relationships and dimensions to control the exact size of the geometry and your Design Intent. Design for change.

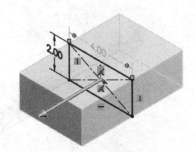

Create features by selecting edges or faces of existing features, such as a Fillet. The Fillet feature rounds sharp corners.

Dimensions drive features. Change a dimension, and you change the size of the part.

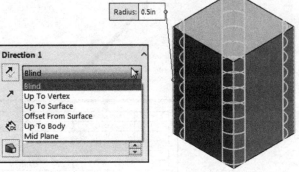

Use Geometric relationships: Vertical, Horizontal, Parallel, etc. and various End Conditions to maintain the Design Intent.

Create a hole that penetrates through a part (Through All). SOLIDWORKS maintains relationships through the change.

The step-by-step approach used in this text allows you to create, edit and modify parts, assemblies and drawings. Change is an integral part of design.

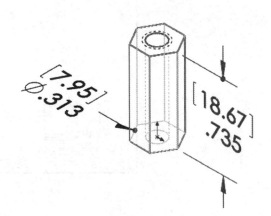

About the Book

You will find a wealth of information in this book. The following conventions are used throughout the text:

- The term document is used to refer to a SOLIDWORKS part, drawing or assembly file.

- The list of items across the top of the SOLIDWORKS interface is the Main menu. Each item in the Main menu has a pull-down menu. When you need to select a series of commands from these menus, the following format is used: Click **Insert**, **Reference Geometry**, **Plane** from the Main bar. The Plane PropertyManager is displayed.

- Screen shots in the book were made using SOLIDWORKS 2018 SP0 running Windows® 10.

- The ANSI overall drafting standard and Third Angle projection is used as the default setting in this text. IPS (inch, pound, second) and MMGS (millimeter, gram, second) unit systems are used.

- Redeem your code on the inside cover of the book. View the provided videos and models to enhance the user experience. All templates, logos and model documents along with additional support materials are available.

The book is designed to expose the new SOLIDWORKS user to many tools, techniques and procedures. It may not always use the most direct tool or process. Learn by doing, not just by reading.

The following command syntax is used throughout the text. Commands that require you to perform an action are displayed in **Bold** text.

Format:	Convention:	Example:
Bold	<ul><li>All commands actions.</li><li>Selected icon button.</li><li>Selected geometry: line, circle.</li><li>Value entries.</li></ul>	<ul><li>Click **Options** ⚙ from the Menu bar toolbar.</li><li>Click the **Extruded Boss/Base** feature.</li><li>Click **Corner Rectangle** ⬜ from the Consolidated Sketch toolbar.</li><li>Select the **centerpoint**.</li><li>Enter **3.0** for Radius.</li></ul>
Capitalized	<ul><li>Filenames.</li><li>First letter in a feature name.</li></ul>	<ul><li>Save the **FLASHLIGHT** assembly.</li><li>Click the **Fillet** feature.</li></ul>

Windows Terminology in SOLIDWORKS

The mouse buttons provide an integral role in executing SOLIDWORKS commands. The mouse buttons execute commands, select geometry, display Shortcut menus and provide information feedback.

A summary of mouse button terminology is displayed below:

Item:	Description:
Click	Press and release the left mouse button.
Double-click	Double press and release the left mouse button.
Click inside	Press the left mouse button. Wait a second, and then press the left mouse button inside the text box. Use this technique to modify Feature names in the FeatureManager design tree.
Drag	Point to an object, press and hold the left mouse button down. Move the mouse pointer to a new location. Release the left mouse button.
Right-click	Press and release the right mouse button. A Shortcut menu is displayed. Use the left mouse button to select a menu command.
Tool Tip	Position the mouse pointer over an Icon (button). The tool name is displayed below the mouse pointer.
Large Tool Tip	Position the mouse pointer over an Icon (button). The tool name and a description of its functionality are displayed below the mouse pointer.
Mouse pointer feedback	Position the mouse pointer over various areas of the sketch, part, assembly or drawing. The cursor provides feedback depending on the geometry.

A mouse with a center wheel provides additional functionality in SOLIDWORKS. Roll the center wheel downward to enlarge the model in the Graphics window. Hold the center wheel down. Drag the mouse in the Graphics window to rotate the model. Review various Windows terminology that describe menus, toolbars and commands that constitute the graphical user interface in SOLIDWORKS.

Visit SOLIDWORKS website: http://www.SOLIDWORKS.com/sw/support/hardware.html to view their supported operating systems and hardware requirements.

This book is designed to expose the new user to numerous tools and procedures. It may not always use the simplest and most direct process.

This book does not cover starting a SOLIDWORKS

Hardware & System Requirements
Research graphics cards hardware, system requirements, and other related topics.

SolidWorks System Requirements
Hardware and system requirements for SolidWorks 3D CAD products.

Graphics Card Drivers
Find graphics card drivers for your system to ensure system performance and stability.

Data Management System Requirements
Hardware and system requirements for SolidWorks Product Data Management (PDM) products.

Anti-Virus
The following Anti-Virus applications have been tested with SolidWorks 3D CAD products.

SolidWorks Composer System Requirements
Hardware and system requirements for SolidWorks Composer and other 3DVIA related products.

Hardware Benchmarks
Applications and references that can help determine hardware performance.

SolidWorks Electrical System Requirements
Hardware and system requirements for SolidWorks Electrical products.

session in detail for the first time. A default SOLIDWORKS installation presents you with several options. For additional information for an Education Edition, visit the following site: http://www.SOLIDWORKS.com/sw/engineering-education-software.htm

The Instructor's information contains over 45 classroom presentations, along with helpful hints, What's new, sample quizzes, avi files of assemblies, projects, and all initial and final SOLIDWORKS model files.

Notes:

Project 1

Overview of SOLIDWORKS® 2018 and the User Interface

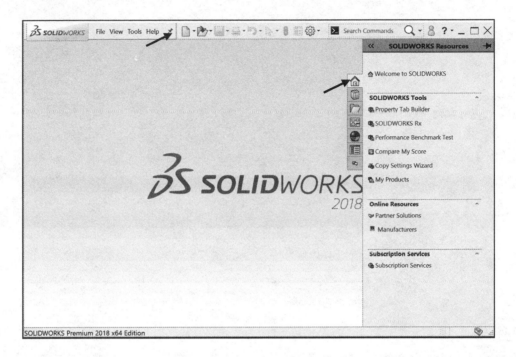

Below are the desired outcomes and usage competencies based on the completion of Project 1.

Desired Outcomes:	Usage Competencies:
• A comprehensive understanding of the SOLIDWORKS® 2018 User Interface (UI) and CommandManager.	• Ability to establish a SOLIDWORKS session. • Aptitude to utilize the following items: *Menu bar toolbar, Menu bar menu, Drop-down menus, Context toolbars, Consolidated drop-down toolbars, System feedback icons, Confirmation Corner, Heads-up View toolbar, Document Properties and more.* • Open a new and existing SOLIDWORKS part. • Knowledge to zoom, rotate and maneuver a three button mouse in the SOLIDWORKS Graphics window.

Notes:

Project 1 - Overview of SOLIDWORKS® 2018 and the User Interface

Project Objective

Provide a comprehensive understanding of the SOLIDWORKS® default User Interface and CommandManager: Menu bar toolbar, Menu bar menu, Drop-down menu, Right-click Pop-up menus, Context toolbars/menus, Fly-out tool button, System feedback icons, Confirmation Corner, Heads-up View toolbar and more.

On the completion of this project, you will be able to:

- Utilize the SOLIDWORKS Welcome dialog box.

- Establish a SOLIDWORKS session.

- Comprehend the SOLIDWORKS User Interface.

- Recognize the default Reference Planes in the FeatureManager.

- Open a new and existing SOLIDWORKS part.

- Utilize SOLIDWORKS Help and SOLIDWORKS Tutorials.

- Zoom, rotate and maneuver a three button mouse in the SOLIDWORKS Graphics window.

What is SOLIDWORKS®?

- SOLIDWORKS® is a mechanical design automation software package used to build parts, assemblies and drawings that takes advantage of the familiar Microsoft® Windows graphical user interface.

- SOLIDWORKS is an easy to learn design and analysis tool (SOLIDWORKS Simulation, SOLIDWORKS Motion, SOLIDWORKS Flow Simulation, Sustainability, etc.), which makes it possible for designers to quickly sketch 2D and 3D concepts, create 3D parts and assemblies and detail 2D drawings.

- Model dimensions in SOLIDWORKS are associative between parts, assemblies and drawings. Reference dimensions are one-way associative from the part to the drawing or from the part to the assembly.

- The book is written for the beginner to intermediate user.

Start a SOLIDWORKS Session

Start a SOLIDWORKS session and familiarize yourself with the SOLIDWORKS User Interface. As you read and perform the tasks in this project, you will obtain a sense of how to use the book and the structure. Actual input commands or required actions in the chapter are displayed in bold.

The book does not cover starting a SOLIDWORKS session in detail for the first time. A default SOLIDWORKS installation presents you with several options. For additional information, visit http://www.SOLIDWORKS.com.

Redeem the code on the inside cover of the book. Download the ENGDESIGN-W-SOLIDWORKS 2018 folder and videos to your local hard drive. View the provided videos to enhance the user experience.

Activity: Start a SOLIDWORKS 2018 Session.

Start a SOLIDWORKS session.

1)   Click **Start** on the Windows Taskbar.

2)   Click **All Programs**.

3)   Click the **SOLIDWORKS 2018** application (or if available, **double-click** the SOLIDWORKS icon on the Desktop). When you open the SOLIDWORKS software, view the Welcome dialog box. The Welcome dialog box provides a convenient means to open documents, view folders, access SOLIDWORKS resources, and stay updated on SOLIDWORKS news. Note: Do not open a document at this time.

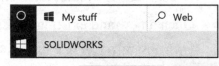

4)   **View** your options. Do not open a document at this time.

If available, double-click the SOLIDWORKS icon on your desktop to start a SOLIDWORKS session.

You can also click **Welcome to SOLIDWORKS** 🏠 (Standard toolbar), **Help** > **Welcome** to SOLIDWORKS, or **Welcome to SOLIDWORKS** on the SOLIDWORKS

Resources 🏠 tab in the Task Pane to open the Welcome dialog box.

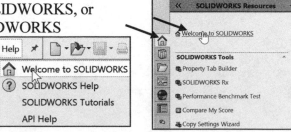

Home Tab

The Home tab lets you open new and existing documents, view recent documents and folders, and access SOLIDWORKS resources.

Sections in the Home tab include *New*, *Recent Documents*, **Recent Folders**, and *Resources*.

Recent Tab

The Recent tab lets you view a longer list of recent documents and folders. Sections in the Recent tab include *Documents* and *Folders*.

The Documents section includes thumbnails of documents that you have opened recently. Click a thumbnail to open the document, or hover over a thumbnail to see the document location and access additional information about the document. When you hover over a thumbnail, the full path and last saved date of the document appears.

Learn Tab

The Learn tab lets you access instructional resources to help you learn more about the SOLIDWORKS software.

Sections in the Learn tab include:

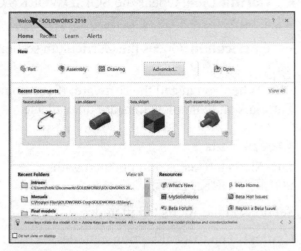

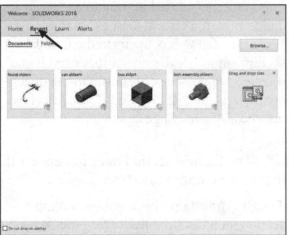

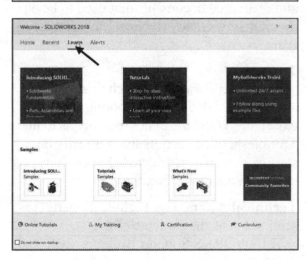

- **Introducing SOLIDWORKS**. Opens the Introducing SOLIDWORKS book.

- **Tutorials**. Opens the step-by-step tutorials in the software.

- **MySolidWorks Training**. Opens the Training section at MySolidWorks.com.

- **Samples**. Opens local folders containing sample models.

- **3DContentCentral**. Opens 3DContentCentral.com.

- **Online Tutorials**. Opens the SOLIDWORKS Tutorials (videos) section at solidworks.com.

- **My Training**. Opens the My Training section at MySolidWorks.com.

- **Certification**. Opens the SOLIDWORKS Certification Program section at solidworks.com.

- **Curriculum**. Opens the Curriculum section at solidworks.com.

When you install the software, if you do not install the Help Files or Example Files, the Tutorials and Samples links are unavailable.

Alerts Tab

The Alerts tab keeps you updated with SOLIDWORKS news.

Sections in the Alerts tab include:

Critical. The Critical section includes important messages that used to appear in a dialog box. If a critical alert exists, the Welcome dialog box opens to the Critical section automatically on startup, even if you selected Do not show at startup in the dialog box. Alerts are displayed until you select Do not show this message again.

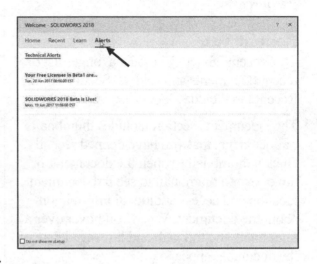

The Critical section does not appear if there are no critical alerts to display.

Troubleshooting. The Troubleshooting section includes troubleshooting messages and recovered documents that used to be on the SOLIDWORKS Recovery tab in the Task Pane.

If the software has a technical problem and an associated troubleshooting message exists, the Welcome dialog box opens to the Troubleshooting section automatically on startup, even if you selected Do not show at startup in the dialog box.

Technical Alerts. The Technical Alerts section opens the contents of the SOLIDWORKS Support Bulletins RSS feed at solidworks.com.

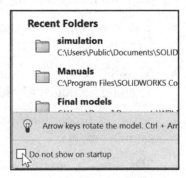

Close the Welcome dialog box.

5)   Click **Close** ✕ from the Welcome dialog box. The SOLIDWORKS Graphics window is
          displayed.

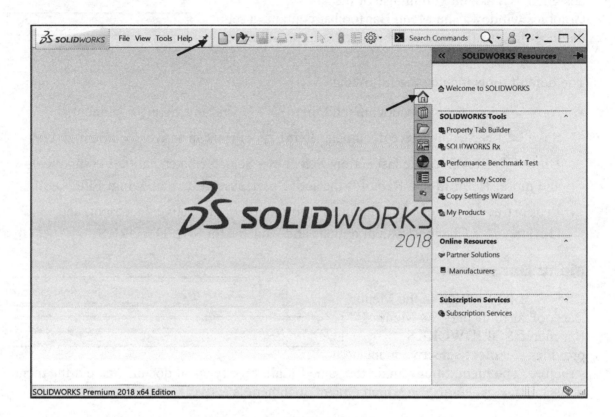

If you do not see this screen, click the SOLIDWORKS Resources ⌂ icon on the right side of
the Graphics window located in the Task Pane.

6)   **Hover** the mouse pointer over the SOLIDWORKS icon as illustrated.

7)   If needed, **Pin** the Menu Bar
          toolbar. View your options.

Menu Bar toolbar

The SOLIDWORKS 2018 (UI) is
designed to make maximum use of the
Graphics window. The Menu Bar toolbar contains a set of
the most frequently used tool buttons from the Standard
toolbar.

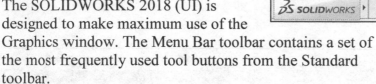

The default tools include the following:

- **New** □ - Creates a new document; **Open** ▷ - Opens an existing document;
  Save 🖫 - Saves an active document; **Print** 🖨 - Prints an active document;
  Undo ↺ - Reverses the last action; **Select** ▷ - Selects Sketch entities, components
  and more; **Rebuild** ❶ - Rebuilds the active part, assembly or drawing; **File**
  Properties 🗏 - Shows the summary information on the active document; and
  Options ⚙ - Changes system options and Add-Ins for SOLIDWORKS.

Menu Bar menu

Click SOLIDWORKS in the Menu
Bar toolbar to display the Menu
Bar menu. SOLIDWORKS
provides a context-sensitive menu

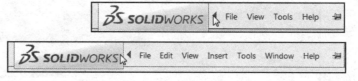

structure. The menu titles remain the same for all three types of documents, but the menu
items change depending on which type of document is active.

Example: The Insert menu includes features in part documents, mates in assembly
documents, and drawing views in drawing documents. The display of the menu is also
dependent on the workflow customization that you have selected. The default menu items
for an active document are *File*, *Edit*, *View*, *Insert*, *Tools*, *Window*, *Help* and *Pin*.

The Pin 📌 option displays the Menu bar toolbar and the Menu bar menu as illustrated.
Throughout the book, the Menu bar menu and the Menu bar toolbar are referred to as the
Menu bar.

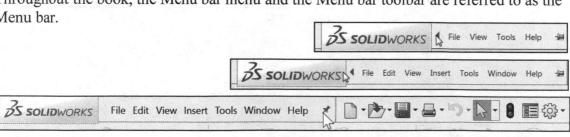

Drop-down menu

SOLIDWORKS takes advantage of the familiar Microsoft® Windows user interface. Communicate with SOLIDWORKS through drop-down menus, Context sensitive toolbars, Consolidated toolbars or the CommandManager tabs.

💡 A command is an instruction that informs SOLIDWORKS to perform a task.

To close a SOLIDWORKS drop-down menu, press the Esc key. You can also click any other part of the SOLIDWORKS Graphics window or click another drop-down menu.

Create a New Part Document

In the next section create a new part document.

Activity: Create a new Part Document.

A part is a 3D model, which consists of features. What are features?

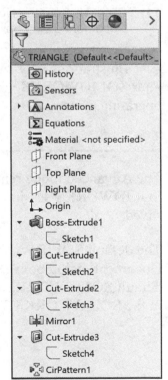

- Features are geometry building blocks.

- Most features either add or remove material.

- Some features do not affect material (Cosmetic Thread).

- Features are created either from 2D or 3D sketched profiles or from edges and faces of existing geometry.

- Features are individual shapes that, combined with other features, makes up a part or assembly. Some features, such as bosses and cuts, originate as sketches. Other features, such as shells and fillets, modify a feature's geometry.

- Features are displayed in the FeatureManager as illustrated (Boss-Extrude1, Cut-Extrude1, Cut-Extrude2, Mirror1, Cut-Extrude3 and CirPattern1).

💡 The first sketch of a part is called the Base Sketch. The Base sketch is the foundation for the 3D model. In this book, we focus on 2D sketches and 3D features.

There are two modes in the New SOLIDWORKS Document dialog box: *Novice* and *Advanced*. The *Novice* option is the default option with three templates. The *Advanced* mode contains access to additional templates and tabs that you create in system options. Use the *Advanced* mode in this book.

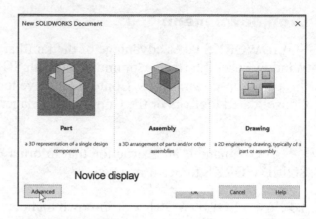

Novice display

Create a new part.

8) Click **New** from the Menu bar. The New SOLIDWORKS Document dialog box is displayed.

Select the Advanced mode.

9) Click the **Advanced** button as illustrated. The Advanced mode is set. Click the Templates tab.

10) Click Part. Part is the default template from the New SOLIDWORKS Document dialog box.

11) Click **OK** from the New SOLIDWORKS Document dialog box.

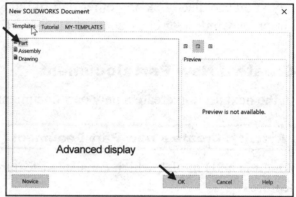

Advanced display

Illustrations may vary depending on your SOLIDWORKS version and operating system.

The Advanced mode remains selected for all new documents in the current SOLIDWORKS session. When you exit SOLIDWORKS, the Advanced mode setting is saved.

The default SOLIDWORKS installation contains two tabs in the New SOLIDWORKS Document dialog box: *Templates* and *Tutorial*. The *Templates* tab corresponds to the default SOLIDWORKS templates. The *Tutorial* tab corresponds to the templates utilized in the SOLIDWORKS Tutorials.

Part1 is displayed in the FeatureManager and is the name of the document. Part1 is the default part window name.

The Part Origin ⫪ is displayed in blue in the center of the Graphics window. The Origin represents the intersection of the three default reference planes: *Front Plane*, *Top Plane* and *Right Plane*. The positive X-axis is horizontal and points to the right of the Origin in the Front view. The positive Y-axis is vertical and points upward in the Front view. The FeatureManager contains a list of features, reference geometry, and settings utilized in the part.

☀ Edit the document units directly from the Graphics window as illustrated.

☀ Dynamic Annotation Views 🔍: Only available with SOLIDWORKS MBD (Model Based Definition). Provides the ability to control how annotations are displayed when you rotate models.

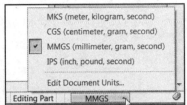

MKS (meter, kilogram, second)
CGS (centimeter, gram, second)
✓ MMGS (millimeter, gram, second)
IPS (inch, pound, second)

Edit Document Units...

Editing Part   MMGS

Main Menu Toolbar

Default CommandManager

Heads-up View Toolbar

Task Pane

Default Part FeatureManager

Origin

Hide/Show FeatureManager

Triad

Model Mode   3D Views   Motion Mode   Units   Tags

View the Default Sketch Planes.

12) Click the **Front Plane** from the FeatureManager.

13) Click the **Top Plane** from the FeatureManager.

14) Click the **Right Plane** from the FeatureManager.

15) Click the **Origin** from the FeatureManager. The Origin is the intersection of the Front, Top and Right Planes.

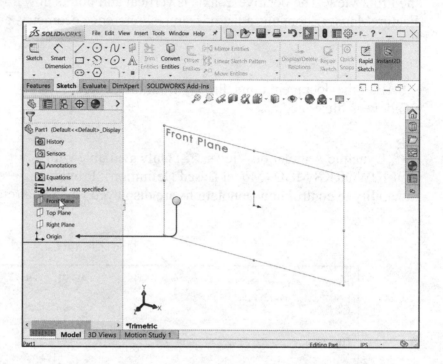

Redeem the code on the inside cover of the book. In the next section, download the ENGDESIGN-W-SOLIDWORKS 2018 folder to your hard drive. Open the Bracket part. Review the features and sketches in the Bracket FeatureManager. Work directly from your hard drive.

Activity: Download the ENGDESIGN-W-SOLIDWORKS folder. Open a Part.

Download the **ENGDESIGN-W-SOLIDWORKS 2018** folder.
Open an existing SOLIDWORKS part.

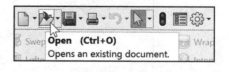

16) **Download** the ENGDESIGN-W-SOLIDWORKS 2018 folder to a local hard drive.

17) Click **Open** from the Menu bar menu.

18) Browse to the **ENGDESIGN-W-SOLIDWORKS 2018\Bracket** folder.

19) Double-click the **Bracket** part. The Bracket part is displayed in the Graphics window.

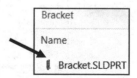

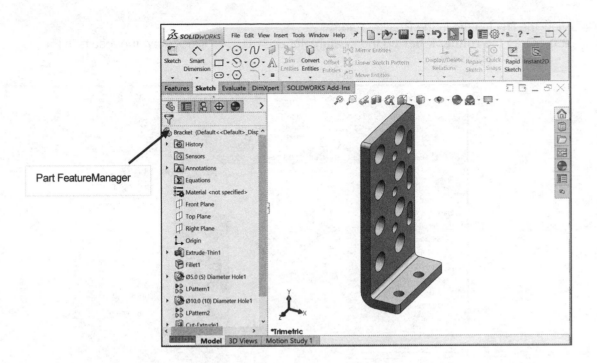

The FeatureManager design tree is located on the left side of the SOLIDWORKS Graphics window. The FeatureManager provides a summarized view of the active part, assembly, or drawing document. The tree displays the details on how the part, assembly or drawing document was created.

Use the FeatureManager rollback bar to temporarily roll back to an earlier state, to absorbed features, roll forward, roll to previous, or roll to the end of the FeatureManager design tree. You can add new features or edit existing features while the model is in the rolled-back state. You can save models with the rollback bar placed anywhere.

In the next section, review the features in the Bracket FeatureManager using the Rollback bar.

Activity: Use the FeatureManager Rollback Bar option.

Apply the FeatureManager Rollback Bar. Revert to an earlier state in the model.

20) Place the **mouse pointer** over the rollback bar in the FeatureManager design tree as illustrated. The pointer changes to a hand 🖐. Note the provided information on the feature. This is called Dynamic Reference Visualization.

21) Drag the **rollback bar** up the FeatureManager design tree until it is above the features you want rolled back, in this case Diameter Hole1.

22) **Release** the mouse button.

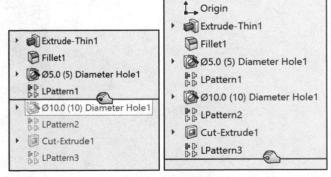

View the first feature in the Bracket Part.

23) Drag the **rollback bar** up the FeatureManager above Fillet1. View the results in the Graphics window.

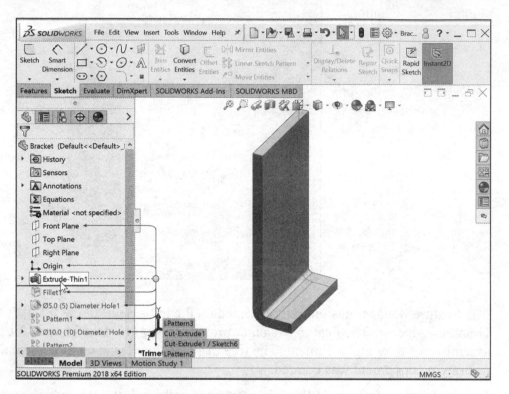

Return to the original Bracket Part FeatureManager.

24) Right-click **Extrude-Thin1** in the FeatureManager. The Pop-up Context toolbar is displayed.

25) Click **Roll to End**. View the results in the Graphics window.

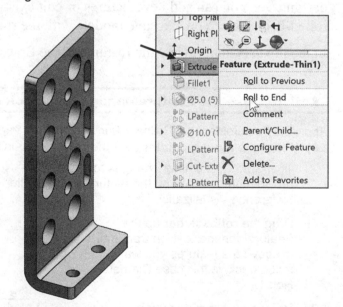

Heads-up View toolbar

SOLIDWORKS provides the user with numerous view options. One of the most useful tools is the Heads-up View toolbar displayed in the Graphics window when a document is active.

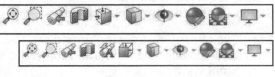

Dynamic Annotation Views : Only available with SOLIDWORKS MBD (Model Based Definition). Provides the ability to control how annotations are displayed when you rotate models.

In the next section, apply the following tools: Zoom to Fit, Zoom to Area, Zoom out, Rotate and select various view orientations from the Heads-up View toolbar.

Activity: Utilize the Heads-up View toolbar.

Zoom to Fit the model in the Graphics window.

26) Click the **Zoom to Fit** icon. The tool fits the model to the Graphics window.

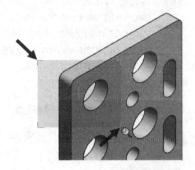

Zoom to Area on the model in the Graphics window.

27) Click the **Zoom to Area** icon. The Zoom to Area icon is displayed.

Zoom in on the top left hole.

28) Window-select the top left corner as illustrated. View the results.

De-select the Zoom to Area tool.

29) Click the **Zoom to Area** icon.

Fit the model to the Graphics window.

30) Press the **f** key.

Rotate the model.

31) Hold the **middle mouse button** down. Drag **upward** ↻, **downward** ↻, to the **left** ↻ and to the **right** ↻ to rotate the model in the Graphics window.

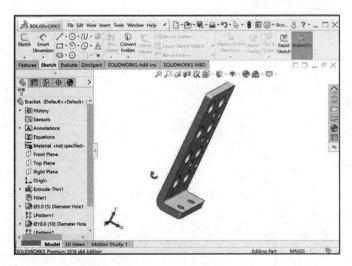

Display a few Standard Views.

32) Click **inside** the Graphics window.

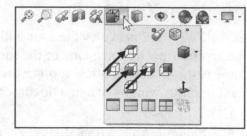

33) Click **Front** from the drop-down Heads-up view toolbar. The model is displayed in the Front view.

34) Click **Right** from the drop-down Heads-up view toolbar. The model is displayed in the Right view.

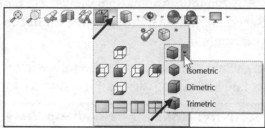

35) Click **Top** from the drop-down Heads-up view toolbar. The model is displayed in the Top view.

Display a Trimetric view of the Bracket model.

36) Click **Trimetric** from the drop-down Heads-up view toolbar as illustrated. Note your options. View the results in the Graphics window.

SOLIDWORKS Help

Help in SOLIDWORKS is context-sensitive and in HTML format. Help is accessed in many ways, including Help buttons in all dialog boxes and PropertyManager (or press F1) and Help ⑦ tool on the Standard toolbar for SOLIDWORKS Help.

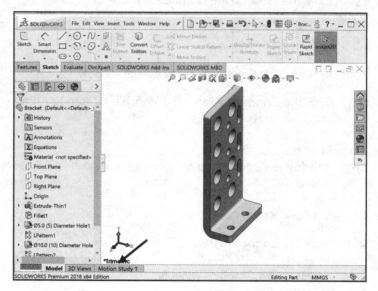

37)   Click **Help** from the Menu bar.

38)   Click **SOLIDWORKS Help**. The SOLIDWORKS Help Home Page is displayed by default. View your options.

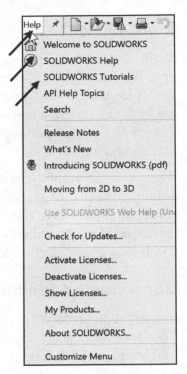

☼ SOLIDWORKS Web Help is active by default under Help in the Main menu.

Close Help. Return to the SOLIDWORKS Graphics window.

39)   **Close ⊠** SOLIDWORKS Home.

SOLIDWORKS Tutorials

Display and explore the SOLIDWORKS tutorials.

40)   Click **Help** from the Menu bar.

41)   Click **SOLIDWORKS Tutorials**. The SOLIDWORKS Tutorials are displayed. The SOLIDWORKS Tutorials are presented by category.

42)   Click the **Getting Started** category. The Getting Started category provides three 30 minute lessons on parts, assemblies, and drawings.

In the next section view the additional User Interface tools.

Activity: Close all Tutorials and Models.

Close SOLIDWORKS Tutorials and models.

43)   **Close ⊠** SOLIDWORKS Tutorials.

44)   Click **Window**, **Close All** from the Menu bar menu.

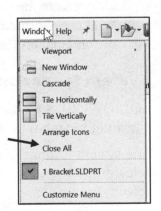

SOLIDWORKS Icon Style

SOLIDWORKS provides a new icon style. It allows vector-based scaling for superior support of high resolution, high pixel density displays. The new icon style standardized the perspective of icons. It also removes non-essential details, and emphasizes primary elements. Consistent visual styling applies to all icons.

Additional User Interface Tools

The book utilizes additional areas of the SOLIDWORKS User Interface. Explore an overview of these tools in the next section.

Right-click

Right-click in the Graphics window on a model, or in the FeatureManager on a feature or sketch to display the Context-sensitive toolbar. If you are in the middle of a command, this toolbar displays a list of options specifically related to that command.

Right-click an empty space in the Graphics window of a part or assembly, and a selection context toolbar above the shortcut menu is displayed. This provides easy access to the most commonly used selection tools.

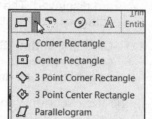

Consolidated toolbar

Similar commands are grouped together in the CommandManager. Example: Variations of the Rectangle sketch tool are grouped in a single fly-out button as illustrated.

If you select the Consolidated toolbar button without expanding:

For some commands such as Sketch, the most commonly used command is performed. This command is the first listed and the command shown on the button.

For commands such as rectangle, where you may want to repeatedly create the same variant of the rectangle, the last used command is performed. This is the highlighted command when the Consolidated toolbar is expanded.

System feedback icon

SOLIDWORKS provides system feedback by attaching a symbol to the mouse pointer cursor.

The system feedback symbol indicates what you are selecting or what the system is expecting you to select.

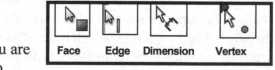

As you move the mouse pointer across your model, system feedback is displayed in the form of a symbol, riding next to the cursor as illustrated. This is a valuable feature in SOLIDWORKS.

Confirmation Corner

When numerous SOLIDWORKS commands are active, a symbol or a set of symbols are displayed in the upper right hand corner of the Graphics window. This area is called the Confirmation Corner.

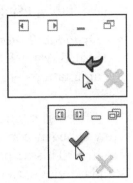

When a sketch is active, the confirmation corner box displays two symbols. The first symbol is the sketch tool icon. The second symbol is a large red X. These two symbols supply a visual reminder that you are in an active sketch. Click the sketch symbol icon to exit the sketch and to save any changes that you made.

When other commands are active, the confirmation corner box provides a green check mark and a large red X. Use the green check mark to execute the current command. Use the large red X to cancel the command.

Confirm changes you make in sketches and tools by using the D keyboard shortcut to move the OK and Cancel buttons to the pointer location in the Graphics window.

Heads-up View toolbar

SOLIDWORKS provides the user with numerous view options from the Standard Views, View and Heads-up View toolbar.

The Heads-up View toolbar is a transparent toolbar that is displayed in the Graphics window when a document is active.

You can hide, move or modify the Heads-up View toolbar. To modify the Heads-up View toolbar, right-click on a tool and select or deselect the tools that you want to display.

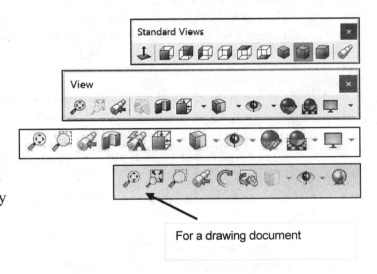

For a drawing document

The following views are available: Note: The available views are document dependent.

- *Zoom to Fit* : Zooms the model to fit the Graphics window.

- *Zoom to Area* : Zooms to the areas you select with a bounding box.

- *Previous View* : Displays the previous view.

- *Section View* : Displays a cutaway of a part or assembly, using one or more cross section planes.

- *Dynamic Annotation Views* : Only available with SOLIDWORKS MBD. Provides the ability to control how annotations are displayed when you rotate models.

The Orientation dialog has an option to display a view cube (in-context View Selector) with a live model preview. This helps the user to understand how each standard view orientates the model. With the view cube, you can access additional standard views. The views are easy to understand and they can be accessed simply by selecting a face on the cube.

To activate the Orientation dialog box, press (Ctrl + spacebar) or click the View Orientation icon from the Heads up View toolbar. The active model is displayed in the View Selector in an Isometric orientation (default view).

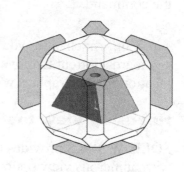

As you hover over the buttons in the Orientation dialog box, the corresponding faces dynamically highlight in the View Selector. Select a view in the View Selector or click the view from the Orientation dialog box. The Orientation dialog box closes and the model rotates to the selected view.

Click the View Selector icon in the Orientation dialog box to show or hide the in-context View Selector.

Press **Ctrl + spacebar** to activate the View Selector.

Press the **spacebar** to activate the Orientation dialog box.

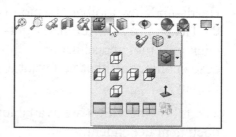

- *View Orientation box* : Provides the ability to select a view orientation or the number of viewports. The available options are *Top*, *Left*, *Front*, *Right*, *Back*, *Bottom*, *Single view*, *Two view - Horizontal*, *Two view - Vertical*, *Four view*. Click the drop-down arrow to access Axonometric views: Isometric, Dimetric and Trimetric.

- *Display Style* : Provides the ability to display the style for the active view. The available options are *Wireframe*, *Hidden Lines Visible*, *Hidden Lines Removed*, *Shaded*, *Shaded With Edges*.

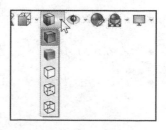

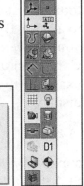

- *Hide/Show Items* ⬥ ⁻: Provides the ability to select items to hide or show in the Graphics window. The available items are document dependent. Note the View Center of Mass ⬥ icon.

- *Edit Appearance* 🌐: Provides the ability to edit the appearance of entities of the model.

- *Apply Scene* 🔮 ⁻: Provides the ability to apply a scene to an active part or assembly document. View the available options.

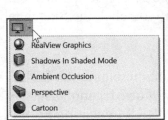

- *View Setting* 🖥 ⁻: Provides the ability to select the following settings: *RealView Graphics*, *Shadows In Shaded Mode, Ambient Occlusion, Perspective* and *Cartoon*.

- *Rotate view* ↻ : Provides the ability to rotate a drawing view. Input Drawing view angle and select the ability to update and rotate center marks with view.

- *3D Drawing View* 🔄: Provides the ability to dynamically manipulate the drawing view in 3D to make a selection.

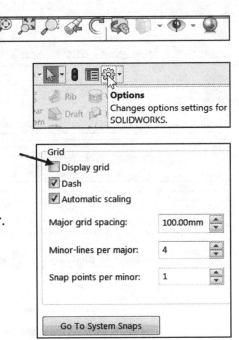

💡 To display a grid for a part, click Options ⚙, Document Properties tab. Click Grid/Snaps, check the Display grid box.

💡 Add a custom view to the Heads-up View toolbar. Press the space key. The Orientation dialog box is displayed. Click the New View 🖊 tool. The Name View dialog box is displayed. Enter a new named view. Click OK.

SOLIDWORKS CommandManager

The SOLIDWORKS CommandManager is a Context-sensitive toolbar that automatically updates based on the toolbar you want to access. By default, it has toolbars embedded in it based on your active document type. When you click a tab below the CommandManager, it updates to display that toolbar. For example, if you click the Sketch tab, the Sketch toolbar is displayed.

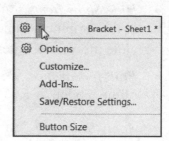

For commercial users, SOLIDWORKS Model Based Definition (MBD) is a separate application. For education users, SOLIDWORKS MBD is included in the SOLIDWORKS Education Edition as an Add In.

Below is an illustrated CommandManager for a default Part document.

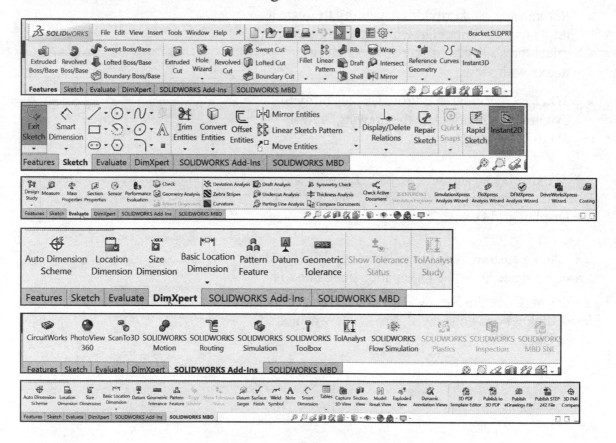

Button sizes. You can set sizes for buttons from the Toolbars tab of the Customize dialog box. To facilitate element selection on touch interfaces such as tablets, you can set up the larger Size buttons and text from the Options menu (Standard toolbar).

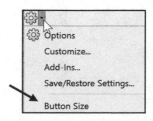

The SOLIDWORKS CommandManager is a Context-sensitive toolbar that automatically updates based on the toolbar you want to access. By default, it has toolbars embedded in it based on your active document type.

💡 For commercial users, SOLIDWORKS Model Based Definition (MBD) is a separate application. For education users, SOLIDWORKS MBD is included in the SOLIDWORKS Education Edition as an Add In.

Below is an illustrated CommandManager for a default Drawing document.

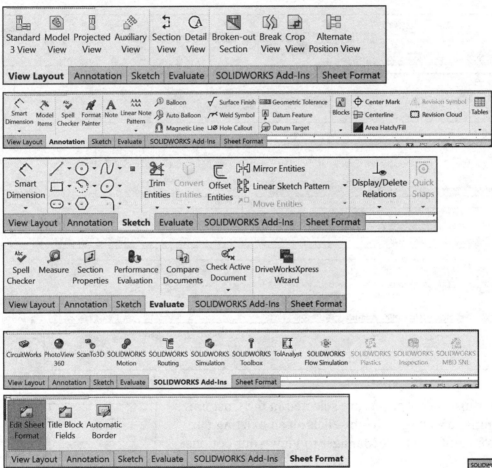

💡 To add a custom tab to your CommandManager, right-click on a tab and click Customize CommandManager from the drop-down menu. The Customize dialog box is displayed. You can also select to add a blank tab as illustrated and populate it with custom tools from the Customize dialog box.

The SOLIDWORKS CommandManager is a Context-sensitive toolbar that automatically updates based on the toolbar you want to access. By default, it has toolbars embedded in it based on your active document type.

🔅 For commercial users, SOLIDWORKS Model Based Definition (MBD) is a separate application. For education users, SOLIDWORKS MBD is included in the SOLIDWORKS Education Edition as an Add In.

Below is an illustrated CommandManager for a default Assembly document.

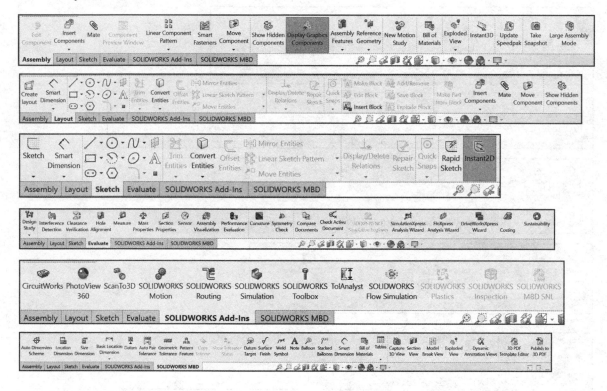

By default, the illustrated options are selected in the Customize box for te CommandManager. Right-click on an existing tab and click Customize CommandManager to view your options.

🔅 You can set the number of mouse gestures to 2, 3, 4, 6, 8, or 12 gestures. If you set the number to 2 gestures, you can orient them vertically or horizontally.

Float the CommandManager. Drag the Features, Sketch or any CommandManager tab. Drag the CommandManager anywhere on or outside the SOLIDWORKS window.

To dock the CommandManager, perform one of the following:

While dragging the CommandManager in the SOLIDWORKS window, move the pointer over a docking icon -

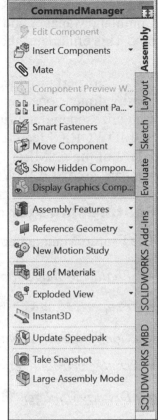

Dock above, Dock left, Dock right and click the needed command.

Double-click the floating CommandManager to revert the CommandManager to the last docking position.

Screen shots in the book were made using SOLIDWORKS 2018 SP0 running Windows® 10.

An updated color scheme for certain icons makes the SOLIDWORKS application more accessible to people with color blindness. Icons in the active PropertyManager use blue to indicate what you must select on the screen: faces, edges, and so on.

Selection Enhancements

Right-click an empty space in the Graphics window of a part or assembly; a selection context toolbar above the shortcut menu provides easy access to the most commonly used selection tools.

- **Box Selection** . Provides the ability to select entities in parts, assemblies, and drawings by dragging a selection box with the pointer.

- **Lasso Selection** . Provides the ability to select entities by drawing a lasso around the entities.

- **Selection Filters** . Displays a list of selection filter commands.

- **Previous Selection** . Displays the previous selection.

- **Select Other** . Displays the Select Other dialog box.

- **Select** . Displays a list of selection commands.

- **Magnified Selection** . Displays the magnifying glass, which gives you a magnified view of a section of a model.

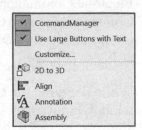

💡 Save space in the CommandManager: right-click in the CommandManager and un-check the Use Large Buttons with Text box. This eliminates the text associated with the tool.

💡 DimXpert provides the ability to graphically check if the model is fully dimensioned and toleranced. DimXpert automatically recognizes manufacturing features. Manufacturing features are not SOLIDWORKS features. Manufacturing features are defined in 1.1.12 of the ASME Y14.5M-1994 Dimensioning and Tolerancing standard. See SOLIDWORKS Help for additional information.

FeatureManager Design Tree

The FeatureManager consists of five default tabs:

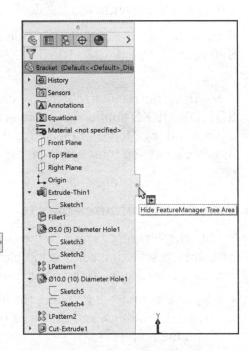

- *FeatureManager design tree* 🔧 tab.

- *PropertyManager* 📋 tab.

- *ConfigurationManager* 🔖 tab.

- *DimXpertManager* ⊕ tab.

- *DisplayManager* ● tab.

Select the Hide FeatureManager Tree Area arrows as illustrated to enlarge the Graphics window for modeling.

💡 The Sensors tool 📟 located in the FeatureManager monitors selected properties in a part or assembly and alerts you when values deviate from the specified limits. There are five sensor types: Simulation Data, Mass properties, Dimensions, Measurement and Costing Data.

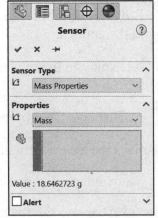

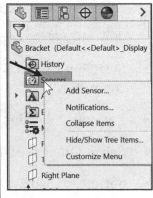

Various commands provide the ability to control what is displayed in the FeatureManager design tree:

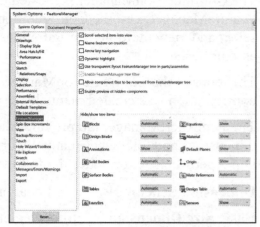

1. Show or Hide FeatureManager items.

💡 Click **Options** ⚙ from the Menu bar. Click **FeatureManager** from the System Options tab. **Customize** your FeatureManager from the Hide/Show Tree Items dialog box.

2. Filter the FeatureManager design tree. Enter information in the filter field. You can filter by *Type of features*, *Feature names*, *Sketches*, *Folders*, *Mates*, *User-defined tags* and *Custom properties*.

💡 Tags are keywords you can add to a SOLIDWORKS document to make them easier to filter and to search. The Tags 🏷 icon is located in the bottom right corner of the Graphics window.

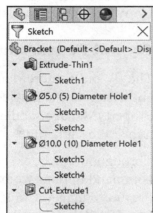

💡 Collapse all items in the FeatureManager, **right-click** and select **Collapse items**, or press the **Shift** + **C** keys.

The FeatureManager design tree and the Graphics window are dynamically linked. Select sketches, features, drawing views, and construction geometry in either pane.

Split the FeatureManager design tree and either display two FeatureManager instances, or combine the FeatureManager design tree with the ConfigurationManager or PropertyManager.

Split line

Move between the FeatureManager design tree, PropertyManager, ConfigurationManager and DimXpertManager by selecting the tabs at the top of the menu.

The ConfigurationManager is located to the right of the FeatureManager. Use the ConfigurationManager to create, select and view multiple configurations of parts and assemblies.

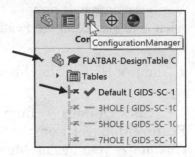

The icons in the ConfigurationManager denote whether the configuration was created manually or with a design table.

The DimXpertManager tab provides the ability to insert dimensions and tolerances manually or automatically. The DimXpertManager provides the following selections: **Auto Dimension Scheme** ⊕, Basic Location Dimension ↦⊣, **Basic Size Dimension** ⤵ **Show Tolerance Status** ±⊚, **Copy Scheme** ⊕ and **TolAnalyst Study** ⊑⊥.

Fly-out FeatureManager

The fly-out FeatureManager design tree provides the ability to view and select items in the PropertyManager and the FeatureManager design tree at the same time.

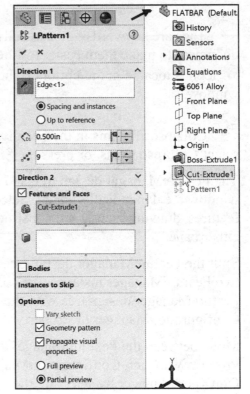

Throughout the book, you will select commands and command options from the drop-down menu, fly-out FeatureManager, Context toolbar or from a SOLIDWORKS toolbar.

🔅 Another method for accessing a command is to use the accelerator key. Accelerator keys are special key strokes, which activate the drop-down menu options. Some commands in the menu bar and items in the drop-down menus have an underlined character.

Press the Alt or Ctrl key followed by the corresponding key to the underlined character activates that command or option.

🔅 Illustrations may vary depending on your SOLIDWORKS version and operating system.

Task Pane

The Task Pane is displayed when a SOLIDWORKS session starts. You can show, hide, and reorder tabs in the Task Pane. You can also set a tab as the default so it appears when you open the Task Pane, pin or unpin to the default location.

The Task Pane contains the following default tabs:

- *SOLIDWORKS Resources* ⌂.

- *Design Library* ⬚ .

- *File Explorer* ▭.

- *View Palette* ▦.

- *Appearances, Scenes and Decals* ●.

- *Custom Properties* ▤ .

- *SOLIDWORKS Forum* ⬚ .

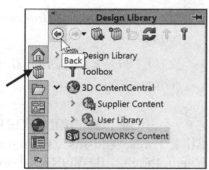

💡 Additional tabs are displayed with Add-Ins.

Use the **Back** and **Forward** buttons in the Design Library tab and the Appearances, Scenes, and Decals tab of the Task Pane to navigate in folders.

SOLIDWORKS Resources

The basic SOLIDWORKS Resources ⌂ menu displays the following default selections:

- *Getting Started.*

- *SOLIDWORKS Tools.*

- *Community.*

- *Online Resources.*

- *Subscription Services.*

- *Tip of the Day.*

Other user interfaces are available during the initial software installation selection: *Machine Design, Mold Design* or *Consumer Products Design.*

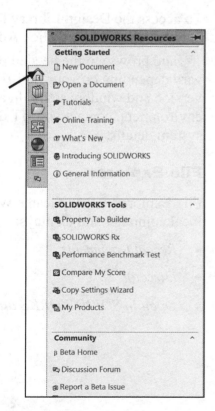

Design Library

The Design Library  contains reusable parts, assemblies, and other elements including library features.

The Design Library tab contains four default selections. Each default selection contains additional sub categories.

The default selections are:

- *Design Library.*
- *Toolbox.*
- *3D ContentCentral (Internet access required).*
- *SOLIDWORKS Content (Internet access required).*

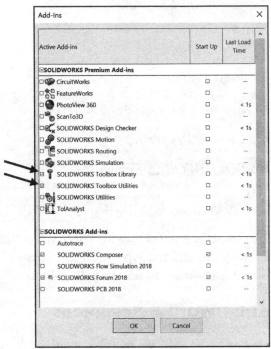

Activate the SOLIDWORKS Toolbox. Click Tools, Add-Ins.., from the Main menu. Check the SOLIDWORKS Toolbox Library from the Add-ins dialog box or click SOLIDWORKS Toolbox from the SOLIDWORKS Add-Ins tab.

To access the Design Library folders in a non-network environment, click Add File Location and browse to the needed path. Paths may vary depending on your SOLIDWORKS version and window setup. In a network environment, contact your IT department for system details.

File Explorer

File Explorer duplicates Windows Explorer from your local computer and displays:

- *Recent Documents.*
- *Directories.*
- *Open in SOLIDWORKS and Desktop folders.*

Search

The SOLIDWORKS Search box is displayed in the upper right corner of the SOLIDWORKS Graphics window (Menu Bar toolbar). Enter the text or key words to search.

New search modes have been added to SOLIDWORKS Search as illustrated.

View Palette

The View Palette 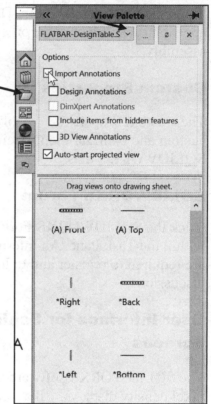 tool located in the Task Pane provides the ability to insert drawing views of an active document, or click the Browse button to locate the desired document.

Click and drag the view from the View Palette into an active drawing sheet to create a drawing view.

The selected model is FLATBAR DesignTable in the illustration.

Appearances, Scenes, and Decals

Appearances, Scenes, and Decals ⬤ provide a simplified way to display models in a photo-realistic setting using a library of Appearances, Scenes, and Decals.

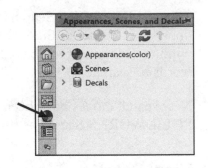

An appearance defines the visual properties of a model, including color and texture. Appearances do not affect physical properties, which are defined by materials.

Scenes provide a visual backdrop behind a model. In SOLIDWORKS they provide reflections on the model. PhotoView 360 is an Add-in. Drag and drop a selected appearance, scene or decal on a feature, surface, part or assembly.

Custom Properties

The Custom Properties 📋 tool provides the ability to enter custom and configuration specific properties directly into SOLIDWORKS files.

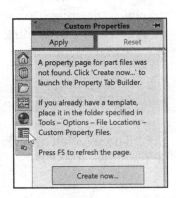

SOLIDWORKS Forum

Click the SOLIDWORKS Forum 🗨 icon to search directly within the Task Pane. An internet connection is required. You are required to register and to login for postings and discussions.

User Interface for Scaling High Resolution Screens

The SOLIDWORKS software supports high-resolution, high-pixel density displays. All aspects of the user interface respond to the Microsoft Windows® display scaling setting. In dialog boxes, PropertyManagers, and the FeatureManager design tree, the SOLIDWORKS software uses your display scaling setting to display buttons and icons at an appropriate size. Icons that are associated with text are scaled to a size appropriate for the text. In addition, for toolbars, you can display Small, Medium, or Large buttons. Click the **Options drop-down arrow** from the Standard Menu bar, and click Button size to size the icons.

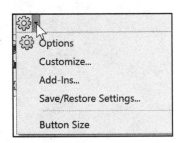

Motion Study tab

Motion Studies are graphical simulations of motion for an assembly. Access the MotionManager from the Motion Study tab. The Motion Study tab is located in the bottom left corner of the Graphics window.

Incorporate visual properties such as lighting and camera perspective. Click the Motion Study tab to view the MotionManager. Click the Model tab to return to the FeatureManager design tree.

The MotionManager displays a timeline-based interface and provides the following selections from the drop-down menu as illustrated:

- *Animation:* Apply Animation to animate the motion of an assembly. Add a motor and insert positions of assembly components at various times using set key points. Use the Animation option to create animations for motion that do **not** require accounting for mass or gravity.

- *Basic Motion:* Apply Basic Motion for approximating the effects of motors, springs, collisions and gravity on assemblies. Basic Motion takes mass into account in calculating motion. Basic Motion computation is relatively fast, so you can use this for creating presentation animations using physics-based simulations. Use the Basic Motion option to create simulations of motion that account for mass, collisions or gravity.

If the Motion Study tab is not displayed in the Graphics window, click **View, MotionManager** from the Menu bar.

3D Views tab

SOLIDWORKS MBD (Model Based Definition) lets you create models without the need for drawings giving you an integrated manufacturing. MBD helps companies define, organize, and publish 3D product and manufacturing information (PMI), including 3D model data in industry standard file formats.

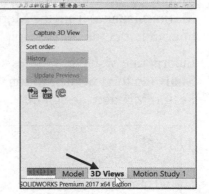

Create 3D drawing views of your parts and assemblies that contain the model settings needed for review and manufacturing. This lets users navigate back to those settings as they evaluate the design.

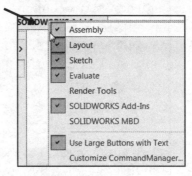

Use the tools in the SOLIDWORKS MBD CommandManager to set up your model with selected configurations, including explodes and abbreviated views, annotations, display states, zoom level, view orientation and section views. Capture those settings so that you and other users can return to them at any time using the 3D view palette.

To access the 3D View palette, click the 3DViews tab at the bottom of the SOLIDWORKS window or the SOLIDWORKS MBD tab in the CommandManager. The Capture 3D View button opens the Capture 3D View PropertyManager, where you specify the 3D view name, and the configuration, display state and annotation view to capture. See SOLIDWORKS help for additional information.

Dynamic Reference Visualization

Dynamic Reference Visualization provides the ability to view the parent relationships between items in the FeatureManager design tree. When you hover over a feature with references in the FeatureManager design tree, arrows display showing the relationships. If a reference cannot be shown because a feature is not expanded, the arrow points to the feature that contains the reference and the actual reference appears in a text box to the right of the arrow. Use Dynamic reference visualization for a part, assembly and ever mates.

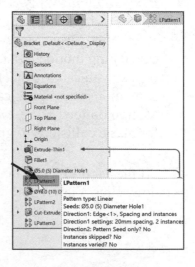

Use Dynamic reference visualization for a part, assembly and ever mates. To display the Dynamic Reference Visualization, click **View ➢ User Interface ➢ Dynamic Reference Visualization** from the Main menu bar.

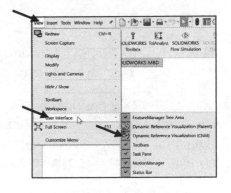

Mouse Movements

A mouse typically has two buttons: a primary button (usually the left button) and a secondary button (usually the right button). Most mice also include a scroll wheel between the buttons to help you scroll through documents and to Zoom in, Zoom out and rotate models in SOLIDWORKS. It is highly recommended that you use a mouse with at least a Primary, Scroll and Secondary button.

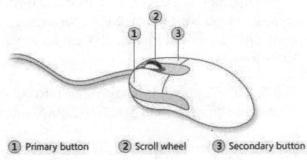

① Primary button   ② Scroll wheel   ③ Secondary button

Single-click

To click an item, point to the item on the screen, and then press and release the primary button (usually the left button). Clicking is most often used to select (mark) an item or open a menu. This is sometimes called single-clicking or left-clicking.

Double-click

To double-click an item, point to the item on the screen, and then click twice quickly. If the two clicks are spaced too far apart, they might be interpreted as two individual clicks rather than as one double-click. Double-clicking is most often used to open items on your desktop. For example, you can start a program or open a folder by double-clicking its icon on the desktop.

Right-click

To right-click an item, point to the item on the screen, and then press and release the secondary button (usually the right button). Right-clicking an item usually displays a list of things you can do with the item. Right-click in the open Graphics window or on a command in SOLIDWORKS, and additional pop-up context is displayed.

Scroll wheel

Use the scroll wheel to zoom-in or to zoom-out of the Graphics window in SOLIDWORKS. To zoom-in, roll the wheel backward (toward you). To zoom-out, roll the wheel forward (away from you).

Summary

The SOLIDWORKS (UI) is designed to make maximum use of the Graphics window for your model. Displayed toolbars and commands are kept to a minimum.

SOLIDWORKS provides a new icon style. It also allows vector-based scaling for superior support of high resolution high pixel density displays. The new icon style standardized the perspective of icons, removes non-essential details, and emphasizes primary elements. Consistent visual styling applies to all icons.

The SOLIDWORKS User Interface and CommandManager consist of the following main options: Menu bar toolbar, Menu bar menu, Drop-down menus, Context toolbars, Consolidated fly-out menus, System feedback icons, Confirmation Corner and Heads-up View toolbar.

The default CommandManager Part tabs control the display of the *Features*, *Sketch*, *Evaluate*, *DimXpert* and *SOLIDWORKS Add-Ins* toolbars.

The FeatureManager consists of five default tabs:

- FeatureManager design tree.

- PropertyManager.

- ConfigurationManager.

- DimXpertManager.

- DisplayManager.

You learned about creating a new SOLIDWORKS part and opening an existing SOLIDWORKS part along with using the Rollback bar to view the sketches and features.

Opening a SOLIDWORKS document from an earlier release can take extra time. After you open and save a file, subsequent opening time returns to normal. Use the SOLIDWORKS Task Scheduler (SOLIDWORKS Professional) to convert multiple files from an earlier version to the SOLIDWORKS 2018 format. Click Windows Start ➤ All Apps ➤ SOLIDWORKS 2018 ➤ SOLIDWORKS Tools 2018 ➤ SOLIDWORKS Task Scheduler.

Templates are part, drawing and assembly documents which include user-defined parameters. Open a new part, drawing or assembly. Select a template for the new document.

- **Parts**. The Parts default template is located in the C:\ProgramData\SolidWorks\\SOLIDWORKS 2018\templates\Part.prtdot folder.

- *Assemblies*. The Assemblies default template is located in the C:\ProgramData\SolidWorks\\SOLIDWORKS 2018\templates\Assembly.asmdot folder.

- *Drawings*. The Drawings default template is located in the C:\ProgramData\SolidWorks\\SOLIDWORKS 2018\templates\Drawing.drwdot folder.

For commercial users, SOLIDWORKS Model Based Definition (MBD) is a separate application. For education users, SOLIDWORKS MBD is included in the SOLIDWORKS Education Edition as an Add In.

In Project 2, obtain the working familiarity of the following SOLIDWORKS sketch and feature tools: *Line, Circle, Centerpoint Straight Slot, Smart Dimension, Extruded Boss/Base, Extruded Cut and Linear Pattern.*

Create three individual parts: AXLE, SHAFT-COLLAR and FLATBAR.

Create the assembly, LINKAGE using the three created parts and the downloaded sub-assembly - AirCylinder.

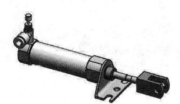

Notes:

Project 2

Fundamentals of Part Modeling

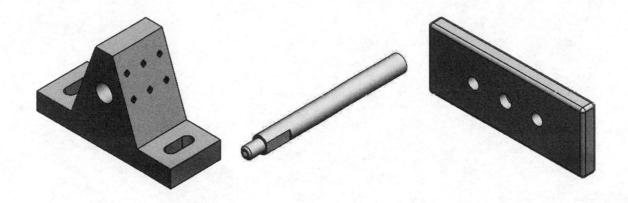

Below are the desired outcomes and usage competencies based on the completion of Project 2.

Project Desired Outcomes:	Usage Competencies:
• Address File Management.	• Create file folders for various Project Models and Document Templates.
• PART-ANSI-MM Part Template.	• Apply and modify System Options and Document Properties.
• PART-ANSI-MM Part Template. 　○ PLATE 　○ ROD 　○ GUIDE	• Create 2D Sketch profiles on the correct Sketch plane. • Understand and apply proper design intent. • Knowledge to create and modify the following SOLIDWORKS 3D features: • Extruded Boss/Base, Extruded Cut, Instant3D, Fillet, Mirror, Chamfer, Hole Wizard (Slot hole), and Linear Pattern.

Notes:

Project 2 - Fundamentals of Part Modeling

Project Objective

Create three folders: PROJECTS, MY-TEMPLATES, and VENDOR COMPONENTS.

Create the PART-MM-ANSI Part template utilized for the parts in this project.

Create three parts:

- PLATE
- ROD
- GUIDE

On the completion of this project, you will be able to:

- Create file folders to manage and organize projects.
- Apply and understand System Options and Document Properties.
- Create a Part template.
- Open, Save and Close Part documents and templates.
- Create 2D sketch profiles on the proper Sketch plane.
- Utilize the following Sketch tools: Line, Corner Rectangle, Circle, Arc, Trim Entities, Convert Entities, Straight Slot and Mirror Entities.
- Apply and edit sketch dimensions.
- Insert the following Geometric relations: Horizontal, Vertical, Symmetric, Equal and Coradial.
- Create and modify 3D parts.
- Create and modify the following SOLIDWORKS features: Extruded Boss\Base, Extruded Cut, Fillet, Mirror, Hole Wizard (Slot hole), Chamfer, and Linear Pattern.

Screen shots and illustrations in the book display the SOLIDWORKS user default setup.

Redeem the code on the inside cover of the book to view the provided videos on 2D Sketching, Sketch Planes, Sketch tools along with 3D Features and Design Intent.

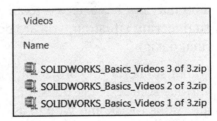

Project Situation

You receive an email or a concept sketch from a customer. The customer is retooling an existing assembly line.

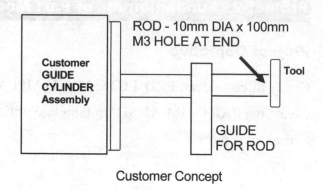

Customer Concept

You are required to design and manufacture a ROD. The ROD part is 10mm in diameter x 100mm in length.

One end of the ROD connects to an existing customer GUIDE CYLINDER assembly.

The other end of the ROD connects to the customer's tool. The ROD contains a 3mm hole and a key-way to attach the tool.

The ROD requires a support GUIDE. The ROD travels through the support GUIDE. The GUIDE-ROD assembly is the finished customer product.

The GUIDE-ROD assembly is a component used in a low volume manufacturing environment. Investigate a few key design issues:

- How will the customer use the GUIDE-ROD assembly?

- How are the parts PLATE, ROD and GUIDE used in the GUIDE-ROD assembly?

- Does the GUIDE-ROD assembly affect other components?

- Identify design requirements for load, structural integrity or other engineering properties.

- Identify cost effective materials.

- How are parts manufactured?

- What are their critical design features?

- How will each part behave when modified?

You may not have access to all of the required design information. Placed in a concurrent engineering situation, you are dependent on others and are ultimately responsible for the final design.

Dimensions for this project are in millimeters. Design information is provided from various sources. Ask questions. Part of the learning experience is to know which questions to ask and who to ask.

The ROD part requires support. During the manufacturing operations, the ROD exhibits unwanted deflection.

The engineering group calculates working loads on test samples. Material test samples include Plain Carbon Steel, AISI 304 steel and a machinable plastic Nylon 101.

The engineering group recommends AISI 304 steel for the ROD and GUIDE parts.

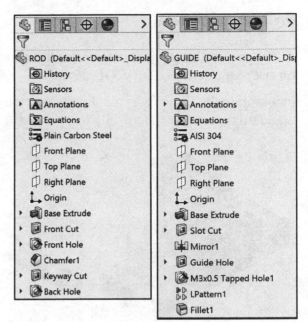

In the real world, there are numerous time constraints. The customer requires a quote, design sketches and a delivery schedule, YESTERDAY. If you wait for all of the required design information, you will miss the project deadline.

You create a rough concept sketch with notes in a design review meeting.

Your colleagues review and comment on the concept sketch.

The ROD cannot mount directly to the customer's GUIDE CYLINDER assembly without a mounting PLATE part.

The PLATE part mounts to the customer's PISTON PLATE part in the GUIDE CYLINDER assembly.

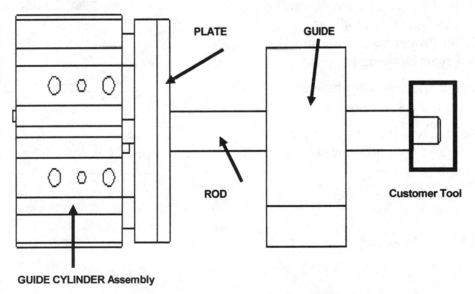

Concept Sketch and Design
Review Notes

Project Overview

A key goal in this project is to create three individual parts: PLATE, ROD and GUIDE for the requested GUIDE-ROD assembly.

Incorporate the three individual parts into the GUIDE-ROD assembly in Project 3 using Standard mates.

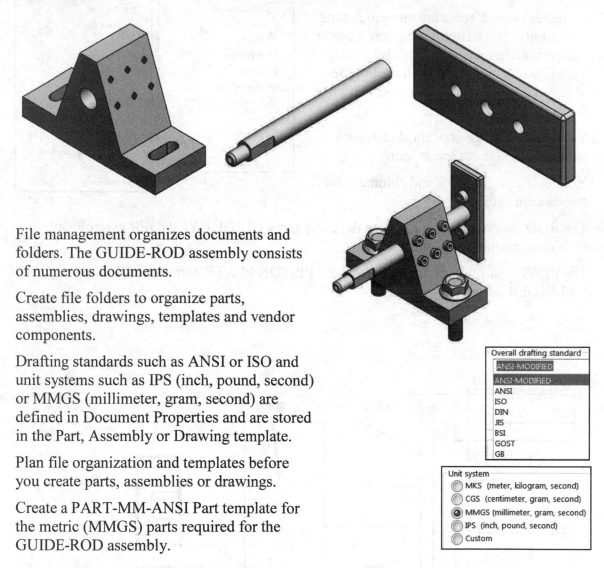

File management organizes documents and folders. The GUIDE-ROD assembly consists of numerous documents.

Create file folders to organize parts, assemblies, drawings, templates and vendor components.

Drafting standards such as ANSI or ISO and unit systems such as IPS (inch, pound, second) or MMGS (millimeter, gram, second) are defined in Document Properties and are stored in the Part, Assembly or Drawing template.

Plan file organization and templates before you create parts, assemblies or drawings.

Create a PART-MM-ANSI Part template for the metric (MMGS) parts required for the GUIDE-ROD assembly.

Overall drafting standard

- ANSI-MODIFIED
- ANSI-MODIFIED
- ANSI
- ISO
- DIN
- JIS
- BSI
- GOST
- GB

Unit system

- MKS (meter, kilogram, second)
- CGS (centimeter, gram, second)
- ● MMGS (millimeter, gram, second)
- IPS (inch, pound, second)
- Custom

File Management

File management organizes parts, assemblies and drawings. Why do you need file management? A large assembly contains hundreds or even thousands of parts.

Parts and assemblies are distributed between team members to save time. Design changes occur frequently in the development process. How do you manage and control changes? Answer: Through file management. File management is a very important tool in the development process.

The GUIDE-ROD assembly consists of many files. Utilize file folders to organize projects, vendor parts and assemblies, templates and various libraries.

Folders exist on the local hard drive, example C:\. Folders also exist on a network drive, example Z:\. The letters C:\ and Z:\ are used as examples for a local drive and a network drive respectfully. The following example utilizes the folder "Documents" to contain the folders for your projects.

Activity: File Management - Create three new sub-folders.

Create three sub-folders in the **ENGDESIGN-W-SOLIDWORKS** folder. In Project 1, you downloaded the **ENGDESIGN-W-SOLIDWORKS** folder to your hard drive. Work directly from your hard drive. Save all models, assemblies and templates to the ENGDESIGN-W-SOLIDWORKS folder and the sub-folders. Note: un-zip files before you try to open any files.

1) Double-click the **ENGDESIGN-W-SOLIDWORKS** folder on your hard drive.

2) Click **New folder** from the Windows Main menu. A new folder icon is displayed.

Create the first sub-folder. The procedure will be different depending on your operating system.

3) Enter **MY-TEMPLATES** for the folder name.

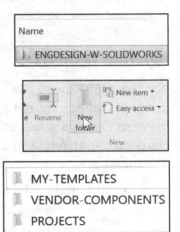

Create the second sub-folder.

4) Click **New folder**.

5) Enter **PROJECTS** for the second sub-folder name.

Create the third sub-folder.

6) Click **New folder** from the Windows Main menu. A new folder icon is displayed.

7) Enter **VENDOR-COMPONENTS** for the third sub-folder name.

Return to the ENGDESIGN-W-SOLIDWORKS folder.

8) Click the **Up arrow** to return to the ENGDESIGN-W-SOLIDWORKS folder.

Utilize the MY-TEMPLATES folder and the PROJECTS folder throughout the text.

Store the Part templates, Assembly templates and Drawing templates in the MY-TEMPLATES folder.

Store the parts, assemblies and drawings that you create in the PROJECTS folder.

Store parts and assemblies that you download from the internet or 3D ContentCentral in the VENDOR-COMPONENTS folder.

🔆 Redeem the code on the inside cover of the book to view the provided videos on the SOLIDWORKS Interface, 2D sketching, sketch planes and sketch tools along with 3D Features and Design Intent.

🔆 Screen shots and illustrations in the book display the SOLIDWORKS user default setup.

Activity: Start a SOLIDWORKS Session. Open a New Part Document.

Start a SOLIDWORKS session.

9) **Start** a SOLIDWORKS 2018 session. The SOLIDWORKS 2018 Welcome dialog box is displayed.

10) **Close** the Welcome dialog box. The SOLIDWORKS program window is displayed.

Create a New Part Document.

11) Click **New** ⬜ from the Menu bar. The New SOLIDWORKS Document dialog box is displayed.

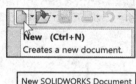

12) Double-click **Part** from the Templates tab (not from the Tutorial tab). View the default Part FeatureManager and empty Graphics window.

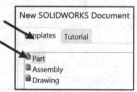

System Options

System Options are stored in the registry of the computer. System Options are not part of the document. Changes to the System Options affect current and future documents. Note: This can be different in a network environment. Ask your IT administrator.

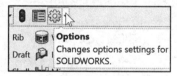

Review and modify the System Options. If you work on a local drive C:\, the System Options are stored on the computer.

If you work on a network drive Z:\, set System Options for each SOLIDWORKS session.

Set the System Options before you start a project. The File Locations Option contains a list of folders referenced during a SOLIDWORKS session.

Add the ENGDESIGN-W-SOLIDWORKS\MY-TEMPLATES folder path name to the Document Templates File Locations list.

Activity: Set System Options

Set System Options.

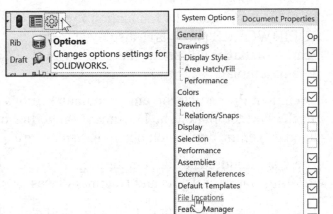

13) Click **Options** ⚙ from the Menu bar. The System Options General dialog box is displayed.

14) Click **File Locations** to set the folder path for custom Document Templates.

15) Click the **Add** button.

16) Select the **ENGDESIGN-W-SOLIDWORKS\MY-TEMPLATES** folder in the Browse For Folder dialog box.

17) Click **OK**.

18) Click **Yes**.

Each folder listed in the System Options, File Locations, Document Templates, Show Folders For option produces a corresponding tab in the New SOLIDWORKS Document dialog box.

The templates used for this project are located in the following folder: ENGDESIGN-W-SOLIDWORKS\MY-TEMPLATES.

The new tab, MY-TEMPLATES is not displayed in the New SOLIDWORKS Document dialog box until a template is added to the folder.

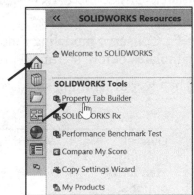

You can also use the SOLIDWORKS Resources tab from the Task Pane as illustrated. Click Property Tab Builder. Property Tab Builder opens. The center pane contains the template you are creating for the tab. The pane on the right is ready to accept page-level attributes.

Part Document Template and Document Properties

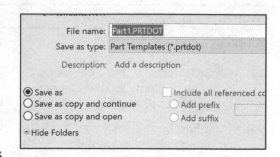

The Part template is the foundation for a SOLIDWORKS part. Part1 displayed in the FeatureManager utilizes the Part #.PRTDOT default template.

Document Properties contain the default settings for the Part template. The Document Properties include the drafting/dimensioning standard, units, dimension decimal display, grids, note font and line styles.

There are hundreds of document properties. Modify the following Document Properties: Drafting standard, Units and Decimal Places.

The Drafting standard determines the display of dimension text, arrows, symbols and spacing. Units are the measurement of physical quantities. Millimeter (MMGS) dimensioning and decimal inch (IPS) dimensioning are the two most common unit types specified for engineering parts and drawings.

Document Properties are stored with the document. Apply the Document Properties to the Part template.

Create a Part template named PART-MM-ANSI from the default Part template. Save the Custom Part template in the ENGDESIGN-W-SOLIDWORKS\MY-TEMPLATES folder. Utilize the PART-MM-ANSI Part template for all metric parts.

Conserve modeling time. Set the Document Properties and create the required templates before starting a project.

The Overall Drafting standard options are ANSI, ISO, DIN, JIS, BSI, GOST and GB.

Overall Drafting standard options:	Abbreviation:	Description:
ANSI ISO DIN JIS BSI GOST GB	ANSI	American National Standards Institute
	ISO	International Standards Organization
	DIN	Deutsche Institute für Normumg (German)
	JIS	Japanese Industry Standard
	BSI	British Standards Institution
	GOST	Gosndarstuennye State Standard (Russian)
	GB	Guo Biao (Chinese)

Display the parts, assemblies and drawings created in the GUIDE-ROD assembly in the ANSI drafting standard.

The Units Document Property assists the designer by defining the Unit System, Length unit, Angular unit, Density unit and Force unit of measurement for the Part, Assembly or Drawing document.

The Decimals option displays the number of decimal places for the Length and Angle unit of measurement.

Activity: Create a Part Document Template. Apply Document Properties.

Set Document Properties. Set overall drafting standard, units, and precision.

19) Click **Options** ⚙ from the Menu bar. The System Options General dialog box is displayed

20) Click the **Document Properties** tab. Click the **Drafting Standard** folder.

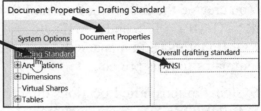

21) Select **ANSI** from the Overall drafting standard drop-down menu. Various Detailing options are available depending on the selected standard.

Set document units and precision.

22) Click **Units**.

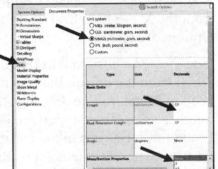

23) Click **MMGS (millimeter, grams, second)** for Unit system.

24) Select **.12** (two decimal places) for Length basic units.

25) Select **None** for Angle decimal places.

26) Click **OK** from the Document Properties - Units dialog box.

Save the Part template.

27) Click **Save As** from the Menu bar.

28) Select **Part Templates *.PRTDOT** from the Save as type box. The default Part Templates folder is displayed.

29) Click the **drop-down arrow** in the Save in box.

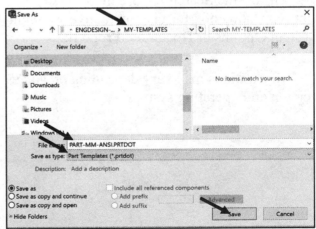

30) Select the **ENGDESIGN-W-SOLIDWORKS\MY-TEMPLATES** folder.

31) Enter **PART-MM-ANSI** in the File name box.

32) Click **Save**. PART-MM-ANSI is displayed in the FeatureManager.

Close the PART-MM-ANSI template.

33) Click **File**, **Close** from the Menu bar. All SOLIDWORKS documents are closed. If required, click **Window**, **Close All** from the Menu bar to close all open documents.

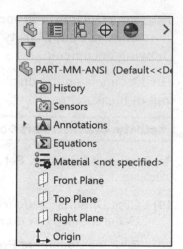

Additional details on System Options and Document Properties are available using the Help option. Index keyword: System Options, Document Templates, units and properties.

Review of the File Folders and Custom Part Template

You created the following folders: PROJECTS, MY-TEMPLATES and VENDOR-COMPONENTS to manage the documents and templates for this project.

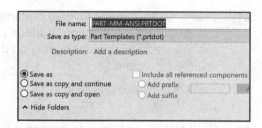

System Options, File Locations directed SOLIDWORKS to open Document Templates from the MY-TEMPLATES folder. System Options are stored in the registry of the computer.

You created a custom Part template, PART-MM-ANSI from the default Part document. Document Properties control the drafting standard, units, tolerance and other properties. Document Properties are stored in the current document. The Part template file extension is .PRTDOT.

If you modify a document property from an Overall drafting standard, a modify message is displayed as illustrated.

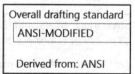

Utilize the Search feature in the Systems Options dialog box for fast interactive searching for System Options or Document Properties.

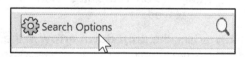

Illustrations may vary depending on your SOLIDWORKS version and operating system.

PLATE Part Overview

Determine the functional and geometric requirements of the PLATE.

Functional:

The PLATE part fastens to the customer's PISTON PLATE part in the GUIDE-CYLINDER assembly.

Fasten the PLATE part to the ROD part with a countersunk screw.

Geometric:

The dimensions of the PISTON PLATE part are 56mm x 22mm.

The dimensions of the PLATE part are 56mm x 22mm.

Locate the 4mm mounting holes with respect to the PISTON PLATE mounting holes. The mounting holes are 23mm apart on center.

The PLATE part requires a Countersink Hole to fasten the ROD part to the PLATE part.

Review the mating part dimensions before creating the PLATE part.

The GUIDE-CYLINDER assembly dimensions referenced in this project are derived from the SMC Corporation of America (www.smcusa.com).

Click **View**, **Hide/Show**, **Origins** from the Menu bar menu to display the Origin in the Graphics window.

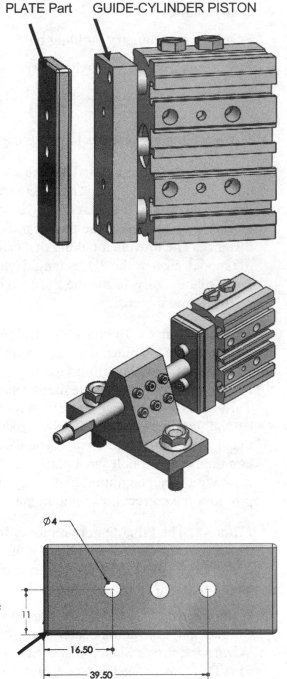

PLATE Part   GUIDE-CYLINDER PISTON

Ø4

11

16.50

39.50

Start the translation of the initial design functional and geometric requirements into SOLIDWORKS features.

What are features?

- Features are geometry building blocks.

- Features can add or remove material.

- Features are created from sketched profiles or from edges and faces of existing geometry.

Utilize the following features to create the PLATE part:

- **_Extruded Boss/Base_** 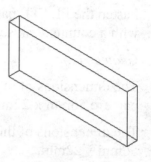: The Extruded Boss/Base feature adds material to the part. The Boss-Extrude1 (Base) feature is the first feature of the PLATE. An extrusion extends a profile along a path normal to the profile plane for a specified distance. The movement along that path becomes the solid 3D model. Sketch the 2D rectangle (close profile) on the Front plane. Fully define the Base Sketch with Geometric relations and dimensions.

- **_Extruded Cut_** 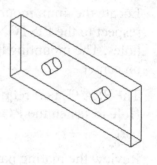: The Extruded Cut feature removes material. The Extruded Cut starts with a 2D circle sketched on the front face. The front face is your Sketch plane. Copy the sketched circle to create the second circle. Utilize the Through All Depth End Condition (Design Intent). The holes extend through the entire Boss-Extrude1 feature. Note: Use the Hole Wizard feature when you need to incorporate complex or blind holes in a part. The Hole Wizard (Call out) information can be imported directly into a drawing and provides the correct hole annotations.

- **_Fillet_** 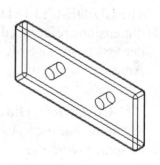: The Fillet feature removes sharp edges of the PLATE. Add Fillets to a solid, not the sketch. Group small corner edge fillets together. In this exercise, group Tangent edge fillets together.

- **_Hole Wizard_** 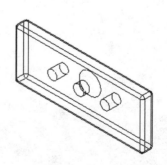: The Hole Wizard feature creates the Countersink hole at the center of the PLATE. The Hole Wizard requires a Sketch plane. Select the back face of the PLATE for your Sketch plane. Note: In this book, we focus on 2D Sketches vs. 3D Sketches.

☼ A goal of this book is to expose the new user to various SOLIDWORKS design tools and features.

Activity: Create the PLATE Part. Use a Custom Part Template.

Create a New part.

34) Click **New** ⬜ from the Menu bar. The New SOLIDWORKS Document dialog box is displayed.

35) Click the **MY-TEMPLATES** tab. This tab was created earlier in the project.

36) Double click the **PART-MM-ANSI** icon. The Part FeatureManager is displayed.

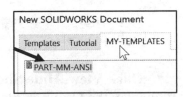

In the New SOLIDWORKS Document, Advanced option, Large icons are displayed by default. Utilize the List option or List Detail option to view the complete template name.

🔆 Set the MY-TEMPLATES tab with **Options** ⚙, **System Options, File Locations, Document Templates** option. The Templates tab is displayed in the New SOLIDWORKS Document Advanced option. If the *Novice* option is displayed, select the *Advanced* button.

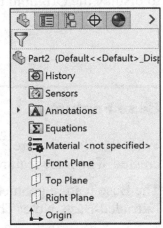

🔆 The first system default Part filename is Part1. The system attaches the .SLDPRT suffix to the created part. The second created part in the same session increments to the filename Part2.

🔆 There are numerous ways to manage a part. Use Project Data Management (PDM) systems to control, manage and document file names and drawing revisions. Use appropriate filenames that describe the part.

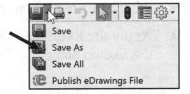

Save the part. Enter a name and description.

37) Click **Save As** 🖫.

38) Select the **ENGDESIGN-W-SOLIDWORKS\PROJECTS** folder.

39) Enter **PLATE** for File name.

40) Enter **PLATE 56MM x 22MM** for Description.

🔆 Create a smart part to create a smart assembly and drawing document.

41) Click **Save**.

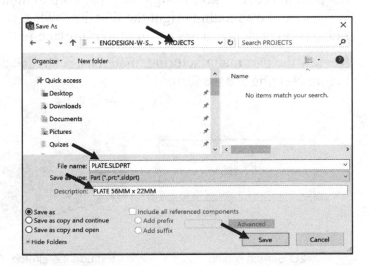

PLATE is displayed in the FeatureManager design tree. The Part 🖢 icon is displayed at the top of the FeatureManager.

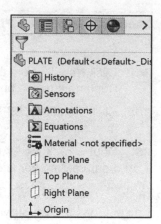

💡 The origin ⥮ represents the intersection of the Front, Top and Right planes.

💡 Click **View**, **Hide/Show**, **Origins** from the Menu bar menu to display the Origin in the Graphics window.

Base Feature

What is a Base feature? The Base feature (Boss-Extrude1) is the first feature that is created. The Base feature is the foundation of the part. Keep the Base feature simple!

The Base feature geometry for the PLATE is an extrusion. Rename the base feature Extruded-Base. Note: The default name is Boss-Extrude1.

How do you create a 3D Extruded Boss/Base feature?

- Select a Sketch plane.

- Sketch a 2D profile on the Sketch plane.

- Apply all needed Geometric relations and dimensions to fully define the sketch.

- Apply the Extruded Boss/Base feature tool. Extend the profile perpendicular (⊥) to the Sketch plane.

💡 It is considered very poor design practice not to fully define a sketch. You will be tempted in order to save design time, but keep in mind that the extra couple of minutes you take to do something right the first time will save you additional time in the end. Fully defined sketches are required to manufacture the part.

Review the part manufacturing, materials, reference planes and orthographic projection before creating a Base feature.

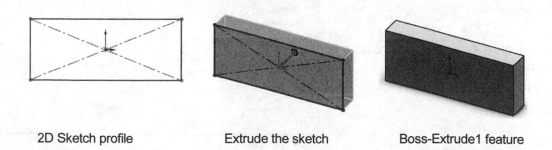

2D Sketch profile   Extrude the sketch   Boss-Extrude1 feature

Machined Part

In earlier conversations with manufacturing, a decision was made that the part would be machined.

Your material supplier stocks raw material in rod, sheet, plate, angle and block forms.

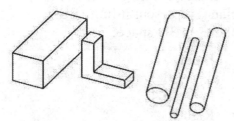

Block, Angle and Rod Stock

You decide to start with a standard plate form. A standard plate form will save time and money.

Select the best profile for the extrusion. The best profile is a simple 2D rectangle.

The boss and square internal cuts are very costly to manufacture from machined bar stock.

As a designer, review your manufacturing options.

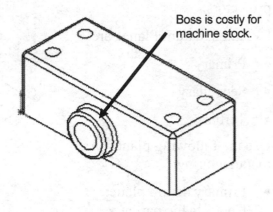

Boss is costly for machine stock.

Utilize standard material thickness and hole sizes in the PLATE design. Utilize standard stock in the ROD design.

Utilize slot cuts in the GUIDE design. Machined parts require dimensions to determine the overall size and shape. Datum planes determine the location of referenced dimensions. Add geometric relations and dimensions to fully define the geometry Extruded Boss/Base feature.

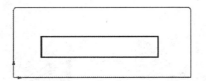

Internal Square Cut (expensive)

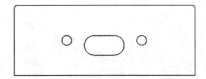

Holes and Slots (less expensive)

Understanding reference planes and views is important as you design machined parts. Before you create the PLATE, review the next topic on Reference Planes and Orthographic Projection.

Reference Planes and Orthographic Projection

The three default ⊥ Reference planes represent infinite 2D planes in 3D space:

- Front
- Top
- Right

Planes have no thickness or mass. Orthographic projection is the process of projecting views onto parallel planes with ⊥ projectors.

Default ⊥ datum Planes are:

- Primary
- Secondary
- Tertiary

Use the following planes in manufacturing:

- Primary datum plane: Contacts the part at a minimum of three points.

- Secondary datum plane: Contacts the part at a minimum of two points.

- Tertiary datum plane: Contacts the part at a minimum of one point.

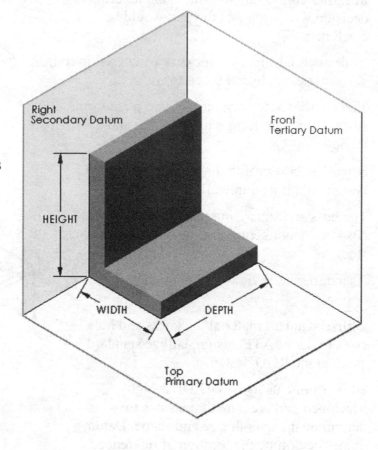

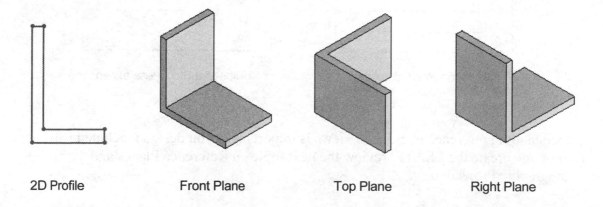

2D Profile   Front Plane   Top Plane   Right Plane

The part view orientation is dependent on the first feature Sketch plane. Compare the available default Sketch planes in the FeatureManager: Front Plane, Top Plane and Right Plane.

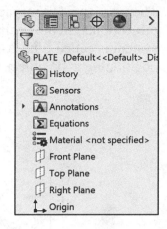

Each Extruded feature above was created with an L-shaped 2D Sketch profile. There are six principle views of Orthographic projection listed in the ASME Y14.3M standards:

- Top

- Front

- Right side

- Bottom

- Rear (Back)

- Left side

SOLIDWORKS Standard view names correspond to these Orthographic projection view names.

ASME Y14.3 Principle View Name:	SOLIDWORKS Standard View:
Front	Front
Top	Top
Right side	Right
Bottom	Bottom
Rear	Back
Left side	Left

The most common standard drawing views in Third angle Orthographic projection are:

- Front

- Top

- Right

- Isometric

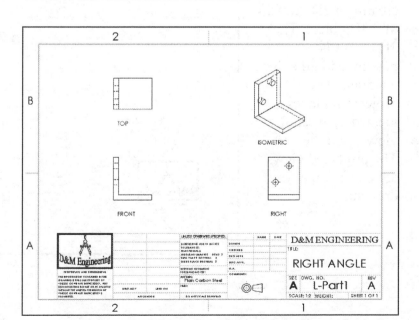

There are two Orthographic projection drawing systems. The first Orthographic projection system is called Third Angle Projection. The second Orthographic projection system is called First Angle Projection. The two drawing systems are derived from positioning a 3D object in the Third or First quadrant as illustrated.

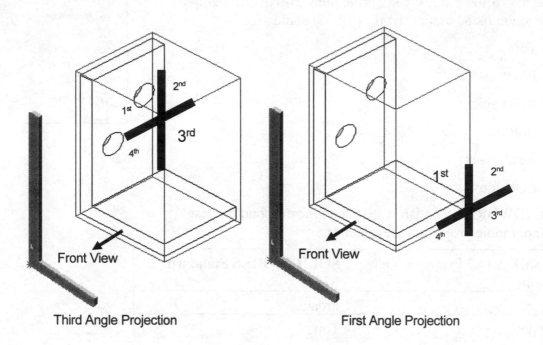

Third Angle Projection

First Angle Projection

Third Angle Projection

Third angle projection is a method of creating a 2D drawing of a 3D object. The 3D object is positioned in the third quadrant. The 2D projection planes are located between the viewer and the part. The projected views are placed on a drawing.

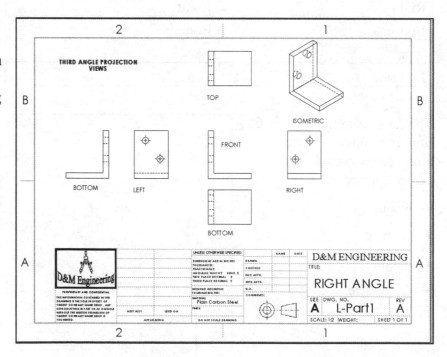

First Angle Projection

First-angle projection is a method of creating a 2D drawing of a 3D object. It is mainly used in Europe and Asia. The 3D object is positioned in the first quadrant. Views are projected onto the planes located behind the part. The projected views are placed on a drawing.

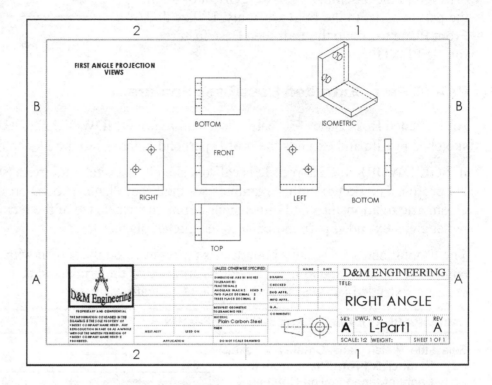

Third Angle Projection is primarily used in the U.S. & Canada and is based on the ASME Y14.3M multi and sectional view drawings standard. Designers should have knowledge and understanding of both systems.

There are numerous multi-national companies. Example: A part is designed in the U.S., manufactured in Japan and destined for a European market.

Third Angle Projection is used in this text. A truncated cone symbol should appear on the drawing to indicate the Projection system.

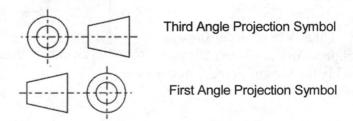

Third Angle Projection Symbol

First Angle Projection Symbol

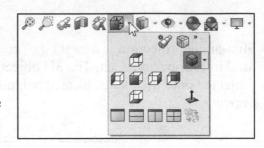

Utilize the Heads-up View toolbar or the Orientation dialog box to orient or set view style for your model.

Before incorporating your design intent into the Sketch plane, ask a question: How will the part be oriented in the assembly? Answer: Orient or align the part to assist in the final assembly. Utilize the Front Plane to create the Extruded Base feature for the PLATE part.

PLATE Part - Extruded Boss/Base Feature

An Extruded Boss/Base feature is a feature in SOLIDWORKS that utilizes a sketched profile and extends the profile perpendicular ($\perp$) to the Sketch plane.

In SOLIDWORKS, a 2D profile is called a sketch. A sketch requires a Sketch plane and a 2D profile. The sketch in this example uses the Front Plane. The 2D profile is a rectangle. Geometric relationships and dimensions define the exact size of the rectangle. The rectangle is extruded perpendicular to the Sketch plane.

The Front Plane's imaginary boundary is represented on the screen with four visible edges. Planes are flat and infinite. Planes are used as the primary sketch surface for creating Extruded Boss/Base and Extruded Cut features.

Activity: PLATE Part - Extruded-Base Feature

Select the Sketch plane. Create a 2D sketch.

42) Right-click **Front Plane** from the FeatureManager. The Context toolbar is displayed.

43) Click **Sketch** from the Context toolbar. The Sketch toolbar is displayed. Front Plane is your Sketch Plane. The grid is de-activated to improve picture clarity.

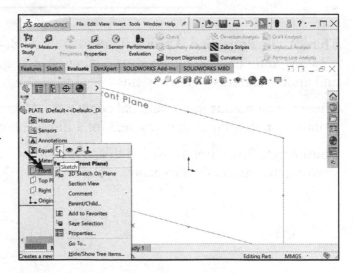

The plane or face you select for the Base sketch determines the orientation of the part.

Click **View, Hide/Show, Origins** from the Menu bar menu to display the Origin in the Graphics window.

You can also click the Front Plane from the
FeatureManager and click the Sketch tab from
the CommandManager. Front Plane is the Sketch
plane and the Sketch toolbar is displayed.

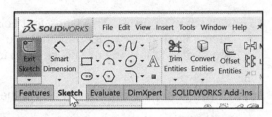

The Origin ↕ represents the intersection of
the Front, Top and Right Planes. The left corner
point of the rectangle is Coincident with the
origin. The Origin is displayed in red.

Sketch a Corner rectangle.

44) Click the **Corner Rectangle** ⬜ tool from the Sketch toolbar. The sketch opens in the Front

view. The mouse pointer displays the Corner Rectangle feedback symbol ⬜. The Front
Plane feedback indicates the current Sketch plane.

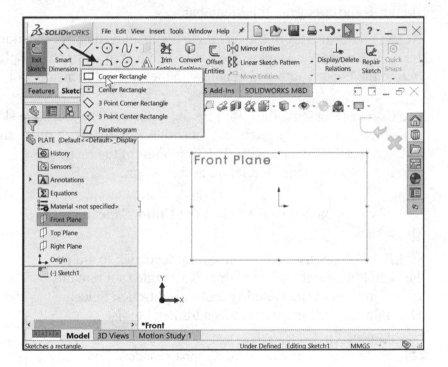

The Rectangle-based tool uses a Consolidated Rectangle PropertyManager. The
SOLIDWORKS application defaults to the last used rectangle type.

45) Click the **Origin** ↳. This is the first point of the rectangle. It is very important that you always reference the sketch to the origin. This helps to fully define the sketch.

46) Drag the **mouse pointer** up and to the right.

47) Release the **mouse button** to create the second point and the rectangle. The first point is Coincident to the Origin.

View the illustrated sketch relations in the Graphics window.

48) Click **View**, **Hide/Show**, **Sketch Relations** from the Menu bar. The sketch relations are displayed in the Graphics window.

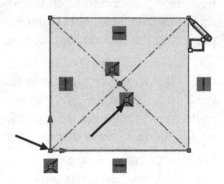

🔆 To deactivate the illustrated sketch relations, click **View**, **Hide/Show**, and uncheck **Sketch Relations** from the Menu bar.

The X-Y coordinates of the rectangle are displayed above the mouse pointer as you drag the mouse pointer up and to the right. The X-Y coordinates display different values. Define the exact width and height with the Smart dimension tool.

The Corner Rectangle tool ⬜ remains selected. The rectangle sketch contains predefined geometric sketch relations: two Horizontal ⁻ relations and two Vertical ⏐ relations. The first point of the rectangle contains a Coincident ⬜ relation at the Origin.

🔆 If you make a mistake, select the **Undo** ↺ icon in the Menu bar.

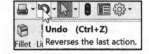

SOLIDWORKS uses color and cursor feedback to aid in the sketching process. The Corner Rectangle tool remains active until you select another tool or right-click Select. The right mouse button contains additional tools.

Deactivate the Corner Rectangle tool.

49) Right-click a **position** in the Graphics window. The Context toolbar is displayed.

50) Click **Select**. The mouse pointer displays the Select icon and the Corner Rectangle Sketch tool is deactivated.

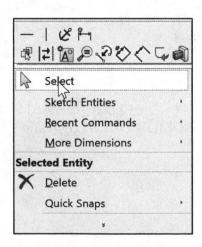

🔆 The phrase "Right-click Select" means **right-click** the mouse in the Graphics window. Click **Select**. This will de-select the current tool that is active.

Size the geometry of the rectangle.

51) Click and drag the **top horizontal line** of the rectangle upward. View the Line Properties PropertyManager.

52) **Right-click** and view your options.

53) Release the **mouse button**.

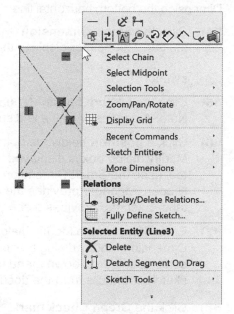

The Line Properties PropertyManager is displayed to the left of the Graphics window.

The selected line displays a Horizontal relation in the Existing Relations box.

The geometry consists of four lines.

Context menus and toolbars save time. Commands and tools vary depending on the mouse position in the SOLIDWORKS window and the active Sketch tool and history.

💡 Rename a feature or sketch. Slowly click the feature or sketch name twice and enter the new name when the old one is highlighted.

💡 Design Intent is how your part reacts as parameters are modified. Example: If you have a hole in a part that must always be .125≤ from an edge, you would dimension to the edge rather than to another point on the sketch. As the part size is modified, the hole location remains .125≤ from the edge.

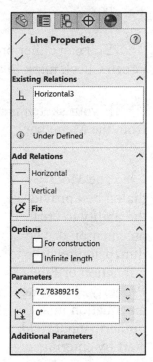

Dimension the bottom horizontal line.

54) Click the **Smart Dimension** tool from the Sketch toolbar. The pointer displays the dimension feedback symbol.

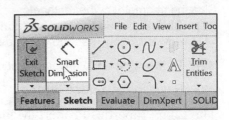

55) Click the **bottom horizontal line** of the rectangle. Note: A dimension value is displayed.

56) Click a **position** below the bottom horizontal line. The Modify dialog box is displayed. The Smart Dimension tool uses the Smart Dimension PropertyManager. The PropertyManager provides the ability to select three tabs. Each tab provides a separate menu.

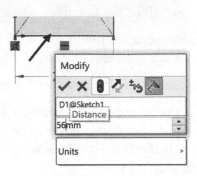

57) Enter **56**mm in the Modify dialog box. Note: The Dimension Modify dialog box provides the ability to select a unit drop-down menu to modify units in a sketch or feature from the document properties.

58) Click the **Green Check mark** in the Modify dialog box.

💡 You can either press the enter key or click the Green Check mark to accept the dimension value.

💡 If your sketch is not correct, select **UNDO** ↰ from the Menu bar.

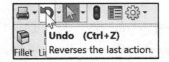

💡 The Modify dialog box provides the ability to create an equation driven dimension relative to a function or File Property. To create an equation in a numeric input field, start by entering an = (equal sign). A drop-down list displays options for Global Variables, functions and file properties.

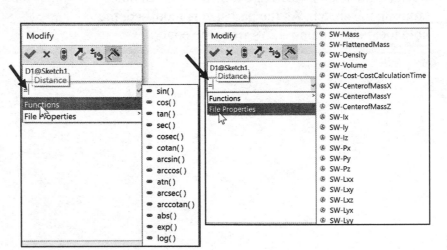

Dimension the vertical line.

59) Click the **left vertical line**.

60) Click a **position** to the left of the vertical line.

61) Enter **22**mm.

62) Click the **Green Check mark** ✔ in the Modify dialog box. The sketch is fully defined. All lines and vertices are displayed in black.

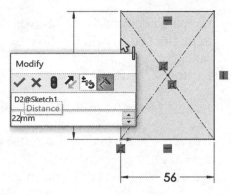

💡 To flip the arrow direction, click the control point as illustrated.

Deactivate the sketch relations and select the dimension value.

63) Click **View**, **Hide/Show**, and uncheck **Sketch Relations** from the Menu bar.

64) Right-click **Select** in the Graphics window.

65) Position the **mouse pointer** over the dimension text. The pointer changes to a linear dimension symbol with a displayed text box. D2@Sketch1 represents the second linear dimension created in Sketch1.

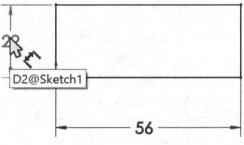

Modify the dimension text.

66) Double-click the **22** dimension text in the Graphics window.

67) Enter **42**mm.

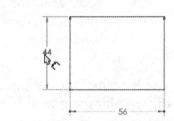

💡 Click the **Spin Box Arrows** ⬍ to increase or decrease dimensional values.

Return to the original vertical dimension.

68) Double-click the **42**mm dimension text in the Graphics window.

69) Enter **22**mm.

70) Click the **Green Check mark** ✔ from the Modify dialog box.

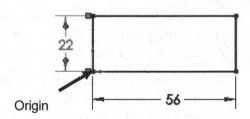

💡 Click the **Undo** ↺ tool to return to the original dimension.

Options in the Modify dialog box:

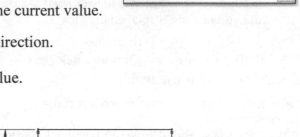

- Green Check mark ✔ - Saves the current value and exits the Modify dialog box.

- Restore ✖ - Restores the original value and exits.

- Rebuild 📍 - Rebuilds the model with the current value.

- Reverse ↗ - Reverse the sense of the direction.

- Reset ⁺ᵗ꜄ - Reset the spin increment value.

- Mark for Drawing ⬦ - Set by default. Inserts part dimensions into the drawing.

- Set Functions and File Properties. Create on the fly Global Variables. Enter an = sign as illustrated to start an equation in the Modify dialog box. View the available drop-down menu for functions, file properties and any created variables.

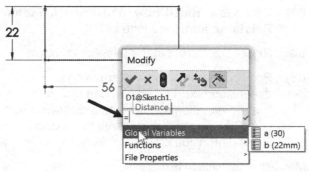

The System displays Under Defined in the Status bar located in the lower right corner of the Graphics window. In an under defined Sketch, the entities that require position, dimensions or Sketch relations are displayed in blue.

In a fully defined sketch, all entities are displayed in black. The Status bar displays Fully Defined. In machining practices, parts require fully defined sketches. *A fully defined sketch has defined positions, dimensions and/or relationships.*

In an over defined sketch, there is geometry conflict with the dimensions and or relationships. In an over defined sketch, entities are displayed in red. SOLIDWORKS provides the SketchXpert tool to correct an over defined sketch. The SketchXpert PropertyManager is displayed by default for an over defined sketch. Click the Diagnose button and accept an option to correct the over defined sketch.

☼ Although dimensions are not required to create features in SOLIDWORKS, dimensions provide location and size information. Models require dimensions for manufacturing. Dimension the rectangle with horizontal and vertical dimensions.

Insert an Extruded Boss/Base feature. This is the first (Base) feature of the model. Utilize the default Blind End Condition.

71) Click the **Features** tab from the CommandManager.

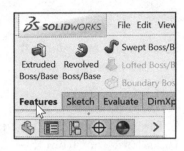

72) Click the **Extruded Boss/Base** ⬒ tool from the Features toolbar. The Boss-Extrude PropertyManager is displayed. The extruded sketch is previewed in a Trimetric view. The preview displays the direction of the extrude feature.

Reverse the direction of the extruded depth.

73) Click the **Reverse Direction** box. Note the location of the Origin in the Graphics window. Blind is the default End Condition for Direction 1.

74) Enter **10**mm for Depth in Direction1.

75) Click **OK** ✔ from the Boss-Extrude PropertyManager. The Boss-Extrude1 feature is displayed in the Graphics window. The name Boss-Extrude1 is displayed in the PLATE FeatureManager.

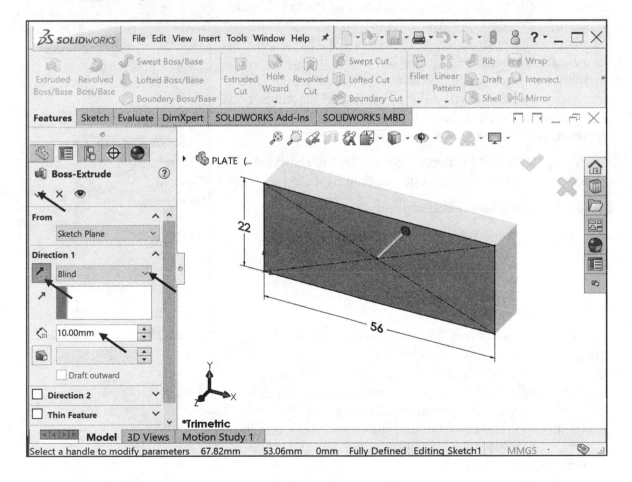

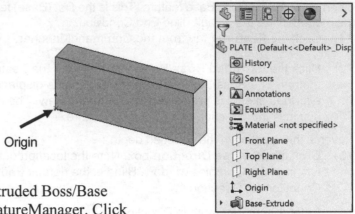

The OK ✔ button accepts and completes the Feature process.

If you exit the sketch before selecting the Extruded Boss/Base tool from the Features toolbar, Sketch1 is displayed in the FeatureManager. To create the Extruded Boss/Base feature, click Sketch1 from the FeatureManager. Click Extruded Boss/Base to create the feature.

For many features (Extruded Boss/Base, Extruded Cut, Simple Hole, Revolved Boss/Base, Revolved Cut, Fillet, Chamfer, Scale, Shell, Rib, Circular Pattern, Linear Pattern, Curve Driven Pattern, Revolved Surface, Extruded Surface, Fillet Surface, Edge Flange and Base Flange) you can enter and modify equations directly in the PropertyManager fields that allow numerical inputs. You can create equations with Global Variables, functions, and file properties without accessing the Equations, Global Variables and Dimensions dialog box.

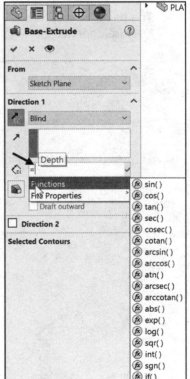

For example, in the PropertyManager for the Extruded Boss/Base feature, you can enter equations in:

- Depth fields for Direction 1 and Direction 2.

- Draft fields for Direction 1 and Direction 2.

- Thickness fields for a Thin Feature with two direction types.

- Offset Distance field.

To create an equation in a numeric input field, start by entering an = (equal sign). A drop-down list displays options for Global Variables, functions and file properties. Numeric input fields that contain equations can display either the equation itself or its evaluated value. You can toggle between the equation and the value by clicking the Equations or Global Variable button that appears at the beginning of the field.

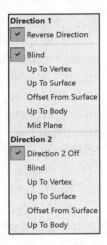

In edit mode, right-click anywhere on an extruded feature and modify the end condition from a pop-up shortcut menu. Click in empty space, on geometry, or on the handle. The pop-up shortcut menu provides the document dependent options for Direction 1 and Direction 2.

Display an Isometric view of the model. Press the **space bar** to display the Orientation dialog box. Click the **Isometric view** icon.

PLATE Part - Modify Dimensions and Rename

Incorporate a design change into the PLATE. Modify dimension values in the Modify dialog box or directly in the Graphics window. Utilize the Rebuild tool to update the Extruded Boss/Base feature.

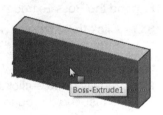

Either double-click the Boss-Extrude1 folder from the FeatureManager or double-click on the feature in the Graphics window to display the model dimensions. Rename entries by selecting on the text (slowly double-click) in the FeatureManager.

Activity: PLATE Part - Modify Dimensions and Rename

Modify the PLATE part.
76) Double-click on the **front face** of Boss-Extrude1 in the Graphics window as illustrated. Boss-Extrude1 is highlighted in the FeatureManager.

Fit the model to the Graphics window.
77) Press the **f** key.

Modify the width dimension.
78) Click the **10**mm dimension in the Graphics window.

79) Enter **5**mm.

Display an Isometric view.
80) Press the **space bar** to display the Orientation dialog box.

81) Click **Isometric view** . You can also access the Isometric view tool from the Heads-up View toolbar.

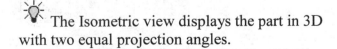

The Isometric view displays the part in 3D with two equal projection angles.

The Instant3D tool provides the ability to drag geometry and dimension manipulator points to resize or create new features in the Graphics window. Use the on-screen ruler to measure and apply modifications. In this book, you will primarily use the PropertyManager and dialog box to modify model dimensions and to create new features. It is very difficult to apply design intent (End Conditions) using the Instant 3D tool. See SOLIDWORKS Help for additional information.

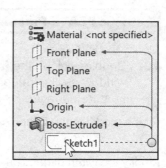

Incorporate the machining process into the PLATE design. Reference dimensions from the three datum planes with the machined Origin in the lower left hand corner of the PLATE.

Maintain the Origin on the front lower left hand corner of the Extruded Boss/Base feature and co-planar with the Front Plane.

Rename Sketch1 and Boss-Extrude1. The first Extruded Boss/Base feature is named Boss-Extrude1 in the FeatureManager. Sketch1 is the name of the sketch utilized to create the Boss-Extrude1 feature.

The Arrow ‣ icon indicates that additional feature information is available.

Expand the Boss-Extrude1 entry from the FeatureManager.

82) Click the **Arrow** ‣ icon of the Boss-Extrude1 feature in the FeatureManager.

Rename Sketch1.

83) Slowly double-click **Sketch1** in the FeatureManager. A white box is displayed around the selected item as illustrated.

84) Enter **Sketch-Base** for the sketch name. The Down arrow ⌄ icon indicates that the feature information is expanded.

85) Click the **Down arrow** ⌄ icon to collapse the Boss-Extrude1 feature.

Rename the Boss-Extrude1 feature.

86) Slowly double-click **Boss-Extrude1** in the FeatureManager. A white box is displayed around the selected item.

87) Enter **Base-Extrude** for the feature name as illustrated.

Save the PLATE.

88) Click **Save** 💾 .

Display Modes, View Modes, View tools and Appearances

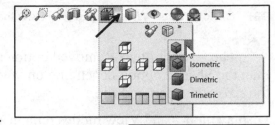

Access the display modes, view modes, view tools and appearances from the Standard Views toolbar and the Heads-up View toolbar. Apply these tools to display the required modes in your document.

The Apply scene 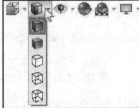 tool adds the likeness of a material to a model in the Graphics window without adding the physical properties of the material. Select a scene from the drop-down menu.

Activity: Display Modes, View Modes, View tools and Appearances

View the display modes from the Heads-up View toolbar.

89) Click **Wireframe** ⌗. View the results.

90) Click **Hidden Lines Visible** ⌗. View the results.

91) Click **Hidden Lines Removed** ⬦. View the results.

92) Click **Shaded** ▮. View the results.

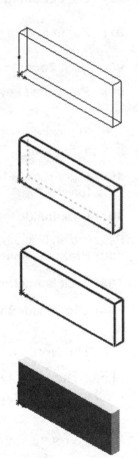

93) Click **Shaded With Edges** 🔲. View the results.

💡 The Hidden Lines Removed option may take longer to display than the Shaded option depending on your computer configuration and the size of the part and file.

Display modes and View modes remain active until deactivated.

Display the View modes.

94) Click **Previous view** to display the previous view of the part in the current window.

95) Click **Zoom to Fit** to display the full size of the part in the current window.

96) Click **Zoom to Area**.

97) **Zoom in** on a corner of the PLATE.

98) Click **Zoom to Area** to deactivate the tool.

99) Press the **f** key to fit the model to the Graphics window.

💡 Press the lower case z key to zoom out. Press the upper case Z key to zoom in.

100) Right-click **Select** in the Graphics window. View the available view tools.

101) Click **inside** the Graphics window.

Return to an Isometric Shaded With Edges view.

102) Press the **space bar** to display the Orientation dialog box.

103) Click **Isometric view** 🔲. View the results.

104) Click **Shaded With Edges** 🔲 from the Heads-up View toolbar.

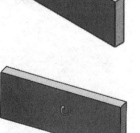

💡 The Normal To view ⚓ tool displays the part ⊥ to the selected plane. The View Orientation tool 🖋 from the Standard Views toolbar creates and displays a custom named view.

💡 To rotate your model, use the middle mouse button. The rotate icon ↻ is displayed.

PLATE Part - Extruded Cut Feature

An Extruded Cut  feature removes material perpendicular ⊥ to the sketch for a specified depth. The Through All End Condition option creates the holes through the entire depth of the PLATE.

Utilize the Circle Sketch tool to create the hole profile on the front face. The first circle is defined with a center point and a point on its circumference. The second circle utilizes the Ctrl-C (Copy) key to create a copy of the first circle.

Geometric relations are constraints between one or more entities. The holes may appear aligned and equal, but they are not! Add two relations: Insert an Equal relation between the two circumferences and a Horizontal relation between the two center points.

The machinist references all dimensions from the reference datum planes to produce the PLATE holes. Reference all three linear dimensions from the Right datum plane (left vertical edge) and the Top datum plane (bottom horizontal edge). Utilize a dimension to define the diameter of the circle.

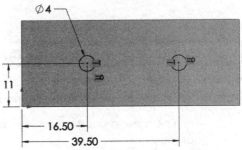

The linear dimensions are positioned to the left and below the Boss-Extrude1 feature. You will learn more about dimensioning to a drawing standard in Project 3.

☀ Position the dimensions off the sketch profile. Place smaller dimensions to the inside. These dimensioning techniques in the part sketch will save time in dimensioning the drawing.

Design intent is not a static concept that controls changing geometry. As the designer, think about how to maintain the design intent with geometric changes to the model. Dimension and geometric relation schemes vary depending on the design intent of the part. In the next section there are alternate dimension schemes to define the position of two circles.

The following two examples are left as an exercise.

Example A: The center points of the holes reference a horizontal centerline created at the Midpoint of the two vertical lines. A dimension is inserted between the two center points.

Example B: The center points of the circles reference the horizontal line. Insert a vertical line at the Midpoint of the two horizontal lines. The center points of the circles utilize a Symmetric relation with the vertical centerline.

Dimension schemes defined in the part can also be redefined in the future drawing.

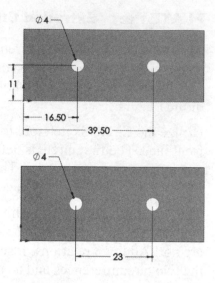

Activity: PLATE Part - Extruded Cut Feature

View the mouse pointer feedback.

105) Position the mouse pointer on the **front face**. View the Select Face icon

. The mouse pointer provides feedback depending on the selected geometry.

106) Position the mouse pointer on the **right edge**. View the Select Edge icon .

Select the Sketch plane for the holes.

107) Right-click the **front face** of the Base-Extrude feature in the Graphics window as illustrated.

Create a sketch.

108) Click **Sketch** from the Context toolbar. The Sketch toolbar is displayed.

Display the Front view.

109) Click **Front view** . The Origin is located in the lower left corner of the part.

110) Click **Hidden Lines Visible** from the Heads-up View toolbar.

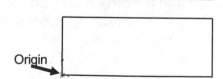

Sketch the first circle.

111) Click the **Circle** Sketch tool. The cursor displays the Circle feedback symbol . The Circle PropertyManager is displayed.

Origin

112) Click a **center point** on the front face diagonally to the right of the Origin as illustrated.

113) Click a **position** to the right of the center point. The circumference and the center point are selected.

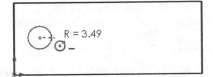

Copy the first circle.

114) Right-click **Select** in the Graphics window to deselect the circle Sketch tool.

115) Hold the **Ctrl** key down.

116) Click and drag the **circumference** of the first circle to the right.

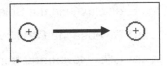

117) Release the **mouse button**.

118) Release the **Ctrl** key. The second circle is displayed.

119) Click a position in the **Graphics window**, off the sketch. The first and second circles are displayed in blue.

Add an Equal relation between the circles.

120) Click the **circumference** of the first circle.

121) Hold the **Ctrl** key down.

122) Click the **circumference** of the second circle.

123) Release the **Ctrl** key. Arc1 and Arc2 are displayed in the Selected Entities text box in the Properties PropertyManager.

124) Click **Equal** from the Add Relations box. The Equal radius/length is displayed in the Existing Relations text box.

125) Click **OK** from the Properties PropertyManager.

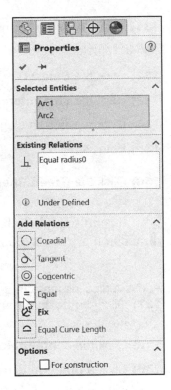

You can right-click the **Make Equal** tool from the Context toolbar.

The SOLIDWORKS default name for curve entities is Arc#. There is an Arc# for each circle. The default name for any point is Point#. The two center points are entities named Point2 and Point4.

To remove unwanted entities from the Selected Entities box, right-click in the **Selected Entities** window. Click **Clear Selections**. The Selected Entities # differs if geometry was deleted and recreated. The Existing Relations box lists Equal radius/length0.

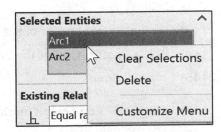

Add a Horizontal relation between the first circle center point and the
second circle center point.

126) Click the first circle **center point**.

127) Hold the **Ctrl** key down.

128) Click the second circle **center point**. The Properties
PropertyManager is displayed. The selected entities are
displayed in the Selected Entities box.

129) Release the **Ctrl** key.

130) Click **Horizontal** from the Add Relations box.

131) Click **OK** ✔ from the Properties PropertyManager.

Add a diameter dimension using the Smart Dimension tool.

132) Click the **Smart Dimension** ✎ Sketch tool.

133) Click the **circumference** of the first circle.

134) Click a **position** diagonally upward.

135) Enter **4**mm.

136) Click the **Green Check mark** ✔ from the Modify dialog box.

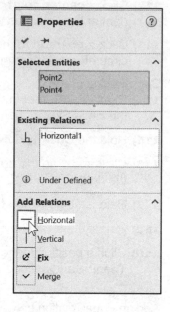

💡 If required, click the dimension arrowhead to toggle the
dimension arrow as illustrated.

Add a vertical dimension.

137) Click the **bottom horizontal** line of the Base-Extrude
feature.

138) Click the **center point** of the first circle.

139) Click a **position** to the left of the profile.

140) Enter **11**mm.

141) Click the **Green Check mark** ✔ from the
Modify dialog box.

Add a horizontal dimension.

142) Click the **left vertical** line of the Base-Extrude
feature.

143) Click the **center point** of the first circle.

144) Click a **position** below the profile.

145) Enter **16.5**mm.

146) Click the **Green Check mark** ✔ from the
Modify dialog box.

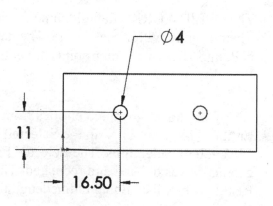

Add a second horizontal dimension.
147) Click the **left vertical** line of the Base-Extrude feature.

148) Click the **center point** of the second circle.

149) Click a **position** below the 16.50 dimension in the Graphics window.

150) Enter **39.5**mm.

151) Click the **Green Check mark** ✓ from the Modify dialog box.

Fit the model to the Graphics window.
152) Press the **f** key.

The drafting standard preference is to place arrows to the inside of the extension lines. Select the arrowhead control blue dots to alternate the arrowhead position.

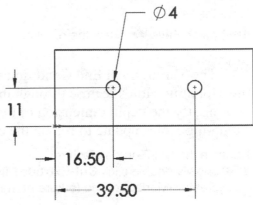

If required, flip the dimension arrows to the inside.
153) Click the **16.50**mm dimension text in the Graphics window.

154) Click the **arrowhead dot** to display the arrows inside the extension lines.

155) If required, repeat the flip dimension arrow procedure for the **11**mm dimension text.

Fit the PLATE to the Graphics window.
156) Press the **f** key.

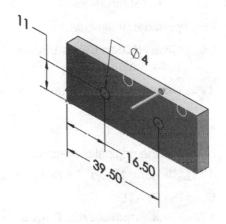

157) Press the **space bar** to display the Orientation dialog box.

158) Click **Isometric view** . View the results.

159) Click **Shaded With Edges** from the Heads-up View toolbar.

Create an Extruded Cut feature.
160) Click the **Features** tab from the CommandManager.

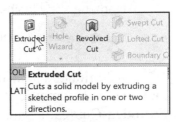

161) Click the **Extruded Cut** tool from the Features toolbar. The Cut-Extrude PropertyManager is displayed. The direction arrow points into the Extruded feature as illustrated.

162) Select **Through All** for End Condition in Direction1 as illustrated. View the End Condition options.

163) Click **OK** ✓ from the Cut-Extrude PropertyManager. Cut-Extrude1 is displayed in the FeatureManager.

Fit the PLATE to the Graphics window.
164) Press the **f** key.

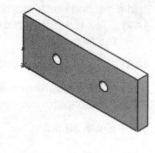

Save the PLATE.
165) Click **Save** 💾. View the results.

💡 The Through All End Condition option in Direction 1 creates the Mounting Holes feature through the Base-Extrude feature. As you modify the depth dimension of the Base-Extrude, the Mounting Holes update to reflect the change.

Rename the Cut-Extrude1 feature.
166) Slowly double click **Cut-Extrude1** from the FeatureManager. The feature name is highlighted.

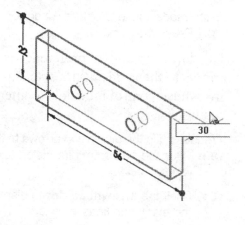

167) Enter **Mounting Holes** for new name.

Modify the Base-Extrude depth.
168) Double-click **Base-Extrude** from the FeatureManager.

Modify the depth dimension.

169) Click **Hidden Lines Visible** ⬚ from the Heads-up View toolbar.

170) Double-click the **5**mm dimension from the Graphics window.

171) Enter **30**mm.

Return to the original dimension.

172) Click the **Undo** ↺ tool from the Menu bar.

Save the PLATE.
173) Click **Shaded With Edges** ⬛ from the Heads-up View toolbar.

174) Click **Save** 💾. View the results.

🔍 Additional details on Circles, Dimension, Geometric Relations and Extruded-Cut feature are available in SOLIDWORKS Help Topics. Index keyword: Circles, Dimension, Add Relations in Sketch and Extruded Cut.

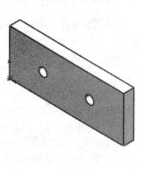

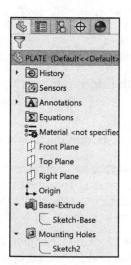

💡 A goal of this book is to expose the new user to various SOLIDWORKS design tools and features.

 Review the Extruded Boss/Base and Extruded Cut
Features

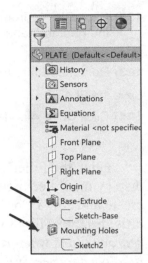

An Extruded Boss/Base feature adds material. An Extruded Cut
feature removes material. Extrude features require:

- Sketch Plane

- Sketch

- Geometric relations and dimensions

- Depth and End Condition

PLATE Part - Fillet Feature

The Fillet feature 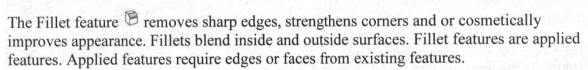 removes sharp edges, strengthens corners and or cosmetically
improves appearance. Fillets blend inside and outside surfaces. Fillet features are applied
features. Applied features require edges or faces from existing features.

There are many fillet options. The Fillet PropertyManager provides the ability to select
either the Manual or FilletXpert tab. Each tab has a separate menu. The Fillet
PropertyManager displays the appropriate selections based on the type of fillet you
create.

In the next activity, select the Manual tab and single edges for the first Fillet. Select the
Tangent propagation option for the second Fillet.

On castings, heat-treated machined parts and plastic molded parts, implement Fillets
into the initial design. If you are uncertain of the exact radius value, input a small test
radius of 1mm. It takes less time for the manufacturing supplier to modify an existing
Fillet dimension than to create a new one.

Activity: PLATE Part - Fillet Feature

Insert a Constant radius edge Fillet.

175) Click **Hidden Lines Visible** from the Heads-up View toolbar.

176) Click the **Fillet** Features tool. The Fillet PropertyManager is
displayed.

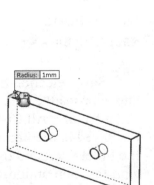

177) Click the **Manual** tab from the Fillet PropertyManager.

178) Click the **Constant Size Fillet** type icon.

179) Enter **1.0**mm in the Fillet Parameters Radius box.

180) Click the **top left corner edge** as illustrated. The fillet option pop-
up toolbar is displayed. Options are model dependent.

181) Select the **Connected to start loop, 3 Edges icon**. The four
selected edges are displayed in the Edges, Faces, Features, and
Loop box.

182) Click **OK** ✔ from the Fillet PropertyManager. Fillet1 is displayed in the FeatureManager.

Rename Fillet1.
183) Rename **Fillet1** to **Small Edge Fillet**.

Save the PLATE.
184) Click **Save** 💾. View the results.

🔅 Click the FilletXpert tab in the Fillet PropertyManager to display the FilletXpert PropertyManager. The FilletXpert can only create and edit a Constant radius fillet type. The FilletXpert provides the ability to *Create multiple fillets, automatically reorder fillets when required, and manage the desired type of fillet corner.* See SOLIDWORKS Help for additional information.

Insert a Constant Size tangent edge Fillet.
185) Click the **front top horizontal edge** as illustrated.

186) Click the **Fillet** 🔘 Features tool. The Fillet PropertyManager is displayed. The Tangent propagation check box is checked.

187) Click the **Manual** tab.

188) Click the **back top horizontal edge**. The selected edges Edge<1>, Edge<2> are displayed in the Items To Fillet box.

189) Repeat the process for the **bottom edges**.

190) Enter **1.0**mm in the Fillet Radius box.

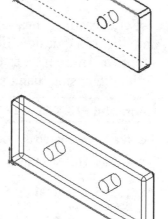

191) Click **OK** ✔ from the Fillet PropertyManager. Fillet2 is displayed in the FeatureManager.

Rename Fillet2.
192) Rename **Fillet2** to **Front-Back Edge Fillet**.

Save the PLATE.
193) Click **Save** 💾.

🔅 Save selection time. Utilize Ctrl-Select and choose all edges to fillet. Select the Fillet feature. All Ctrl-Select edges are displayed in the Items to Fillet box, or apply the FilletXpert and use the connected to start the loop option.

💡 Minimize the number of Fillet radius sizes created in the FeatureManager. Combine Fillets and Rounds that have a common radius. Select all Fillet/Round edges. Add edges to the Items to Fillet list in the Fillet PropertyManager. Select the small edges, and then fillet the larger edges.

PLATE Part - Hole Wizard Feature

The PLATE part requires a Countersink hole. Apply the Hole Wizard 🗗 feature. The Hole Wizard creates simple and complex Hole features by stepping through a series of options (Wizard) to define the hole type and hole placement. The Hole Wizard requires a face or Sketch plane to position the Hole feature. Select the back face.

Activity: PLATE Part - Hole Wizard Feature

Rotate the view.
194) Click and drag the **mouse pointer** using the middle mouse key
to the left to rotate the part as illustrated. The Rotate icon ↻ is displayed.

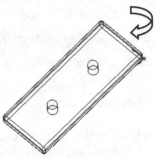

Display the Back view.
195) Click **Back view** 🗗 from the Heads-up View toolbar. The Origin is displayed at the lower right corner.

Insert the Countersink hole.
196) Click the **Hole Wizard** 🗗 Feature tool. The Hole Specification PropertyManager is displayed.

197) Click the **Type** tab.

198) Click the **Countersink** icon.

199) Select **ANSI Metric** for Standard.

200) Select **Flat Head Screw - ANSI B18.6.7M** for Type.

201) Select **M4** for Size.

202) Select **Through All** End Condition. Accept the default settings.

203) Click the **Positions** tab.

Select the Sketch plane for the Hole Wizard.
204) Click the **middle back face** of Base-Extrude as illustrated. The
Point ▱ tool is selected.

205) Click **again to place** the center point of the hole.

206) Right-click **Select** in the Graphics window to de-select the Point tool.

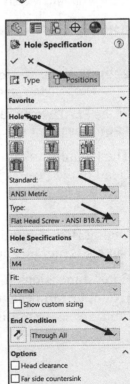

Add a horizontal dimension.

207) Click the **Smart Dimension** ↖ Sketch tool. The Point tool is de-selected and the Smart Dimension icon is displayed.

208) Click the **center point** of the Countersink hole.

209) Click the far **right vertical line**.

210) Click a **position** below the bottom horizontal edge.

211) Enter **56/2**. The dimension value 28mm is calculated.

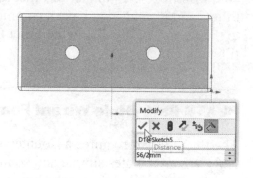

212) Click the **Green Check mark** ✔ from the Modify dialog box.

Add a vertical dimension.

213) Click the **center point** of the Countersink hole.

214) Click the **bottom horizontal line**.

215) Click a **position** to the right of the profile.

216) Enter **11**mm.

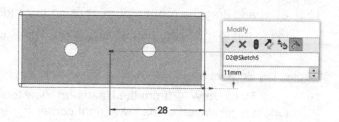

217) Click the **Green Check mark** ✔ from the Modify dialog box.

218) Click **OK** ✔ from the Dimension PropertyManager.

219) Click **OK** ✔ from the Hole Position PropertyManager. View the results.

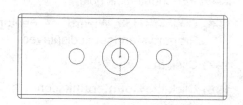

Fit the PLATE to the Graphics window.
220) Press the **f** key.

The Countersink hole is named CSK for M4 Flat Head in the FeatureManager.

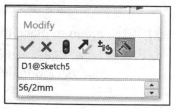

☼ Enter dimensions as a formula in the Modify Dialog Box. Example 56/2 calculates 28. Note you can also enter an equation.

Enter dimensions for automatic unit conversion. Example 2.0in calculates 50.8mm when primary units are set to millimeters (1in. = 25.4mm).

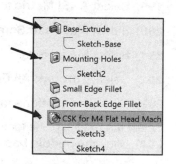

Display an Isometric view of the PLATE.
221) Press the **space bar** to display the Orientation dialog box.

222) Click **Isometric view** . You can also access the
Isometric view tool from the Heads-up View toolbar.

223) Click the **Shaded With Edges** icon.

Save the PLATE.
224) Click **Save** .

Close the PLATE.
225) Click **File**, **Close** from the Menu bar.

Display in the Shaded Isometric view before you save
and close the part. A bitmap image of the model is saved.

Additional details on the Hole Wizard feature are available in SOLIDWORKS
Help. Keywords: Hole Wizard, holes, simple and complex holes.

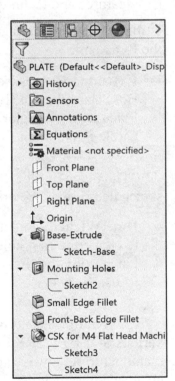

 Review the PLATE Part

The PLATE part utilized an Extruded Boss/Base feature. An
Extruded Boss/Base feature consisted of a rectangular profile
sketched on the Front Plane. You added linear dimensions to
correspond to the overall size of the PLATE based on the
GUIDE-CYLINDER assembly, Piston Plate.

The two holes utilized an Extruded Cut feature. You sketched
two circles on the front face and added Equal and Horizontal
relations. You utilized Geometric relations in the sketch to
reduce dimensions and to maintain the geometric requirements
of the part. Linear dimension defined the diameter and position
of the circles.

You created fully defined sketches in both the Extruded Base
and Extruded Cut feature to prevent future rebuild problems and
obtain faster rebuild times. The Fillet feature inserted the edge
Fillets and tangent edge Fillets to round the corners of the Base-
Extrude feature.

The Hole Wizard feature created the M4 Flat Head Countersink
on the back face of the PLATE. You renamed all features in the
PLATE FeatureManager.

ROD Part Overview

Recall the functional requirements of the customer:

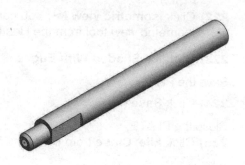

- The ROD is part of a sub-assembly that positions materials onto a conveyor belt.

- The back end of the ROD fastens to the PLATE.

- The front end of the ROD mounts to the customer's components.

- The customer supplies the geometric requirements for the keyway cut and hole.

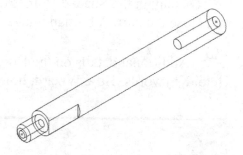

The ROD utilizes an Extruded Boss/Base feature with a circular profile sketched on the Front Plane. The ROD also utilizes the Hole Wizard feature for a simple hole and the Extruded Cut feature. Explore new features and techniques with the ROD.

Utilize the following features to create and modify the ROD part:

- **Extruded Boss/Base** 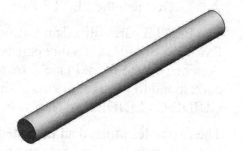: Create the Extruded Boss/Base feature (Boss-Extrude1) on the Front Plane with a circular sketched profile. Use the Circle Sketch tool.

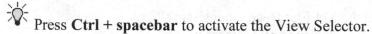 The plane or face you select for the Base sketch determines the orientation of the part.

- **Hole Wizard** : Utilize the Hole Wizard feature to create the first hole on the front face of the ROD. Later, copy the first hole to the back face of the ROD.

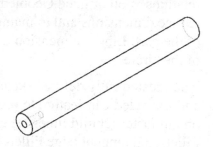

A goal of this text is to expose the new user to various SOLIDWORKS design tools and features. For manufacturing, always apply the Hole Wizard feature to create a complex hole.

Press **Ctrl + spacebar** to activate the View Selector.

- **Chamfer** 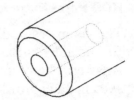: Create the Chamfer feature from the circular edge of the front face. The Chamfer feature removes material along an edge or face. The Chamfer feature assists the ROD by creating beveled edges for ease of movement in the GUIDE-ROD assembly.

- **Extruded Cut** : Create the Extruded Cut feature from a 2D sketch. Convert the edge of the Boss-Extrude1 (Base Extrude) feature to form a Keyway. Utilize the Trim Entities sketch tool to delete sketched geometry.

- **Hole Wizard** : Create the back hole with the Copy/Paste tool. Copy the front hole to the back face. Modify the Hole Wizard hole diameter from 3mm to 4mm.

- **Extruded Cut** : Apply the Rollback bar and Edit Feature function to implement a customer design change. Add a new Extruded Cut feature to the front face.

- **Redefine** the Chamfer feature and Keyway cut feature. Modify the Sketch Plane.

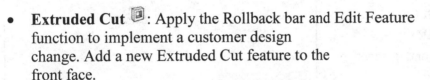

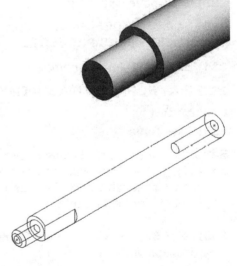

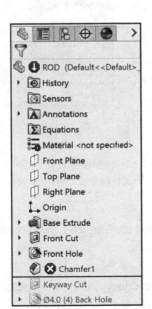

ROD Part - Extruded Boss/Base Feature

The geometry of the Boss-Extrude1 (Base) feature is a cylindrical extrusion. The Extruded Boss/Base feature is the foundation for the ROD. What is the Sketch plane? Answer: The Front Plane is the Sketch plane for this model. What is the shape of the sketched 2D profile? Answer: A circle.

Activity: ROD Part - Extruded Base Feature

Create a New part.

226) Click **New** ☐ from the Menu bar.

227) Click the **MY-TEMPLATES** tab. The New SOLIDWORKS Document dialog displays the PART-MM-ANSI Part Template. The MY-TEMPLATES tab was created earlier in the project.

228) Double-click **PART-MM-ANSI**.

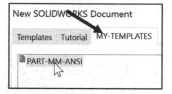

💡 If the MY-TEMPLATES tab is not displayed, click **Options** ⚙, **File Locations** from the Menu bar. Add the full pathname to the **MY-TEMPLATES** folder in the Document Templates box.

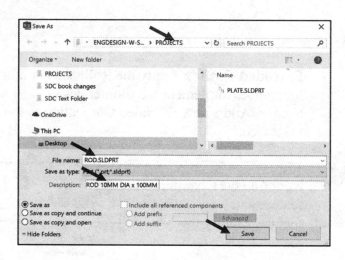

Save the part. Enter name and description.

229) Click **Save** 💾.

230) Select **ENGDESIGN-W-SOLIDWORKS\PROJECTS**.

231) Enter **ROD** for File name.

232) Enter **ROD 10MM DIA x 100MM** for Description.

233) Click **Save**.

Select the Sketch plane. Create Sketch1.

234) Right-click **Front Plane** from the FeatureManager. Front Plane is your Sketch plane. The Base Sketch plane determines the orientation of the part.

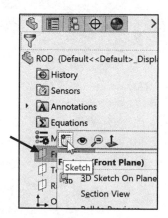

235) Click **Sketch** 📝 from the Context toolbar. The Sketch toolbar is displayed.

💡 To de-activate the display Reference planes, right-click on the **selected plane** in the FeatureManager, click **Hide** 🏷 from the Context toolbar or click **View**, **Hide/Show** and uncheck **Planes** from the Menu bar.

Sketch a circle Coincident at the Origin. Only a single dimension is required to fully define the Base Sketch.

236) Click the **Circle** 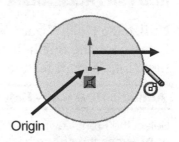 Sketch tool. The Circle PropertyManager is displayed. Use the Origin in the sketch.

237) Click the **Origin** . The Origin is Coincident with the center point of the circle.

238) Click a **position** to the right of the Origin.

Add a dimension using the Smart Dimension tool.

239) Right-click the **Smart Dimension** Sketch tool. The

Smart Dimension icon is displayed.

240) Click the **circumference** of the circle.

241) Click a **position** diagonally off the profile.

242) Enter **10**mm.

243) Click the **Green Check mark** from the Modify dialog box.

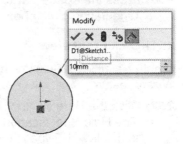

If required, click the arrowhead dot to toggle the direction of the dimension arrow.

Insert an Extruded Boss/Base feature.

244) Click the **Extruded Boss/Base** feature tool from the Features tab in the CommandManager. Blind is the default End Condition in Direction 1.

245) Enter **100**mm for Depth. Accept the default settings. Note the direction of the extrude feature.

246) Click **OK** from the Boss-Extrude PropertyManager. Boss-Extrude1 is displayed in the FeatureManager. Note the location of the Origin.

Fit the model to the Graphics window.
247) Press the **f** key.

Rename Boss-Extrude1.
248) Rename **Boss-Extrude1** to **Base Extrude**.

Save the ROD.
249) Click **Save** . View the results.

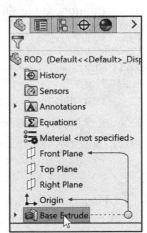

ROD Part - HOLE Wizard Feature

Utilize the Hole Wizard 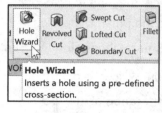 feature to create the first hole on the front face of the ROD. Later, copy the first hole to the back face of the ROD. What is the Sketch Plane for the first Hole Wizard feature? Answer: Front face of the ROD.

Activity: ROD Part - Hole Wizard Feature

Select the Front view.

250) Press the **space bar** to display the Orientation dialog box.

251) Click **Front view**.

Create a simple ANSI Metric hole using the Hole Wizard feature.

252) Click the **Hole Wizard** feature tool. The Hole Specification PropertyManager is displayed. Type is the default tab.

253) Click the **Type** tab. Click **Hole** for Hole Type. Select **ANSI Metric** for Standard.

254) Select **Drill sizes** for Type.

255) Select **3.0** for Hole Size.

256) Select **Blind** for End Condition.

257) Enter **10mm** for Depth.

258) Click the **Positions** tab.

259) Click a position to the right of the **Origin** as illustrated. Do not click the Origin. The Point tool icon is displayed.

260) Click the **Origin** to place the center point of the hole at the Origin. Note the Geometric relation.

De-select the Point tool.

261) Right-click **Select** in the Graphics area. Note: If you Right-click on the Origin, it will exit directly out of the PropertyManager.

Exit out of the Dimension/Hole Position.

262) Click **OK** from the Dimension/Hole Position PropertyManager. View the results.

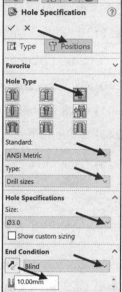

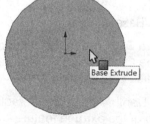

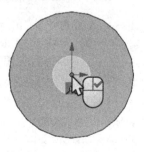

Utilize the Search feature in the Systems Options dialog box to quickly locate information for System Options or Document Properties.

263) Expand Diameter Hole1 in the FeatureManager. View the two sketches. Both sketches are fully defined.

Display an Isometric view with Hidden Lines Visible.

264) Press the **space bar** to display the Orientation dialog box.

265) Click **Isometric view** .

266) Click **Hidden Lines Visible** from the Heads-up View toolbar.

Rename Diameter Hole1.

267) Rename **Diameter Hole1** to **Front Hole**.

Save the ROD.

268) Click **Save** .

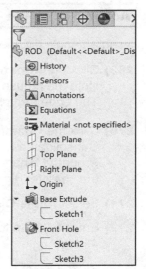

The Hole Wizard feature creates either a 2D or 3D sketch for the placement of the hole in the FeatureManager. You can consecutively place multiple holes of the same type. The Hole Wizard creates 2D sketches for holes unless you select a non-planar face or click the 3D Sketch button in the Hole Position PropertyManager.

ROD Part - Chamfer Feature

The Chamfer feature removes material along an edge or face. The Chamfer feature assists the ROD by creating beveled edges for ease of movement in the GUIDE-ROD assembly. The various Chamfer features are:

- *Angle Distance*. Selected by default. Set the Distance and Angle.

- *Distance Distance*. Enter values for both distances on either side of the selected chamfer edges or Equal Distance and specify a single value.

- *Offset Face*. Arc solved by offsetting the faces adjacent to selected edges. The software calculates the intersection point of the offset faces, then calculates the normal from that point to each face to create the chamfer.

- *Face Face*. Creates symmetric, asymmetric, hold line, and chord width chamfers.

- *Vertex*. Enter values for the three distances on each side of the selected vertex or click Equal Distance and specify a single value.

Chamfer the front circular edge and utilize the Angle distance option in the next activity.

Selection order does not matter when creating a Chamfer or Fillet feature. Example 1: Select edges and faces. Select the feature tool. Example 2: Select the feature tool. Select edges and faces.

Activity: ROD Part - Chamfer Feature

Insert an Angle distance Chamfer feature.

269) Click the **front outer circular edge** of the ROD. Note the mouse icon feedback symbol for an edge. The edge turns blue.

270) Click the **Chamfer** feature tool from the Consolidated toolbar. The Chamfer PropertyManager is displayed. The arrow indicates the chamfer direction.

271) Select **Angle Distance** for Chamfer type.

272) Enter **1.00**mm for Distance. Accept the default 45deg Angle.

273) Click **OK** ✔ from the Chamfer PropertyManager. Chamfer1 is displayed in the FeatureManager.

Save the ROD.

274) Click **Save** 🖫.

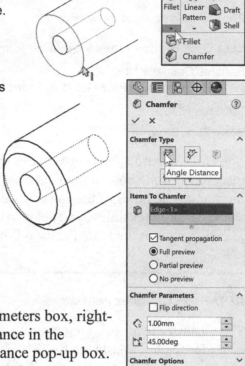

💡 To deselect geometry inside the Chamfer Parameters box, right-click **Clear Selections**. Enter values for Angle distance in the Chamfer Property Manager or inside the Angle distance pop-up box.

💡 Design Intent is how your part reacts as parameters are modified. Example: If you have a hole in a part that must always go through, apply the Through All End Condition. As the part thickness changes, the hole will always go through the part.

ROD Part-Extruded Cut Feature & Convert Entities Sketch tool

The Extruded Cut 🔲 feature removes material from the front of the ROD to create a keyway. A keyway locates the orientation of the ROD into a mating part in the assembly.

Utilize the Convert Entities Sketch tool in the Sketch toolbar to extract existing geometry to the selected Sketch plane. The Trim Entities Sketch tool deletes sketched geometry. Utilize the Power Trim option from the Trim PropertyManager.

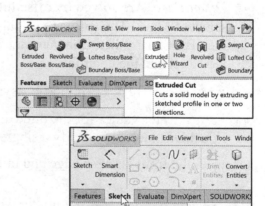

Activity: ROD Part-Extruded Cut Feature and Convert Entities Sketch tool

Select the Sketch plane for the keyway. The Sketch plane is the front circular face of the ROD.

275) Click **Shaded With Edges** 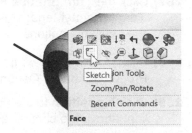 from the Heads-up View toolbar.

276) Right-click the **front circular face** of the ROD. Base Extrude is highlighted in the FeatureManager.

Create a sketch.

277) Click **Sketch** from the Context toolbar.

Zoom in on the front face.

278) Click **Zoom to Area** from the Heads-up View toolbar.

279) Zoom in on the front face. Click **Zoom to Area** to deactivate from the Heads-up View toolbar.

Convert the outside edge to the Sketch plane.

280) Click the **outside front circular edge** of the Base Extrude  as illustrated.

281) Click the **Convert Entities** Sketch tool from the Sketch toolbar. The system extracts the outside edge and positions it on the Sketch plane.

Display a Front view.

282) Press the **space bar** to display the Orientation dialog box.

283) Click **Front view** . The front face of the ROD is displayed.

Sketch a vertical line.

284) Click the **Line** Sketch tool. The Insert Line PropertyManager is displayed.

285) Sketch a **vertical line** as illustrated. The end points of the line extend above and below the diameter of the circle. Insert a Vertical relation if needed.

De-select the Line Sketch tool.

286) Right-click **Select** in the Graphics window.

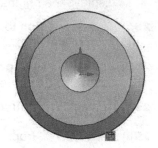

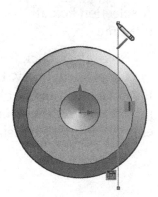

💡 Model about the Origin; this provides a point of reference for your dimensions to fully define the sketch.

Trim the sketch.

287) Click the **Trim Entities** ✂ Sketch tool. The Trim PropertyManager is displayed. Note the available options.

288) Click **Power trim** ⬛ from the Trim PropertyManager Options box.

289) Click a **position** to the right of the vertical line, above the circular edge.

290) Drag the **mouse pointer** to intersect the vertical line above the circular edge.

291) Release the **mouse button** when you intersect the vertical line. The line is removed.

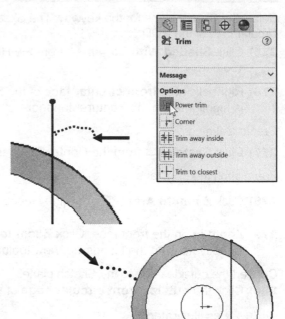

The Power trim option trims the vertical line to the converted circular edge and adds a Coincident relation between the endpoint of the vertical line and the circle.

292) Click a **position** to the left of the circular edge.

293) Click and drag the **mouse pointer** to intersect the left circular edge.

294) Release the **mouse button**. The left circular edge displays the original profile.

295) Click a **position** to the right of the vertical line, below the circular edge.

296) Drag the **mouse pointer** to intersect the vertical line below the circular edge.

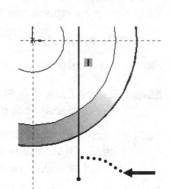

297) Release the **mouse pointer**.

298) Click **OK** ✔ from the Trim PropertyManager.

A selection box allows selection of multiple entities at one time.

To Box-select, click the upper left corner of the box in the Graphics window. Drag the mouse pointer to the lower right corner. A blue box is displayed. Release the mouse pointer. If you create the box from left to right, the entities completely inside the box are selected.

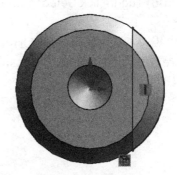

Box-select from left to right.

299) Click a **position** to the upper left corner of the profile as illustrated.

300) Drag the **mouse pointer** diagonally to the lower right corner.

301) Release the **mouse pointer**. The Properties PropertyManager is displayed. The Arc and Line sketch entity are displayed in the Selected Entities dialog box.

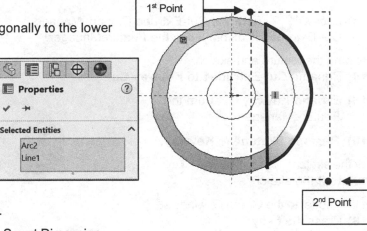

302) Click **OK** ✔ from the Properties PropertyManager.

Insert a vertical dimension using the Smart Dimension tool.

303) Click the **Smart Dimension** ⤢ Sketch tool. The Smart Dimension ⤢ icon is displayed.

304) Click the **vertical line** as illustrated.

305) Click a **position** to the right of the profile.

306) Enter **6.0**mm.

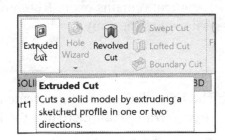

307) Click the **Green Check mark** ✔ from the Modify dialog box. The vertical line endpoints are coincident with the circle outside edge. The profile is a single closed loop.

Fit the sketch to the Graphics window.

308) Press the **f** key.

💡 Although dimensions are not required to create features in SOLIDWORKS, dimensions provide location and size information. Models require dimensions for manufacturing.

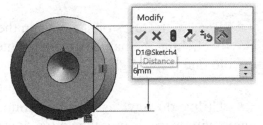

Insert an Extruded Cut feature.

309) Click the **Extruded Cut** ▣ feature tool. The Cut-Extrude PropertyManager is displayed.

Display an Isometric view.

310) Press the **space bar** to display the Orientation dialog box.

311) Click **Isometric view** ▣. View the arrow direction.

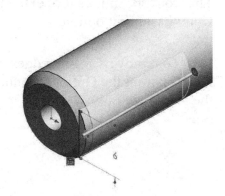

312) Enter **15**mm for the Depth in Direction 1. Accept the default settings. Blind is the default End Condition.

313) Click **OK** ✔ from the Cut-Extrude PropertyManager. Cut-Extrude1 is displayed in the FeatureManager.

Rename the feature and sketch.
314) Rename **Cut-Extrude1** to **Keyway Cut**.

315) **Expand** Keyway Cut from the FeatureManager.

316) Rename **Sketch4** to **Keyway**.

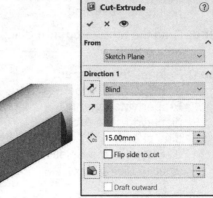

Exit the sketch.
317) **Exit** the sketch.

Fit the model to the Graphics window.
318) Press the **f** key.

Save the ROD.
319) Click **Save** 💾.

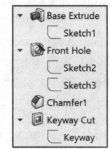

💡 Rename features or sketches in the FeatureManager by slowly clicking the name twice and entering the new name when the old one is highlighted.

💡 Utilize the Save as copy and continue command to save the document to a new file name without replacing the active document. Utilize the Save as copy and open command to save the document to a new file name that becomes the active document. The original document remains open. References to the original document are not automatically assigned to the copy.

The Extruded Cut feature created the Keyway Cut by converting existing geometry. The Hole Wizard feature created the Front Simple Hole on the front face.

The Convert Entities tool extracts feature geometry. Copy features utilizing the Copy/Paste Edit option.

💡 To remove Tangent edges, click **Display** from the Options menu, and check the **Removed** box.

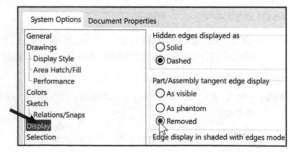

ROD Part - View Orientation, Named View & Viewport Option

The View Orientation defines the preset position of the ROD in the Graphics window. It is helpful to display various views when creating and editing features. The View Orientation options provide the ability to perform the following tasks:

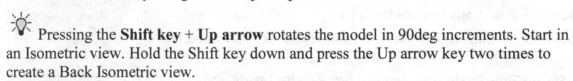

- Create a named view.

- Select a standard view.

- Combine views by using the Viewport option.

☼ Pressing the **Shift key** + **Up arrow** rotates the model in 90deg increments. Start in an Isometric view. Hold the Shift key down and press the Up arrow key two times to create a Back Isometric view.

Activity: ROD Part - View Orientation and Named View

Rotate the ROD.
320) Hold the **Shift** key down.

321) Press the **Up arrow** key twice.

322) Release the **Shift** key.

Insert a new view.
323) Press the **space bar** on your keyboard or click **View Orientation** ⚲ from the Standard Views toolbar. The Orientation dialog box is displayed.

324) Click the **Push Pin** 📌 icon from the Orientation dialog box to maintain the displayed menu.

325) Click the **New View** ⚲ icon in the Orientation dialog box.

326) Enter **back iso** in the View name text box.

327) Click **OK** from the Named View dialog box.

Close the Orientation menu.
328) Close ✖ the dialog box.

Save the ROD in the new view.
329) Click **Save** 💾. View the results.

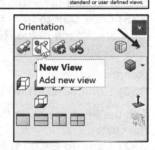

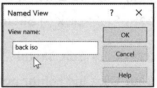

ROD Part - Copy/Paste Feature

The Copy and Paste feature allows you to copy selected sketches and features from one face to another face or from one model to different models.

The ROD requires an additional hole on the back face. Copy the Front Hole feature to the back face. Edit the sketch and modify the dimensions and geometric relations.

Activity: ROD Part - Copy/Paste Feature

Copy the Front Hole to the Back Face.

330) Click **Back view** either from the Heads-up View toolbar or from the Orientation dialog box.

331) Click **Front Hole** Front Hole from the FeatureManager.

332) Click **Edit**, **Copy** from the Menu bar.

333) Click the **back face** of the ROD as illustrated. DO NOT select the center point.

334) Click **Edit**, **Paste** from the Menu bar. View the results.

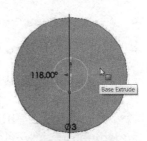

The Copy Confirmation dialog box is displayed. The box states that there are external constraints in the feature being copied. The external constraint is the Coincident geometric relation used to place the Front Hole.

If needed, delete the old constraint.

335) Click **Delete** from the Copy Confirmation dialog box. The back face of the ROD contains a copy of the Front Hole feature.

Rename the feature.

336) Rename Ø3.0 (#) Diameter Hole# to **Ø4.0 (4) Back Hole**.

Edit the Sketch. Locate the Back Hole. Insert a Coincident relation to the Origin.

337) **Expand Ø4.0 (4) Back Hole** for the FeatureManager.

338) Right-click **(-) Sketch5**.

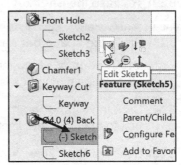

339) Click **Edit Sketch** from the Context toolbar.

340) Click the **center point** of the 3mm circle.

341) Hold the **Ctrl** key down.

342) Click the **Origin** of the ROD as illustrated.

343) Release the **Ctrl** key. The Properties PropertyManager is displayed. The selected sketch entities are displayed in the Selected Entities box.

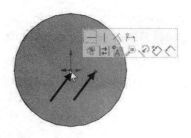

344) Click **Coincident** ⋏ from the Add Relations box. The center point of the back hole and the origin are Coincident.

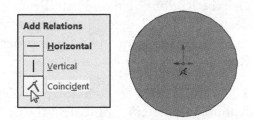

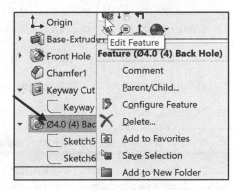

345) Click **OK** ✔ from the Properties PropertyManager.

Exit the Sketch.

346) Click **Exit Sketch** 📝 from the Context toolbar. Sketch5 is fully defined.

Display an Isometric view. Modify the diameter dimension of the hole. Edit the Hole Wizard feature.

347) Click **Isometric view** 🔲.

348) Right-click **Ø4.0 (4) Back Hole** from the FeatureManager.

349) Click **Edit Feature** 📑 from the Context toolbar. The Hole Specification PropertyManager is displayed.

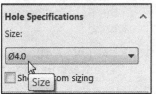

350) Select **Ø4.0** for Hole Size.

351) Click **OK** ✔ from the Hole Specification PropertyManager.

Save the ROD.

352) Click **Save** 💾.

💡 The Confirmation Corner in the sketch Graphics window indicates that a sketch is active.

ROD Part - Design Changes with Rollback Bar

You are finished for the day. The phone rings. The customer voices concern about the GUIDE-ROD assembly. The customer provided incorrect dimensions for the mating assembly to the ROD. The ROD must fit into a 7mm hole. The ROD fastens to the customer's assembly with a 4mm Socket head cap screw.

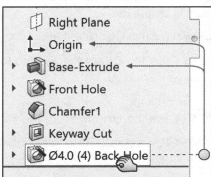

You agree to make the changes at no additional cost since the ROD has not been machined. You confirm the customer's change in writing.

The customer agrees but wants to view a copy of the GUIDE-ROD assembly design by tomorrow. You are required to implement the design change and to incorporate it into the existing part.

Use the Rollback bar and Edit Feature function to implement a
design change. The Rollback bar provides the ability to
redefine a feature in any state or order. Reposition the
Rollback bar in the FeatureManager.

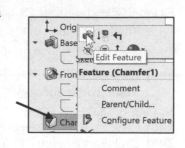

The Edit Feature 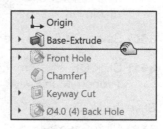 function provides the ability to redefine
feature parameters. Implement the design change.

- Modify the ROD with the Extruded Cut feature to address
  the customer's new requirements.

- Edit the Chamfer feature to include the new edge.

- Redefine a new Sketch plane for the Keyway Cut feature.

In the next section, develop rebuild errors and correct the rebuild errors.

Activity: ROD Part - Design Changes with Rollback Bar in the FeatureManager

Position the Rollback bar in the FeatureManager.

353) Place the **mouse pointer** over the Rollback bar at the bottom of
the FeatureManager. The mouse pointer displays a symbol of a
hand.

354) Drag the **Rollback** bar upward below the Base Extrude feature.
The Base Extrude feature is displayed.

Select the Sketch plane and display the Sketch toolbar.

355) Click **Front view** .

356) Right-click the **Front face** of the Base Extrude feature. This is
your Sketch plane. Base Extrude is highlighted in the
FeatureManager.

357) Click **Sketch** from the Context toolbar. The Sketch toolbar
is displayed.

Sketch a circle Coincident about the Origin.

358) Click the **Circle** Sketch tool. The Circle PropertyManager
is displayed.

359) Click the **Origin** of the Base Extrude feature.

360) Click a **position** to the right of the Origin as
illustrated.

Add a diameter dimension.

361) Right-click the **Smart Dimension** tool.

362) Click the **circumference** of the circle. Click a
position diagonally off the profile. Enter **7mm**.

363) Click the **Green Check mark** from the Modify
dialog box.

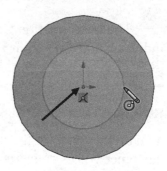

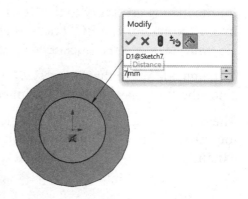

Fit the ROD to the Graphics window.
364) Press the **f** key.

Insert an Extruded Cut feature.
365) Click the **Extruded Cut** feature tool. The Cut-Extrude PropertyManager is displayed.

366) Click the **Flip side to cut** box. The direction arrow points outward. Blind is the default End Condition.

367) Enter **10.00**mm for Depth. Accept the default settings.

368) Click **OK** ✔ from the Cut-Extrude PropertyManager.

Display an Isometric view.
369) Press the **space bar** to display the Orientation dialog box.

370) Click **Isometric view**.

Rename the feature.
371) Rename the **Cut-Extrude#** feature to **Front Cut**.

Save the ROD.
372) Click **Save**.

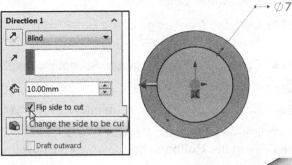

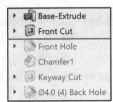

ROD Part - Recover from Rebuild Errors

Rebuild errors can occur when using the Rollback function. A common error occurs when an edge or face is missing. Redefine the edge for the Chamfer feature. When you delete sketch references, Dangling dimensions, sketches and plane errors occur. Redefine the face for the Keyway Sketch plane and the Keyway Sketch.

☀ Dangling indicates sketch geometry that cannot be resolved, for example, deleting an entity that was used to define another sketch entity.

Activity: ROD Part - Recover from Rebuild Errors

Redefine the Chamfer feature.
373) Drag the **Rollback** bar downward below the Chamfer1 feature. A red x ⬡ ⊗ Chamfer1 is displayed next to the name of the Chamfer1 feature. A red arrow is displayed ⬡ ⬇ ROD (Default next to the ROD part icon. Red indicates a model rebuild error. When you created the Front Cut feature, you deleted the original edge from the Base Extrude feature.

Insert the Chamfer feature on the Front Cut edge.
374) Right-click **Chamfer1** from the FeatureManager.

375) Click **Edit Feature** . The Chamfer1 PropertyManager is displayed. **Missing**Edge<1> is displayed in the Chamfer Parameters box.

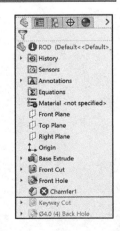

376) Right-click **inside** the Edges and Faces or Vectors box.

377) Click **Clear Selections**.

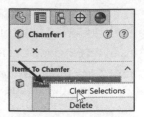

378) Click the **front edge** of the Base Extrude feature as illustrated. Edge<2> is displayed in the Items to Chamfer box.

379) Enter **1.0**mm for Distance. Accept the default settings.

Update the Chamfer feature.

380) Click **OK** ✔ from the Chamfer1 PropertyManager.

Redefine the face for the Keyway Sketch plane.

381) Drag the **Rollback** bar downward below the Keyway Sketch. The Rod Rebuild Error box is displayed. The ⚠ Keyway Sketch indicates a dangling Sketch plane. The Front Cut deleted the original Sketch plane.

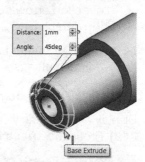

Redefine the Sketch plane and sketch relations to correct the rebuild errors.

382) Right-click the **Keyway** sketch from the FeatureManager.

383) Click **Edit Sketch Plane** 🖉 from the Context toolbar. The Sketch Plane PropertyManager is displayed. **Delete** any selected entities.

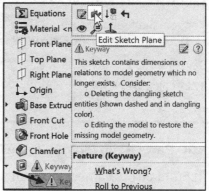

384) Click the **large circular face** of the Front Cut as illustrated. Face<2> is displayed in the Sketch Plane/Face box.

385) Click **OK** ✔ from the Sketch Plane PropertyManager. The SOLIDWORKS dialog box is displayed.

386) Click the **Stop and Repair** button. The What's Wrong dialog box is displayed. View the description.

387) Click **Close** from the What's Wrong dialog box.

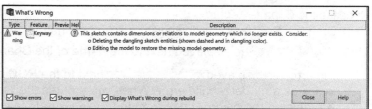

The Keyway Sketch moved to the new Sketch plane. To access the Edit Sketch Plane option, right-click the Sketch entry in the FeatureManager.

Redefine the sketch relations and dimensions to correct the two build errors. Dangling sketch entities result from a dimension or relation that is unresolved. An entity becomes dangling when you delete previously defined geometry.

💡 The Display/Delete Relations sketch tool lists dangling geometry.

Edit the Keyway Cut sketch.
388) Right-click **Keyway Cut** from the FeatureManager.

389) Click **Edit Sketch** ✎ from the Context toolbar.

Display the Sketch relations. Delete the On Edge 0 relation.

390) Click the **Display/Delete Relations** ⊥₀ Sketch tool. The Display/Delete Relations PropertyManager is displayed.

391) Right-click the **On Edge 0** relation.

392) Click **Delete**. The Dangling status is replaced by Satisfied. The sketch is under defined and is displayed in blue.

393) Click **OK** ✔ from the Display/Delete Relations PropertyManager.

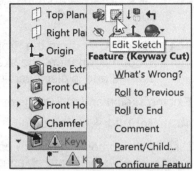

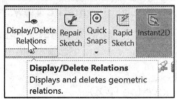

💡 When you select a relation from the Relations list box, the appropriate sketch entities are highlighted in the Graphics window. If Sketch Relations (**View**, **Hide/Show**, **Sketch Relations**) is selected, all icons are displayed, but the icons for the highlighted relations are displayed in a different color.

Display a Front view.
394) Press the **space bar** to display the Orientation dialog box.

395) Click **Front view** ⬦.

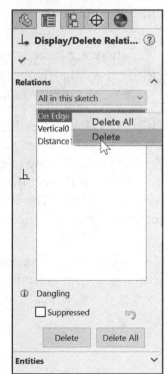

Add a Coradial relation.

396) Click the **right blue arc** as illustrated.

397) Hold the **Ctrl** key down.

398) Click the **outside circular edge**.

399) Release the **Ctrl** key. The Properties PropertyManager is displayed.

400) Click **Coradial** from the Add Relations box.

401) Click **OK** ✔ from the Properties PropertyManager. The sketch is fully defined and is displayed in black.

Exit the sketch and rebuild the model.

402) **Rebuild** 🔘 the model.

Display an Isometric view. Roll back the Rollback bar in the FeatureManager.

403) Click **Isometric view** 🔲.

404) Drag the **Rollback** bar downward below the Back Hole feature in the FeatureManager.

Save the ROD.

405) Click **Save** 💾.

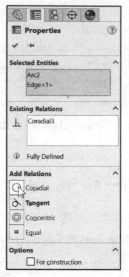

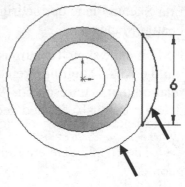

💡 There are three ways to add geometric relations. Utilize the Add Relations Sketch tool ⊥ , use the Properties PropertyManager, or use the Context toolbar.

Utilize the Properties PropertyManager 📋 **Properties** or the Context toolbar when you select multiple sketch entities in the Graphics window. Select the first entity. Hold the Ctrl key down and select the remaining geometry to create Sketch relations. The Properties and Context toolbar technique requires fewer steps.

💡 View the mouse pointer icon to confirm selection of the correct point, edge, face, etc.

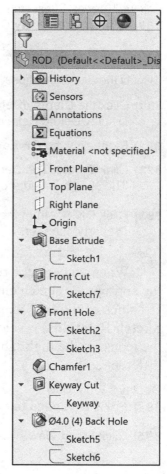

Modify the depth of the Back Hole.

406) Right-click Ø4.0 (4) **Back Hole** from the FeatureManager.

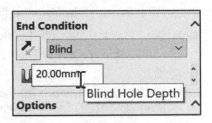

407) Click **Edit Feature** from the Context toolbar. The Hole Specification PropertyManager is displayed.

408) Select **Blind** for End Condition. Enter **20.00**mm for Blind Hole Depth.

409) Click **OK** from the Hole Specification PropertyManager.

Save the ROD part.

410) Click **Save**.

A machinist rule of thumb states that the depth of a hole does not exceed 10 times the diameter of the hole. The purpose of this book is to expose various SOLIDWORKS design tools and features.

A Countersunk screw fastens the PLATE to the ROD. The depth of the PLATE and the depth of the Back Hole determine the length of the Countersunk screw. Utilize standard length screws.

ROD Part - Edit Part Appearance

Parts are shaded gray by system default. The Edit Part Color option provides the ability to modify part, feature and face color. In the next activity, modify the part color and the machined faces.

Activity: ROD Part - Edit Part Appearance

Edit the part color.

411) Click **Shaded with Edges** from the Heads-up View toolbar.

412) Right-click the **ROD Part** icon at the top of the FeatureManager.

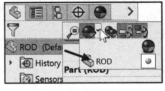
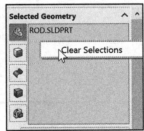

413) Click the **Appearances** drop-down arrow as illustrated. Click ROD. The

414) Click **ROD** as illustrated. The Color PropertyManager is displayed.

415) Right-click **ROD** inside the Selection Geometry box.

416) Click **Clear Selections**.

Modify machined surfaces on the ROD part.

417) Click **Select Features** in the Selection Geometry box.

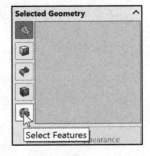

418) Click **Keyway Cut** from the fly-out FeatureManager.

419) Click **Front Hole** from the fly-out FeatureManager.

420) Click **Back Hole** from the fly-out FeatureManager. The selected entities are displayed in the Selected Entities box.

421) Click a **color swatch** for the selected faces. Example: Blue. The ROD faces are displayed in the selected color.

422) Click **OK** ✔ from the Color PropertyManager.

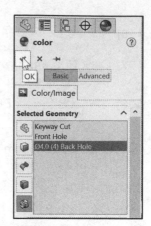

Expand the Show Display Pane.
423) Click the **Show Display Pane arrow**. View the results.

Collapse the Hide Display Pane.
424) Click the **Hide Display Pane arrow** to collapse the FeatureManager.

Display an Isometric view.
425) Click **Isometric view** ⬛.

The ROD part is complete.
Save the ROD.
426) Click **Save** 💾.

Close the ROD.
427) Click **File**, **Close** from the Menu bar. View the results.

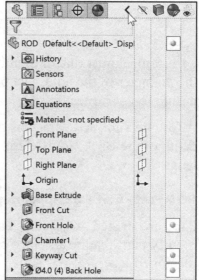

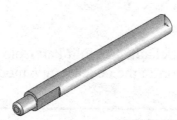

Standardize the company part colors. Example: Color red indicates a cast surface. Color blue indicates a machined surface.

Review of the ROD Part

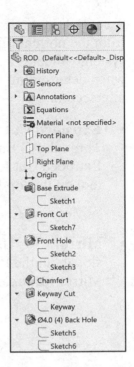

The ROD part utilized an Extruded Boss/Base feature with a circular profile sketched on the Front Plane. Linear and diameter dimensions corresponded to the customer's requirements.

The Chamfer feature removed the front circular edge. The Chamfer feature required an edge, distance and angle parameters. The Extruded Cut feature utilized a converted edge of the Base Extrude feature to form the Keyway Cut. Reuse geometry with the Convert Entity Sketch tool.

You utilized the Hole Wizard feature for a simple hole on the front face to create the Front Hole. You reused geometry. The Front Hole was copied and pasted to the back face to create the Back Hole. The Back Hole dimensions were modified. The Display\Delete Sketch Tool deleted dangling dimensions of the copied feature. You inserted an Extruded Cut feature by moving the Rollback bar to a new position in the FeatureManager. You recovered from the errors that occurred with Edit Feature, Edit Sketch, and Edit Sketch plane tools by redefining geometry.

GUIDE Part Overview

The GUIDE part supports the ROD. The ROD moves linearly through the GUIDE. Two slot cuts provide flexibility in locating the GUIDE in the final assembly.

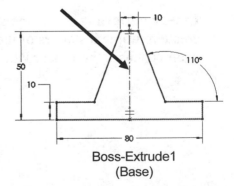

The GUIDE supports a small sensor mounted on the angled right side. You do not have the information on the exact location of the sensor. During the field installation, you receive additional instructions.

Create a pattern of holes on the angled right side of the GUIDE to address the functional requirements.

Address the geometric requirements with the GUIDE part:

- **Extruded Boss/Base** : Use the Extruded Boss/Base feature to create a symmetrical sketched profile for the GUIDE part.

Utilize a centerline and the Dynamic Mirror Sketch tool. The centerline acts as the mirror axis. The copied entity becomes a mirror image of the original across the centerline.

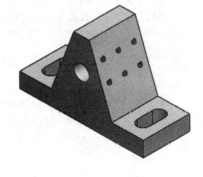

Boss-Extrude1
(Base)

A centerline is a line with a property that makes it exempt from the normal rules that govern sketches. When the system validates a sketch, it does not include centerlines when determining if the profile is disjointed or self-intersecting.

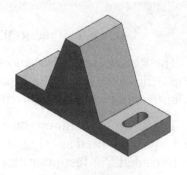

- **Extruded Cut** : Use the Extruded Cut feature with the Through All End Condition to create the first slot. The Extruded Cut feature utilizes the Straight Slot Sketch tool on the top face to form the slot.

- **Mirror** : Use the Mirror feature to create a second slot about the Right Plane. Reuse geometry with the Mirror feature.

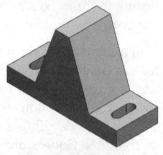

- **Extruded Cut** : Use the Extruded Cut feature with the Through All End Condition to create the Guide Hole. The ROD glides through the Guide Hole. Note: The purpose of this book is to expose users to various SOLIDWORKS design tools and features.

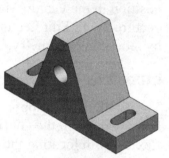

- **Hole Wizard** : Use the Hole Wizard feature to create an M3 tapped hole, seed feature for the Linear Pattern.

- **Linear Pattern** : Use the Linear Pattern feature to create multiple instances of the M3 tapped hole. Reuse geometry with the Linear Pattern feature.

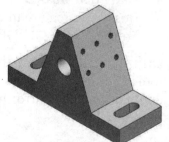

GUIDE Part - Extruded Boss/Base Feature and Dynamic Mirror

The GUIDE utilizes an Extruded Boss/Base feature. What is the Sketch plane? Answer: The Front Plane is the Sketch plane. The Mid Plane End Condition extrudes the sketch symmetric about the Front Plane.

How do you sketch a symmetrical 2D profile? Answer: Use a sketched centerline and the Dynamic Mirror Sketch tool.

Sketching lines and centerlines produce different behavior depending on how you sketch. Review sketching line techniques before you begin the GUIDE part.

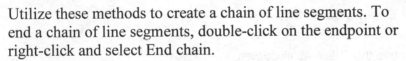

There are two methods to create a line:

- Method 1: Click the first point of the line. Drag the mouse pointer to the end of the line and release. Utilize this technique to create an individual line segment.

- Method 2: Click the first point of the line. Release the left mouse button. Move to the end of the line and click again. The endpoint of the first line segment becomes the start point of the second line segment. Move to the end of the second line and click again.

Utilize these methods to create a chain of line segments. To end a chain of line segments, double-click on the endpoint or right-click and select End chain.

Utilize Method 1 to create the individual centerline.

Utilize Method 2 to create the chained profile lines.

Activity: GUIDE Part - Extruded Base Feature and Dynamic Mirror

Create a New part.

428) Click **New** ⬜ from the Menu bar.

429) Click the **MY-TEMPLATES** tab.

430) Double-click **PART-MM-ANSI** in the New SOLIDWORKS Document Template dialog box.

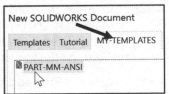

Save the part. Enter name and description.

431) Click **Save** 💾.

432) Select **ENGDESIGN-W-SOLIDWORKS\PROJECTS**.

433) Enter **GUIDE** for the File name.

434) Enter **GUIDE SUPPORT** for the Description.

435) Click **Save**. The GUIDE FeatureManager is displayed.

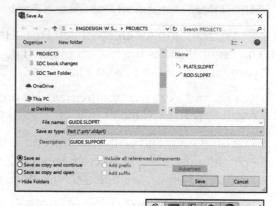

Create the first sketch. Select the Front Plane.

436) Right-click **Front Plane** from the GUIDE FeatureManager. This is your Sketch Plane.

437) Click **Sketch** 🖾 from the Context toolbar. The Sketch toolbar is displayed.

Sketch a centerline.

438) Click the **Centerline** ✏ Sketch tool from the Consolidate toolbar. The Insert Line PropertyManager is displayed.

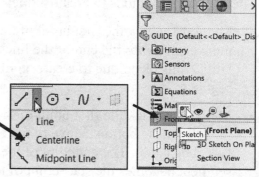

💡 A centerline is displayed with a series of long and short dashes.

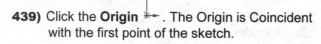

439) Click the **Origin** ↳. The Origin is Coincident with the first point of the sketch.

440) Drag and click a **point** above the Origin as illustrated.

Activate the Mirror Sketch tool.

441) Click **Tools, Sketch Tools, Dynamic Mirror** from the Menu bar.

442) Click the **vertical centerline**. The Line Properties PropertyManager is displayed.

Sketch the profile.

443) Click the **Line** ✏ Sketch tool. The Insert Line PropertyManager is displayed.

444) Sketch the first horizontal line from the **Origin** to the right as illustrated. The mirror of the sketch is displayed on the left side of the centerline. Note: Click the endpoint of the line.

445) Sketch the **four additional line segments** to complete the sketch as illustrated. The last point is coincident with the centerline. The profile is continuous; there should be no gaps or overlaps.

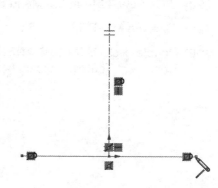

446) Double-click the **end of the line** to end the Sketch.

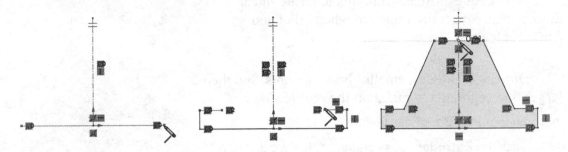

Add angular, vertical and horizontal dimensions.
447) Click the **Smart Dimension** Sketch tool.

Create an angular dimension.
448) Click the **right horizontal line**.

449) Click the **right-angled line**.

450) Click a **position** between the two lines.

451) Enter **110**.

452) Click the **Green Check mark**.

Create two vertical dimensions.
453) Click the **left small vertical line**.

454) Click a **position** to the left of the profile.

455) Enter **10**mm. Click the **Green Check mark** ✔.

456) Click the **bottom horizontal line**.

457) Click the **top horizontal line**.

458) Click a **position** to the left of the profile.

459) Enter **50**mm.

460) Click the **Green Check mark** ✔.

Create two linear horizontal dimensions.
461) Click the **top horizontal line**.

462) Click a **position** above the profile.

463) Enter **10**mm.

464) Click the **Green Check mark** ✔.

465) Click the **bottom horizontal line**.

466) Click a **position** below the profile.

467) Enter **80**mm.

468) Click the **Green Check mark** ✔.

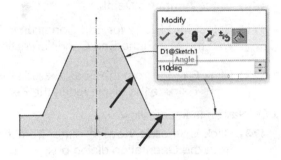

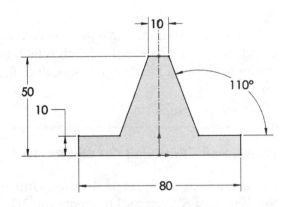

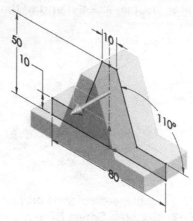

💡 Select edges instead of points to create linear dimensions. Points are removed when Fillet and Chamfer features are added.

💡 Dimension angles, smaller line segments, and then larger line segments to maintain the profile shape.

Insert an Extruded Boss/Base feature.

469) Click the **Extruded Boss/Base** feature tool from the Features tab in the CommandManager. The Boss-Extrude PropertyManager is displayed. The direction arrow points to the front.

470) Enter **30**mm for Depth.

471) Select **Mid Plane** for End Condition in Direction1. Think Symmetry with the Mid Plane End Condition option.

472) Click **OK** ✔ from the Boss-Extrude PropertyManager. Boss-Extrude1 is displayed in the FeatureManager.

Display an Isometric view.

473) Click **Isometric view** 🔲 either from the Heads-up View toolbar or from the Orientation dialog box.

Rename the feature.

474) Rename **Boss-Extrude1** to **Base Extrude**.

Save the GUIDE.

475) Click **Save** 💾.

💡 There are two Mirror Sketch tools in SOLIDWORKS: *Mirror* and *Dynamic Mirror*. Mirror requires a predefined profile and a centerline. Select **Mirror**, select the **profile** and the **centerline**. Dynamic Mirror requires a centerline. As you sketch, entities are created about the selected centerline.

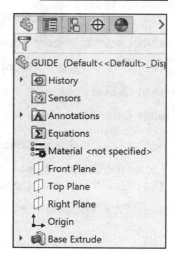

💡 Think design intent. When do you use the various End Conditions and Geometric sketch relations? What are you trying to do with the design? How does the component fit into an assembly?

GUIDE Part - Extruded Cut Slot Profile

Create an Extruded Cut 🔲 feature. Utilize the Straight Slot Sketch tool to create the 2D sketch profile for the slot.

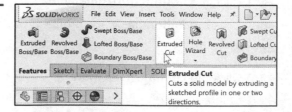

Activity: GUIDE Part - Extruded Cut Slot Profile

Display a Top view. Create the Slot sketch on the top right face of the model.

476) Click **Top view** ⬚.

477) Right-click the **top right face** of the GUIDE in the Graphics window.

478) Click **Sketch** ⬚ from the Context toolbar. The Sketch toolbar is displayed.

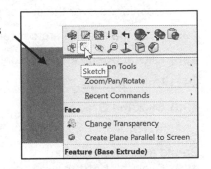

Apply the Straight Slot Sketch tool. The Straight Slot Sketch tool requires three points,

479) Click the **Straight Slot** Sketch ⬚ tool from the Consolidated Slot toolbar as illustrated. The Slot PropertyManager is displayed.

480) Click the **first point** as illustrated.

481) Click the **second point** directly above the first point.

482) Click the **third point** directly to the right of the second point. The Slot Sketch is created.

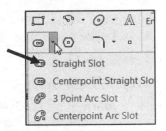

💡 Create a slot without using the Slot Sketch tool by sketching two lines, two 180-degree tangent arcs and apply the needed geometric relations.

Add dimensions.

483) Click the **Smart Dimension** ⬚ Sketch tool.

484) Click the **Origin** ⬚ as illustrated.

485) Click the **center point** of the top arc. Note: When the mouse pointer is positioned over the center point, the center point is displayed in red.

486) Click a **position** above the profile.

487) Enter **30**mm.

488) Click the **Green Check mark** ✓.

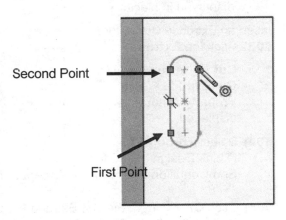

Second Point

First Point

Create a vertical dimension.

489) Click the **top horizontal edge** of the GUIDE.

490) Click the **center point** of the top arc.

491) Click a **position** to the right of the profile.

492) Enter **10**mm.

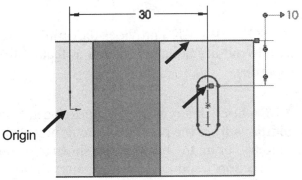

Origin

493) Click the **Green Check mark** ✔ .

Create a dimension between the two arc center points.

494) Click the top arc **center point**. When the mouse pointer is positioned over the center point, the center point is displayed in red.

495) Click the second arc **center point**.

496) Click a **position** to the right of the profile.

497) Enter **10**mm.

498) Click the **Green Check mark** ✔ .

Create a radial dimension.

499) Click the **bottom arc** radius.

500) Click a **position** diagonally below the profile.

501) Enter **3**mm.

502) Click the **Green Check mark** ✔ . The Sketch is fully defined and is displayed in black.

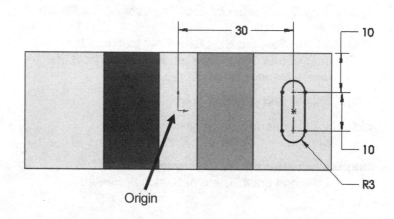

Insert an Extruded Cut feature.

503) Click the **Extruded Cut** 🔲 feature tool. The Cut-Extrude PropertyManager is displayed.

504) Select **Through All** (Think Design Intent) for End Condition in Direction1. Accept all default settings.

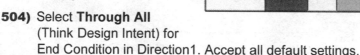

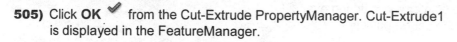

505) Click **OK** ✔ from the Cut-Extrude PropertyManager. Cut-Extrude1 is displayed in the FeatureManager.

☀ SOLIDWORKS provides the ability to select a unit drop-down menu to modify units in a sketch or feature from the set document properties.

☀ Right-click anywhere on an extruded feature; view the end conditions from the pop-up shortcut menu. Click in empty space, on geometry, or on the handle. The shortcut menu provides the options for Direction 1 and Direction 2. Note: Options are document dependent.

Display an Isometric view.

506) Click **Isometric view** 🧊 either from the Heads-up View toolbar or the Orientation dialog box.

Rename the feature.

507) Rename **Cut-Extrude1** to **Slot Cut**.

Save the GUIDE.

508) Click **Save** 💾.

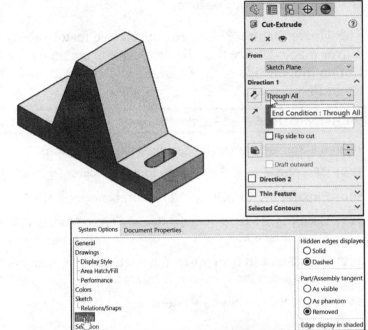

💡 To remove Tangent edges, click **Display** from the Options menu, and check the **Removed** box.

SOLIDWORKS contains additional methods to create sketch slots. Here are three examples that you can explore as an exercise:

- Sketch a rectangle. Delete two opposite line segments. Insert two arcs. Extrude the profile.

- Sketch a single vertical centerline. Utilize Offset Entities, Bidirectional option and Cap Ends option. Extrude the profile.

- Utilize a Design Library Feature.

The machined, metric, straight slot feature provides both the Slot sketch and the Extruded Cut feature. You define the references for dimensions, overall length and width of the Slot. Explore the SOLIDWORKS Design Library later in the book.

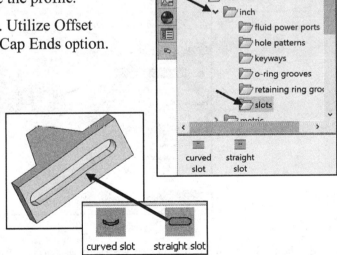

💡 Slots are available in the Hole Wizard. You can create regular slots as well as counterbore and countersink slots. You also have options for position and orientation of the slot. If you have hardware already mated in place, the mates will not be broken if you switch from a hole to a slot.

GUIDE Part - Mirror Feature

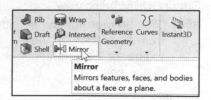

The Mirror feature ![icon] mirrors a selected feature about a mirror plane. Utilize the Mirror feature to create the Slot Cut on the left side of the GUIDE. The mirror plane is the Right Plane.

Activity: GUIDE Part - Mirror Feature with Geometric Pattern Option

Insert a Mirror feature with the Geometric Pattern options.

509) Click the **Mirror** ![icon] feature tool. The Mirror PropertyManager is displayed.

510) **Expand** the GUIDE fly-out FeatureManager in the Graphics window.

511) Click **Right Plane** from the fly-out FeatureManager. Right Plane is displayed in the Mirror Faces/Plane box.

512) Click **Slot Cut** from the fly-out FeatureManager for the Features to Mirror. Slot Cut is displayed in the Features to Mirror box.

513) Check the **Geometry Pattern** box.

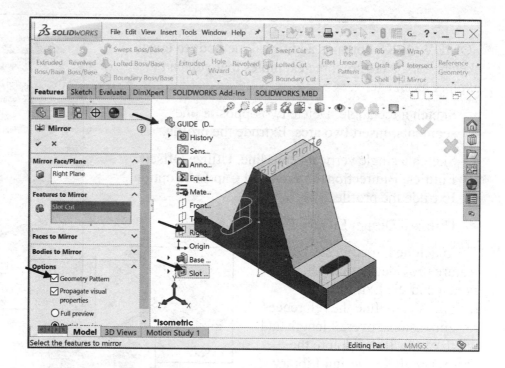

514) Click **OK** ![icon] from the Mirror PropertyManager. Mirror1 is displayed in the FeatureManager. Save the GUIDE part.

515) Click **Save** ![icon].

Fit the GUIDE part to the Graphics window.
516) Press the **f** key.

Save rebuild time--check the Geometry pattern option in the Mirror Feature PropertyManager. Each instance is an exact copy of the faces and edges of the original feature.

Additional details on Geometry Pattern, Mirror Feature and Sketch Mirror are available in SOLIDWORKS Help. Index keywords: Geometry, Mirror Features and Mirror Entities.

GUIDE Part - Holes

The Extruded Cut feature removes material from the front face of the GUIDE to create a Guide Hole. The ROD moves through the Guide Hole.

Create the tapped holes with the Hole Wizard feature and the Linear Pattern feature. Apply dimensions to the tapped holes relative to the Guide Hole.

Activity: GUIDE Part - Holes

Create a sketch on the front face.
517) Right-click the **front face** of the GUIDE in the Graphics window. This is your Sketch Plane.

518) Click **Sketch** from the Context toolbar.

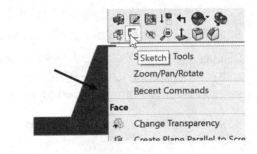

519) Click **Normal To view** from the Heads-up View toolbar. The front face of the GUIDE is displayed.

520) Click the **Circle** Sketch tool. The Circle PropertyManager is displayed.

521) Sketch a **circle** in the middle of the GUIDE as illustrated.

522) Right-click **Select** to deselect the Circle sketch tool.

523) If needed, add a **Vertical Relation** between the center of the circle and the Origin.

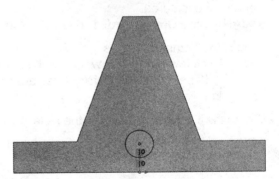

Add dimensions.

524) Click the **Smart Dimension** ✐ Sketch tool.

525) Click the circle **circumference**.

526) Click a **position** below the horizontal line.

527) Enter **10**mm.

528) Click the **Green Check mark** ✔.

529) Click the **center point** of the circle.

530) Click the **Origin**.

531) Click a **position** to the right of the profile.

532) Enter **29**mm.

533) Click the **Green Check mark** ✔. The sketch is fully defined.

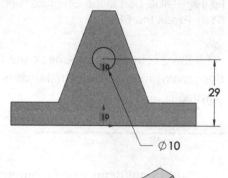

Create an Extruded Cut feature.

534) Click the **Extruded Cut** ▣ feature tool. The Cut-Extrude PropertyManager is displayed.

535) Select **Through All** for End Condition in Direction1. Accept the default conditions.

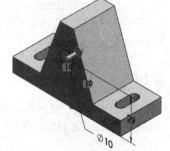

536) Click **OK** ✔ from the Cut-Extrude PropertyManager. Cut-Extrude1 is displayed in the FeatureManager.

537) Press the **space bar** to display the Orientation dialog box.

Display an Isometric view with Hidden Lines Removed.

538) Click **Isometric view** 🔲 either from the Heads-up View toolbar or from the Orientation dialog box.

539) Click **Hidden Lines Removed** ▢ from the Heads-up View toolbar.

Rename the feature.

540) Rename **Cut-Extrude1** to **Guide Hole**.

Display the Temporary Axes.

541) Click **View**, **Hide/Show**, check **Temporary Axes** from the Menu bar. The Temporary Axes are displayed in the Graphics window.

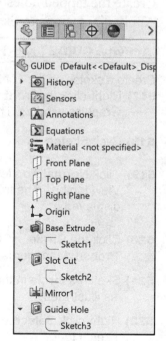

Create the Tapped Hole. Utilize the Hole Wizard.

542) Click the **Hole Wizard** 🕮 feature tool. The Hole Specification PropertyManager is displayed.

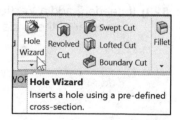

543) Click the **Type** tab.

544) Click **Straight Tap** for Hole Specification.

545) Select **ANSI Metric** for Standard.

546) Select **Bottoming Tapped Hole** for Type.

547) Select **M3x0.5** for Size.

548) Select **Blind** for End Condition. Accept the default conditions.

549) Click the **Positions** tab.

550) Click the **right angled face** below the Guide Hole Temporary Axis as illustrated. The Point icon is displayed.

551) Click **again** to place the center of the hole.

Display a Normal To view.

552) Click **Normal To view** from the Heads-up View toolbar to view the Sketch plane.

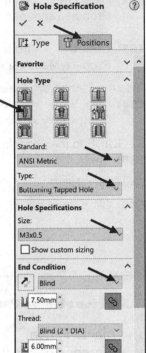

Dimension the tapped hole relative to the Guide Hole axis. Add the horizontal dimension.

Insert Dimensions.

553) Click the **Smart Dimension** Sketch tool. The Point tool is de-selected. The Smart Dimension icon is displayed.

554) Click the **center point** of the tapped hole. Note: When the mouse pointer is positioned over the center point, the center point is displayed in red.

555) Click the **right vertical edge**. Click a **position** above the profile. Enter **25**mm.

556) Click the **Green Check mark** .

Add a vertical dimension.

557) Click the **center point** of the tapped hole.

558) Click the horizontal **Temporary Axis**.

559) Click a **position** to the right of the profile.

560) Enter **4**mm.

561) Click the **Green Check mark** .

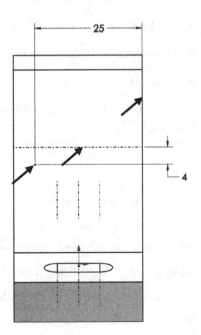

Display the tapped hole.

562) Click **OK** ✔ from the Dimension PropertyManager.

563) Click **OK** ✔ from the Hole Position PropertyManager.

Hide the Temporary Axes.
564) Click **View**, **Hide/Show**, and uncheck **Temporary Axes** from the Menu bar.

Fit GUIDE to the Graphics window.
565) Press the **f** key.

Save the GUIDE.
566) Click **Save** 💾.

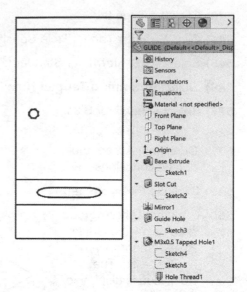

GUIDE Part-Linear Pattern Feature

The Linear Pattern ⬚⬚ feature creates multiple copies of a feature in an array. A copy of a feature is called an instance. A Linear Pattern feature requires:

- One or two directions.

- A Feature, Face or Body to pattern.

- Distance between instances.

- Number of instances.

Insert a Linear Pattern feature of tapped holes on the right-angled face of the GUIDE.

Activity: GUIDE Part-Linear Pattern Feature with Geometry Pattern option

Display a Right view.
567) Click **Right view** 🗔.

Insert a Linear Pattern feature with the Geometry Pattern option.

568) Click the **Linear Pattern** ⬚⬚ feature tool. The Linear Pattern PropertyManager is displayed.

569) **Expand** GUIDE from the fly-out FeatureManager.

570) Click inside the **Features to Pattern** box.

571) Click **M3X0.5 Tapped Hole1** from the fly-out FeatureManager.

572) Click inside the **Pattern Direction** box for Direction 1.

573) Click the **top horizontal edge** for Direction 1. Edge<1> is displayed in the Pattern Direction box. The direction arrow points to the right.

574) Enter **10**mm in the Direction 1 Spacing box. The direction arrow points to the right. If required, click the Reverse Direction button.

575) Enter **3** in the Number of Instances box for Direction 1.

576) Click inside the **Pattern Direction** box for Direction 2.

577) Click the **left vertical edge** for Direction 2. The direction arrow points upward. If required, click the Reverse Direction button.

578) Enter **12**mm in the Direction 2 Spacing box.

579) Enter **2** in the Number of Instances box.

580) Check **Geometry Pattern** in the Options box.

Display the Linear Pattern feature.

581) Click **OK** ✔ from the Linear Pattern PropertyManager. LPattern1 is displayed in the FeatureManager.

Display an Isometric view - Shaded With Edges.
582) Press the **space bar** to display the Orientation dialog box.

583) Click **Isometric view** 🔲.

584) Click **Shaded With Edges** 🔲 from the Heads-up View toolbar.

Save the GUIDE.
585) Click **Save** 💾.

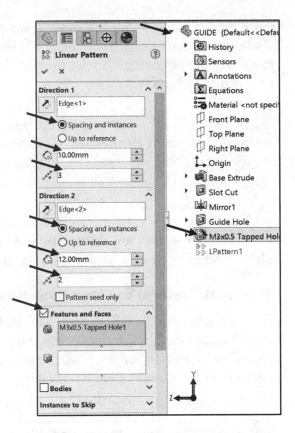

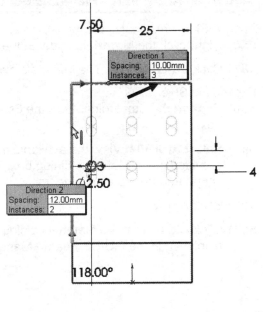

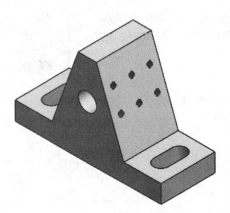

Reuse geometry. The Linear Pattern feature with the Geometry Pattern option saves rebuild time. Rename the seed feature in the part pattern. The seed feature is the first feature for the pattern, Example: M3x0.5 Tapped Hole1. Assemble additional parts to the seed feature in the assembly.

Save time when sketching by utilizing the Grid. Click **Options**, **Document Properties**, **Grid/Snap** to control display and spacing settings. The Display grid check box turns on/off the Grid in the sketch.

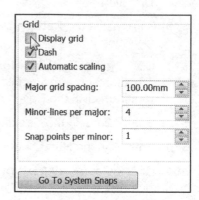

Hide geometric relations to clearly display profile lines in the sketch. Check **View**, **Hide/Show**, **Sketch Relations** to hide/display geometric relations in the sketch.

GUIDE Part - Material and Mass Properties

Apply material to the GUIDE part. The Material dialog box helps you manage physical materials. You can work with pre-defined materials, create custom materials, apply materials to parts, and manage favorites.

Calculate the Mass Properties of the GUIDE part using the Mass Properties tool.

Activity: GUIDE Part - Materials Editor and Mass Properties

Apply AISI 304 material to the model.

586) Right-click the **Materials** folder in the FeatureManager.

587) Click **Edit Material**. The Materials dialog box is displayed.

Expand the Steel folder.

588) **Expand** the Steel folder from the SOLIDWORKS Materials box.

589) Select **AISI 304**. View the available materials and information provided in the Materials dialog box. The Materials dialog box is updated from previous years.

590) Click **Apply**.

591) Click **Close** from the Materials dialog box. AISI 304 is displayed in the GUIDE FeatureManager.

Calculate the Mass Properties of the GUIDE.
592) Click the **Evaluate** tab from the CommandManager.

593) Click the **Mass Properties** ⚖ tool. The Mass Properties dialog box is displayed.

Set the units to display.
594) Click the **Options** button.

595) Click the **Use custom settings** button.

596) Enter **4** in the Decimal box.

597) Click **OK** from the Mass/Section Property Options dialog box. Review the Mass properties of the GUIDE.

598) Click **Close** from the Mass Properties dialog box.

Save the GUIDE.
599) Click **Save** 💾.

Close the GUIDE.
600) Click **File**, **Close** from the Menu bar.

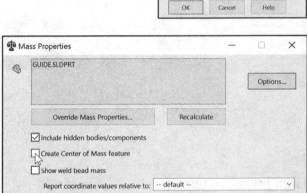

Add or create a Center of Mass (COM) ⊕ point directly in the Mass Properties box or from the Main menu for a part or assembly. The position of the COM point ⊕ updates when the model's center of mass changes. For example, the position of the COM point updates as you add, modify, delete, or suppress features or components. The COM point can be useful when you are designing assemblies requiring balanced mass to avoid excessive vibration.

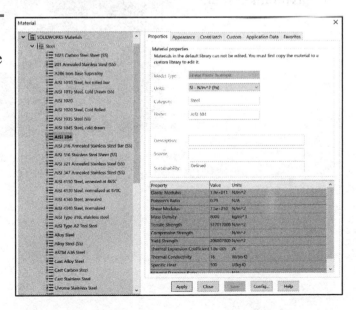

💡 A material that is selected in the Material dialog box propagates to other SOLIDWORKS applications: Mass Properties, SOLIDWORKS Simulation, PhotoWorks and PhotoView to name a few. Specify a material early in the design process. Define a material for the SW library or define your own material.

Manufacturing Considerations

CSI Mfg.
(www.compsources.com)
Southboro, MA, USA
provides solutions to their
customers by working with
engineers to obtain their
design goals and to reduce
cost.

Customer Focus
Courtesy of CSI Mfg (www.compsources.com)
Southboro, MA

High precision turned
parts utilize a variety of
equipment such as single
and multi-spindle lathes
for maximum output.

MultiDECO 20/8b, 8 spindle, full CNC 23 axis automatic
Courtesy of Tornos USA (www.tornosusa.com)

Obtain knowledge about the manufacturing process from your colleagues before you design the part.

Ask questions on material selection, part tolerances, machinability and cost. A good engineer learns from past mistakes of others.

Review the following design guidelines for Turned Parts.

Examples of Cost Effective Materials
- AISI 303
- BRASS
- AL 2011

Guideline 1: Select the correct material.

Consider part functionality, operating environment, cost, "material and machine time," machinability and surface finish. Utilize standard stock sizes and material that is readily available. Materials are given a machinability rating. Brass (100), AISI 303(65) and AISI 304(30).

Example: Using AISI 304 vs. AISI 303. AISI 303 is easier to machine due to its higher sulfur content.

Guideline 2: Minimize tool changes. Maximize feature manufacturability.

The Extruded Cut, Hole, Chamfer and Fillet features require different tooling operations. Will changes in form or design reduce the number of operations and the cost of machining?

Example:

- Avoid Fillets on inside holes to remove sharp edges. Utilize the Chamfer feature to remove sharp edges.

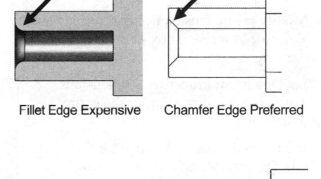

Fillet Edge Expensive   Chamfer Edge Preferred

- Utilize common size holes for fewer tool changes.

- Leave space between the end of an external thread and the shoulder of the rod. Rule of thumb: 1-2 pitch of thread, minimum.

- Leave space at the end of an internal thread for the tool tip endpoint.

💡 To display a cosmetic thread, right-click the Annotations folder in the FeatureManager and click Details. Check the Cosmetic threads box and the Shaded cosmetic thread box. Click OK. Select Shaded With Edges for display modes. View the results.

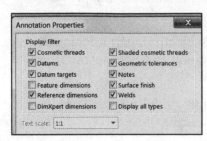

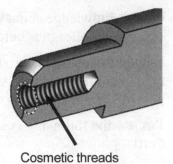

Cosmetic threads

Guideline 3: Optimize the design for production. Provide dimension, tolerance and precision values that can be manufactured and inspected.

- Check precision on linear, diameter and angular dimensions. Additional non-needed decimal places increase cost to a part.

Example:

The dimension 45 at +/- .5° indicates the dimension is machined between 44.5° and 45.5.

The dimension 45.00 at +/- .05° indicates the dimension is machined between 44.95 and 45.05.

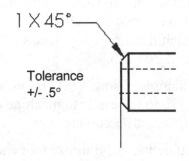

This range of values adds additional machining time and cost to the final product.

Modify the dimension decimal display in the Tolerance/Precision box in the part, assembly or drawing.

Set Document Properties, Dimensions, Tolerance and Precision for the entire document.

In Project 3, create drawings for the GUIDE part. Modify the dimension values and tolerance on the ROD and GUIDE to create a clearance fit.

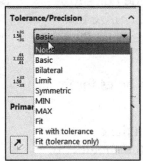

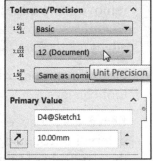

Sketch Entities and Sketch Tools

The Sketch Entities and Sketch Tools entries in the tools menu contain additional options not displayed in the default Sketch toolbar. Each option contains information in Help, SOLIDWORKS Help Topics.

The most common errors are open contours (gaps) or overlapping contours. Edit the sketch and correct the error or click the Show the problem using Check Sketch for Feature message.

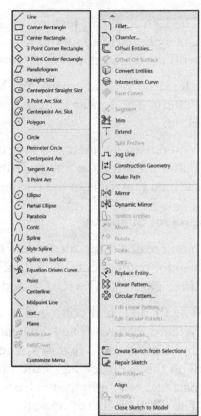

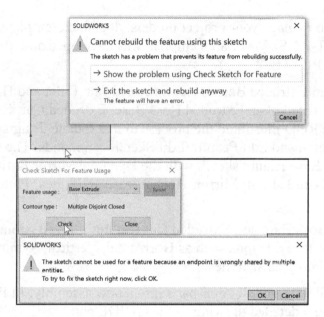

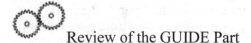 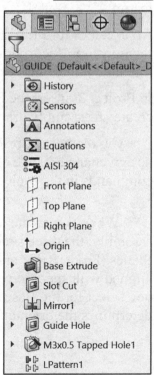

Review of the GUIDE Part

The GUIDE part utilized an Extruded Boss/Base feature. The 2D sketched profile utilized the Dynamic Mirror Sketch tool to build symmetry into the part.

The linear and diameter dimensions corresponded to the ROD and GUIDE-CYLINDER assembly.

The Extruded Cut feature created the Slot-Cut. The Mirror feature created a second Slot-Cut symmetric about the Right Plane. The Guide Hole utilized the Extruded Cut feature with the Through All End Condition.

The Hole Wizard created the M3 x0.5 Tapped Hole, the seed feature in the Linear Pattern. The Linear Pattern feature created multiple instances of the M3 x0.5 Tapped Hole.

You reused geometry to save design time and rebuild time. The Mirror feature, Linear Pattern feature, Geometry Pattern option and Sketch Mirror are time saving tools. You defined materials in the part used by other software applications.

Project Summary

You created three file folders to manage your project models, document templates and vendor components. The PART-ANSI-MM Part Template was developed and utilized for three parts: ROD, GUIDE and PLATE.

The PLATE part consisted of an Extruded Boss/Base, Extruded Cut, Fillet and Hole Wizard features. The ROD consisted of an Extruded Boss/Base, Extruded Cut, Chamfer, and Hole Wizard features. The ROD illustrated the process to manipulate design changes with the Rollback bar, Edit Sketch and Edit Feature/Edit Sketch commands. The GUIDE part utilized an Extruded Boss/Base feature sketch with the Dynamic Mirror Entities tool. The GUIDE also utilized the Extruded Cut, Mirror, Hole Wizard and Linear Pattern features.

All sketches for the Extruded Boss/Base and Extruded Cut features utilized geometric relationships. Incorporate geometric relations such as Horizontal, Vertical, Symmetric, Equal, Collinear and Coradial into your sketches.

In Project 3, insert the ROD, GUIDE and PLATE parts into a new assembly. In Project 4, create an assembly drawing and a detailed drawing of the GUIDE part.

Are you ready to start Project 3? Stop! Examine and create additional parts. Perform design changes. Take chances, make mistakes and have fun with the various features and commands in the project exercises.

When you create a new part or assembly, the three standard default Planes (Front, Right and Top) are aligned with specific views. The Plane you select for the Base sketch determines the orientation of the part.

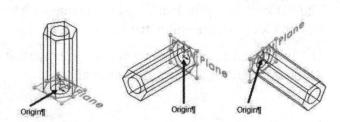

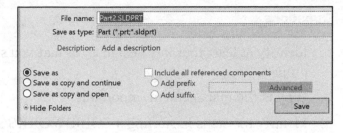

- Utilize the Save As/Save as command to save the file in another file format.

- Utilize the Save as copy and continue command to save the document to a new file name without replacing the active document.

- Utilize the Save as copy and open command to save the document to a new file name that becomes the active document. The original document remains open. References to the original document are not automatically assigned to the copy.

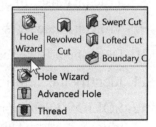

A goal of this book is to expose various SOLIDWORKS design tools and features. For manufacturing, use the Hole Wizard feature to create a simple or complex hole along with the Advance Hole feature.

Redeem the code on the inside cover of the book. View the provided videos to enhance the user experience.

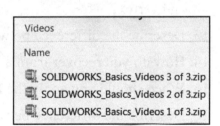

- Start a SOLIDWORKS session.

- SOLIDWORKS Interface.

- Create 2D Sketches, Sketch Planes and utilize various Sketch tools.

- Create 3D Features and apply Design Intent.

- Create an Assembly.

- Create fundamental Drawings Part 1 & Part 2.

Templates are part, drawing and assembly documents which include user-defined parameters. Open a new part, drawing or assembly. Select a template for the new document.

- *Parts*. The Parts default template is located in the C:\ProgramData\SolidWorks\SOLIDWORKS 2018\templates\Part.prtdot folder.

- *Assemblies*. The Assemblies default template is located in the C:\ProgramData\SolidWorks\SOLIDWORKS 2018\templates\Assembly.asmdot folder.

- *Drawings*. The Drawings default template is located in the C:\ProgramData\SolidWorks\SOLIDWORKS 2018\templates\Drawing.drwdot folder.

Questions

1. Identify at least four key design areas that you should investigate before starting a project.

2. Why is file management important?

3. Identify the steps in starting a SOLIDWORKS session.

4. Identify the three default Reference planes in SOLIDWORKS.

5. Describe a Base sketch. Provide an example.

6. True or False: The plane or face you select for the Base sketch determines the orientation of the part.

7. True or False: In a fully defined sketch, all sketch entities are displayed in black.

8. True or False: In an over defined sketch, all sketch entities are displayed in blue.

9. True or False: It is considered very poor design practice not to fully define a sketch.

10. Describe a Base feature. Provide an example.

11. Identify the Drafting Standard which is used in the United States.

12. SOLIDWORKS provides system feedback by attaching a symbol to the mouse pointer cursor. The system feedback symbol indicates what you are selecting or what the system is expecting you to select. What is the feedback symbol for a face, edge and vertex?

13. Describe the difference between First Angle Projection and Third Angle Projection.

14. How do you recover from a Rebuild error?

15. When should you apply the Hole Wizard feature vs. the Extruded Cut feature?

16. Describe the difference between an Extruded Boss/Base feature and an Extruded Cut feature.

17. Describe a Fillet feature. Provide an example.

18. Name the command keys used to Copy sketched geometry.

19. Describe the difference between the Edit Feature and Edit Sketch tool.

20. Describe a Chamfer feature. Provide an example.

21. Describe the Rollback bar function. Provide an example.

22. Identify three types of Geometric Relations that you can apply to a sketch.

23. Describe the procedure in creating a Part Document template.

24. Describe the procedure to modify material in an existing part.

25. Describe the procedure to set the Overall drafting standard, units and precision for a Part document.

26. Describe the procedure to sketch an Elliptical profile.

27. Identify the name of the following feature tool icons.

A   B   C   D   E   F

A		B		C		D
E		F				

28. Identify the name of the following Sketch tool icons.

A   B   C   D   E   F   G   H   I   J

A		B		C		D
E		F		G		H
I		J				

29. Identify the surfaces with the appropriate letter that would appear in the FRONT view, TOP view and RIGHT view.

FRONT view surfaces: _____

TOP view surfaces: _____

RIGHT view surfaces:_____

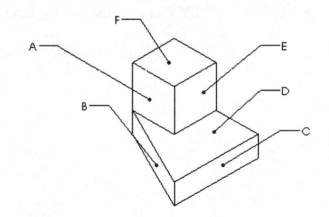

30. Identify the Angle of Projection for the provided illustration.

Angle of Projection_____.

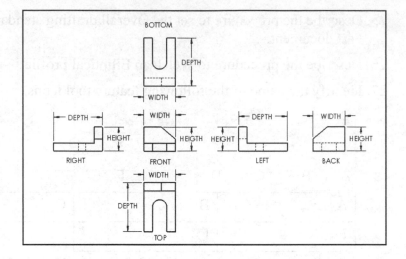

Exercises

Exercise 2.1 and Exercise 2.2

A fast and economical way to manufacture parts is to utilize Steel Stock, Aluminum Stock, Cast Sections, Plastic Rod and Bars and other stock materials.

Stock materials are available in different profiles and various dimensions from tool supply companies.

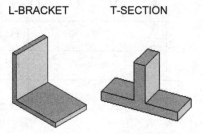

L-BRACKET   T-SECTION

- Create a new L-BRACKET part and a T-SECTION part. Each part contains an Extruded Base feature. Apply units from the below tables.

- Utilize the Front Plane for the Sketch plane.

Exercise 2.1: L-BRACKET

Ex.:	A:	B:	C:	LENGTH:	Units:
1.1a	3	3	5/8	10	IN
1.1b	4	4	3/4	10	
1.1c	6	4	1/2	25	
1.1d	8	6	3/4	25	
1.1e	12	12	3/2	25	
1.1f	7	4	3/4	30	
1.1g	75	75	10	250	MM
1.1h	75	100	10	250	
1.1i	200	100	25	500	

Exercise 2.2: T-SECTION

Ex.:	A:	B:	C:	LENGTH:	Units:
1.2a	3	3	5/8	30	IN
1.2b	4	4	3/4	30	
1.2c	6	6	1	30	
1.2d	75	75	10	250	MM
1.2e	75	100	10	250	
1.2f	200	100	25	500	

Exercise 2.2a: PART DOCUMENT TEMPLATES

Create a Metric (MMGS) Part document template.

- Use the ISO drafting standard and the MMGS unit system.

- Set the Drawing/Units and precision to the appropriate values.

- Name the Template PART-MM-ISO.

Exercise 2.2b: PART DOCUMENT TEMPLATES

Create an English (IPS) Part document template.

- Use the ANSI drafting standard and the IPS unit system.

- Set Drawing/Units and precision to the appropriate values.

- Name the Template PART-IN-ANSI.

Exercise 2.3: Identify the Sketch plane for the Boss-Extrude1 feature. View the Origin location. Simplify the number of features.

A: Top Plane

B: Front Plane

C: Right Plane

D: Left Plane

Correct answer _____.

Origin

Exercise 2.4: Identify the Sketch plane for the Boss-Extrude1 feature. View the Origin location. Simplify the number of features.

A: Top Plane

B: Front Plane

C: Right Plane

D: Left Plane

Correct answer _____.

Origin

Exercise 2.5: Identify the Sketch plane for the Boss-Extrude1 feature. View the Origin location. Simplify the number of features.

A: Top Plane

B: Front Plane

C: Right Plane

D: Left Plane

Correct answer _____.

Origin

Exercise 2.6: Identify the Sketch plane for the Boss-Extrude1 feature. View the Origin location. Simplify the number of features.

A: Top Plane

B: Front Plane

C: Right Plane

D: Left Plane

Correct answer _____.

Origin

Exercise 2.7: Identify the material category for 6061 Alloy.

A: Steel

B: Iron

C: Aluminum Alloys

D: Other Alloys

E: None of the provided

Correct answer _____.

Exercise 2.8: AXLE

Create an AXLE part as illustrated with dual units. Overall drafting standard - ANSI. IPS is the primary unit system.

- Utilize the Front Plane for the Sketch plane.

- Utilize the Mid Plane End Condition. The AXLE is symmetric about the Front Plane. Note the location of the Origin.

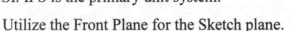
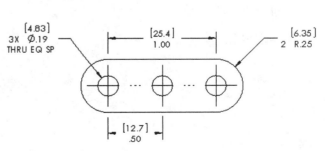

Exercise 2.9: SHAFT COLLAR

Create a SHAFT COLLAR part as illustrated with dual system units IPS (inch, pound, second) and MMGS (millimeter, gram, second). Overall drafting standard - ANSI.

- Utilize the Front Plane for the Sketch plane. Note the location of the Origin.

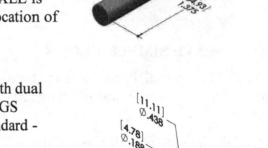

Exercise 2.10: FLAT BAR - 3 HOLE

Create the FLAT BAR - 3 HOLE part as illustrated with dual system units IPS (inch, pound, second) and MMGS (millimeter, gram, second). Overall drafting standard - ANSI.

- Utilize the Front Plane for the Sketch plane.

- Utilize the Center point Straight Slot Sketch tool.

- Utilize a Linear Pattern feature for the three holes. The FLAT BAR - 3 HOLE part is stamped from 0.060in [1.5mm] Stainless Steel.

Exercise 2.11: FLAT BAR - 9 HOLE

Create the the FLAT BAR - 9 HOLE part. Overall drafting standard - ANSI.

- The dimensions for hole spacing, height and end arcs are the same as the FLAT BAR - 3 HOLE part.

- Utilize the Front Plane for the Sketch plane.

- Utilize the Centerpoint Straight Slot Sketch tool.

- Utilize the Linear Pattern feature to create the hole pattern. The FLAT BAR - 9 HOLE part is stamped from 0.060in [1.5mm] 1060 Alloy.

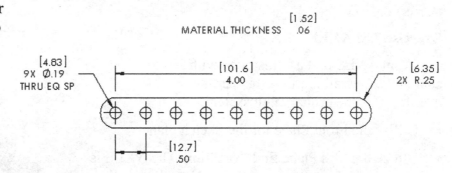

Exercise 2.12: SIMPLE BLOCK

Create the illustrated part. Note the location of the Origin. Overall drafting standard - ANSI.

- Calculate the overall mass of the illustrated model. Apply the Mass Properties tool.

- Think about the steps that you would take to build the model.

- Review the provided information carefully.

- Units are represented in the IPS (inch, pound, second) system.

- A = 3.50in, B = .70in

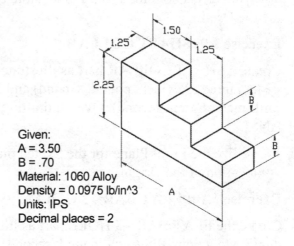

Given:
A = 3.50
B = .70
Material: 1060 Alloy
Density = 0.0975 lb/in^3
Units: IPS
Decimal places = 2

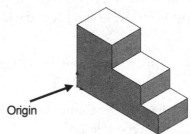

Origin

Exercise 2.13: SIMPLE T

Create the illustrated part. Note the location of the Origin. Overall drafting standard - ANSI.

- Calculate the overall mass of the illustrated model. Apply the Mass Properties tool.

- Think about the steps that you would take to build the model.

- Review the provided information carefully. Units are represented in the IPS (inch, pound, second) system.

- A = 3.00in, B = .75in

Given:
A = 3.00
B = .75
Material: Copper
Density = 0.321 lb/in^3
Units: IPS
Decimal places = 2

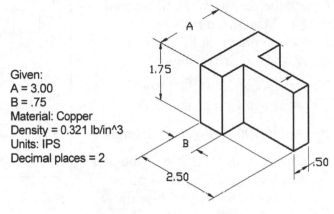

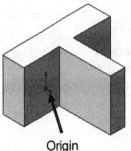

Origin

Exercise 2.14: SIMPLE BLOCK STEP

Create the illustrated part. Note the location of the Origin. Overall drafting standard - ANSI.

- Calculate the volume of the part and locate the Center of mass relative to the origin with the provided information.

Given:
A = 3.30
B = 2.00
Material: 2014 Alloy
Density = .101 lb/in^3
Units: IPS
Decimal places = 2

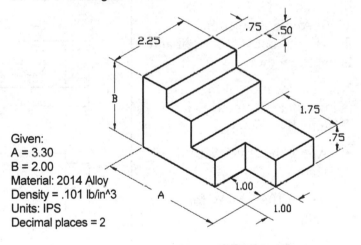

- Apply the Mass Properties tool.

- Think about the steps that you would take to build the model.

- Review the provided information carefully.

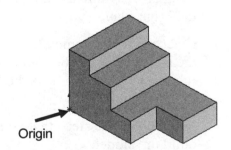

Origin

Exercise 2.15: DRAWING 1

Create the part from the illustrated ANSI - MMGS Third Angle Projection drawing: Front, Top, Right and Isometric views.

Note the location of the Origin (shown in an Isometric view).

- Apply 1060 Alloy for material.

- Calculate the Volume of the part and locate the Center of mass relative to the origin. Use the Mass Properties tool.

- Think about the steps that you would take to build the model. The part is symmetric about the Front Plane.

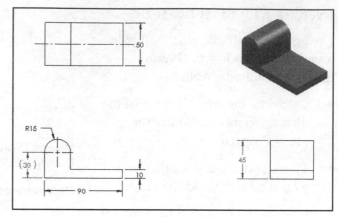

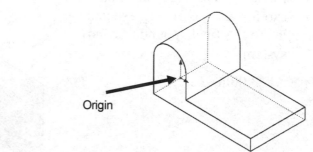

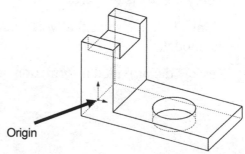

Origin

Exercise 2.16: DRAWING 2

Create the part from the illustrated ANSI - MMGS Third Angle Projection drawing: Front, Top, Right and Isometric views.

Note the location of the Origin (shown in an Isometric view). Assume Symmetry.

- Apply the Hole Wizard feature.

- Apply 1060 Alloy for material.

- The part is symmetric about the Front Plane.

- Calculate the Volume of the part and locate the Center of mass relative to the origin. Use the Mass Properties tool.

- Think about the steps that you would take to build the model.

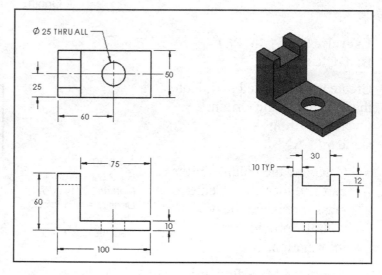

Origin

Exercise 2.17: Slider Part

Create the part from the illustrated ANSI - MMGS Third Angle Projection drawing: Front, Top, Right and Isometric view.

Note: The location of the Origin displayed in the Isometric view.

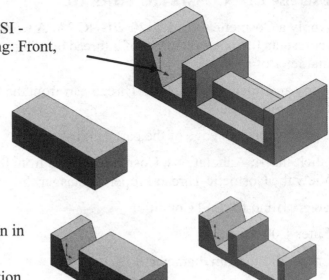

- Apply Cast Alloy steel for material.

- The part is symmetric about the Front Plane.

- Apply Mid Plane for End Condition in Boss-Extrude1.

- Apply Through All for End Condition in Cut-Extrude1.

- Apply Through All for End Condition in Cut-Extrude2.

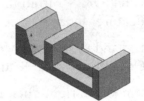

- Apply Up to Surface for End Condition in Boss Extrude2

- Calculate the Volume of the part and locate the Center of mass.

- Calculate the Volume of the part and locate the Center of mass relative to the origin. Use the Mass Properties tool.

Think about the steps that you would take to build the model. Do you need the Right view for manufacturing? Does it add any valuable information?

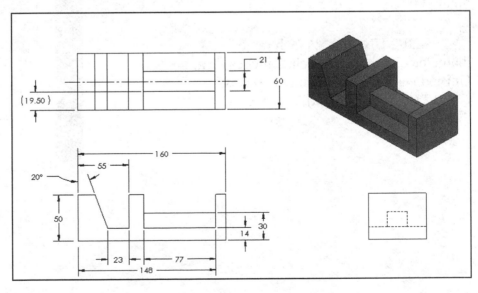

Exercise 2.18: COSMETIC THREAD

Apply a Cosmetic thread ¼-20-2 UNC 2A. A Cosmetic thread represents the inner diameter of a thread on a boss or the outer diameter of a thread.

Copy and open the Cosmetic Thread part from the Chapter 2 Homework folder.

Click the bottom edge of the part as illustrated.

Click Insert, Annotations, Cosmetic Thread from the Menu bar menu. View the Cosmetic Thread PropertyManager. Edge<1> is displayed.

Select Blind for End Condition.

Enter 1.00 for depth.

Enter .200 for min diameter.

Enter ¼-20-2 UNC 2A in the Thread Callout box.

Click OK ✔ from the Cosmetic Thread FeatureManager.

Expand the FeatureManager. View the Cosmetic Thread feature.

If needed, right-click the Annotations folder, and click Details.

Check the Cosmetic threads and Shaded cosmetic threads box.

Click OK from the Annotation Properties dialog box.

🔅 ¼-20-2 UNC 2A: ¼ inch major diameter - 20 threads / inch, 2 inches long, Unified National Coarse thread series, Class 2 (General Thread), A -External threads.

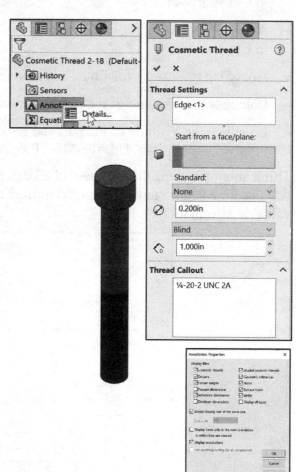

Exercise 2.19: HOLE BLOCK

Create the Hole-Block using the Hole Wizard feature as illustrated. Create the Hole-Block part on the Front Plane. The Hole Wizard 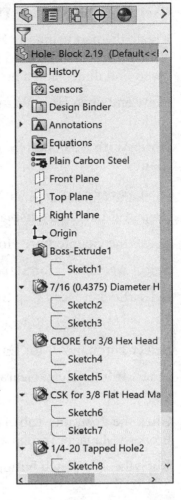 feature creates either a 2D or 3D sketch for the placement of the hole in the FeatureManager. You can consecutively place multiple holes of the same type. The Hole Wizard creates 2D sketches for holes unless you select a non-planar face or click the 3D Sketch button in the Hole Position PropertyManager.

Hole Wizard creates two sketches, a sketch of the revolved cut profile of the selected hole type and a sketch of the center placement for the profile. Fully define the sketch for the center placement. Both sketches should be fully defined.

Create a rectangular prism 2 inches wide by 5 inches long by 2 inches high. On the top surface of the prism, place four holes, 1 inch apart.

- Hole #1: Hole Type: Fractional Drill Size, 7/16 diameter, End Condition: Blind, 0.75 inch deep.
- Hole #2: Counterbore hole Type: for 3/8 inch diameter Hex bolt, End Condition: Through All.
- Hole #3: Countersink hole Type: for 3/8 inch diameter Flat head screw (100), 1.5 inch deep.
- Hole #4: Straight Tapped hole Type, Size ¼-20, 1 inch Blind hole deep.

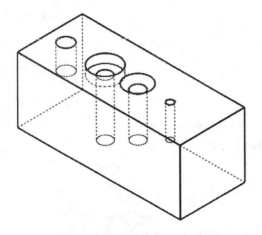

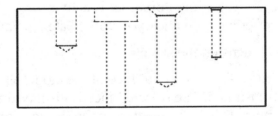

Exercise 2.20: 3D SKETCH/HOLE WIZARD

Create the part using the Hole Wizard feature. Apply the 3D sketch placement method as illustrated in the FeatureManager. Insert and dimension a hole on a cylindrical face.

Copy and open Hole Wizard 2.20 from the Chapter 2 Homework folder.

Note: With a 3D sketch, press the Tab key to move between planes.

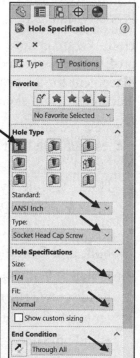

Click the Hole Wizard 📷 Features tool. The Hole Specification PropertyManager is displayed.

Select the Counterbore Hole Type.

Select ANSI Inch for Standard.

Select Socket Head Cap Screw for fastener Type.

Select 1/4 for Size. Select Normal for Fit.

Select Through All for End Condition.

Enter .100 for Head clearance in the Options box.

Click the Positions Tab. The Hole Position PropertyManager is displayed.

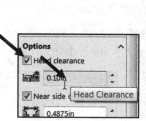

Click the 3D Sketch button. SOLIDWORKS displays a 3D interface

with the Point ^{XY} tool active.

🔆 When the Point tool is active, wherever you click you will create a point.

Dimension the sketch.

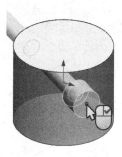

Click the cylindrical face of the model as illustrated. The selected face is displayed in blue. This indicates that an OnSurface sketch relations will be created between the sketch point and the cylindrical face. The hole is displayed in the model.

Insert a .25in dimension between the top face and the sketch point.

Locate the point angularly around the cylinder. Apply construction geometry.

Display the Temporary Axes.

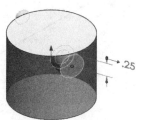

Click the Line ✎ Sketch tool. Note: 3D sketch is still activated.

Ctrl+click the top flat face of the model. This moves the red space handle origin to the selected face. This also constrains any new sketch entities to the top flat face. Note the mouse pointer icon.

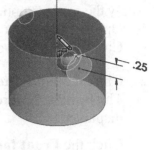

Move the mouse pointer near the center of the activated top flat face as illustrated. View the small black circle. The circle indicates that the end point of the line will pick up a Coincident relation.

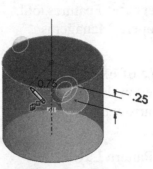

Click the center point of the circle.

Sketch a line so it picks up the **AlongZ sketch relation. The cursor displays the relation to be applied. This is a very important step!** If needed insert an AlongZ relation.

Create an AlongY sketch relation between the center point of the hole on the cylindrical face and the endpoint of the sketched line as illustrated. The sketch is fully defined.

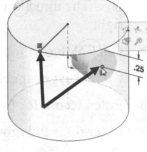

Add Relations

- ⊥ **Along X**
- ⊥ Along Y
- ⊥ Along Z
- ⋏ Coincident
- ⊗ Fix

Click OK ✔ from the Properties PropertyManager. Click OK ✔ from the Hole Position PropertyManager.

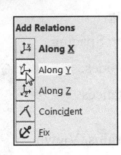

Expand the FeatureManager and view the results. The two sketches are fully defined. One sketch is the hole profile, the other sketch is to define the position of the feature.

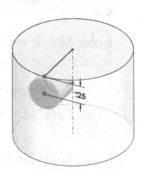

Close the model.

💡 You can create a second sketched line and insert an angle dimension between the two lines. This process is used to control the position of the centerpoint of the hole on the cylindrical face as illustrated. Insert an AlongY sketch relation between the centerpoint of the hole on the cylindrical face and the end point of the second control line. Control the hole position with the angular dimension. **Note: With a 3D sketch, press the Tab key to move between planes.**

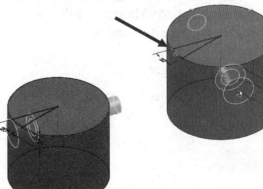

Exercise 2.21: Fill Pattern

Create a Polygon Layout Fill Pattern feature. Apply the seed cut option.

1. Open **Fill Pattern 2.21** from the Chapter 2 Homework folder.

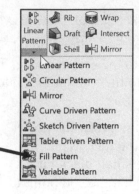

2. Click the **Fill Pattern** 🖼 Features tool. The Fill Pattern PropertyManager is displayed.

3. Click the **Front face** of Extrude1. Face<1> is displayed in the Fill Boundary box. The direction arrow points to the right.

4. Click **Polygon** for Pattern Layout.

5. Click **Target spacing**.

6. Enter **15**mm for Loop Spacing.

7. Enter **6** for Polygon sides.

8. Enter **10**mm for Instances Spacing.

9. Enter **10**mm for Margins. View the direction arrow.

10. Click **Create seed cut**.

11. Click **Circle** for Features to Pattern.

12. Enter **4**mm for Diameter.

13. Click **OK** ✔ from the Fill Pattern PropertyManager. Fill Pattern1 is created and is displayed in the FeatureManager.

14. **View** the results.

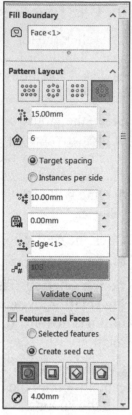

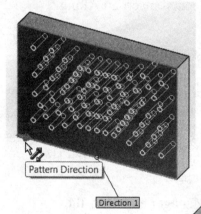

Exercise 2.22: Fill Pattern

Create a Circular Layout Fill Pattern feature.

1. **Copy** the Chapter 2 Homework folder to your hard drive.

2. Open **Fill Pattern 2.22** from the Chapter 2 Homework folder.

3. Click the **Fill Pattern**  Features tool. The Fill Pattern PropertyManager is displayed.

4. Click the **top face** of Boss-Extrude1. Face<1> is displayed in the Fill Boundary box. The direction arrow points to the back. Click **Circular** for Pattern Layout.

5. Enter **.10**in for Loop Spacing.

6. Click **Target spacing**.

7. Enter **.10**in Instance Spacing.

8. Enter **.05**in for Margins. Edge<1> is selected for Pattern Direction.

9. Click **inside** the Features to Pattern box.

10. Click **Boss-Extrude2** from the fly-out FeatureManager. Boss-Extrude2 is displayed in the Features to Pattern box.

11. Click **OK** ✔ from the Fill Pattern PropertyManager. Fill Pattern1 is created and is displayed in the FeatureManager.

12. **View** the results.

Notes:

Project 3
Fundamentals of Assembly Modeling

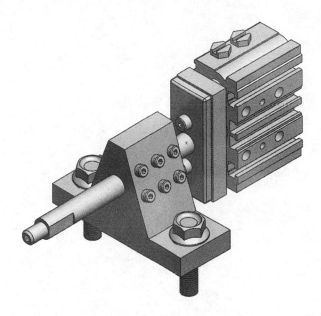

Below are the desired outcomes and usage competencies based on the completion of Project 3.

Project Desired Outcomes:	Usage Competencies:

• Two assemblies:
   ○ GUIDE-ROD assembly
   ○ CUSTOMER assembly
	• Insert and edit components. Apply Standard mates and Mechanical mates using the Bottom-up assembly design approach. Utilize the standard and Quick mate procedure.

• Use an assembly from SMC USA. ○ GUIDE-CYLINDER	• Use and assemble components from 3D ContentCentral.
• Flange bolt part from the SOLIDWORKS Design Library.	• Assemble components from the SOLIDWORKS Design Library.

• Two parts:
   ○ 4MMCAPSCREW
   ○ 3MMCAPSCREW
	• Apply the Revolved Base, Extruded Cut, Chamfer, Component Pattern features along with the Copy, Edit and suppress tools.
	• Apply and edit material.

Notes:

Project 3 - Fundamentals of Assembly Modeling

Project Objective

Provide an understanding of the Bottom-up assembly design approach. Insert existing parts into an assembly. Orient and position the components in the assembly using Standard and Mechanical Mates. Utilize the standard and Quick mate procedure.

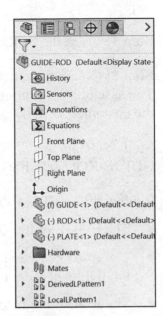

Create the GUIDE-ROD assembly. Utilize the ROD, GUIDE and PLATE parts. The ROD, GUIDE and PLATE parts were created in project 2.

Create the CUSTOMER assembly. The CUSTOMER assembly consists of two sub-assemblies: GUIDE-ROD and GUIDE-CYLINDER.

Insert the Flange bolt from the SOLIDWORKS Design Library into the GUIDE-ROD assembly. Utilize a Revolved Base feature to create the 4MMCAPSCREW. Copy the 4MMCAPSCREW part and modify dimensions to create the 3MMCAPSCREW part.

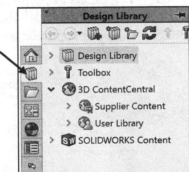

On the completion of this project, you will be able to:

- Understand the Assembly FeatureManager Syntax.

- Insert parts/sub-components into an assembly.

- Insert and edit Standard and Mechanical mates.

- Utilize the Standard and Quick mate procedure.

- Rename parts and copy assemblies with internal references.

- Save an assembly with references.

- Incorporate design changes into an assembly.

- Insert a sub-component from the SOLIDWORKS Design Library.

- Modify, Edit and Suppress features in an assembly.

- Recover from Mate errors in an assembly.

- Suppress/Un-suppress component features.

- Create an Isometric Exploded view of an assembly.

- Create a Section view of an assembly.

Project Situation

The PLATE, ROD and GUIDE parts were created in Project 2. Perform the following steps:

Step 1: Insert the ROD, GUIDE and PLATE into a GUIDE-ROD assembly.

Step 2: Obtain the customer's GUIDE-CYLINDER assembly using 3D ContentCentral. The assembly is obtained to guarantee proper fit between the GUIDE-ROD assembly and the customer's GUIDE-CYLINDER assembly.

Step 3: Create the CUSTOMER assembly. The CUSTOMER assembly combines the GUIDE-ROD assembly with the GUIDE-CYLINDER assembly.

Review the GUIDE-ROD assembly design constraints:

- The ROD requires the ability to travel through the GUIDE.

- The ROD keyway is parallel to the right surface of the GUIDE. The top surface of the GUIDE is parallel to the work area.

- The ROD mounts to the PLATE. The GUIDE mounts to a flat work surface.

The PISTON PLATE is the front plate of the GUIDE-CYLINDER assembly. The PLATE from the GUIDE-ROD assembly mounts to the PISTON PLATE. Create a rough sketch of the conceptual assembly.

Rough Sketch of Design Situation: CUSTOMER assembly

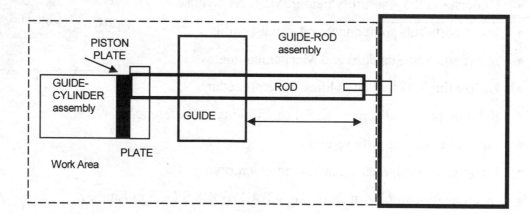

An assembly combines two or more parts. In an assembly, parts are referred to as components. Design constraints directly influence the assembly design process. Other considerations indirectly affect the assembly design, namely cost, manufacturability and serviceability.

Project Overview

Translate the rough conceptual sketch into a SOLIDWORKS assembly.

The GUIDE-ROD is the first assembly.

Determine the first component of the assembly.

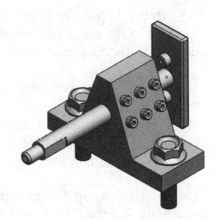

The first component is the GUIDE. The GUIDE remains stationary. The GUIDE is a fixed component.

The action of assembling components in SOLIDWORKS is defined as mates.

Mates are relationships between components that simulate the construction of the assembly in a manufacturing environment. Mates affect/restrict the *six degrees* of freedom in an assembly.

The CUSTOMER assembly combines the GUIDE-CYLINDER assembly with the GUIDE-ROD assembly.

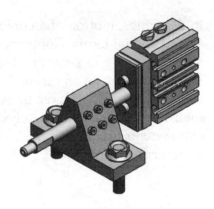

Assembly Modeling Approach

In SOLIDWORKS, components and their assemblies are directly related through a common file structure. Changes in the components directly affect the assembly and vice versa. Create assemblies using the Bottom-up assembly approach, Top-down assembly approach or a combination of both methods. This project focuses on the Bottom-up assembly approach.

The Bottom-up approach is the traditional method that combines individual components. Based on design criteria, the components are developed independently. The three major steps in a Bottom-up assembly approach are Create each component independent of any other component in the assembly, insert the components into the assembly, and mate the components in the assembly as they relate to the physical constraints of your design.

In the Top-down assembly approach, major design requirements are translated into assemblies, sub-assemblies and components.

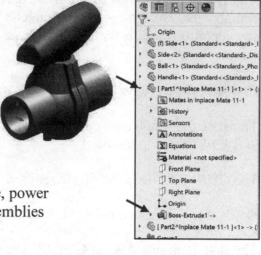

In the Top-down approach, you do not need all of the required component design details. Individual relationships are required.

Example: A computer. The inside of a computer can be divided into individual key sub-assemblies such as a motherboard, disk drive, power supply, etc. Relationships between these sub-assemblies must be maintained for proper fit.

Use the Bottom-up design approach for the GUIDE-ROD assembly and the CUSTOMER assembly.

Linear Motion and Rotational Motion

In dynamics, motion of an object is described in linear and rotational terms. Components possess linear motion along the x, y and z-axes and rotational motion around the x, y and z-axes. In an assembly, each component has six degrees of freedom: three translational (linear) and three rotational. Mates remove degrees of freedom. All components are rigid bodies. The components do not flex or deform.

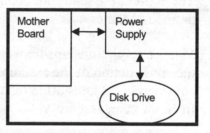

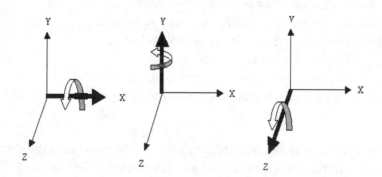

GUIDE-ROD assembly

The GUIDE-ROD assembly consists of six components.

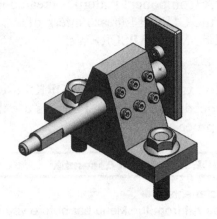

- GUIDE

- ROD

- PLATE

- Flange bolt

- 4MMCAPSCREW

- 3MMCAPSCREW

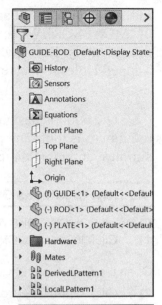

The first component is the GUIDE. The GUIDE is the fixed component in the GUIDE-ROD assembly.

The second component is the ROD. The ROD translates linearly through the GUIDE.

The third component is the PLATE. The PLATE is assembled to the ROD and the GUIDE-CYLINDER assembly.

The fourth component is the Flange bolt. Obtain the Flange bolt from the SOLIDWORKS Design Library.

The fifth component is the 4MM CAPSCREW. Create the 4MM CAPSCREW component with a Revolved Base feature. A Revolved Base feature requires a centerline for an axis and a sketched profile. The profile is rotated about the axis to create the feature.

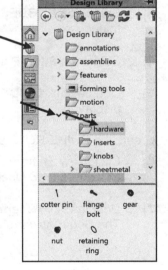

The sixth component is the 3MMCAPSCREW. Create the 3MMCAPSCREW from the 4MMCAPSCREW. Reuse geometry to save time.

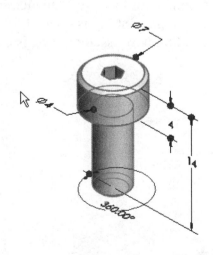

Insert a Component Pattern feature of the 3MMCAPSCREW part. A Component Pattern is created in the assembly. Reference the GUIDE Linear Pattern of Tapped Holes to locate the 3MMCAPSCREWs.

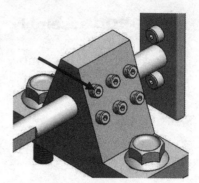

💡 The book is design to expose the SOLIDWORKS user to numerous tools and procedures. It may not always use the simplest and most direct process.

Activity: Create the GUIDE flange bolt Assembly

Close all SOLIDWORKS documents.

1)   Click **Windows**, **Close All** from the Menu bar before you begin this project.

Open the GUIDE part.

2)   Click **Open** 📂 from the Menu bar.

3)   Select the **ENGDESIGN-W-SOLIDWORKS\PROJECTS** folder.

4)   Select the **Filter Parts (*.prt; *sldprt)** button.

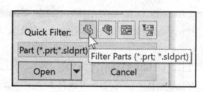

5)   Double-click **GUIDE**. The GUIDE FeatureManager is displayed.

Create the GUIDE-ROD assembly.

6)   Click **New** 🗋 from the Menu bar. The New SOLIDWORKS Documents dialog box is displayed.

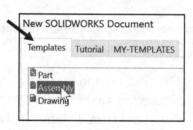

7)   Double-click **Assembly** from the default Templates tab. The Begin Assembly PropertyManager is displayed.

💡 The Begin Assembly PropertyManager and the Insert Component PropertyManager are displayed when a new or existing assembly is opened, if the Start command when creating new assembly box is checked.

When a part is inserted into an assembly, it is called a component. The Begin Assembly PropertyManager is displayed to the left of the Graphics window. GUIDE is listed in the Part/Assembly to Insert box.

Insert the GUIDE and fix it to the assembly Origin.
8)   Double-click **GUIDE** in the Open documents box.

9)   Click **OK** ✔ from the Begin Assembly PropertyManager. The GUIDE part icon is displayed in the assembly FeatureManager. It is fixed to the Assembly origin.

The GUIDE name is added to the assembly FeatureManager with the symbol (f).

The symbol (f) represents a fixed component. A fixed component cannot move and is locked to the assembly Origin. The first component in an assembly should be fixed, fully defined, or mated to an axis.

To fix the first component to the Origin in an assembly, click OK ✔ from the Begin Assembly PropertyManager or click the Origin in the Graphics window.

☀ To remove the fixed state (f), right-click the **fixed component name** in the FeatureManager. Click **Float**. The component is free to move.

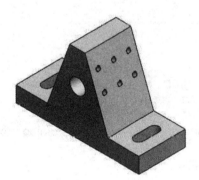

Your default system document templates may be different if you are a new user of SOLIDWORKS vs. an existing user who has upgraded from a previous version.

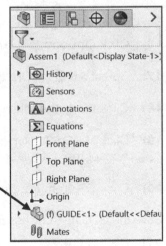

Save and name the assembly document.

10) Click **Save As** from the Menu bar.

11) Select the **ENGDESIGN-W-SOLIDWORKS\PROJECTS** folder.

12) Enter **GUIDE-ROD** for File name.

13) Click **Save**.

Close the GUIDE part.

14) Click **Window**, **GUIDE** from the Menu bar.

15) Click **File**, **Close** from the Menu bar. The GUIDE-ROD assembly is the open document.

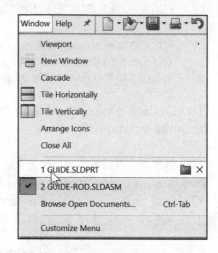

🔆 Select the Ctrl-Tab keys to quickly alternate between open SOLIDWORKS documents. Select inside Close ☒ to close the current SOLIDWORKS document. The outside Close ☒ exits SOLIDWORKS.

Set Document Properties. Set drafting standard and units.

16) Click **Options** ⚙ ˅, **Document Properties** tab.

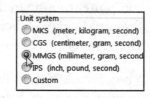

17) Select **ANSI** for Drafting Standard.

18) Click the **Units** folder.

19) Click **MMGS** (millimeter, gram, second) for Unit system. Accept the default settings.

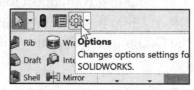

20) Click **OK** from the Document Properties - Units dialog box.

Save the GUIDE-ROD assembly.

21) Click **Save** 💾.

To customize the CommandManager, right-click on an existing tab. Click Customize Command Manager. Check the option to display in the CommandManager. The options you select are based on the tool types you require for the design.

🔆 View the provided video on creating an assembly to enhance your experience in this section.

GUIDE-ROD Assembly-Insert Component

The first component is the foundation of the assembly. The GUIDE is the first component in the GUIDE-ROD assembly. The ROD is the second component in the GUIDE-ROD assembly. Add components to assemblies utilizing the following techniques:

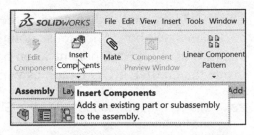

- Utilize the Insert Components 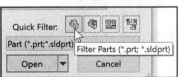 Assembly tool.

- Utilize Insert, Component from the Menu bar.

- Drag a component from Windows Explorer (3D ContentCentral) into the Assembly.

- Drag a component from the SOLIDWORKS Design Library into the Assembly.

- Drag a component from an Open part file into the Assembly.

Activity: GUIDE-ROD Assembly - Insert Component

Insert the ROD component into the assembly. Use the Insert Components tool.

22) Click the **Assembly** tab from the CommandManager.

23) Click the **Insert Components** 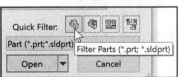 Assembly tool. The Insert Component PropertyManager is displayed.

24) Click the **BROWSE** button from the Open documents box.

25) Browse to the **ENGDESIGN-W-SOLIDWORKS\PROJECTS** folder.

26) Select the **Filter Parts (*.prt; *sldprt)** button to view only parts in the folder.

27) Double-click the **ROD** part. The ROD is displayed in the Graphics window.

The ROD part icon is displayed in the Open documents text box. The mouse pointer displays the ROD component when positioned inside the GUIDE-ROD Graphics window.

28) Click a **position** to the left of the GUIDE as illustrated.

Deactivate the Origin display in the assembly if needed.
29) Click **View**, **Hide/Show**, and uncheck **Origins** from the Main menu.

Re-position the ROD component in the Graphics window.
30) Click and drag the **ROD** in the Graphics window.

31) Move the **ROD** on the right side of the GUIDE as illustrated.

Fit the model to the Graphics window.
32) Press the **f** key.

The component movement in the assembly is determined by its degrees of freedom.

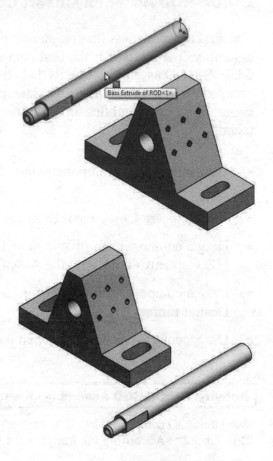

💡 Degree of freedom is geometry that is not defined by dimensions or relations and is free to move. In a 2D sketch, there are three degrees of freedom: movement along the X axis, Y axis and rotation about the Z axis (the axis normal to the Sketch plane).

Save the GUIDE-ROD assembly.
33) Click **Save** 💾.

34) Click **Rebuild and Save** the document. The ROD component is displayed in the GUIDE-ROD FeatureManager. The ROD part is free to move and rotate.

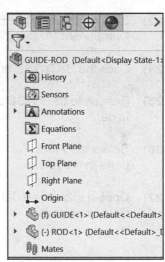

Review the FeatureManager Syntax.

35) **Expand** (f) GUIDE<1> in the FeatureManager. The features are displayed in the FeatureManager. Note: Base Extrude, Slot Cut, Guide Hole and M3x0.5 Tapped Hole1 contain additional sketches. Features were renamed in Project 2. Example: Boss-Extrude1 was renamed to Base Extrude.

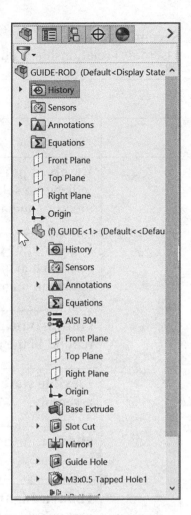

In Project 2, you applied AISI 304 as a material to the GUIDE part.

36) **Expand** (-) ROD<1> in the FeatureManager. The features are displayed in the FeatureManager.

37) Click the **Drop down arrow** ⌄ icon to the left of the GUIDE entry and ROD entry to collapse the list.

An Arrow › icon indicates that additional feature information is available. The Drop down arrow ⌄ icon indicates that the feature list is fully expanded. Manipulating the FeatureManager is an integral part of the assembly. In the step-by-step instructions, expand and collapse are used as follows:

 Expand - Click the Arrow › icon.

 Collapse - Click the Drop down arrow ⌄ icon.

FeatureManager Syntax

How do you distinguish the difference between an assembly and a part in the FeatureManager? Answer: The assembly icon contains a blue square block and an upside down yellow "T" extrusion. The part icon contains an upside down yellow "T" extrusion. Entries in the FeatureManager design tree have specific definitions. Understanding syntax and states saves time when creating and modifying assemblies. Review the columns below.

First Column: › (f) GUIDE<1> (Default<<Default>. A Resolved component (not in lightweight state) displays an arrow › icon. The arrow icon indicates that additional feature information is available. A Drop down arrow ⌄ icon displays the fully expanded feature list.

Second Column: ▸ (f) GUIDE<1> (Default<<Default> . Identifies a component's (part or assembly) relationship with other components in the assembly.

	Component or Part States:
	State:
	Resolved part. A part icon indicates a resolved state. A blue part icon indicates a selected, resolved part. The component is fully loaded into memory and all of its features and mates are editable.
	Lightweight part. A blue feather on the part icon indicates a lightweight state. When a component is lightweight, only a subset of its model data is loaded in memory.
	Out-of-Date part. An eye is displayed on the part icon. The part needs a rebuild in the assembly.
	Flexible state. A blue box on the part icon indicates a flexible part in an assembly.
	Hidden. A transparent icon indicates that the component is active but invisible.
	Hidden Lightweight. A clear feather over a clear part icon indicates the part is hidden and lightweight.
	Smart Component. A lightning bolt on the part icon indicates that the component is a Smart Component.
	Smart Fastener. A lightning bolt is displayed on the Fastener folder icon.
	Resolved assembly. Resolved (or unsuppressed) is the normal state for assembly components. A resolved assembly is fully loaded in memory, fully functional, and fully accessible.
	Inserted New Component. A new inserted component into an active assembly.

Third Column: ▸ 🧊 (f) GUIDE<1> (Default<<Default>. The part is fixed (f). You can fix the position of a component so that it cannot move with respect to the assembly Origin. By default, the first part in an assembly is fixed; however, you can float it at any time.

It is recommended that at least one assembly component is either fixed or mated to the assembly planes or Origin. This provides a frame of reference for all other mates and helps prevent unexpected movement of components when mates are added. The Component Properties are:

Component Properties in an assembly:	
Symbol:	**Relationship:**
(-)	A minus sign (–) indicates that the part or assembly is under-defined and requires additional information.
(+)	A plus sign (+) indicates that the part or assembly is over-defined.
None	The Base component is mated to three assembly reference planes. It is fully defined.
(f)	A fixed symbol (f) indicates that the part or assembly does not move.
(?)	A question mark (?) indicates that additional information is required on the part or assembly.

Fourth Column: ▸ 🧊 (f) GUIDE<1> (Default<<Default>. Name of the part.

Fifth Column: ▸ 🧊 (f) GUIDE<1> (Default<<Default>. The symbol <#> indicates the particular inserted instance of a component. The symbol <1> indicates the first inserted instance of a component in the assembly. If you delete a component and reinsert the same component again, the <#> symbol increments by one.

Fifth Column: ▸ 🧊 MOTHERBOARD<1> -> (Default The Resolved state displays the icon with an External reference symbol, "- >". The state of External references is displayed as follows:

- If a part or feature has an external reference, its name is followed by - >. The name of any feature with external references is also followed by - >.

- If an external reference is currently out of context, the feature name and the part name are followed by - >?

- The suffix - >* means that the reference is locked.

- The suffix - >x means that the reference is broken.

🔆 There are modeling situations in which unresolved components create rebuild errors. In these situations, issue the forced rebuild, Ctrl+Q. The Ctrl+Q option rebuilds the model and its features. If the mates still contain rebuild errors, resolve all the components below the entry in the FeatureManager that contains the first error.

Mate Types

Mates provide the ability to create geometric relationships between assembly components. Mates define the allowable directions of rotational or linear motion of the components in the assembly. Move a component within its degrees of freedom in the Graphics window to view the behavior of an assembly.

Mates are solved together as a system. The order in which you add mates does not matter. All mates are solved at the same time. You can suppress mates just as you can suppress features.

The Mate PropertyManager provides the ability to select either the **Mates** or **Analysis** tab. Each tab has a separate menu. The **Analysis** tab requires the ability to run SOLIDWORKS Motion. The Analysis tab is not covered in this book. The Mate PropertyManager displays the appropriate selections based on the type of mate you create. The components in the GUIDE-ROD assembly utilize Standard Mate types.

Review **Standard**, **Advanced**, and **Mechanical** Mate types.

Standard Mates:

Components are assembled with various mate types. The Standard Mate types are:

- **Coincident**: Locates the selected faces, edges, or planes so they use the same infinite line. A Coincident mate positions two vertices for contact.

- **Parallel**: Locates the selected items to lie in the same direction and to remain a constant distance apart.

- **Perpendicular**: Locates the selected items at a 90° angle to each other.

- **Tangent**: Locates the selected items in a tangent mate. At least one selected item must be a conical, cylindrical or spherical face.

- **Concentric**: Locates the selected items so they can share the same center point.

- **Lock Mate**: Maintains the position and orientation between two components.

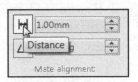

- **Distance Mate**: Locates the selected items with a specified distance between them. Use the drop-down arrow box or enter the distance value directly.

- **Angle Mate**: Locates the selected items at the specified angle to each other. Use the drop-down arrow box or enter the angle value directly.

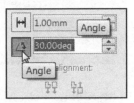

There are two Mate Alignment options. The Aligned option positions the components so that the normal vectors from the selected faces point in the same direction. The Anti-Aligned option positions the components so that the normal vectors from the selected faces point in opposite directions.

Use for positioning only. When selected, components move to the position defined by the mate, but a mate is not added to the FeatureManager design tree. A mate appears in the Mates box so you can edit and position the components, but nothing appears in the FeatureManager design tree when you close the Mate PropertyManager.

Advanced Mates:

The Advanced Mate types are:

- **Profile Center**: Mate to Center automatically center-aligns common component types such as rectangular and circular profiles to each other and fully defines the components.

- **Symmetric**: Positions two selected entities to be symmetric about a plane or planar face. A Symmetric Mate does not create a Mirrored Component.

- **Width**: Centers a tab within the width of a groove. You can select geometry to drive the limits of width mates, eliminating the need for numerical input.

- **Path Mate**: Constrains a selected point on a component to a path.

- **Linear/LinearCoupler**: Establishes a relationship between the translation of one component and the translation of another component.

- **Distance (Limit)**: Locates the selected items with a specified distance between them. Use the drop-down arrow box or enter the distance value directly.

- **Angle**: Locates the selected items at the specified angle to each other. Use the drop-down arrow box or enter the angle value directly.

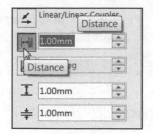

Mechanical Mates:

The Mechanical Mate types are:

- **Cam**: Forces a plane, cylinder, or point to be tangent or coincident to a series of tangent extruded faces.

- **Slot**: Mate bolts to straight or arced slots and you can mate slots to slots. Select an axis, cylindrical face, or a slot to create slot mates.

- **Hinge**: Limits the movement between two components to one rotational degree of freedom. It has the same effect as adding a Concentric mate plus a Coincident mate.

- **Gear**: Forces two components to rotate relative to one another around selected axes.

- **Rack Pinion**: Provides the ability to have Linear translation of a part; rack causes circular rotation in another part, pinion, and vice versa.

- **Screw**: Constrains two components to be concentric and adds a pitch relationship between the rotation of one component and the translation of the other.

- **Universal Joint**: The rotation of one component (the output shaft) about its axis is driven by the rotation of another component (the input shaft) about its axis.

Mates reflect the physical behavior of a component in an assembly. In this project, the two most common Mate types are Concentric and Coincident.

Quick Mate:

Quick Mate is a procedure to mate components together. No command (click Mate from the Assembly CommandManager) is required. Hold the Ctrl key down, make your selections. Release the Ctrl key, and a Quick Mate pop-up menu is displayed below the context toolbar. Select your mate and you are finished.

Utilize the Quick Mate procedure for Standard mates, Cam mate, Profile Center mate, Slot mate, Symmetric mate and Width mate. To activate the Quick Mate functionality, click Tools, Customize. On the toolbars tab, under Context toolbar settings, select Show Quick Mates. Quick Mate is selected by default.

GUIDE-ROD Assembly - Mate the ROD Component

Recall the initial assembly design constraints:

- The ROD requires the ability to travel through the GUIDE.

- The face of the Keyway in the ROD is parallel to the right face of the GUIDE.

Utilize Concentric and Parallel mates between the ROD and GUIDE. The Concentric mate utilizes the cylindrical face of the shaft with the cylindrical face of the GUIDE Hole. The Parallel mate utilizes two planar faces from the ROD Keyway Cut and the right face of the GUIDE.

Concentric Mate - 2 Cylindrical   Parallel - 2 Planar faces

The Concentric and Parallel mate provides the ability for the ROD to translate linearly through the GUIDE Hole. The ROD does not rotate.

You can use the following steps to create a Mate using the Mate tool:

- Click the Mate tool ✎ from the Assembly toolbar.

- Select the geometry from the first component (usually the part or plane).

- Select the geometry from the second component (usually the assembly or plane).

- Select the Mate type.

- Click OK ✔ to create the Mate. View the created mate.

Activity: GUIDE-ROD Assembly - Mate the ROD Component

Insert a Concentric mate between two faces. Use the Quick mate procedure.

38) Click the **inside cylindrical face** of the Guide Hole as illustrated. Note the icon feedback symbol.

39) Hold the **Ctrl** key down.

40) Click the **cylindrical face** of the ROD.

41) Release the **Ctrl** key. The Mate Pop-up menu is displayed.

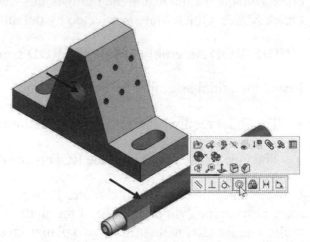

42) Click **Concentric** 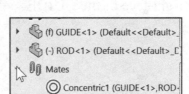 from The Mate Pop-up menu. A Concentric mate locates the selected items so they can share the same center point.

The ROD is Concentric with the GUIDE. The ROD has the ability to move and rotate while remaining concentric to the GUIDE hole.

Move and rotate the ROD in the Graphics window.

43) Click and drag the **ROD** in a horizontal direction. The ROD travels linearly in the GUIDE.

44) Click and drag the **ROD** in a vertical direction. The ROD rotates in the GUIDE.

45) Rotate the **Rod** until the Keyway cut is approximately parallel to the right face of the GUIDE. Display origins in the Assembly.

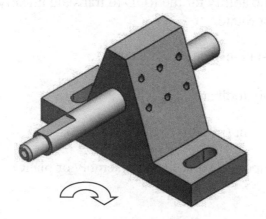

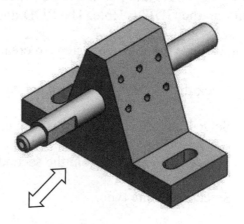

Recall the second assembly design constraint. The flat end of the ROD must remain parallel to the right surface of the GUIDE.

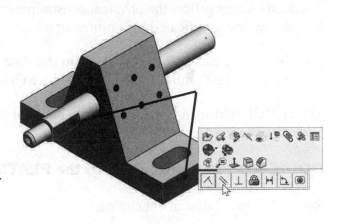

Insert a Parallel mate between two faces.
46) Hold the **Ctrl** key down.

47) Click the **Keyway face** of the ROD.

48) Click the **flat right face** of the GUIDE.

49) Release the **Ctrl** key. The Mate Pop-up menu is displayed.

50) Click **Parallel** ⬉ from the Mate Pop-up menu.

Move the ROD in the Graphics window.
51) Click and drag the **ROD** in a horizontal direction and position it approximately in the center of the GUIDE.

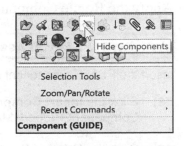

Hide the GUIDE component in the Graphics window.
52) Right-click on the front face of the **GUIDE** in the Graphics window.

53) Click **Hide Components** from the Context toolbar. The GUIDE component is not displayed in the Graphics window. Note the change in color in the FeatureManager.

Display the Mate types.
54) **Expand** the Mates folder from the FeatureManager.

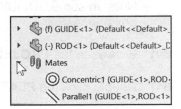

Display the full Mate names.
55) Drag the **vertical FeatureManager border** to the right. View the two created mates: Concentric1 and Parallel1.

Save the GUIDE-ROD assembly.

56) Click **Save** 💾.

57) Click **Rebuild and Save** the document.

🔆 Use the Alt key to temporarily hide a face when you need to select an obscured face for mates.

The ROD mates reflect the physical constraints in the GUIDE-ROD assembly. The ROD is under defined, indicated by a minus sign (-).

The Concentric mate allows the ROD to translate freely in the Z direction through the Guide Hole. The Parallel Mate prevents the ROD from rotating in the Guide Hole.

The GUIDE part icon is displayed with no color in the FeatureManager to reflect the Hide state.

GUIDE-ROD Assembly - Mate the PLATE Component

Recall the initial design constraints.

- The ROD is fastened to the PLATE.

- The PLATE part mounts to the GUIDE-CYLINDER PISTON PLATE part.

Apply the Rotate Component ⟲ tool to position the PLATE before applying the Mates.

Activity: GUIDE-ROD Assembly - Mate the PLATE Component

Insert the PLATE component into the assembly from the PROJECTS folder.

58) Click the **Insert Components** 📁 Assembly tool. The Insert Component PropertyManager is displayed.

59) Click the **BROWSE** button from the Open documents box. Note: In some SOLIDWORKS versions, the Open document box opens.

60) Browse to the **PROJECTS** folder.

61) Double-click the **PLATE** part.

62) Click a **position** behind the ROD in the Graphics window as illustrated. The PLATE component is added to the GUIDE-ROD FeatureManager.

Display an Isometric view.
63) Click **Isometric view** 🔲.

Fit the model to the Graphics window.
64) Press the **f** key. View the results in the Graphics window.

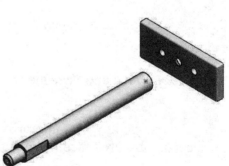

Rotate the PLATE.

65) Click the **Rotate Component** 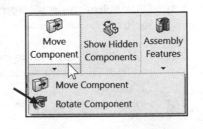 Assembly tool. The Rotate Component PropertyManager is displayed. View your options.

66) Click and drag the **front face** of the PLATE downward until the PLATE rotates approximately 90°.

67) Click **OK** ✔ from the PropertyManager to deactivate the tool.

68) Click a **position** in the Graphics window, to the right of the PLATE, to deselect any faces or edges.

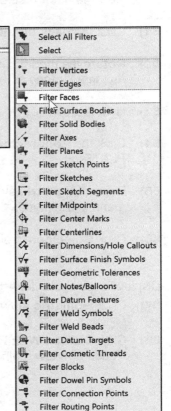

If needed, deactivate the Origins.
69) Click **View**, **Hide/Show**, uncheck **Origins** from the Menu bar.

Use Selection filters to select difficult individual features such as *faces*, *edges* and *points*. Utilize the Filter Faces tool to select the hidden ROD Back Hole face from the Selection Filter toolbar.

Display the Selection Filter toolbar. Activate the Filter Faces tool.
70) **Right-click** in the Graphics window.

71) Click the **Selection Filters** icon. The Selection Filter toolbar is displayed.

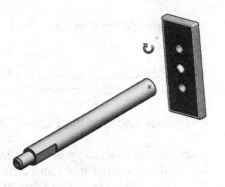

72) Click **Filter Faces** from the Selection Filter drop-down menu as illustrated. The Selection Filter icon is displayed in your mouse pointer ▽.

To deactivate the Filter Faces tool, click Clear All Filters from the Selection Filter toolbar.

🔅 The book is design to expose the SOLIDWORKS user to numerous tools and procedures. It may not always use the simplest and most direct process.

Insert a Concentric mate between two faces.

73) Click **WireFrame** 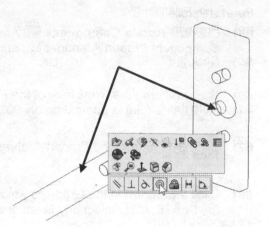 from the Heads-up View toolbar.

74) Click the **center inside cylindrical face** from the PLATE Countersink hole.

75) Hold the **Ctrl** key down.

76) Click the **cylindrical face** of the ROD.

77) Release the **Ctrl** key. The Mate Pop-up menu is displayed.

78) Click **Concentric** from the Mate Pop-up menu. Concentric2 is created.

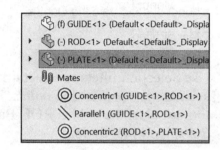

79) Click and drag the **PLATE** behind the ROD.

Insert a Coincident mate between two faces.
80) Use the **middle mouse** button to rotate the model to view the **back face** of the ROD.

81) Click the **back circular face** of the ROD.

82) **Rotate** the model to view the front rectangular face of the PLATE.

83) Hold the **Ctrl** key down.

84) Click the **front rectangular face** of the PLATE.

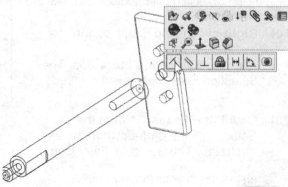

85) Release the **Ctrl** key. The Mate Pop-up menu is displayed.

86) Click **Coincident** from the Mate Pop-up menu.

Display an Isometric view - Shaded with Edges.
87) Click **Isometric view** .

88) Click **Shaded With Edges** from the Heads-up View toolbar. View the results.

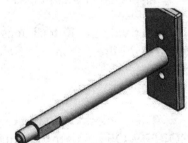

Insert a Parallel mate between two faces.

89) Press the **Shift + z** keys to zoom in on the ROD.

90) Click the ROD **Keyway Cut** flat face.

91) Hold the **Ctrl** key down.

92) Click the PLATE right **rectangular face** as illustrated.

93) Release the **Ctrl** key.

94) Click **Parallel** from the Mate Pop-up Menu. Parallel2 is created.

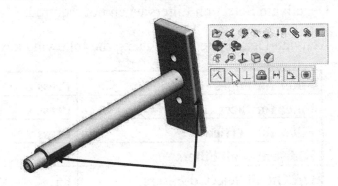

Clear all filters in SOLIDWORKS.

95) **Right-click** in the Graphics window.

96) Click the **Clear All Filters** icon.

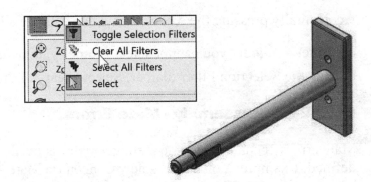

Save the GUIDE-ROD assembly.

97) Click **Save** . View the results.

Mates reflect the physical relations between the PLATE and the ROD. The Concentric mate aligns the PLATE Countersink Hole and the ROD cylindrical face. The Coincident mate eliminates translation between the PLATE front face and the ROD back face.

The Parallel mate removes PLATE rotation about the ROD axis. Create the Parallel mate.

A Distance mate of 0 provides additional flexibility over a Coincident mate. A Distance mate value can be modified. Utilize a Coincident mate when mating faces remain coplanar.

Use the Pack and Go tool to save and gather all related files for a model design (parts, assemblies, drawings, references, design tables, Design Binder content, decals, appearances, scenes, and SOLIDWORKS Simulation results) into a single folder or zip file. It is one of the best tools to utilize when you are trying to save a large assembly or drawing with references and SOLIDWORKS Toolbox components.

The mouse pointer displays the Filter ⍦ icon when the Selection Filter is activated. Deactivate Selection Filters when not required.

Activate/Deactivate Filters using the following keys:

Filter for edges	Press e
Filter for faces	Press x
Filter for vertices	Press v
Hide/Show all Filters	F5
Off/On all Selected Filters	F6

Accidentally pressing the e, x or v keys can activate a Filter. If the mouse pointer displays the Filter ⍦ icon, you cannot select geometry, dimensions or text. Press the F5 key to display the Selection Filter toolbar. Select Clear All Filters ⍦.

GUIDE-ROD Assembly - Mate Errors

Mate errors occur when component geometry is over defined. Example: You added a new Concentric Mate between the PLATE bottom Mounting Hole and the ROD cylindrical face.

The ROD back hole cannot physically exist with a Concentric Mate to both the PLATE middle CSK Hole and bottom Mounting Hole.

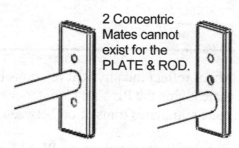

2 Concentric Mates cannot exist for the PLATE & ROD.

- Review the design intent. Know the behavior of the components in the assembly.

- Review the messages and symbols in the FeatureManager.

- Utilize Delete, Edit Feature and Undo commands to recover from Mate errors.

To view the mates for more than one component, hold the Ctrl key down, select the components, then right-click, and click the View Mates 🐾 tool.

A Mate problem displays the following icons:

⚠ **Warning.** The mate is satisfied but is involved in over defining the assembly.

⊗ **Error.** The mate is not satisfied.

Insert the second Concentric mate which will create a Mate error in the following steps.

Activity: GUIDE-ROD Assembly - Mate Errors

Insert a Concentric mate between two faces.
98) Click the **outside cylindrical face** of the ROD.

99) Hold the **Ctrl** key down.

100) Click the **bottom Mounting Hole inside cylindrical face** of the PLATE.

101) Release the **Ctrl** key.

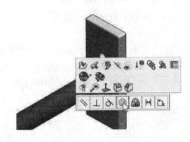

102) Click **Concentric** from the Mate Pop-up Menu. Adding a Concentric mate would over-define the assembly. A Mate error message is displayed. The components cannot be moved to a position that satisfies this mate.

103) Click **Cancel** from the dialog box.

Review the created Mates.
104) **Expand** the Mates folder in the FeatureManager. View the created mates and their icons.

If you delete a Mate and then recreate it, the Mate number will be different. View the mate symbols in the Mates folder.

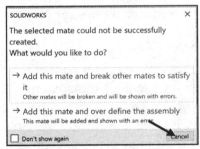

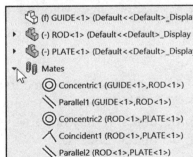

Review the mates for a component.

105) Right-click **ROD** from the Graphics window.

106) Click **View Mates**  from the Context toolbar. The Mates dialog box is displayed. Note: The GUIDE is Hidden in the FeatureManager.

Components involved in the mate system for the selected components are vaguely transparent in the Graphics window. Components not involved are hidden.

107) Close the pop-up mate dialog box to return to the GUIDE-ROD FeatureManager.

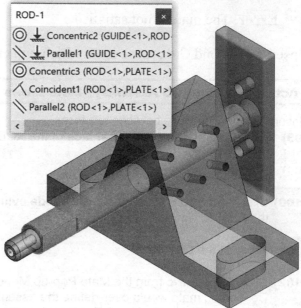

Organize the Mates names. Rename Mate names with descriptive names for clarity.

You can set an option in the Mate PropertyManager so that the first component you select from becomes transparent. Then selecting from the second component is easier, especially if the second component is behind the first. The option is supported for all mate types except those that might have more than one selection from the first component (width, symmetry, linear coupler, cam, and hinge).

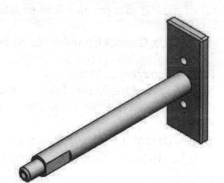

The book is designed to expose the SOLIDWORKS user to numerous tools and procedures. It may not always use the simplest and most direct process.

GUIDE-ROD Assembly - Collision Detection

The Collision Detection assembly function detects collisions between components as they move or rotate. A collision occurs when geometry on one component coincides with geometry on another component. Place components in a non-colliding position, then test for collisions.

Activity: GUIDE-ROD Assembly - Collision Detection

Display the GUIDE and apply the Collision Detection tool.

108) Right-click **GUIDE** ▸ (f) GUIDE<1> (Default· from the FeatureManager.

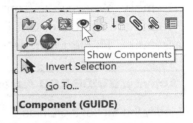

109) Click **Show Components** 👁 from the Context toolbar. The GUIDE is displayed in the Graphics window.

110) Click **Shaded with Edges** from the Heads-up View toolbar.

Move the PLATE behind the GUIDE.

111) Click the **Move Component** Assembly tool. The Move Component PropertyManager is displayed.

112) Drag the **PLATE** backward until the PLATE clears the GUIDE. The PLATE is free to translate along the Z-axis.

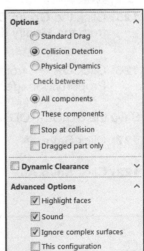

Apply the Collision Detection tool.

113) Click the **Collision Detection** checkbox. Un-check the Stop at collision box if needed.

114) Check the **All components** box.

115) Check **Highlight faces**, **Sound** and **Ignore complex surfaces** from the Advanced Options box.

116) Drag the **PLATE** forward. The GUIDE back, top and angled right faces turn blue when the PLATE front face collides with the back face of the GUIDE. The ROD is blue due to an interference with the GUIDE hole.

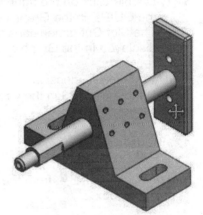

Return the PLATE to the original position.

117) Drag the **PLATE** backward until the ROD is approximately halfway through the GUIDE.

118) Click **OK** ✔ from the Move Component PropertyManager.

Save the GUIDE-ROD assembly.
119) Click **Save** 💾.

120) Click **Rebuild and Save** the document.

GUIDE-ROD Assembly - Modify Component Dimension

Modify part dimensions in the assembly. Utilize Rebuild to update the part and the assembly. You realize from additional documentation that the Slot in the GUIDE is 4mm. Modify the right Slot Cut feature dimensions in the GUIDE-ROD assembly. Rebuild the assembly. The left Mirror Slot Cut and right Slot Cut update with the new value.

Activity: GUIDE-ROD Assembly - Modify Component Dimension

Modify the Slot of the Guide.
121) Double-click on the right **Slot Cut** of the GUIDE in the Graphics window. The Slot Cut dimensions are displayed in the Graphics window.

Modify the radial dimension.
122) Double-click **R3** in the Graphics window.

123) Enter **4**mm.

124) Click **Rebuild** 🔓 from the Modify dialog box.

125) Click the **Green Check mark** ✔ from the Modify dialog box.

126) Click **OK** ✔ from the Dimension PropertyManager.

Save the GUIDE-ROD assembly.
127) Click **Save** 💾.

128) Click **Save All**.

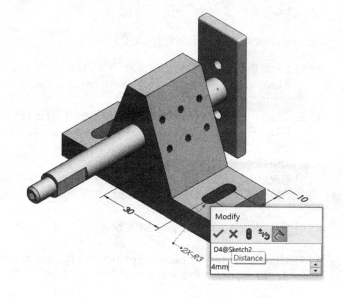

🔍 Additional details on Assembly, Mates, Mate Errors, Collision Detection, Selection Filters are available in SOLIDWORKS Help.

Keywords: Standard Mates, Mate PropertyManager, Quick Mate, Mates (Diagnostics), Collision Detection, Design Methods in Assembly and Selection Filters.

SOLIDWORKS Design Library

A parts library contains components used in a design creation. The SOLIDWORKS Design Library 🗌 tab in the Task Pane provides a centralized location for reusable elements such as parts, assemblies and sketches. It does not recognize non-reusable elements such as SOLIDWORKS drawings, text files or other non-SOLIDWORKS files.

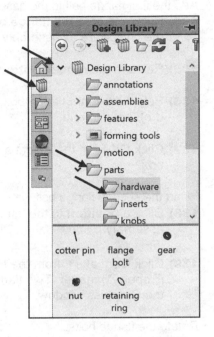

The Design Library consists of annotations, assemblies, features, forming tools, motion, parts, routing, smart components, Toolbox (Add-in) and 3D ContentCentral (models from suppliers). SOLIDWORKS Add-ins are software applications.

Your company issued a design policy. The policy states that you are required to only use parts that are presently in the company's parts library. The policy is designed to lower inventory cost, purchasing cost and design time.

In this project, the SW Design Library parts simulate your company's part library. Utilize a hex flange bolt located in the Hardware folder. Specify a new folder location in the Design Library to locate the components utilized in this project.

🔅 The Design Library saves time locating and utilizing components in an assembly. Note: In some network installations, depending on your access rights, additions to the Design Library are only valid in new folders.

Activity: GUIDE-ROD Assembly - SOLIDWORKS Design Library

Open and obtain components from the SOLIDWORKS Design Library.

129) Click the **Design Library** 🗌 tab from the Task Pane as illustrated. The Design Library menu is displayed in the Graphics window.

Pin the Design Library (Auto Show) to remain open.

130) Click **Pin** 🗕.

131) Expand Design Library.

Select the Flange bolt from the parts folder.

132) Expand the parts folder.

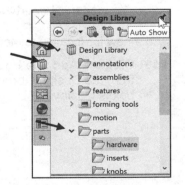

133) Double-click the **hardware** folder. The hardware components are displayed. The flange bolt icon represents a family of similar shaped components in various configurations.

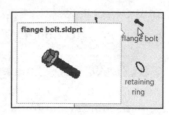

Add the first flange bolt to the assembly.

134) Click and drag the **flange bolt** icon to the right of the GUIDE-ROD assembly in the Graphics window.

135) Release the **mouse button**.

136) Select the 8mm flange bolt, **M8-1.25 x 30**, from the drop down menu.

137) Click **OK** from the Select a Configuration dialog box.

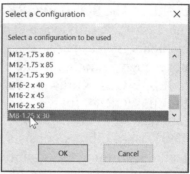

Insert the second flange bolt.

138) Click a **position** to the left of the GUIDE-ROD assembly.

139) Click **Cancel** ✖ from the Insert Component PropertyManager. Two flange bolts are displayed in the Graphics window.

Rotate the flange bolts.

140) Click the **shaft** of the first flange bolt.

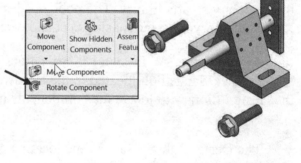

141) Click the **Rotate Component** 🔄 Assembly tool. The Rotate Component PropertyManager is displayed.

142) Click and drag the **shaft** of the first flange bolt in a vertical direction as illustrated. The flange bolt rotates.

143) Click and drag the **shaft** of the second flange bolt in a vertical direction.

144) Click **OK** ✔ from the Rotate Component PropertyManager.

Add a new file location to the Design Library.

145) Click the **Add File Location** 🎁 tool from the Design Library. The Choose Folder dialog box is displayed.

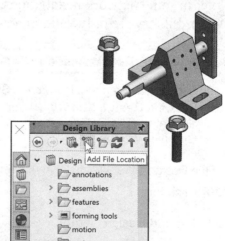

146) Select the **PROJECTS** folder.

147) Click **OK** from the Choose Folder dialog box. The PROJECTS folder is added to the Design Library.

148) Click the **PROJECTS** folder to display your parts and assemblies.

Add a new folder to the Design Library Parts.
149) Click the **PROJECTS** folder in the Design Library.

150) Click **Create New Folder**.

151) Enter **MY-PLATES** for folder name.

152) Click the **MY-PLATES** folder. The folder is empty.

Un-pin the Design Library.
153) Click **Pin** .

Display an Isometric view. Save the GUIDE-ROD assembly.
154) Click **Isometric view** from the Heads-up View toolbar.

155) Click **Save** .

156) Click **Yes** to rebuild. The two flange bolts are displayed in the FeatureManager.

Caution: Do not drag the PLATE part into the MY-PLATES folder at this time. The GUIDE-ROD assembly references the PLATE in the PROJECTS folder. Insert parts into the Design Library when the assembly is closed. This action is left as an exercise.

There are files of type part (*.sldprt) and Library feature part (*.sldlfp). The flange bolt is a Library feature part. The PLATE is a part. The PLATE cannot be saved as a Library feature part because it contains a Hole Wizard feature. Save the PLATE as a part (*.sldprt) when the assembly is not opened.

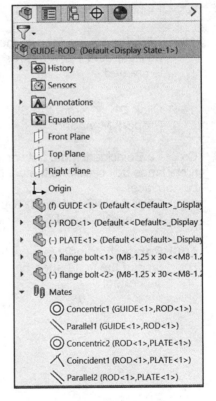

GUIDE-ROD Assembly - Insert Mates for Flange Bolts

Insert a Slot mate for the first flange bolt. You can mate bolts to straight or arced slots and you can mate slots to slots. Select an axis, cylindrical face or slot to create a slot mate.

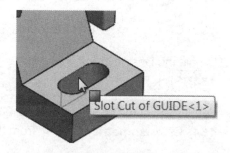

157) Click the **Mate** Assembly tool. The Mate PropertyManager is displayed.

158) Pin the Mate PropertyManager. Click the **Keep Visible** icon.

159) Expand the **Mechanical Mates** section.

160) Select **Slot** for mate type.

161) Select **Center in Slot** from the Constraint drop-down menu. This option centers the component in the slot. For Mate Selections, select a slot face and the feature to mate.

162) Click the **cylindrical face** of the first Flange bolt as illustrated.

163) Click the **inside face** of the first slot as illustrated. View the results. Slot1 is created.

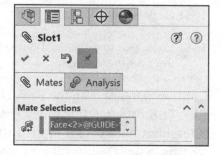

164) Click **OK** from the Slot PropertyManager.

Create a Coincident mate between the first Flange bolt head and the right slot cut top face.

165) Click the **right slot cut top face** as illustrated.

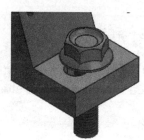

166) **Rotate** the model to view the bottom face of the first flange bolt as illustrated.

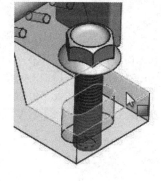

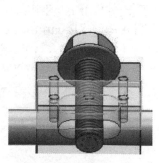

167) Click the **bottom face** of the first flange bolt as illustrated.

168) Click **Coincident** mate.

169) Click **OK** ✔ from the Coincident PropertyManager.

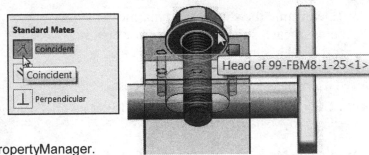

170) Click **OK** ✔ from the Mate PropertyManager.

171) **Rotate** the model and view the results in the Graphics window.

View the created Mates.

172) **Expand** the Mates folder in the Assembly PropertyManager. View the created mates.

Display the Mates in the FeatureManager to check that the components and the Mate types correspond to your design intent.

Display an Isometric view.

173) Press the **space bar** to display the Orientation dialog box.

174) Click **Isometric view** 🧊. The first flange bolt is free to rotate about its centerline in the Slot Cut.

Insert a Parallel mate between two faces.

175) Click the **front face** of the hex head as illustrated.

176) Hold the **Ctrl** key down.

177) Click the **front face** of the GUIDE as illustrated.

178) Release the **Ctrl** key. The Mate pop-up menu is displayed.

179) Click **Parallel** ⟍ from the Mate Pop-up menu.

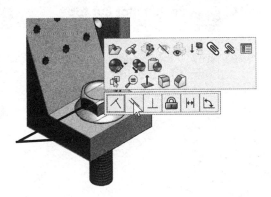

🔆 If you delete a Mate and then recreate it, the Mate numbers will be different (increase).

💡 Determine the static and dynamic behavior of mates in each sub-assembly before creating the top level assembly.

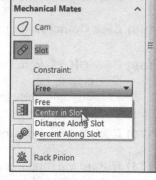

Insert a Slot mate for the second flange bolt.

180) Click the **Mate** ✎ Assembly tool. The Mate PropertyManager is displayed.

181) Pin the Mate PropertyManager. Click the **Keep Visible** icon if needed.

182) Expand the **Mechanical Mates** section.

183) Select **Slot** 𝒪 for mate type.

184) Select **Center in Slot** from the Constraint drop-down menu. This option centers the component in the slot.

Mirror1 of GUIDE<1>

185) Click the **cylindrical face** of the second Flange bolt as illustrated.

186) Click the **inside face** of the second Slot as illustrated. View the results. Slot2 is created.

187) Click **OK** ✔ from the Slot2 PropertyManager.

Create a Coincident mate between the second Flange bolt head and the right slot cut top face.

188) Click the **left slot cut top face** as illustrated.

189) **Rotate** the model to view the bottom face of the second flange bolt as illustrated.

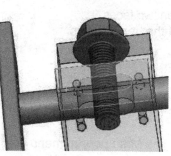

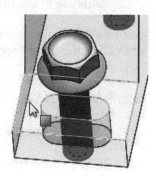

190) Click the **bottom face** of the second flange bolt as illustrated.

191) Click **Coincident** mate.

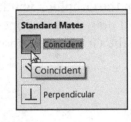

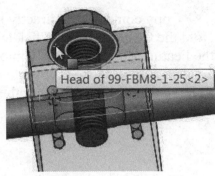

Head of 99-FBM8-1-25<2>

192) Click **OK** ✔ from the Coincident PropertyManager.

193) Click **OK** ✔ from the Mate PropertyManager.

194) Rotate the model and view the results in the Graphics window.

Display an Isometric view.
195) Press the **space bar** to display the Orientation dialog box.

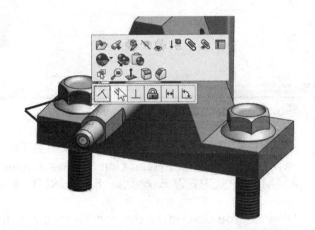

196) Click **Isometric view** ⬛. The second flange bolt is free to rotate about its centerline in the Slot Cut.

Insert a Parallel mate between two faces.
197) Click the **front face** of the hex head as illustrated.

198) Hold the **Ctrl** key down.

199) Click the **front face** of the GUIDE as illustrated.

200) Release the **Ctrl** key. The Mate pop-up menu is displayed.

201) Click **Parallel** ⟍ from the Mate Pop-up menu.

View the created Mates.
202) Expand the Mates folder in the Assembly PropertyManager. View the created mates.

🔆 Later, delete the second flange bolt and apply a Linear Component Pattern tool to create the second flange bolt from the first flange bolt.

Save the GUIDE-ROD assembly.
203) Click **Save** 💾.

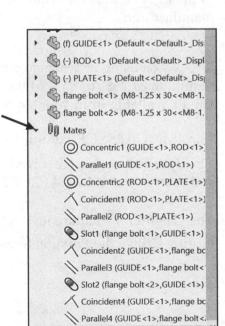

▸ 🧲 (f) GUIDE<1> (Default<<Default>_Dis
▸ 🧲 (-) ROD<1> (Default<<Default>_Displ
▸ 🧲 (-) PLATE<1> (Default<<Default>_Dis
▸ 🧲 flange bolt<1> (M8-1.25 x 30<<M8-1.
▸ 🧲 flange bolt<2> (M8-1.25 x 30<<M8-1.
▾ 📎 Mates
　⊚ Concentric1 (GUIDE<1>,ROD<1>
　⟍ Parallel1 (GUIDE<1>,ROD<1>)
　⊚ Concentric2 (ROD<1>,PLATE<1>)
　⟀ Coincident1 (ROD<1>,PLATE<1>)
　⟍ Parallel2 (ROD<1>,PLATE<1>)
　🟠 Slot1 (flange bolt<1>,GUIDE<1>)
　⟀ Coincident2 (GUIDE<1>,flange bc
　⟍ Parallel3 (GUIDE<1>,flange bolt<
　🟠 Slot2 (flange bolt<2>,GUIDE<1>)
　⟀ Coincident4 (GUIDE<1>,flange bc
　⟍ Parallel4 (GUIDE<1>,flange bolt<

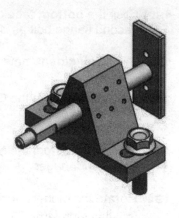

💡 Copy components directly in the Graphics window.
Hold the Ctrl key down. Click and drag the component from
the FeatureManager directly into the Graphics window to
create a new instance (copy).
Release the Ctrl key. Release the
mouse button.

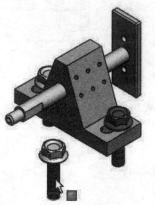

Socket Head Cap Screw Part

The PLATE mounts to the PISTON PLATE of the GUIDE CYLINDER assembly with
two M4x0.7 Socket Head Cap Screws. Create a simplified version of the
4MMCAPSCREW based on the ANSI B 18.3.1M-1986 standard.

How do you determine the overall length of the 4MMCAPSCREW? Answer: The depth
of the PLATE plus the required blind depth of the PISTON PLATE provided by the
manufacturer.

💡 When using fasteners to connect two plates, a design rule of thumb is to use a
minimum of 75% to 85% of the second plate's blind depth. Select a common overall
length available from your supplier.

The 4MMCAPSSCREW utilizes the Revolved Base, Chamfer and Extruded Cut feature.

The Base feature is a Revolved feature. The Revolved Base 🥄 feature creates the head
and shaft of the 4MMCAPSCREW.

The Chamfer 🔲 feature inserts two end cuts. The Extruded Cut 🔳 feature utilizes a hex
profile. Utilize the Polygon Sketch ⬡ tool to create the hexagon.

Activity: Create Socket Head Cap Screw - 4MMCAPSCREW Part

Create the 4MMCAPSCREW.

204) Click **New**  from the Menu bar.

205) Click the **MY-TEMPLATES** tab from the New SOLIDWORKS Document dialog box.

206) Double-click **PART-MM-ANSI**. The Part FeatureManager is displayed.

Save the part. Enter name and description.

207) Click **Save** .

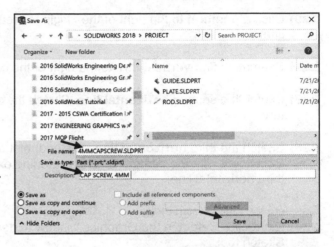

208) Select the **ENGDESIGN-W-SOLIDWORKS\PROJECTS** folder.

209) Enter **4MMCAPSCREW** for File name.

210) Enter **CAP SCREW, 4MM** for Description.

211) Click **Save**. The 4MMCAPSCREW FeatureManager is displayed.

Create the Base sketch on the Front Plane.

212) Right-click **Front Plane** from the FeatureManager.

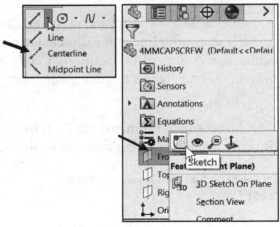

213) Click **Sketch** from the Context toolbar.

214) Click the **Centerline** Sketch tool. The Insert Line PropertyManager is displayed.

215) Click **Front view** from the Heads-up View toolbar.

Sketch a vertical centerline.

216) Click the **Origin** in the Graphics window. The Origin is coincident with the first point.

217) Click a **position** directly above the Origin as illustrated (approximately 25mm).

Sketch the illustrated profile of the 4MMCAPSCREW using the Line Sketch tool.

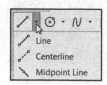

218) Click the **Line** ⟋ Sketch tool. The Insert Line PropertyManager is displayed. Note: You can right-click the Line sketch tool from the Sketch Entities box.

219) Click the **Origin** ⬐.

220) Click a **position** to the right of the Origin to create a horizontal line.

221) Sketch the first **vertical line** to the right of the centerline.

222) Sketch the second **horizontal line**. Sketch the second **vertical line**.

223) Sketch the third **horizontal line**. The endpoint of the line is Coincident with the centerline. The centerline extends above the third horizontal line.

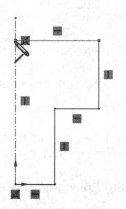

Deselect the Line Sketch tool.
224) Right-click **Select** in the Graphics window to deselect the Line Sketch tool.

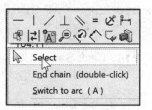

A diameter dimension for the revolved sketch requires a centerline, profile line and a dimension position to the left of the centerline. A dimension position directly below the bottom horizontal line creates a radial dimension.

💡 Insert smaller dimensions first, then larger dimensions to maintain the shape of the sketch profile.

Fit the sketch to the Graphics window.
225) Press the **f** key.

Insert dimensions on the Base Sketch.
226) Click the **Smart Dimension** ⟨ from the Sketch tool.

Add a bottom diameter dimension.
227) Click the **centerline**.

228) Click the **first vertical line**. Click a **position** below and to the left of the Origin to create a diameter dimension.

229) Enter **4**mm.

230) Click the **Green Check mark** ✔ from the Modify dialog box.

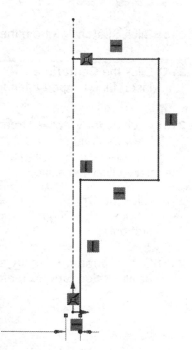

Add a vertical dimension.
231) Click the **second vertical line**.

232) Click a **position** to the right of the profile.

233) Enter **4**mm.

234) Click the **Green Check mark** ✔ from the Modify dialog box.

Add a top diameter dimension.
235) Click the **centerline**.

236) Click the **second vertical line**.

237) Click a **position** to the left of the Origin and above the second horizontal line to create a diameter dimension.

238) Enter **7**mm.

239) Click the **Green Check mark** ✔ from the Modify dialog box.

Create an overall vertical dimension.
240) Click the **top horizontal line**.

241) Click the **Origin**.

242) Click a **position** to the right of the profile.

243) Enter **14**mm.

244) Click the **Green Check mark** ✔.

Deselect the Smart Dimension Sketch tool.
245) Right-click **Select**.

246) Click the **centerline** in the Graphics window for axis for revolution. The Line Properties PropertyManager is displayed.

Create the first feature. Insert a Revolved Base feature.
247) Click **Revolved Boss/Base** 🍥 from the Features toolbar.

248) Click **Yes** to the question, "The sketch is currently open. A non-thin revolution feature requires a closed sketch. Would you like the sketch to be automatically closed?" The Revolve PropertyManager is displayed.

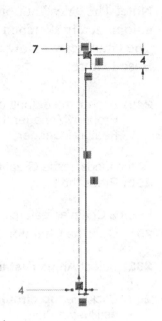

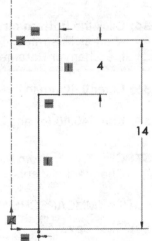

Note: The "Yes" button causes a vertical line to be automatically sketched from the top left point to the Origin. The Graphics window displays a preview of the Revolved Base feature.

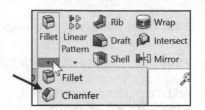

249) Accept the default options. Click **OK** ✔ from the Revolve PropertyManager. Revolve1 is displayed in the FeatureManager.

Fit the model to the Graphics window.
250) Press the **f** key.

Insert a Chamfer feature.
251) Click the **Chamfer** ⬡ feature tool.

252) Select **Angle Distance** for Chamfer type.

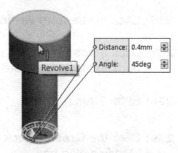

253) Click the **top circular edge** of the 4MMCAPSCREW head as illustrated.

254) Click the **bottom circular edge** of the 4MMCAPSCREW shaft as illustrated. The selected entities are displayed in the Chamfer Parameters box.

255) Enter **0.40**mm in the Distance box.

256) Enter **45.00** for angle. Accept the default settings.

257) Click **OK** ✔ from the Chamfer PropertyManager. Chamfer1 is displayed in the FeatureManager.

Save the 4MMCAPSCREW part.
258) Click **Save** 💾.

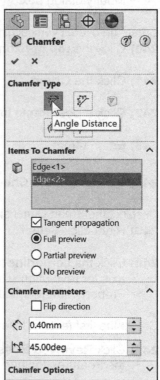

🔆 Angle distance values for the Chamfer feature can be entered directly in the Pop-up box from the Graphics window.

Rotate the view. Create the Hex Extruded Cut feature.

259) Rotate the **middle mouse button** to display the top circular face of 4MMCAPSCREW.

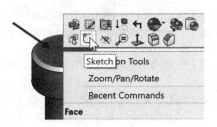

Insert a sketch.

260) Right-click the **top circular face** of Revolve1 for the Sketch plane as illustrated.

261) Click **Sketch** from the Context toolbar.

Display a Top view.

262) Click **Top view** .

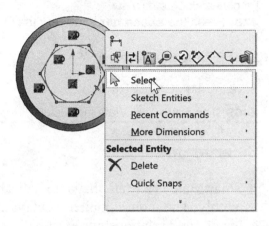

Sketch a hexagon.

263) Click the **Polygon** Sketch tool. The Polygon PropertyManager is displayed.

264) Click the **Origin** . Click a **position** to the right as illustrated.

Insert a Horizontal relation.

265) Right-click **Select** to deselect the Polygon Sketch tool.

266) Click the **Origin** .

267) Hold the **Ctrl** key down.

268) Click the **right point** of the hexagon.

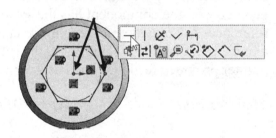

269) Release the **Ctrl** key. The Properties PropertyManager is displayed.

270) Right-click **Make Horizontal** from the Pop-up Context toolbar. Click **OK** from the Properties PropertyManager.

Add a dimension using the Smart Dimension tool.

271) Click the **Smart Dimension** Sketch tool.

272) Click the **inscribed circle**.

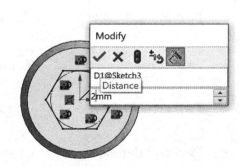

273) Click a position **diagonally** to the right of the profile. Enter **2**mm.

274) Click the **Green Check mark** .

Insert an Extruded Cut feature.

275) Click the **Extruded Cut** 🔲 feature tool. The Cut-Extrude PropertyManager is displayed.

276) Enter **4**mm for the Depth. Accept the default settings.

277) Click **OK** ✔ from the Cut-Extrude PropertyManager. Cut-Extrude1 is displayed in the FeatureManager.

Display an Isometric view.
278) Press the **space bar** to display the Orientation dialog box.

279) Click **Isometric view** 🔲.

Save the 4MMCAPSCREW.
280) Click **Save** 💾. View the results.

SmartMates

A SmartMate is a Mate that automatically occurs when a component is placed into an assembly. The mouse pointer displays a SmartMate feedback symbol when common geometry and relationships exist between the component and the assembly.

SmartMates are Concentric or Coincident. A Concentric SmartMate assumes that the geometry on the component has the same center as the geometry on an assembled reference. A Coincident Plane SmartMate assumes that a plane on the component lies along a plane on the assembly. As the component is dragged into place, the mouse pointer provides feedback such as:

Mating Entities:	*Type of Mate*:	*Icon Feedback*:
• Two linear edges	Coincident	
• Two planner faces	Coincident	
• Two vertices	Coincident	
• Two conical faces	Concentric	
• Two circular edges	Coincident/Concentric	

Coincident/Concentric SmartMate

The most common SmartMate between a screw/bolt and a hole is the Coincident/Concentric SmartMate. The following technique utilizes two windows. The first window contains the 4MMCAPSCREW part and the second window contains the GUIDE-ROD assembly. Zoom in on both windows to view the Mate reference geometry. Drag the part by the shoulder edge into the assembly window. View the mouse pointer for Coincident/Concentric feedback 👆🔩. Release the mouse pointer on the circular edge of the PLATE Mounting Hole.

Activity: Insert - Coincident/Concentric SmartMate

Display the 4MMCAPSCREW part and the GUIDE-ROD assembly.
281) Click **Window**, **Tile Horizontally** from the Menu bar.

282) Zoom in on the PLATE to view the Mounting Holes.

Insert the first 4MMCAPSCREW.
283) Click and drag the **circular edge** of the 4MMCAPSCREW part into the GUIDE-ROD assembly Graphics window.

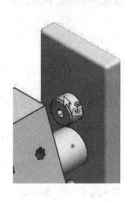

284) Release the mouse pointer on the **top circular edge** of the PLATE. The mouse pointer displays the Coincident/Concentric circular edges feedback icon.

Insert the second 4MMCAPSCREW.
285) Click and drag the **circular edge** of the 4MMCAPSCREW part into the GUIDE-ROD assembly Graphics window.

286) Release the mouse pointer on the **bottom circular edge** of the PLATE. The mouse pointer displays the Coincident/Concentric circular edges feedback 👆🔩 icon.

287) Maximize the GUIDE-ROD assembly.

Fit the GUIDE-ROD assembly to the Graphics window.
288) Press the **f** key.

💡 **Zoom in** before dragging a component into the assembly to select the correct circular edge for a Coincident/Concentric Mate.

The circular edge of the 4MMCAPSCREW produces both the Coincident/Concentric Mates. The cylindrical face of the 4MMCAPSCREW produces only a Concentric Mate. View the mouse pointer for the correct feedback.

Insert a Parallel mate for the first 4MMCAPSCREW.

289) Click the **Mate** ✎ Assembly tool. The Mate PropertyManager is displayed.

290) **Expand** the GUIDE-ROD icon from the fly-out FeatureManager.

291) Click **Right Plane** of the 4MMCAPSCREW<1> in the fly-out FeatureManager as illustrated.

292) Click **Right Plane** of the 4MMCAPSCREW<2> in the fly-out FeatureManager. The selected planes are displayed in the Mate Selections box.

293) Click **Parallel** ⟋.

294) Click **OK** ✔ from the Mate PropertyManager.

Fit the assembly to the Graphics window. Display an Isometric view.
295) Press the **f** key.

296) Click **Isometric view** 🔲.

Click the GUIDE-ROD assembly.
297) Click **Save** 💾.

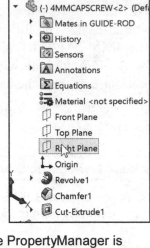

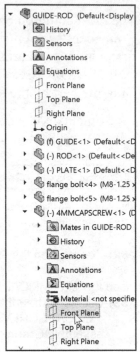

The ROD, PLATE, and 4MMCAPSCREWS are free to translate along the z-axis. Their component status remains under defined, (-) in the GUIDE-ROD FeatureManager.

The GUIDE is fixed (f) to the assembly Origin. The flange bolts are fully defined since they are mated only to the GUIDE.

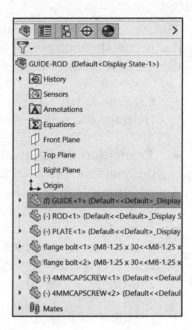

💡 Understand how Mates reflect the physical behavior in an assembly. Move and rotate components to test the mate behavior. The correct mate selection minimizes rebuild time and errors in the assembly.

Tolerance and Fit

The ROD travels through the GUIDE in the GUIDE-ROD assembly. The shaft diameter of the ROD is 10mm. The hole diameter in the GUIDE is 10mm. A 10mm ROD cannot be inserted into a 10mm GUIDE hole without great difficulty. (Press Fit.)

Note: The 10mm dimension is the nominal dimension. The nominal dimension is approximately the size of a feature that corresponds to a common fraction or whole number.

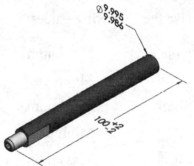

Tolerance is the difference between the maximum and minimum variation (Limits) of a nominal dimension and the actual manufactured dimension.

Example: A ROD has a nominal dimension of 100mm with a tolerance of ± 2mm (100mm ± 2mm). This translates to a part with a possible manufactured dimension range between 98mm (lower limit) to 102mm (upper limit). The total ROD tolerance is 4mm.

Note: Design rule of thumb: Design with the maximum permissible tolerance. Tolerance flexibility saves in manufacturing time and cost.

The assembled relationship between the ROD and the GUIDE is called the fit. The fit is defined as the tightness or looseness between two components. This project discusses three major types of fits:

- **Clearance fit**: The shaft diameter is less than the hole diameter. See SOLIDWORKS Help for additional information.

- **Transition fit**: Clearance or interference can exist between the shaft and the hole. See SOLIDWORKS Help for additional information.

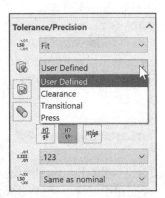

- **Press fit**: The shaft diameter is larger than the hole diameter. The difference between the shaft diameter and the hole diameter is called interference. See SOLIDWORKS Help for additional information.

You require a Clearance fit between the shaft of the ROD and the Guide Hole of the GUIDE. There are multiple categories for Clearance fits. Dimension the GUIDE hole and ROD shaft for a Sliding Clearance fit. All below dimensions are in millimeters.

Use the following values:

Hole	Maximum	10.015mm.
	Minimum	10.000mm.
Shaft	Maximum	9.995mm.
	Minimum	9.986mm.
Fit	Maximum	10.015 - 9.986 = .029 Max. Hole - Min. Shaft.
	Minimum	10.000 - 9.995 = .005 Min. Hole - Max. Shaft.

Calculate the maximum variation:

Hole Max: 10.015 - Hole Min: 10.000 = .015 Hole Max. Variation.

Select features from the FeatureManager and the Graphics window. In the next activity, locate feature dimensions with the FeatureManager for the GUIDE.

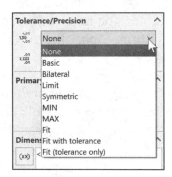

Locate feature dimensions in the Graphics window for the ROD. Select the dimension text, and then apply the Tolerance/Precision through options in the Dimension PropertyManager.

Activity: Address - Tolerance and Fit

Locate the dimension in the FeatureManager.

298) Expand GUIDE from the GUIDE-ROD FeatureManager.

299) Double-click **Guide Hole** from the FeatureManager to display the dimensions.

Add the maximum and minimum Guide Hole dimensions.

300) Click the Ø**10** dimension. The Dimension PropertyManager is displayed.

301) Select **Limit** from the Tolerance/Precision box.

302) Select **.123** for three place Precision.

303) Enter **0.015**mm for Maximum Variation.

304) Enter **0.000**mm for Minimum Variation.

305) Click **OK** ✔ from the Dimension PropertyManager.

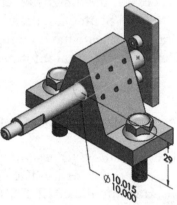

Add maximum and minimum Shaft dimensions.

306) Double-click the **Base-Extrude** (**ROD**) component from the Graphics window.

307) Click the Ø**10** diameter dimension.

308) Select **Limit** from the Tolerance/Precision box.

309) Select **.123** for three place Precision.

310) Enter **-0.005**mm for Maximum Variation.

311) Enter **-0.014**mm for Minimum Variation.

312) Click **OK** ✔ from the Dimension PropertyManager.

Save the GUIDE-ROD assembly.

313) Click **Save** 💾.

314) Click **Save All**.

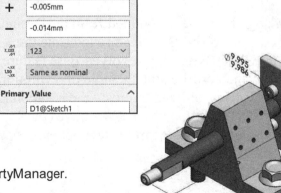

ISO symbol Hole/Shaft Classification is applied to an individual dimension for Fit, Fit with tolerance, or Fit (tolerance only) types. Classification can be User Defined, Clearance, Transitional or Press.

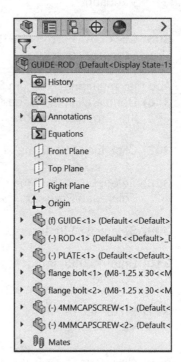

For a hole or shaft dimension, select a classification from the list. The Hole/Shaft designation for a Sliding Fit is H7/g6.

The values for Maximum and Minimum tolerances are calculated automatically based on the diameter of the Hole/Shaft and the Fit Classification.

Utilize Hole/Shaft Classification early in the design process. If the dimension changes, then the tolerance updates. The Hole/Shaft Classification propagates to the details in the drawing. Create the drawing in Project 4. See SOLIDWORKS Help for additional information.

Review of the GUIDE-ROD Assembly

The GUIDE-ROD assembly combined the GUIDE, ROD and PLATE components. The GUIDE was the first component inserted into the GUIDE-ROD assembly.

Mates removed degrees of freedom. Concentric, Coincident and Parallel Mates were utilized to position the ROD and PLATE with respect to the GUIDE.

The flange bolts were obtained from the Design Library. You utilized a Revolved Base feature to create the 4MMCAPSCREW. The Revolved Base feature contained an axis, sketched profile and an angle of revolution. The Polygon Sketch tool was utilized to create the hexagon Extruded Cut feature. The 4MMCAPSCREW utilized the Concentric/Coincident SmartMate option.

Exploded View

The Exploded View illustrates how to assemble the components in an assembly. Create an Exploded View with four steps in the ROD-GUIDE assembly. Click and drag components in the Graphics window.

The Manipulator icon indicates the direction to explode. Select an alternate component edge for the Explode direction. Drag the component in the Graphics window or enter an exact value in the Explode distance box. In this activity, manipulate the top-level components in the assembly.

In the project exercises, create exploded views for each sub-assembly and utilize the Re-use sub-assembly Explode option in the top level assembly.

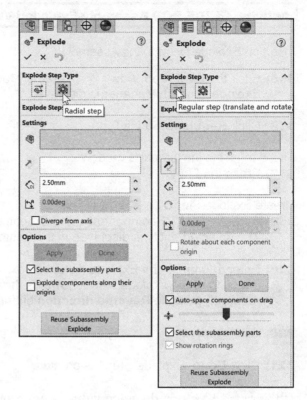

Access the Explode view option as follows:

- Right-click the configuration name in the ConfigurationManager.

- Select the Exploded View tool in the Assembly toolbar.

- Select Insert, Exploded View from the Menu bar.

The Exploded View feature uses the Explode PropertyManager as illustrated. There are two Explode Step Types: Regular step (translate and rotates) and Radial step. The Radial step Type provides the ability to explode components aligned radially/cylindrically about an axis. You can also explode radially by diverging along an axis.

You can create exploded view steps that rotate a component with or without linear translation. Use the rotation and translation handles of the triad. You can include rotation and translation in the same explode step. You can also edit the explode step translation distance and rotation angle values in the PropertyManager.

Activity: GUIDE-ROD Assembly-Exploded View

Insert an Exploded view in the Assembly.

315) Click the **Exploded View** 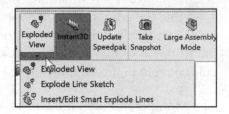 Assembly tool. The Explode PropertyManager is displayed.

316) Click **Regular step** for Type.

Fit the model to the Graphics window.
317) Press the **f** key.

Create Explode Step 1.
318) Click the **PLATE** component in the Graphics window. The selected entity is displayed in the Settings box.

319) Enter **100**mm in the Explode distance box. The direction of the explode view is towards the back. If required, click the **Reverse direction** button.

320) Click **Apply**.

321) Click **Done**. Explode Step1 is created.

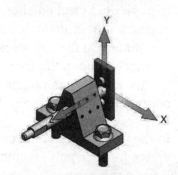

If the exact Explode distance value is not required, click and drag the Manipulator handle as illustrated in the next step.

Create Explode Step2.
322) Click the **ROD** component from the Graphics window.

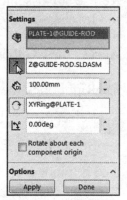

Fit the model to the Graphics window.
323) Press the **f** key.

324) Drag the **blue/orange manipulator handle** backward to position between the PLATE and the GUIDE as illustrated.

325) Click **Done** from the Settings box. Explode Step2 is created.

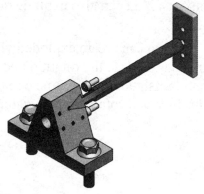

Create Explode Step3.
326) Click the **left flange bolt** from the Graphics window.

327) Drag the **vertical manipulator handle** upward above the GUIDE.

328) Click **Done** from the Settings box. Explode Step3 is created.

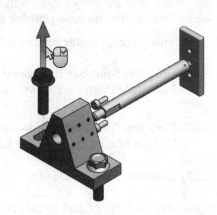

Create Explode Step4.
329) Click the **right flange bolt** from the Graphics window.

330) Drag the **vertical manipulator handle** upward above the GUIDE.

331) Click **Done** from the Settings box. Explode Step4 is created.

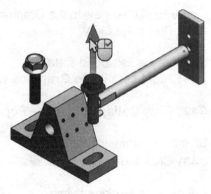

332) Click **OK** ✔ from the Explode PropertyManager.

Display an Isometric view. Save the model.
333) Click **Isometric view** ⬛.

334) Click **Save** 💾.

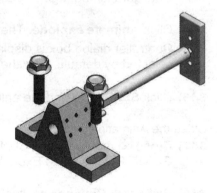

Split the FeatureManager to view the Exploded Steps.

335) Position the **mouse pointer** at the top of the FeatureManager.

The mouse pointer displays the Split bar ⸬.

336) Drag the **Split bar** half way down to display two FeatureManager windows.

337) Click the **ConfigurationManager** 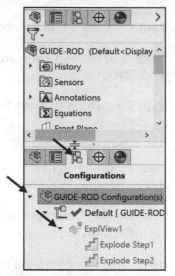 tab to display the Default configuration in the lower window.

338) **Expand** Default <Display State-1>.

339) **Expand** ExplView1 to display the four Exploded Steps.

Fit the Exploded view to the Graphics window.

340) Press the **f** key.

Remove the Exploded state.

341) Right-click in the **Graphics window**.

342) Click **Collapse** from the Pop-up menu.

Display an Isometric view.

343) Click **Isometric view**.

Animate the Exploded view.

344) Right-click **ExplView1** from the ConfigurationManager.

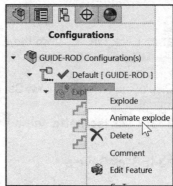

345) Click **Animate explode**. The Animation Controller dialog box is displayed. Play ▶ is selected by default. View the animation.

346) Click **Stop** ■ to end the animation.

Close the Animation Controller.

347) **Close** ☒ the Animation Controller. The GUIDE-ROD is in the Exploded state.

348) Right-click **Collapse** as illustrated.

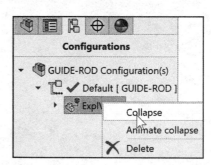

Display the FeatureManager. Return to a single FeatureManager.
349) Drag the **Split bar** upward to display one FeatureManager window.

350) Click the **Assembly FeatureManager** tab.

Save the GUIDE-ROD assembly.
351) Click **Save** .

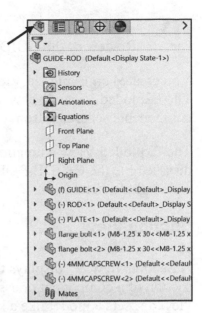

Click the Motion Study # tab at the bottom of the Graphics window. Select Basic Motion from the MotionManager. Click the Animation Wizard to create a simple animation. Click Play to view the Animation Wizard. Click the Model tab at the bottom of the Graphics window to return to the SOLIDWORKS screen.

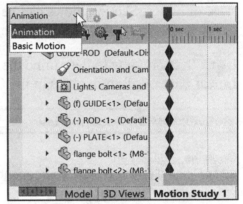

Note: The SOLIDWORKS Animator application is required to record an AVI file through the Animation Controller. Play the animation files through the Windows Media Player. The time required to create the animation file depends on the number of components in the assembly and the options selected.

Stop the animation before closing the Animation Controller toolbar to avoid issues. Utilize SOLIDWORKS Animator for additional control over the Explode/Collapse motion in the assembly.

Create the Exploded steps in the order that you would disassemble the assembly. Collapse the assembly to return to the original assembled position. Your animations will appear more realistic based on the order of the Explode steps.

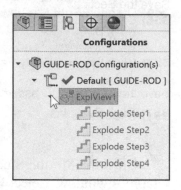

Review of the GUIDE-ROD assembly Exploded View

You created an Exploded View in the GUIDE-ROD assembly. The Exploded View displayed the assembly with its components separated from one another.

The Exploded View animation illustrated how to assemble and disassemble the GUIDE-ROD assembly through the collapse and explode states.

Section View

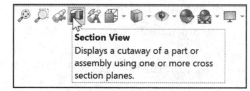

A Section view displays the internal cross section of a component or assembly. The Section view dissects a model like a knife slicing through a stick of butter. Section views can be performed anywhere in a model. The location of the cut corresponds to the Section plane. A Section plane is a planar face or reference plane.

Use a Section view to determine the interference between the 10mm Guide Hole and the Linear Pattern of 3mm holes in the GUIDE.

Activity: GUIDE-Section View

Open the GUIDE part.
352) Right-click the **GUIDE** component in the Graphics window.

353) Click **Open Part** from the shortcut toolbar. The GUIDE part is displayed in the Graphics window.

Insert a Section View on the Front Plane of the GUIDE.
354) Click **Front Plane** from the GUIDE FeatureManager.

355) Click **Section View** from the Heads-up View toolbar in the Graphics window. The Section View PropertyManager is displayed. View your options.

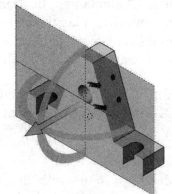

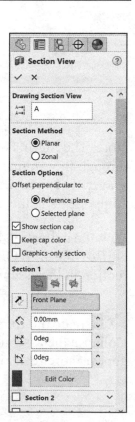

Save the Section View.

356) Click **Save** from the Section View PropertyManager. The Save As dialog box is displayed. SectionView is the default view name.

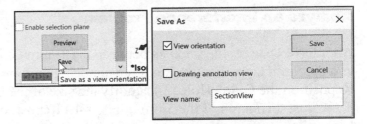

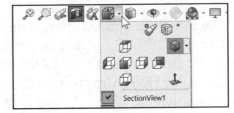

357) Click **Save**.

☀ Click the drop-down arrow from the View Orientation tool as illustrated. SectionView1 is saved as a custom view.

Display and Pin the View Orientation dialog box.

358) Press the **space** bar. View the Orientation box with the new view.

359) Close ✖ the Orientation dialog box.

Return to a full standard view.

360) Click **Front view** ⬜ from the Heads-up View toolbar.

361) Click **Section view** ▥ from the Heads-up View toolbar to return to a full view.

362) Click **Hidden Lines Visible** ⬓ from the Heads-up View toolbar.

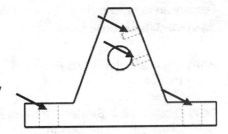

Save the GUIDE part.

363) Click **Save** 💾.

The Section view ▥ tool detects potential problems before manufacturing. The Section view tool determines the interference between the 10mm Guide hole and the Linear Pattern of 3mm holes in the GUIDE.

The GUIDE, ROD and PLATE components all share a common file structure with the GUIDE-ROD assembly. What component do you modify? How will changes affect other components? Let's analyze the interference problem and determine a solution.

Analyze an Interference Problem

An interference problem exists between the 10mm Guide Hole and the 3mm tapped holes. Review your design options: 1.) Reposition the Guide Hole, 2.) Modify the size of the Guide Hole, 3.) Adjust the length of the 3mm holes or 4.) Reposition the Linear Pattern feature.

The first three options affect other components in the assembly. The GUIDE-CYLINDER assembly and PLATE determine the Guide Hole location. The ROD diameter determines the size of the Guide Hole. The position sensor requires the current depth of the 3mm Holes. Reposition the Linear Pattern feature by modifying the first 3mm Thru Hole.

Activity: GUIDE-Analyze an Interference Problem

Modify the Thru Hole dimensions.
364) **Expand** the M3x0.5 Tapped Hole1 feature from the FeatureManager.

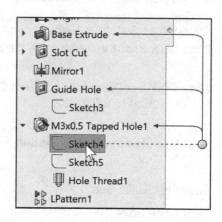

Display an Isometric view - Shaded With Edges.
365) Click **Isometric view** .

366) Click **Shaded With Edges** from the Heads-up View toolbar.

367) Double-click **Sketch4** from the FeatureManager to display the position dimensions.

368) Double-click the **4**mm dimension created from the temporary axis of the Guide Hole.

369) Enter **6**mm.

Rebuild the model.
370) Click **Rebuild** from the Modify dialog box. Click the **Green Check mark** from the Modify dialog box.

371) Click **Exit Sketch**.

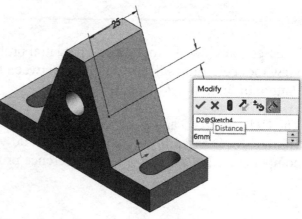

Close the GUIDE part.
372) Click **File**, **Close** from the Menu bar.

373) Click **Yes** to Save changes to the GUIDE.

374) Click **Wireframe** from the Heads-up View toolbar.

375) Click **Front view** 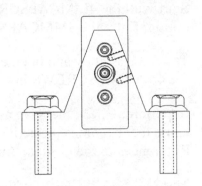. The GUIDE-ROD assembly updates to display the changes to the GUIDE part.

Return to an Isometric view - Shaded With Edges.

376) Click **Isometric view** 🔲.

377) Click **Shaded With Edges** 🔲.

Save the GUIDE-ROD assembly.

378) Click **Save** 💾.

🔅 SOLIDWORKS provides an Interference Detection tool and a Clearance Verification tool to analyze assemblies. See SOLIDWORKS help for additional information.

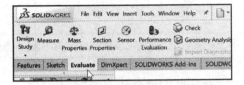

🔅 Analyze issues at the part level first. Working at the part level reduces rebuild time and complexity. Return to the assembly and review the modifications.

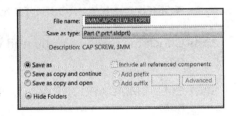

The last component to insert to the GUIDE-CYLINDER is the 3MMCAPSCREW. Utilize the Save As/Save as copy and open option to copy the 4MMCAPSCREW to create the 3MMCAPSCREW part.

Save As/Save as Copy Options

Conserve design time and cost. Modify existing parts and assemblies to create new parts and assemblies. Utilize the Save As/Save as command to save the file in another file format.

Utilize the Save as copy and continue command to save the document to a new file name without replacing the active document. Utilize the Save as copy and open command to save the document to a new file name that becomes the active document. The original document remains open. References to the original document are not automatically assigned to the copy.

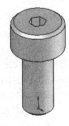

The 4MMCAPSCREW was created earlier. The GUIDE requires 3MMCAPSCREWs to fasten the sensor to the 3mm Tapped Hole Linear Pattern.

Start with the 4MMCAPSCREW. Utilize the Save As/Save as copy and open command option. Enter the 3MMCAPSCREW for the new file name.

The Save as copy and open command prevents the 3MMCAPSCREWS from replacing the 4MMCAPSCREWS in the GUIDE-CYLINDER assembly.

Important: Check the Save as copy and open command to save the document to a new file name that becomes the active document. The original document remains open. References to the original document are not automatically assigned to the copy.

The 3MMCAPSCREW is the new part name. Modify the dimensions of the Revolved Base feature to create the 3MMCAPSCREW.

Activity: 4MMCAPSCREW-Save As/Save as copy and open Option

Open the 4MMCAPSCREW part.

379) Right-click the **4MMCAPSCREW front face** component in the GUIDE-ROD assembly Graphics window.

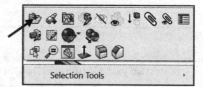

380) Click **Open Part** from the Context toolbar. The 4MMCAPSCREW part is displayed.

Apply the Save as copy and open.

381) Click **File**, **Save As** from the Menu bar. A SOLIDWORKS dialog box is displayed.

382) Click **Save as copy and open** from the SOLIDWORKS dialog box.

383) Select the **ENGDESIGN-W-SOLIDWORKS\PROJECTS** folder.

384) Check the **Save as copy and open** box.

385) Enter **3MMCAPSCREW** for File name.

386) Enter **CAP SCREW, 3MM** for Description.

387) Click **Save**. The 3MMCAPSCREW part is the active document. Note: The 4MMCAPSCREW document is still open.

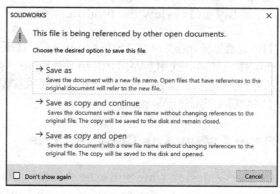

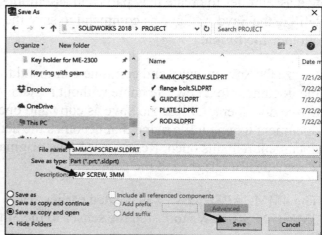

Modify the 3MMCAPSCREW part.
388) Double-click **Revolve1** in the FeatureManager. Dimensions are displayed.

Fit the model to the Graphics window.
389) Press the **f** key.

Modify the Revolve dimensions.
390) Double-click the vertical dimension **4**mm.

391) Enter **3**mm.

392) Double-click the depth dimension **14**mm.

393) Enter **9**mm.

394) Double-click the diameter dimension **4**mm.

395) Enter **3**mm.

396) Double-click the diameter dimension **7**mm.

397) Enter **5.5**mm.

Save the 3MMCAPSCREW part.
398) Click **SAVE** .

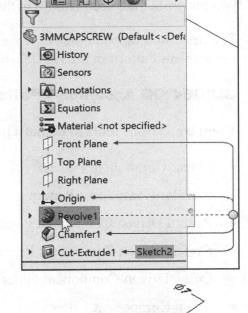

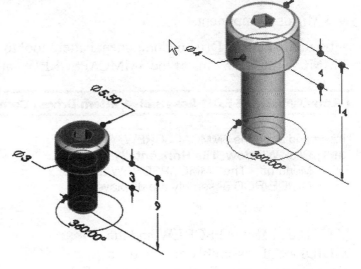

The GUIDE Linear Pattern of 3mm tapped holes requires six 3MMCAPSCREWs. Do you remember the seed feature in the Linear Pattern?

The first 3mm tapped hole is the seed feature. The seed feature is required for a Component Pattern in the GUIDE-ROD assembly.

GUIDE-ROD Assembly - Pattern Driven Component Pattern

There are various tools to define a pattern in an assembly:

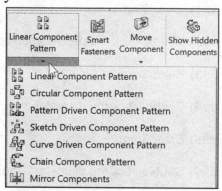

- Linear Component Pattern

- Circular Component Pattern

- Pattern Driven Component Pattern

- Sketch Driven Component Pattern

- Curve Driven Component Pattern

- Chain Component Pattern

- Mirror Components

Utilize the Pattern Driven Component Pattern tool to create instances of the 3MMCAPSCREW. Insert the 3MMCAPSCREW part into the GUIDE-ROD assembly.

Activity: GUIDE-ROD Assembly-Pattern Driven Component Pattern

---

Insert and mate the 3MMCAPSCREW.
399) Click **Window**, **Tile Horizontally** from the Menu bar. The 3MMCAPSCREW and the GUIDE-ROD assembly are displayed.

☼ The 3MMCAPSCREW and the GUIDE-ROD assembly are the open documents. Close all other documents. Select the Close ☒ icon. Click Window, Tile Horizontally again to display the two open documents.

400) **Zoom in** on the bottom left circular edge of the GUIDE left Tapped Hole.

401) Click and drag the **bottom circular edge** of the 3MMCAPSCREW into the GUIDE-ROD assembly.

402) Release the mouse button on the **bottom left circular edge** of the GUIDE left Tapped Hole. The mouse pointer displays the Coincident/Concentric Circular edges feedback symbol. The 3MMCAPSCREW part is positioned in the bottom left Tapped Hole.

Insert a Pattern Driven Component Pattern.
403) Maximize the GUIDE-ROD assembly.

Fit the GUIDE-ROD assembly to the Graphics window.
404) Press the **f** key.

405) Click the **Pattern Driven Component Pattern** tool from the Consolidated Assembly toolbar. The Pattern Driven PropertyManager is displayed. 3MMCAPSCREW<1> is displayed in the Components to Pattern box.

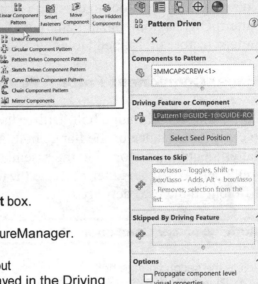

406) Click inside the **Driving Feature or Component** box.

407) Expand GUIDE in the GUIDE-ROD fly-out FeatureManager.

408) Click **LPattern1** under GUIDE<1> from the fly-out FeatureManager. LPattern1@GUIDE-1 is displayed in the Driving Feature or Component to Pattern box.

409) Click **OK** from the Pattern Driven PropertyManager. The DerivedLPattern1 is displayed in the FeatureManager.

Display an Isometric view.
410) Click **Isometric view**.

Save the GUIDE-ROD assembly.
411) Click **Save**.

Close all models.
412) Click **Window**,

413) Close All from the Menu bar.

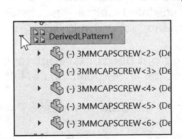

Reuse geometry. Utilize patterns early in the design process. A Linear Pattern feature in the part is utilized as a Pattern Driven Component Pattern in the assembly.

Review the Pattern Driven Component Pattern feature for the 3MMCAPSCREWs

You utilized the Save As/Save as copy and open option to copy the 4MMCAPSCREW to the 3MMCAPSCREW. The 3MMCAPSCREW was mated to the GUIDE 3MM Tapped Hole. You utilized the Pattern Driven Component Pattern tool to create an array of 3MMCAPSCREWs. The Pattern Driven Component Pattern was derived from the GUIDE Linear Pattern feature of Tapped Holes.

Linear Component Pattern Feature

Delete the second flange bolt. Utilize the Linear Component Pattern feature to create an instance of the flange bolt. Additionally, the Linear Component Pattern feature developed in the assembly is called LocalLPattern1. Note: If you delete a mate and then recreate the mate, your instance number will be different in the next activity.

Activity: Linear Component Pattern

Delete the second flange bolt.

414) Right-click the **flange bolt <2> (second flange both in the assembly [left side])** component in the FeatureManager.

415) Click **Delete**.

416) Click **Yes** to confirm.

Insert a Linear Component Pattern feature.

417) Click the **Linear Component Pattern** tool from the Consolidated toolbar. The Linear Pattern PropertyManager is displayed.

418) Click the **front horizontal edge** for Direction 1. The Direction arrow points to the left.

419) Click inside the **Components to Pattern** box.

420) Click the **right flange bolt** in the Graphics window as illustrated.

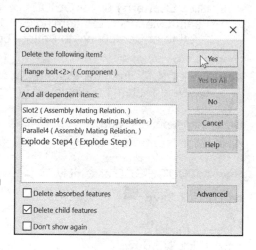

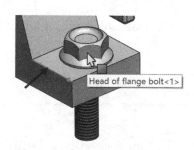

421) Enter **60**mm for Spacing.

422) Enter **2** for Number of Instances.

423) Click **OK** ✔ from the Linear Pattern PropertyManager. LocalLPattern1 is displayed in the FeatureManager.

Display an Isometric view. Save the GUIDE-ROD assembly.
424) Click **Isometric view** 🟦.

425) Click **Save** 💾.

💡 Before you redefine or edit a Mate, save the assembly. One Mate modification can lead to issues in multiple components that are directly related. Understand the Mate Selections syntax. The Mate PropertyManager lists the Mate Type, geometry selected, and component reference (part/assembly and instance number).

The 3mm Tapped Holes utilized the Pattern Driven Component Pattern feature. The flange bolts utilized the Linear Component feature.

Utilize the Pattern Driven Component Pattern feature when a part contains a pattern to reference. Utilize the Linear Component feature when a part does not contain a pattern reference.

Folders and Suppressed Components

The Assembly FeatureManager entries increase as components are inserted into the Graphics window of an assembly. Folders reduce the length of the Assembly FeatureManager in the part and assembly. Folders also organize groups of similar components. Organize hardware in the assembly into folders.

Suppress features, parts and assemblies that are not displayed. During model rebuilding, suppressed features and components are not calculated. This saves rebuilding time for complex models. The names of the suppressed features and components are displayed in light gray.

Create a folder in the FeatureManager named Hardware. Drag all individual bolts and screws into the Hardware Folder. The DerivedLPattern1 feature and the LocalLPattern1 feature cannot be dragged into a folder. Suppress the Hardware folder. Suppress the DerivedLPattern1 and LocalLPattern1 feature.

Activity: Folders and Suppressed Components

Create a new folder in the FeatureManager.

426) Right-click ▸ 🔩 flange bolt<1> (M8-1.25 x 30<<M8-1.25 x in the GUIDE-ROD FeatureManager.

427) Click **Create New Folder**. Folder1 is displayed in the FeatureManager.

428) Enter **Hardware** for folder name.

429) Click and drag ▸ 🔩 flange bolt<1> (M8-1.25 x 30<<M8-1.25 x from the FeatureManager into the Hardware folder. The mouse pointer displays the Move Component ⬦ icon.

430) Repeat the above process for the **4MMCAPSCREW<1>**, **4MMCAPSCREW<2>** and **3MMCAPSCREW<1>** components.

431) **Expand** the Hardware folder. View the components.

Suppress the Folder and Patterns.
432) Right-click the **Hardware** folder in the FeatureManager.

433) Click **Suppress**.

434) Right-click **DerivedLPattern1** in the FeatureManager.

435) Click **Suppress**.

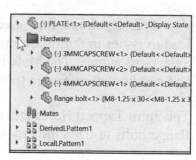

436) Right-click **LocalLPattern1**.

437) Click **Suppress**.

Display an Isometric view.
438) Click **Isometric view** .

Save the GUIDE-ROD assembly.
439) Click **Save** . View the FeatureManager.

Close all parts and assemblies.
440) Click **Window**, **Close All** from the Menu bar.

Standardized folder names, such as Hardware or Fillets, help colleagues recognize folder names. Place a set of continuous features or components into an individual folder.

Make-Buy Decision-3D ContentCentral

In a make-buy decision process, a decision is made on which parts to manufacture and which parts to purchase.

In assembly modeling, a decision is also made on which parts to design and which parts to obtain from libraries and the World Wide Web. SOLIDWORKS contains a variety of designed parts in their Design Library.

The SOLIDWORKS Toolbox is a library of feature based design automation tools for SOLIDWORKS. The Toolbox uses Window's drag and drop functionality with SmartMates. Fasteners are displayed with full thread details. Un-suppress the correct feature to obtain full thread display in the Graphics window.

Activate the SOLIDWORKS Toolbox. Click the **SOLIDWORKS Add-Ins** tab. Click **SOLIDWORKS Toolbox**.

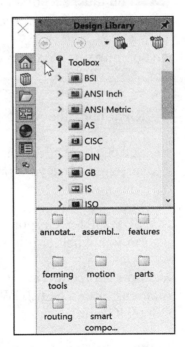

SOLIDWORKS SmartFastener uses the Hole Wizard tool to automatically SmartMate the corresponding Toolbox fasteners in an assembly. The fastener is sized and inserted into the assembly based on the Dimensioning Standard of the Hole.

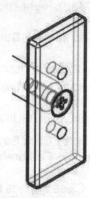

SOLIDWORKS provides a tool known as 3D ContentCentral. 3D ContentCentral provides access to 3D models from component suppliers and individuals in all major CAD formats.

3D ContentCentral provides the following access:

- 🌐 3D ContentCentral : Links to supplier Web sites with certified 3D models. You can search the 3D ContentCentral site, view, configure, and evaluate models online, and download models for your documents.

- 🌐 User Library : Links to models from individuals using 3D PartStream.NET.

💡 You must accept a license agreement when you first open 3D ContentCentral before you can access the contents.

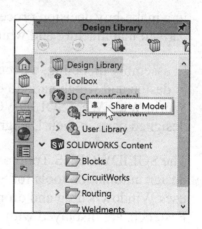

To submit your model to the 3D ContentCentral user library:

- Right-click the 3D ContentCentral 🌐 icon.

- Select Share a Model. Follow the directions on the screen to enter your profile information and to upload your model.

SOLIDWORKS Corporation will review your model and inform you whether it is accepted for sharing.

Vendors utilize this service to share model information with their customers. SOLIDWORKS users share information through 3D ContentCentral.

In the next section, create the CUSTOMER Assembly from the MGPM12-1010 assembly. The MGPM12-1010 assembly was downloaded from 3D ContentCentral. The manufacturer of the MGPM12-1010 assembly is SMC of USA. Utilize the assembly model that is located in the **ENGDESIGN-W-SOLIDWORKS\VENDOR COMPONENTS** folder for the next section.

CUSTOMER Assembly

Three documents for this project are contained in the **ENGDESIGN-W-SOLIDWORKS\VENDOR COMPONENTS** folder. Copy the folder directly to your hard drive. Work directly from your hard drive.

✦ MGPM12-1010
✷ MGPM12-1010_MGPRod(12M)
✦ MGPM12-1010_MGPTube(12M)

The MGPM12-1010_MGPTube(12M) part and MGPM12-1010_MGPRod(12M) part contain references to the MGPM12-1010 assembly.

💡 When parts reference an assembly, open the assembly first. Then open the individual parts from within the Assembly FeatureManager.

Create the CUSTOMER assembly. The CUSTOMER assembly combines the GUIDE-ROD assembly and the GUIDE-CYLINDER (MGPM12-10) assembly. The GUIDE-ROD assembly is fixed to the CUSTOMER assembly Origin.

Activity: Create the CUSTOMER Assembly

Copy the ENGDESIGN-W-SOLIDWORKS\VENDOR COMPONENTS folder.
441) Copy the **ENGDESIGN-W-SOLIDWORKS\VENDOR COMPONENTS** folder to your hard drive (PROJECTS folder). Work from your hard drive.

✦ MGPM12-1010
✷ MGPM12-1010_MGPRod(12M)
✦ MGPM12-1010_MGPTube(12M)

Open the MGPM12-1010 assembly from your hard drive.
442) Double-click the **MGPM12-1010** assembly from the ENGDESIGN-W-SOLIDWORKS\VENDOR-COMPONENTS folder. The MGPM12-1010 FeatureManager is displayed.

443) If needed, click **Yes** to rebuild.

Fit the model to the Graphics window.
444) Press the **f** key.

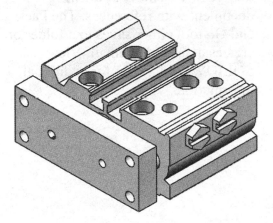

Open the GUIDE-ROD assembly.

445) Double-click the **GUIDE-ROD** assembly from the PROJECTS folder.

Create the CUSTOMER assembly.

446) Click **New**  from the Menu bar.

447) Double-click **Assembly** from the Templates tab. The Begin Assembly PropertyManager is displayed.

448) Double-click **GUIDE-ROD** from the Open documents box.

449) Click **OK** ✔ from the Begin Assembly PropertyManager to fix the GUIDE-ROD to the Origin.

Save the assembly.

450) Click **Save** 💾.

451) Click **Save All**.

452) Select **ENGDESIGN-W-SOLIDWORKS\PROJECTS** for folder.

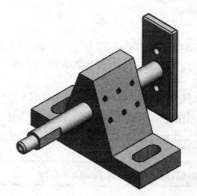

453) Enter **CUSTOMER** for File name.

Utilize the **Pack and Go** option to save an assembly or drawing document with references. The Pack and Go tool either saves to a folder or creates a zip file to email. View SOLIDWORKS help for additional information.

454) Enter **GUIDE-ROD AND GUIDE-CYLINDER ASSEMBLY** for Description.

455) Click **Save**.

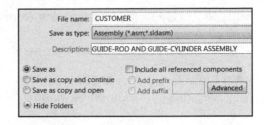

GUIDE-ROD is the first component in the CUSTOMER assembly. Insert the second component.

Activity: CUSTOMER Assembly-Insert Component

Insert the MGPM12-1010 assembly.

456) Click the **Insert Components** Assembly tool. The Insert Component PropertyManager is displayed.

457) Click **MGPM12-1010** from the Part/Assembly box.

458) Click a **position** in the Graphics window behind the GUIDE-ROD assembly as illustrated. The MGPM12-1010<1> component is added to the CUSTOMER FeatureManager.

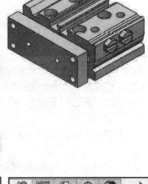

Fit the model to the Graphics window.
459) Press the **f** key.

Reposition the MGPM12-1010<1> component.
460) Click the **MGPM12-1010<1>** component in the FeatureManager. It's highlighted in the Graphics window.

461) Click the **Rotate Component** Assembly tool. The Rotate Component PropertyManager is displayed.

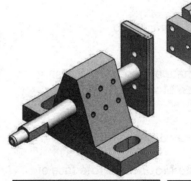

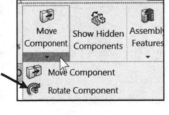

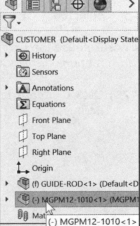

462) Select **By Delta XYZ** from the Rotate box.

463) Enter **90** in the delta Z box.

464) Click **Apply**. The MGPM12-1010<1> component rotated 90 degrees in the Graphics window.

465) Click **OK** ✔ from the Rotate Component PropertyManager.

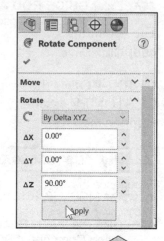

Fit the model to the Graphics window.
466) Press the **f** key.

467) Click and drag the **MGPM12-1010 assembly** behind the GUIDE-ROD assembly. The PISTON PLATE is behind the GUIDE-ROD assembly.

Zoom in on the PISTON PLATE component.

468) Click **Zoom to Area** from the Heads-up View toolbar.

469) **Zoom in** on the PISTON PLATE and the top MountHoles face of the PLATE as illustrated.

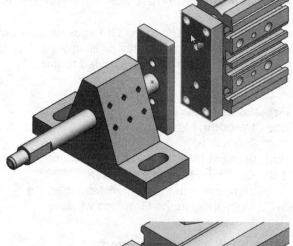

470) Click **Zoom to Area** to deactivate.

Insert a Concentric mate between two faces.
471) Click the **inside MountHoles** face of the PLATE.

472) Hold the **Ctrl** key down.

473) Click the **inside Mounting Hole** face of the PISTON PLATE. Both faces are selected.

474) Release the **Ctrl** key. The Mate pop-up menu is displayed.

475) Click **Concentric** from the Mate pop-up menu.

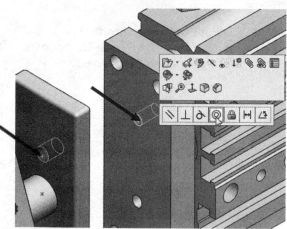

The bottom hole feature contains the Slot Cut and a Mounting Hole. The Mounting Hole is hidden. Utilize the Select Other option to view your options.

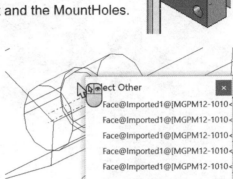

Insert a Concentric mate between two faces. Utilize the Select Other tool to view your options.

476) Click the **Mate** ✎ Assembly tool. The Mate PropertyManager is displayed.

477) Click **View**, **Hide/Show**, and check **Temporary Axes** from the Menu bar.

Experience the Select Other tool. Insert a Concentric mate.

478) Click **Zoom to Area** 🔍. **Zoom in** on the bottom Slot and the MountHoles. Do not select the Slot.

479) Click **Zoom to Area** 🔍 to deactivate.

480) Right-click a **position** behind the Slot Cut on the Piston Plate.

481) Click **Select Other**. The Select Other dialog box is displayed. The Select Other box lists geometry in the selected region of the Graphics window. There are edges, faces and axes. The list order and geometry entries depend on the selection location of the mouse pointer. Your geometry may vary.

482) Clear **all Selections** in the Mate entities box.

483) Position the **mouse pointer** over the Mounting Hole in the Piston Plate. Click the inside **Mounting Hole** face.

484) Click the **bottom inside Mounting Hole face** of the PLATE.

485) Click **Concentric**.

486) Click **OK** ✔ from the Concentric Mate PropertyManager.

487) Click **OK** ✔ from the Mate PropertyManager.

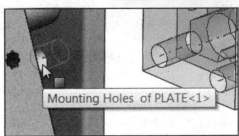

Mounting Holes of PLATE<1>

Deactivate the Temporary Axes.
488) Click **View**, **Hide/Show**, un-check **Temporary Axes** from the Menu bar.

Fit the model to the Graphics window.
489) Press the **f** key.

Insert a Coincident mate between two faces.
490) Click and drag the **PISTON PLATE** backward to create a gap.

491) Click the **front face** of the PISTON PLATE. **Rotate** the view until you can see the back face of the PLATE.

492) Hold the **Ctrl** key down. Click the **back face** of the PLATE.

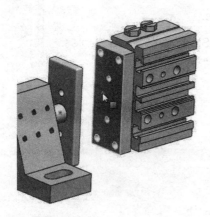
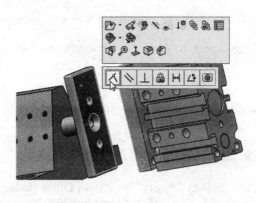

493) Release the **Ctrl** key. The Mate pop-up menu is displayed.

494) Click **Coincident** from the Mate pop-up menu.

Un-suppress all components and Sub-assemblies.
495) Right-click the **GUIDE-ROD** component in the CUSTOMER FeatureManager.

496) Click **Open Sub-assembly** from the Context toolbar. The GUIDE-ROD Assembly FeatureManager is displayed.

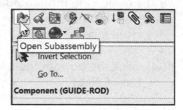

497) Right-click the **Hardware** folder in the FeatureManager.

498) Click **UnSuppress**.

499) Right-click **DerivedLPattern1** in the FeatureManager. Click **UnSuppress**.

500) Right-click **LocalLPattern1** in the FeatureManager. Click **UnSuppress**.

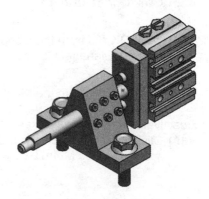

Return to the CUSTOMER assembly.
501) Press **Ctrl+Tab**.

502) Select **CUSTOMER assembly**.

Display an Isometric view.

503) Click **Isometric view** ⬢.

Save the CUSTOMER assembly in the ENGDESIGN-W-SOLIDWORKS/PROJECTS file folder.

504) Click **Save** 💾.

Copy the CUSTOMER Assembly - Apply Pack and Go

Copying an assembly in SOLIDWORKS is not the same as copying a document in Microsoft Word. The CUSTOMER assembly contains references to its parts and other sub-assemblies. Your task is to provide a copy of the CUSTOMER assembly to a colleague for review. Copy the CUSTOMER assembly and all of the components (references) into a different file folder named CopiedModels. Reference all component file locations to the new file folder. Apply the SOLIDWORKS Pack and Go tool.

Activity: CUSTOMER Assembly - SOLIDWORKS Pack and Go tool

Apply the SOLIDWORKS Pack and Go tool to save the CUSTOMER assembly to a new file folder.

505) Click **File**, **Pack and Go** from the Menu bar. Note the save options: Save to folder or Save to Zip file.

506) Click the **Browse** button.

507) Select the **My Documents** folder.

508) Click the **Make New Folder** button.

509) Enter **CopiedModels** for Folder Name.

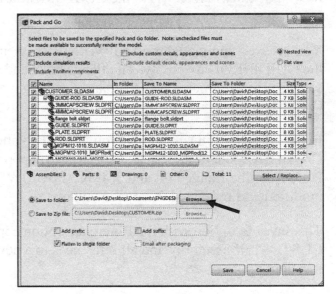

510) Click **OK** from the Browse For Folder dialog box.

Save the model.
511) Click **Save**. The CUSTOMER assembly remains open.

Close all files.
512) Click **Windows**, **Close All** from the Menu bar.

513) Click **Yes**. View the saved assembly in the CopiedModels folder.

    Review of the CUSTOMER Assembly

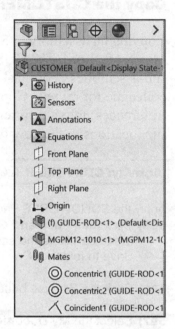

The CUSTOMER assembly combined the GUIDE-ROD assembly and the MGPM12-1010 assembly. In the design process, you decided to obtain the MGPM12-1010 assembly in SOLIDWORKS format from 3D ContentCentral.

The GUIDE-ROD assembly is the first component inserted into the CUSTOMER assembly. The GUIDE-ROD assembly is fixed (f) to the CUSTOMER assembly Origin.

You inserted and mated the MGPM12-1010 assembly to the GUIDE-ROD assembly. The flange bolts and cap screws utilized SmartMates to create Concentric and Coincident mates.

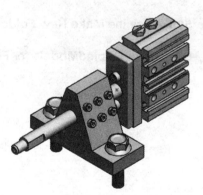

In the GUIDE-ROD assembly, you utilized a Pattern Driven Component Pattern and a Linear Component Pattern for the flange bolt and 3MMCAPSCREW. You modified and redefined the flange bolt mates.

The Save As option with the Reference option copied the CUSTOMER assembly and all references to a new folder location.

Point at the Center of Mass

Add a center of mass (COM) point to parts, assemblies or drawings.

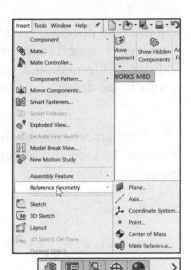

COM points added in component documents also appear in the assembly document. In a drawing document of parts or assemblies that contain a COM point, you can display and reference the COM point.

Add COM to a part or assembly by clicking **Center of Mass** (Reference Geometry toolbar) or **Insert, Reference Geometry, Center of Mass** ⊕ or checking the **Create Center of Mass feature** box in the Mass Properties dialog box.

The center of mass of the model is displayed in the Graphics window and in the FeatureManager design tree just below the origin.

The position of the **COM** point ⊕ updates when the model's center of mass changes. The COM point can be suppressed and unsuppressed for configurations.

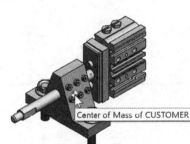

Center of Mass of CUSTOMER

You can measure distances and add reference dimensions between the COM point and entities such as vertices, edges and faces.

💡 If you want to display a reference point where the CG was located at some particular point in the FeatureManager, you can insert a Center of Mass Reference Point. See SOLIDWORKS Help for additional information.

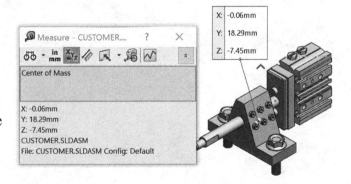

Add a center of mass (COM) point to a drawing view. The center of mass is a selectable entity in drawings, and you can reference it to create dimensions.

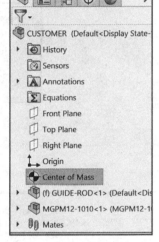

In a drawing document, click **Insert**, **Model Items**. The Model Items PropertyManager is displayed. Under Reference Geometry, click the **Center of Mass** icon. Enter any needed additional information. Click **OK** from the Model Items PropertyManager. View the results in the drawing.

The part or assembly **needs to have a COM before** you can view the COM in the drawing. To view the center of mass in a drawing, click **View**, **Hide/Show**, **Center of mass**.

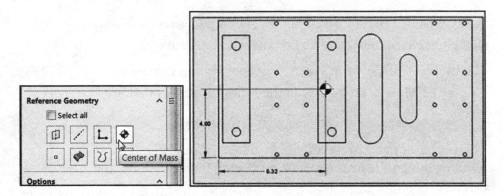

Project Summary

You created the GUIDE-ROD assembly from the following parts: GUIDE, ROD, PLATE, Flange Bolt, 3MMCAPSCREW and 4MMCAPSCREW. The components were oriented and positioned in the assembly using Concentric, Coincident, Parallel and Concentric/Coincident SmartMates.

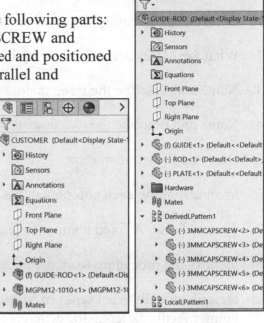

Mates were redefined in the GUIDE-ROD assembly to center the Flange Bolts in the Slot Cut of the GUIDE. In the GUIDE-ROD assembly, you utilized the Pattern Driven Component Pattern feature for the 3MMCAPSCREW and the Local Component Pattern feature for the Flange Bolt.

The CUSTOMER assembly contained the MGPM12-1010 assembly and the GUIDE-ROD assembly. The assemblies utilized a Bottom-up design approach. In a Bottom-up design approach, you possess all the required design information for the individual components.

Project 3 is completed. In Project 4, create an assembly drawing and a detailed drawing of the GUIDE.

Think design intent. When do you use the various End Conditions and Geometric sketch relations? What are you trying to do with the design? How does the component fit into an assembly?

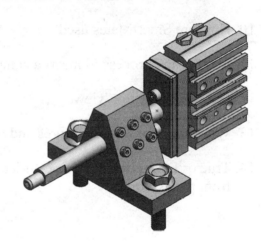

☀ Use the Defeature tool to remove details from a part or assembly and save the results to a new file in which the details are replaced by dumb solids (that is, solids without feature definition or history). You can then share the new file without revealing all the design details of the model.

☀ Use the Alt key to temporarily hide a face when you need to select an obscured face for mates.

Questions

1. Describe an assembly or sub-assembly.

2. What are Mates and why are they important in assembling components?

3. Name and describe the three major types of Fits.

4. Name and describe the two assembly modeling techniques in SOLIDWORKS.

5. Describe Dynamic motion.

6. In an assembly, each component has_____# degrees of freedom? Name them.

7. True or False. A fixed component cannot move and is locked to the Origin.

8. Identify the procedure to create a Revolved Base feature.

9. Describe the different types of SmartMates. Utilize online help to view the mouse pointer feedback icons for different SmartMate types.

10. How are SmartMates used?

11. Identify the process to insert a component from the Design Library.

12. Describe a Section view.

13. What are Suppressed features and components? Provide an example.

14. True or False. If you receive a Mate Error you should always delete the component from the assembly.

Exercises

Exercise 3.1: Weight-Hook Assembly

Create the Weight-Hook assembly. The Weight-Hook assembly has two components: WEIGHT and HOOK.

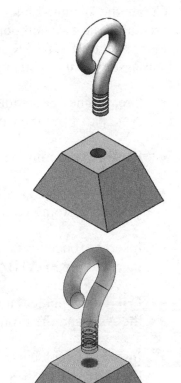

- Create a new assembly document. Copy and insert the WEIGHT part from the Chapter 3 Homework folder.

- Fix the WEIGHT to the Origin as illustrated in the Assem1 FeatureManager.

- Copy and insert the HOOK part from the Chapter 3 Homework folder into the assembly.

- Insert a Concentric mate between the inside top cylindrical face of the WEIGHT and the cylindrical face of the thread. Concentric is the default mate.

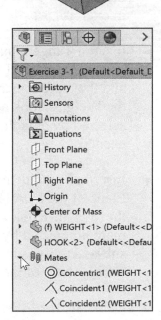

- Insert the first Coincident mate between the top edge of the circular hole of the WEIGHT and the top circular edge of Sweep1, above the thread. The HOOK can rotate in the WEIGHT.

Fix the position of the HOOK. Insert the second Coincident mate between the Right Plane of the WEIGHT and the Right Plane of the HOOK. Coincident is the default mate.

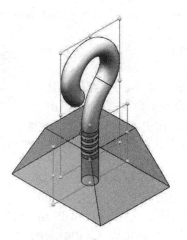

- Expand the Mates folder and view the created mates. Note: Do not display Tangent Edges or Origins in the final assembly.

- Calculate the Mass and Volume of the assembly.

- Identify the Center of Mass for the assembly.

Exercise 3.2: Weight-Link Assembly

Create the Weight-Link assembly. The Weight-Link
assembly has two components and a sub-assembly: Axle
component, FLATBAR component and the Weight-Hook
sub-assembly that you created in Exercise 3.1.

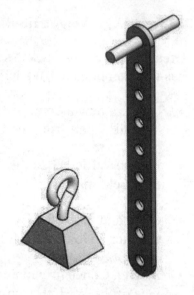

- Create a new assembly document. Copy and insert the
  Axle part from the Chapter 3 Homework folder.

- Fix the Axle component to the Origin of the assembly.

- Copy and insert the FLATBAR part from the
  Chapter 3 Homework folder.

- Insert a Concentric mate between the Axle cylindrical
  face and the FLATBAR inside face of the top circle.

- Insert a Coincident mate between the Front Plane of
  the Axle and the Front Plane of the FLATBAR.

- Insert a Coincident mate between the Right Plane of
  the Axle and the Top Plane of the FLATBAR. Position
  the FLATBAR as illustrated.

- Insert the Weight-Hook sub-assembly that you created
  in exercise 3.1.

- Insert a Tangent mate between the inside
  bottom cylindrical face of the FLATBAR
  and the top circular face of the HOOK in
  the Weight-Hook assembly. Tangent mate is
  selected by default. ClickFlip Mate Alignment if
  needed.

- Insert a Coincident mate between the Front
  Plane of the FLATBAR and the Front Plane of
  the Weight-Hook sub-assembly. Coincident
  mate is selected by default. The Weight-Hook
  sub-assembly is free to move in the bottom
  circular hole of the FLATBAR.

- Calculate the Mass and Volume of the assembly.

- Insert the Center of Mass icon into the assembly.

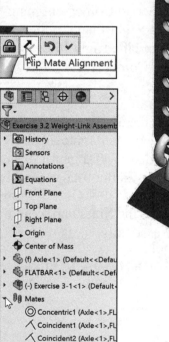

Exercise 3.3: Counter Weight Assembly

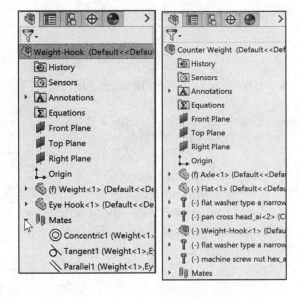

Create the Counter Weight assembly as illustrated using SmartMates and Standard mates. Copy all files from the Chapter 3 Homework\Counter-Weight folder to your hard drive. The Counter Weight consists of the following items:

- Weight-Hook sub-assembly.

- Weight.

- Eye Hook.

- Axle component.

- Flat component.

- Flat Washer Type A from the SOLIDWORKS Toolbox.

- Pan Cross Head Screw from the SOLIDWORKS Toolbox.

- Flat Washer Type A from the SOLIDWORKS toolbox.

- Machine Screw Nut Hex from the SOLIDWORKS Toolbox.

Fix the Axle component to the Origin of the assembly.

Insert other components and sub-assemblies. Apply all needed mates. Use SmartMates with the Flat Washer Type A Narrow_AI, Machine Screw Nut Hex_AI and the Pan Cross Head_AI components.

Use a Distance mate to fit the Axle in the middle of the Flat. Note a Symmetric mate could replace the Distance mate. Think about the design of the assembly.

The symbol (f) represents a fixed component. A fixed component cannot move and is fixed to the assembly Origin.

Exercise 3.4: Binder Clip Assembly

- Create a simple Gem binder clip.

- Create an ANSI - IPS model.

- Create two components - BASE and HANDLE.

- Apply material to each component and address all needed mates. Think about where you would start. Think about how the binder clip assembly would move.

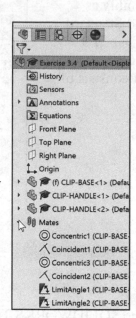

- What would the Base Sketch for each component be?

- Approximate the dimensions from a small or large Gem binder clip.

- View the sample Assembly FeatureManager. Your Assembly FeatureManager can (should) be different. This is just one way to create the assembly.

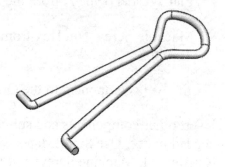

Below is a sample assembly of the Binder Clip from my Freshman Engineering class.

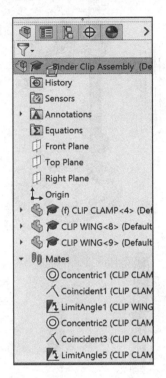

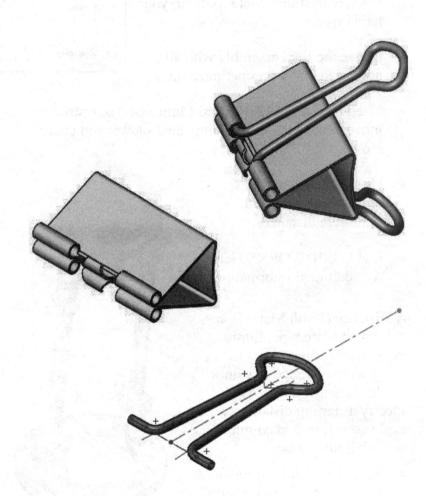

Exercise 3.5: Limit Mate Assembly (Advanced Mate Type)

- Copy all files from the Chapter 3 Homework\Limit Mate folder to your hard drive.

- Create the final assembly with all needed mates for proper movement.

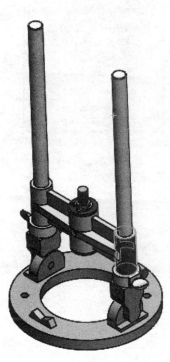

- Insert a Distance (Advanced Limit Mate) to restrict the movement of the Slide Component - lower and upper movement.

- Use the Measure tool to obtain maximum and minimum distances.

- Use SOLIDWORKS Help for additional information.

A Distance (Limit Mate) is an Advanced Mate type. Limit mates allow components to move within a range of values for distance and angle. You specify a starting distance or angle as well as a maximum and minimum value.

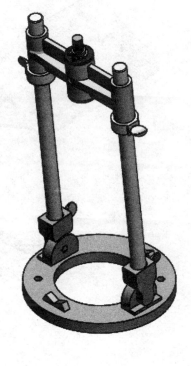

- Save the model and move the slide to view the results in the Graphics window.

- Think about how you would use this mate type in other assemblies.

Exercise 3.6: Screw Mate Assembly (Mechanical Mate Type)

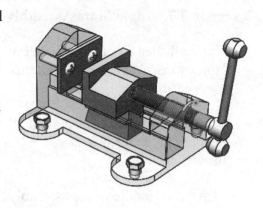

- Copy all files from the Chapter 3 Homework\Screw Mate folder to your hard drive. Create the final assembly with all needed mates for proper movement.

- Insert a Screw mate between the inside Face of the Base and the Face of the vice. A Screw is a Mechanical Mate type.

- Calculate the mass and volume of the assembly.

A Screw mate constrains two components to be concentric. It adds a pitch relationship between the rotation of one component and the translation of the other. Translation of one component along the axis causes rotation of the other component according to the pitch relationship. Likewise, rotation of one component causes translation of the other component. Use SOLIDWORKS Help if needed.

Use the Select Other tool (See SOLIDWORKS Help) to select the proper inside faces and to create the Screw mate for the assembly.

- Rotate the handle and view the results. Think about how you would use this mate type in other assemblies.

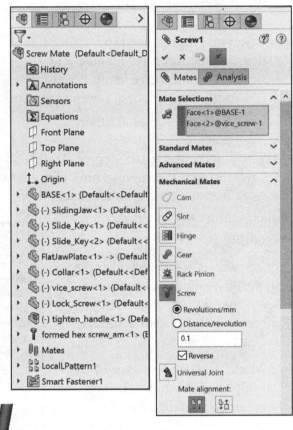

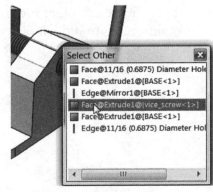

Exercise 3.7: Angle Mate Assembly

- Copy all files from the Chapter 3 Homework\Angle Mate folder to your hard drive.

- Create the final assembly with all needed mates for proper movement.

- Move the Handle in the assembly. The Handle is free to rotate. Set the angle of the Handle.

- Insert an Angle mate (165 degrees) between the Handle and the Side of the valve using Planes. An Angle mate places the selected items at the specified angle to each other.

- The Handle has a 165-degree Angle mate to restrict flow through the valve. Think about how you would use this mate type in other assemblies.

Exercise 3.8: Angle Mate Assembly (Cont:)

Create two end caps (lids) for the ball valve using the Top-down Assembly method. Note: The Reference - In-Content symbols in the FeatureManager.

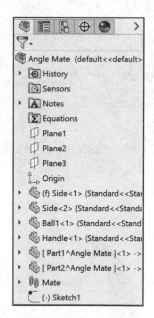

- Modify the Appearance of the body to observe the change - enhance visualization.

- Apply the Select-other tool to obtain access to hidden faces and edges.

Exercise 3.9: 4 Bar Linkage Assembly

Create the 4 bar linkage assembly as illustrated. The four bar linkage assembly has five simple components. Create the five simple components. Assume dimensions.

View the avi file from the Chapter 3 Homework folder for proper movement.

Insert all needed mates.

In an assembly, fix (f) the first component to the origin or fully define it to the three default sketch planes of the Assembly.

Insert additional components and insert needed mates to simulate the movement of a 4 bar linkage assembly.

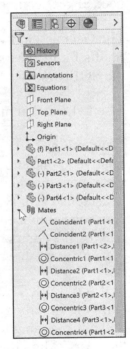

Read the section on Coincident, Concentric and Distance mates in
SOLIDWORKS Help.

Create a base with text for extra credit.

Below are sample models from my Freshman Engineering class. Note the different designs to maintain the proper movement of the 4 bar linkage.

Exercise 3.9A: Gear Mate Assembly

Create the Gear Mate assembly as illustrated.

View the provided ppt and avi in the Chapter 3 Homework\Gear Mate folder.

Create the final assembly.

View the Gear avi file for the proper model movement.

Use the SOLIDWORKS toolbox to insert the needed gear components.

The Gear assembly consists of a Base Plate, two Shafts for the gears, and two different gears from the SOLIDWORKS toolbox.

Insert all needed mates for proper movement.

Gear mate: Forces two components to rotate relative to one another around selected axes. The Gear mate provides the ability to establish gear type relations between components without making the components physically mesh.

SOLIDWORKS provides the ability to modify the gear ratio without changing the size of the gears. Align the components before adding the Mechanical gear mate.

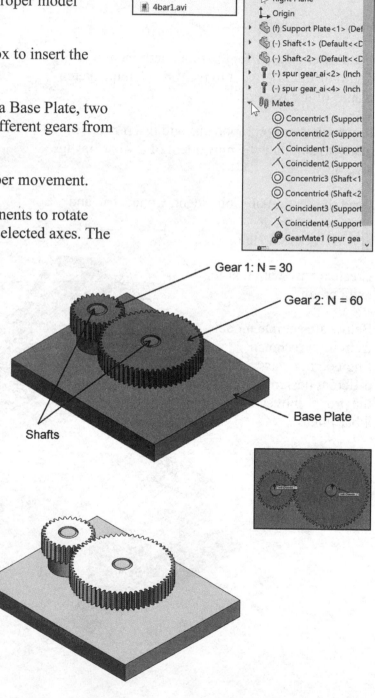

Gear 1: N = 30

Gear 2: N = 60

Base Plate

Shafts

Exercise 3.9B: Cam Assembly

Copy all files from the Chapter 3 Homework\Cam Mate folder.

Create the final assembly with all needed mates for proper movement.

Open the assembly. View the model. The model contains two cams.

Insert a tangent Cam follower mate.

1. Click the **Mate** ✎ tool from the Assembly tab.

2. **Pin** the Mate PropertyManager.

3. **Expand** the Mechanical Mates box from the Mate PropertyManager.

4. Click **Cam Follower** ⌀ mate.

5. Click inside of the **Cam Path** box.

6. Click the **outside face** of cam-s<1> as illustrated.

7. Click the **face of Link-c<1>** as illustrated. Face<2>@Link-c-1 is displayed.

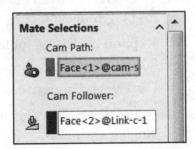

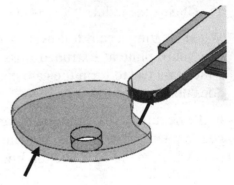

8. Click **OK** ✔ from the CamMateTangent1 PropertyManager.

Insert a Coincident Cam follower mate.

9. Click **Cam Follower** ⬯ mate from the Mechanical Mates box.

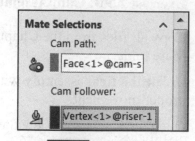

10. Click the **outside face** of cam-s2<1>.

11. Click the vertex of **riser<1>** as illustrated. Vertex<1>@riser-1 is displayed.

12. Click **OK** ✔ from the CamMateCoincident1 PropertyManager.

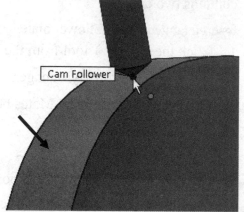

13. **Un-pin** the Mate PropertyManager.

14. Click **OK** ✔ from the Mate PropertyManager.

15. **Expand** the Mates folder. View the new created mates.

16. Display an **Isometric** view.

17. **Rotate** the cams. View the results in the Graphics window.

18. **Close** the model.

When creating a cam follower mate, make sure that your Spline or Extruded Boss/Base feature, which you use to form the cam contact face, does nothing but form the face.

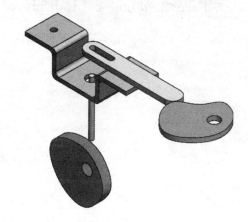

In the next tutorial of the spur gear and rack, the gear's pitch circle must be tangent to the rack's pitch line, which is the construction line in the middle of the tooth cut.

Exercise 3.9C: Slot Assembly

Copy all files from the Chapter 3 Homework\Slot Mate folder.

Open the Slot Mate assembly.

Insert a Slot mate (center in slot option) between the first Flange bolt and the right side slot.

Insert a second Flange bolt into the assembly.

Insert the second slot mate (center in slot option) between the second Flange bolt and the left side slot.

Insert a Coincident mate between the top face of the GUIDE and the bottom face of the second Flange bolt cap.

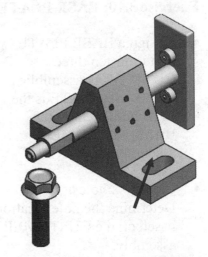

Expand the Mates folder. View the created mates.

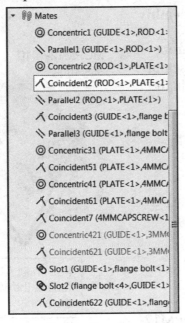

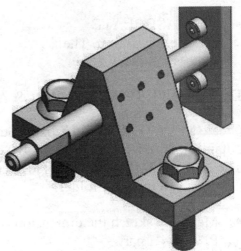

Exercise 3.10: BASE PLATE

- Design a BASE PLATE part to fasten three CUSTOMER assemblies. The BASE PLATE is the first component in the BASE-CUSTOMER assembly:

- Create six holes. Determine the hole location based on the CUSTOMER assembly.

- The BASE PLATE thickness is 10mm. The material is 1060 Alloy.

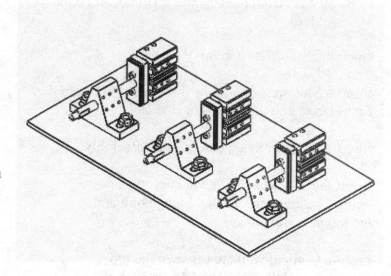

Exercise 3.11: PLATE4H-Part & Assembly

Create a PLATE4H part. Utilizes four front outside holes of the MGPM12-1010 assembly. The MGPM12-1010 zip file is located in the Chapter 3 Homework folder.

- Manually sketch the dimensions of the PLATE4H part.

- Dimension the four outside holes.

- The Material thickness is 10mm.

- Create the new assembly, PLATE4H-GUIDECYLINDER.

- Mate the PLATE4H part to the MGPM12-1010 assembly.

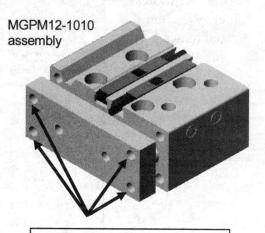

MGPM12-1010 assembly

Fasten PLATE4H to the front face of the MGPM12-1010 assembly

Exercise 3.11A: LINKAGE Assembly

In the Project 2 exercises, you created four parts for the LINKAGE assembly.

- Axle part

- SHAFT COLLAR part

- FLAT BAR - 3 HOLE part

- FLAT BAR - 9 HOLE part

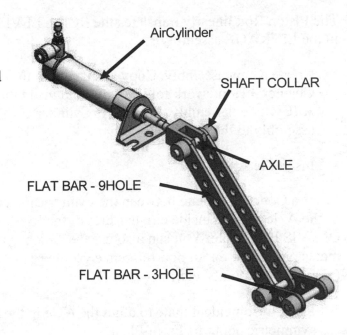

Create the LINKAGE assembly with the SMC AirCylinder and the four parts from Project 2.

The AirCylinder is the first sub-assembly in the LINKAGE assembly. When compressed air goes in to the air inlet, the Piston Rod is pushed out.

Without compressed air, the Piston Rod is returned by the force of the spring.

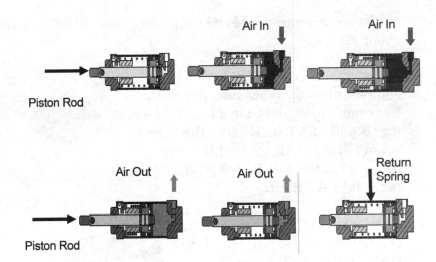

The Piston Rod linearly translates the ROD CLEVIS in the LINKAGE assembly.

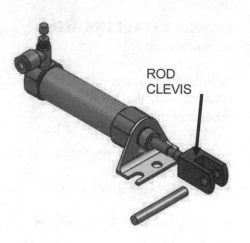

ROD CLEVIS

- Create a new assembly. Copy all files from the Chapter 3 Homework folder. Open and insert the AirCylinder assembly. Fix the AirCylinder assembly to the Origin.

- Insert the Axle.

Insert a Concentric mate between the cylindrical face of the Axle and the inside circular face of the ROD CLEVIS hole. Note: You can also use the Quick mates procedure for all non-reference geometry (planes, axes, points, etc.).

- Insert a Coincident mate to align the Axle in the ROD CLEVIS holes. The Axle is symmetric about its Front Plane.

- Insert the first FLAT BAR - 9 HOLE component as illustrated.

- Utilize a Concentric mate and Coincident mate. The FLAT BAR - 9 HOLE rotates about the Axle.

- Insert the second FLAT BAR - 9 HOLE component on the other side. Repeat the Concentric mate and Coincident mate for the second FLAT BAR - 9 HOLE. The second FLAT BAR - 9 HOLE is free to rotate about the AXLE, independent from the first FLAT BAR.

- Insert a Parallel mate between the two top narrow faces of the FLAT BAR - 9 HOLE. The two FLAT BAR - 9 HOLEs rotate together.

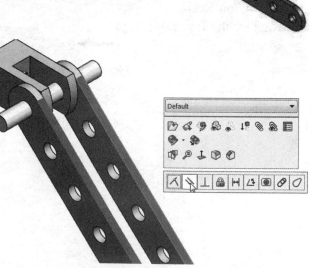

- Insert the first SHAFT-COLLAR component. Insert a Concentric and Coincident mate.

- Insert the second SHAFT-COLLAR component. Repeat the Concentric and Coincident mate.

- Insert the second instance of the Axle. Mate the AXLE to the bottom hole of the FLAT BAR - 9 HOLE. Insert a Concentric and Coincident mate.

- Insert the two FLAT-BAR - 3 HOLE components, the third Axle, and four SHAFT-COLLAR components with the needed mates to complete the assembly.

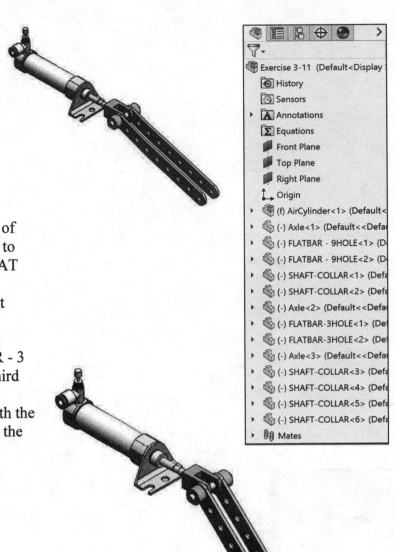

Notes:

Project 4

Fundamentals of Drawing

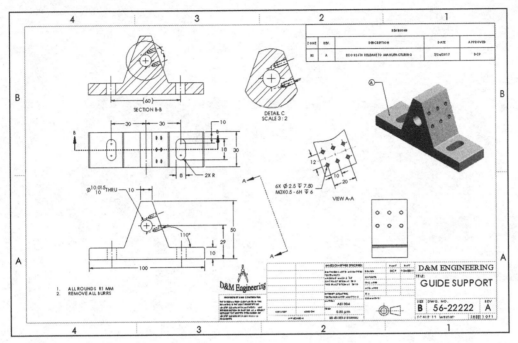

Below are the desired outcomes and usage competencics based on the completion of Project 4.

Project Desired Outcomes:	Usage Competencies:
• B-ANSI-MM Drawing Template.	• Generate a Drawing Template with Document Properties and Sheet Properties.
• CUSTOM-B Sheet Format.	• Produce a Sheet Format with Custom Sheet Properties, Title block, Company logo and more.
• GUIDE Drawing.	• Create Standard Orthographic, Auxiliary, Detail and Section views. • Insert, create, and modify dimensions and annotations.
• GUIDE-ROD Drawing with a Bill of Materials.	• Knowledge to develop and incorporate a Bill of Materials with Custom Properties.

Notes:

Project 4 - Fundamentals of Drawing

Project Objective

Provide an understanding of Drawing Templates, Part drawings, Assembly drawings, details and annotations.

Create a B-ANSI-MM Drawing Template. Create a CUSTOM-B Sheet Format. The Drawing Template contains Document Property settings. The Sheet Format contains a Company logo, Title block, Revision table and Sheet information.

Create the GUIDE drawing. Display Standard Orthographic, Section, Auxiliary and Detail drawing views. Insert, create and modify part and component dimensions.

Create an Isometric exploded GUIDE-ROD assembly drawing with a Bill of Materials. Obtain knowledge to develop and incorporate a Bill of Materials with Custom Properties.

On the completion of this project, you will be able to:

- Create a new Drawing Template.

- Generate a customized Sheet Format with Custom Properties.

- Open, Save and Close Drawing documents.

- Produce a Bill of Materials with Custom Properties.

- Insert and position views on a Multi Sheet drawing.

- Set the Dimension Layers.

- Insert, move and modify dimensions from a part into a drawing view.

- Insert Annotations: Center Mark, Centerline, Notes, Hole Callouts and Balloons.

- Use Edit Sheet Format and Edit Sheet mode.

- Insert a Revision table.

- Modify dimensioning scheme.

- Create parametric drawing notes.

- Link notes in the Title block to SOLIDWORKS properties.

- Rename parts and drawings.

- Insert a Center of Mass point.

Project Situation

The individual parts and assembly are completed. What is the next step? You are required to create drawings for various internal departments, namely production, purchasing, engineering, inspection and manufacturing. Each drawing contains unique information and specific footnotes. Example: A manufacturing drawing would require information on assembly, Bill of Materials, fabrication techniques and references to other relative documents.

Project Overview

Generate two drawings in this project:

- GUIDE drawing with a customized Sheet Format.

- GUIDE-ROD assembly drawing with a Bill of Materials.

The GUIDE drawing contains three Standard Orthographic views and an Isometric view. Do you remember what the three Principle Orthographic Standard views are? They are Top, Front and Right side (Third Angle Projection).

Three new views are introduced in this project: Detailed view, Section view and Auxiliary view. Orient the views to fit the drawing sheet. Incorporate the GUIDE dimensions into the drawing with inserted Annotations.

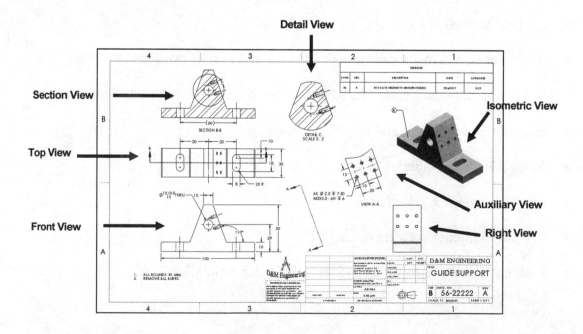

The GUIDE-ROD assembly drawing contains an Isometric Exploded view.

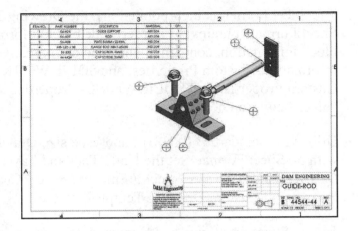

The drawing contains a Bill of Materials with balloon text with bent leaders and magnetic lines.

Both drawings utilize a custom Sheet Format containing a Company logo, Title block and Sheet information.

There are two major design modes used to develop a SOLIDWORKS drawing:

- Edit Sheet Format.

- Edit Sheet.

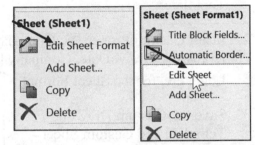

The Edit Sheet Format mode provides the ability to:

- Define the Title block size and text headings.

- Incorporate a company logo.

- Add a picture, design, company text and more.

The Edit Sheet mode provides the ability to:

- Add or modify drawing views.

- Add or modify drawing view dimensions.

- Add or modify text and more.

Drawing Template and Sheet Format

The foundation of a SOLIDWORKS drawing document is the Drawing Template. Drawing size, drawing standards, company information, manufacturing, and or assembly requirements, units and other properties are defined in the Drawing Template.

The Sheet Format is incorporated into the Drawing Template. The Sheet Format contains the border, Title block information, revision block information, company name, and or company logo information, Custom Properties, and SOLIDWORKS Properties. Custom Properties and SOLIDWORKS Properties are shared values between documents.

Utilize the standard B (ANSI) Landscape size Drawing Template with no Sheet Format. Set the Units, Font and Layers. Modify a B (ANSI) Landscape size Sheet Format to create a Custom Sheet Format and Custom Drawing Template.

1. Set Sheet Properties and Document Properties for the Drawing Template.

2. Insert Custom Properties: CompanyName, Revision, Number, DrawnBy, DrawnDate, Company Logo, Third Angle Projection Logo, etc. for the Sheet Format.

3. Save the Custom Drawing Template and Custom Sheet Format in the MY-TEMPLATE file folder.

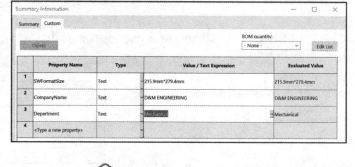

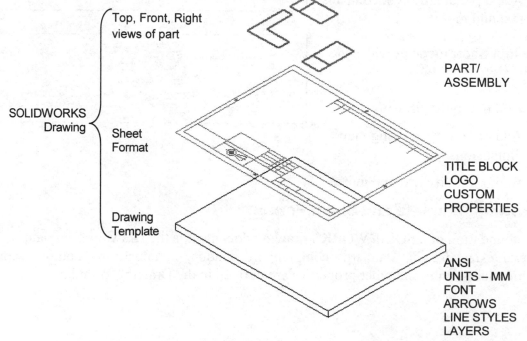

Views from the part or assembly are inserted into the SOLIDWORKS Drawing.

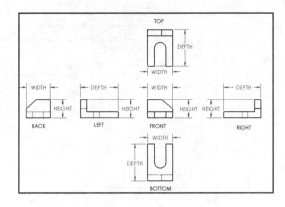

A Third Angle Projection scheme is illustrated in this project. For non-ANSI dimension standards, the dimensioning techniques are the same, even if the displayed arrows and text size are different.

For printers supporting millimeter paper sizes, select A3 (ANSI) Landscape (420mm x 297mm).

The default Drawing Templates with Sheet Format displayed contain predefined Title block Notes linked to Custom Properties and SOLIDWORKS Properties.

View the provided videos on Drawing Fundamentals to enhance your experience in this section.

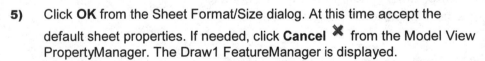
Activity: Create a Drawing Template

Close all documents.
1)   Click **Window, Close All** from the Menu bar.

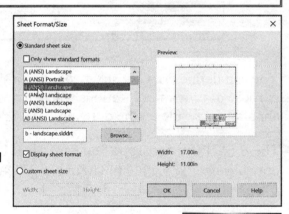

Create a B (ANSI) Landscape, Third Angle Projection drawing document.

2)   Click **New** ⬜ from the Menu bar. Double-click **Drawing** from the Templates tab.

3)   If needed **uncheck** the **Only show standard formats** box.

4)   Select **B (ANSI) Landscape** from the Standard sheet size box.

5)   Click **OK** from the Sheet Format/Size dialog. At this time accept the default sheet properties. If needed, click **Cancel** ✖ from the Model View PropertyManager. The Draw1 FeatureManager is displayed.

If the Start command when creating new drawing option is checked, the Model View PropertyManager is selected by default.

Draw1 is the default drawing name. Sheet1 is the default first sheet name.

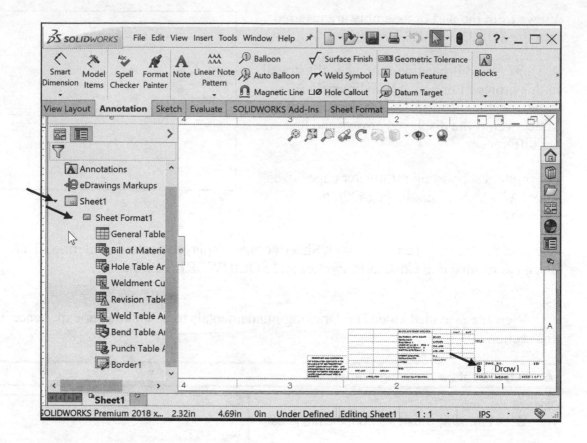

You can define drawing sheet zones on a sheet format for the purpose of providing locations where drawing views and annotations reside on the drawing.

The B (ANSI) Landscape Standard Sheet border defines the drawing size, 17″ x 11″ (431.8mm x 279.4mm).

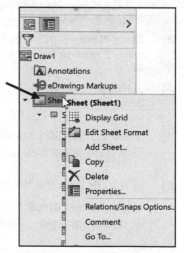

Expand Sheet1 in the FeatureManager. Right-click **Sheet 1** and view your options. The purpose of this book is to expose the new user to various tools and methods.

Use annotation notes and balloons to identify which drawing zone they are in. As you move an annotation in the Graphics area, the drawing zone updates to the current zone. You can add the current zone to an annotation by clicking an open space within the drawing view's bounding box while typing the annotation.

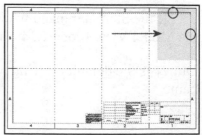

Set Drawing Size, Sheet Properties and Document Properties for the Drawing Template. Sheet Properties control Sheet Size, Sheet Scale and Type of Projection. Document Properties control the display of dimensions, annotations and symbols in the drawing.

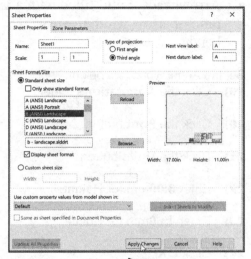

Set Sheet Properties. View your options in the Sheet Properties dialog box.

6)   Right-click in the **Graphics window**.

7)   Click **Properties** ▤. The Sheet Properties dialog box is displayed. If needed, expand the drop-down menu to view the Properties command.

8)   Select Sheet Scale **1:1**.

9)   Select **B (ANSI) Landscape** from the Standard sheet size box.

10) If needed uncheck the **Display sheet format** box.

11) Select **Third Angle** for Type of projection.

12) Click **Apply Changes** from the Sheet Properties dialog box.

Set Document Properties. Set drafting standard, units, and precision.

13) Click **Options** ⚙, **Document Properties** tab from the Menu bar.

14) Select **ANSI** for Overall drafting standard.

15) Click the **Units** folder.

16) Select **MMGS** for Unit system.

17) Select **.12** for basic unit length decimal place.

18) Select **None** for basic unit angle decimal place.

Detailing options provide the ability to address dimensioning standards, text style, center marks, extension lines, arrow styles, tolerance and precision.

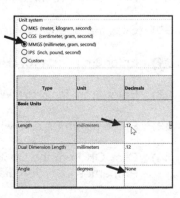

There are numerous text styles and sizes available in SOLIDWORKS. Companies develop drawing format standards and use specific text height for Metric and English drawings.

Numerous engineering drawings use the following format:

- Font: Century Gothic - All capital letters.

- Text height: .125in. or 3mm for drawings up to B (ANSI) Size, 17in. x 22in.

- Text height: .156in. or 5mm for drawings larger than B (ANSI) Size, 17in x 22in.

- Arrowheads: Solid filled with a 1:3 ratio of arrow width to arrow height.

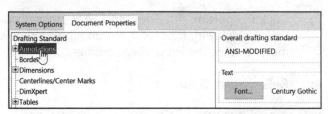

Set the Annotations font height.
19) Click the **Annotations** folder.

20) Click the **Font** button.

21) Click the **Units** button.

22) Enter **3.00**mm for Height.

23) Enter **1.00**mm for Space.

24) Click **OK**.

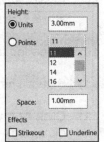

Change the Arrow Height.
25) Click the **Dimensions** folder from the Document Properties column as illustrated.

26) Enter **1**mm for arrow Height.

27) Enter **3**mm for arrow Width.

28) Enter **6**mm for arrow Length.

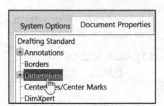

Set Section/Views size.
29) **Expand** the Views folder from the Document Properties column.

30) Click the **Section** folder.

31) Enter **2**mm for arrow Height.

32) Enter **6**mm for arrow Width.

33) Enter **12**mm for arrow Length.

34) Click **OK** from the Document Properties - Section dialog box.

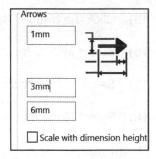

Drawing Layers organize dimensions, annotations and geometry. Create a new drawing layer to contain dimensions and notes. Create a second drawing layer to contain hidden feature dimensions.

Select the Show/Hide eye to turn On/Off Layers. Dimensions placed on the hidden layers are turned on and off for clarity and can be recalled for parametric annotations.

Display the Layer toolbar.

35) **Right-click** in the gray area to the right of the word Help in the Menu bar. Check **Layer** if the Layer toolbar is not active. The Layer toolbar is displayed. Click and drag the Layer toolbar to the center of the Graphics window.

36) Click the **Layer Properties** file folder from the Layer toolbar. The Layers dialog box is displayed.

Create the Dimension Layer.

37) Click the **New** button. Enter **Dim** in the Name column.

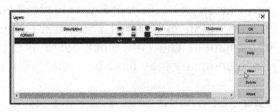

38) **Double-click** under the Description column. Enter **Dimensions** in the Description column.

Create the Notes Layer.

39) Click the **New** button. Enter **Notes** for Name. **Double-click** under the Description column.

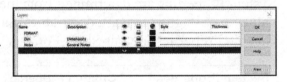

40) Enter **General Notes** for Description.

Create the Hidden Dims Layer.

41) Click the **New** button. Enter **Hidden Dims** for Name. **Double-click** under the Description column. Enter **Hidden Insert Dimensions** for Description.

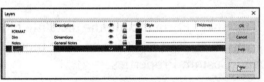

Dimensions placed on the Hidden Dims Layer are not displayed on the drawing until the Hidden Dims Layer status is On. Set the Layer Color to locate dimensions on this layer easily.

Turn the Hidden Insert Dimension Layer Off.

42) Click **On/Off**. The row is displayed in black.

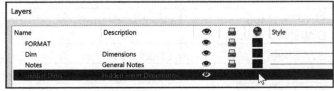

Set the Layer Color.

43) Click the **small black square** in the Hidden Dims row as illustrated.

Select a Color Swatch from the Color dialog box.
44) Select **Dark Blue**.

45) Click **OK**.

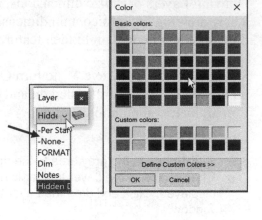

46) Click **OK** from the Layers dialog box.

The current Layer is Hidden Dims. Set the current Layer to None before saving the Drawing Template.

Set None for Layer.
47) Click the **Layer drop-down arrow**.

48) Click **None**. None is displayed in the Layer toolbar.

The Drawing Template contains the drawing Size, Document Properties and Layers. The Overall drafting standard is ANSI and the Units are in millimeters. The current Layer is set to None. The Drawing Template requires a Sheet Format. The Sheet Format contains Title block information. The Title block contains vital part or assembly information. Each company may have a unique version of a Title block.

Sheet Format and Title block

The Sheet Format contains the Title block, Revision block, Company logo, Custom Properties, Zones, etc. The Title block contains text fields linked to System Properties and Custom Properties.

System Properties are determined from the SOLIDWORKS documents. Custom Property values are assigned to named variables. Save time. Utilize System Properties and define Custom Properties in your Sheet Formats.

System Properties Linked to fields in default Sheet Formats:	Custom Properties of drawings linked to fields in default Sheet Formats:		Custom Properties of parts and assemblies linked to fields in default Sheet Formats:
SW-File Name (in DWG. NO. field):	CompanyName:	EngineeringApproval:	Description (in TITLE field):
SW-Sheet Scale:	CheckedBy:	EngAppDate:	Weight:
SW-Current Sheet:	CheckedDate:	ManufacturingApproval:	Material:
SW-Total Sheets:	DrawnBy:	MfgAppDate:	Finish:
	DrawnDate:	QAApproval:	Revision:
	EngineeringApproval:	QAAppDate:	

Utilize the standard landscape B (ANSI) Sheet Format (17in. x 11in.) or the standard-A (ANSI) Sheet Format (420mm x 297mm) to create a Custom Sheet Format.

Activity: Sheet Format and Title block

Display the standard B (ANSI) Landscape Sheet Format.
49) Right-click in the **Graphics window**. **Expand** the drop-down menu.

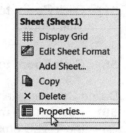

50) Click **Properties**. The Sheet Properties dialog box is displayed.

51) Click the **Standard sheet size** box.

52) Select **B (ANSI) Landscape**.

53) Check the **Display sheet format** box. The default Sheet Format, b - landscape.slddrt is displayed.

54) Click **Apply Changes** from the Sheet Properties dialog box.

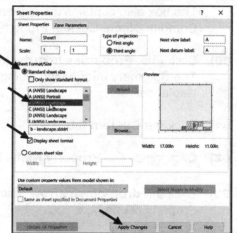

The default Sheet Format is displayed in the Graphics window. The FeatureManager displays Draw1, Sheet 1.

A SOLIDWORKS drawing contains two edit modes:

1. Edit Sheet.

2. Edit Sheet Format.

Insert views and dimensions in the Edit Sheet mode.

Modify the Sheet Format text, lines or Title block information in the Edit Sheet Format mode.

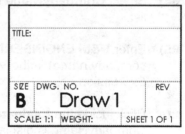

The CompanyName Custom Property is located in the Title block above the TITLE box. There is no value defined for CompanyName. A small text box indicates an empty field. Define a value for the Custom Property CompanyName. Example: D&M ENGINEERING.

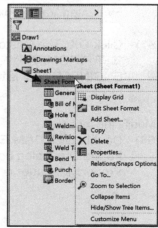

A goal of this book is to expose the user to various tools and methods. See SOLIDWORKS Help for additional information.

Activate the Edit Sheet Format mode.

55) Right-click in the **Graphics window**.

56) Click **Edit Sheet Format**. The Title block lines turn blue.

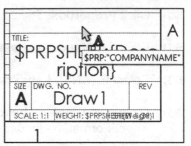

View the right side of the Title block.

57) Click **Zoom to Area** . **Zoom in** on the Sheet Format Title block.

58) Click **Zoom to Area** to deactivate.

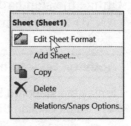

Define CompanyName Custom Property.

59) Position the **mouse pointer** in the middle of the box above the TITLE box. The mouse pointer displays Sheet Format1. The box also contains the hidden text, linked to the CompanyName Custom Property.

60) Click **File**, **Properties** from the Menu bar. The Summary Information dialog box is displayed.

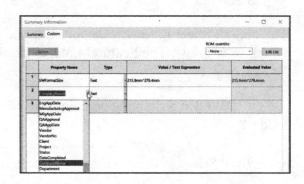

61) Click the **Custom** tab.

62) Click inside the **Property Name** box.

63) Click the **drop-down arrow** in the Property Name box.

64) Select **CompanyName** from the Property list.

65) Enter **D&M ENGINEERING** (or your company name) in the Value/Text Expression box.

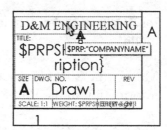

66) Click inside the **Evaluated Value** box. The CompanyName is displayed in the Evaluated Value box.

67) Click **OK**. The Custom Property "$PRP:COMPANYNAME," Value "D&M ENGINEERING" is displayed in the Title block.

Modify the font size.

68) Double-click **D&M ENGINEERING**. The Formatting dialog box and the Note PropertyManager is displayed.

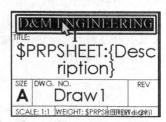

69) Click the **drop-down arrows** to set the Text Font and Height.

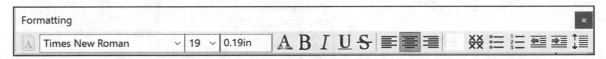

70) Click the **Style buttons** and **Justification buttons** to modify the selected text.

71) Click **OK** ✔ from the Note PropertyManager. View the results.

💡 Click a position outside the selected text box to save and exit the text.

The Tolerance block is located in the Title block. The Tolerance block provides information to the manufacturer on the minimum and maximum variation for each dimension on the drawing. If a specific tolerance or note is provided on the drawing, the specific tolerance or note will override the information in the Tolerance block.

General tolerance values are based on the design requirements and the manufacturing process.

💡 Create Sheet Formats for different parts types. Example: sheet metal parts, plastic parts and high precision machined parts. Create Sheet Formats for each category of parts that are manufactured with unique sets of Title block notes.

Modify the Tolerance block in the Sheet Format for ASME Y14.5-2009 machined, millimeter parts. Delete unnecessary text. The FRACTIONAL text refers to inches. The BEND text refers to sheet metal parts. The Three Decimal Place text is not required for this millimeter part.

Modify the Tolerance Note.

72) Double-click the text **INTERPRET GEOMETRIC TOLERANCING PER:**

73) Enter **ASME Y14.5** as illustrated.

74) Click **OK** ✔ from the Note PropertyManager.

75) Double-click inside the **Tolerance block** text. The Formatting dialog box and the Note PropertyManager is displayed.

76) Delete the text **INCHES**.

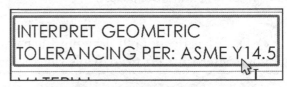

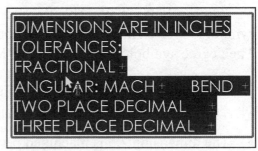

77) Enter **MILLIMETERS**.

78) Delete the line **FRACTIONAL +-**.

79) Delete the text **BEND +-**.

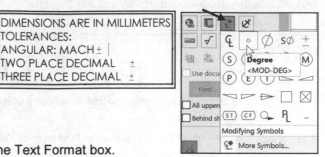

Enter ANGULAR tolerance.
80) Click a **position** at the end of the line.

81) Enter **0**. Click **Add Symbol** from the Text Format box.

82) Select **Degree** from the Modifying Symbols library.

83) Enter **30′** for minutes of a degree.

Modify the TWO and THREE PLACE DECIMAL LINES.
84) Delete the **TWO** and **THREE PLACE DECIMAL lines**.

85) Enter **ONE PLACE DECIMAL +- 0.5**.

86) Enter **TWO PLACE DECIMAL +- 0.15**.

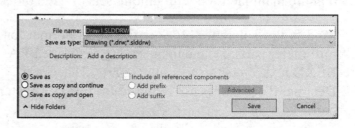

87) Click **OK** ✔ from the Note PropertyManager.

88) Right-click **Edit Sheet** in the Graphics window.

Save Draw1. Fit the drawing to the Graphics window.
89) Press the **f** key.

90) Click **Save** . Accept the default name. View the results.

💡 Draw1 is the default drawing file name. This name is temporary. In the next activity, invoke Microsoft Word. Always save before selecting another software application.

Various symbols are available through the Symbol button in the Text dialog box. The ± symbol is located in the Modify Symbols list. The ± symbol is displayed as <MOD-PM>. The degree symbol ° is displayed as <MOD-DEG>.

Interpretation of tolerances is as follows:

- The angular dimension 110 is machined between 109.5 and 110.5.

- The dimension 2.5 is machined between 2.0 and 3.0.

- The Guide Hole dimension 10.000/10.015 is machined according to the specific tolerance on the drawing.

Company Logo

A Company logo is normally located in the Title block of the drawing. You can create your own Company logo or copy and paste an existing picture.

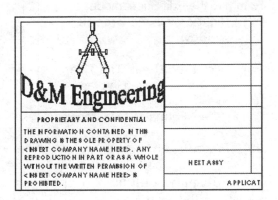

☀ Redeem the code on the inside cover of the book. Download the model files. The COMPASS.jpeg file is located in the LOGO folder. Copy all files from the LOGO folder to your hard drive. Insert the provided Company logo in the Edit Sheet Format mode.

☀ If you have your own Logo, skip the process of copying and applying the LOGO folder and file found below.

Activity: Company Logo

Insert a Company Logo.

91) Copy the **ENGDESIGN-W-SOLIDWORKS\LOGO** folder to your hard drive. If you have your own Logo, skip the following process of copying and applying the LOGO folder and file.

92) Right-click **Edit Sheet Format** in the Graphics window.

93) Click **Insert**, **Picture** from the Menu bar. The Open dialog box is displayed.

94) Double-click the **Logo.jpg** file. The Sketch Picture PropertyManager is displayed.

95) Drag the picture handles to size the **picture** to the left side of the Title block. Note: Text was added to the picture. Un-check the **Enable scale tool** box and the **Lock aspect ratio** box.

96) Click **OK** ✔ from the Sketch Picture PropertyManager.

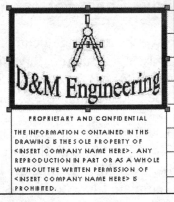

☀ Text can be added to create a custom logo. You can insert a picture or an object.

☀ View the provided videos on Drawing Fundamentals to enhance your experience in this project.

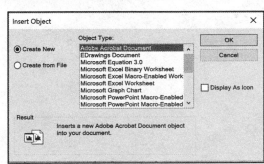

Return to the Edit Sheet mode.

97) Right-click in the **Graphics window**.

98) Click **Edit Sheet**. The Title block is displayed in black.

Fit the Sheet Format to the Graphics window.

99) Press the **f** key.

Draw1 displays Editing Sheet1 in the Status bar. The Title block is displayed in black when in Edit Sheet mode.

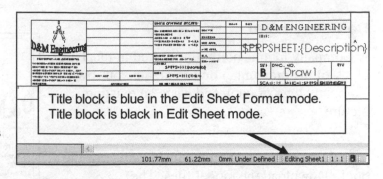

Title block is blue in the Edit Sheet Format mode.
Title block is black in Edit Sheet mode.

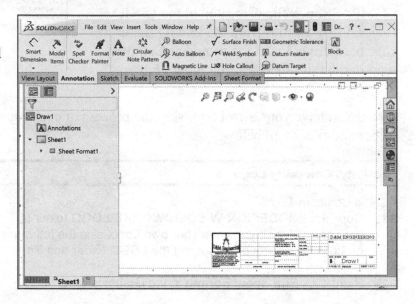

Save Sheet Format and Save As Drawing Template

Save the drawing document in the Graphics window in two forms: Sheet Format and Drawing Template. Save the Sheet Format as a custom Sheet Format named CUSTOM-B. Use the CUSTOM-B (ANSI) Sheet Format for the drawings in this project. The Sheet format file extension is .slddrt.

The Drawing Template can be displayed with or without the Sheet Format. Combine the Sheet Format with the Drawing Template to create a custom Drawing Template named B-ANSI-MM. Utilize the File, Save As option to save a Drawing Template. The Drawing Template file extension is .drwdot.

Select the Save as type option first, then select the Save in folder to avoid saving in default SOLIDWORKS installation directories.

The System Options, File Locations, Document Templates option is only valid for the current session of SOLIDWORKS in some *network locations*. Set the File Locations option in order to view the MY-TEMPLATES tab in the New Document dialog box.

Activity: Save Sheet Format. Utilize the Save As Option for the Drawing Template.

Save the Sheet Format.

100) Click **File**, **Save Sheet Format** from the Menu bar. The Save Sheet Format dialog box is displayed. The file extension for Sheet Format is .slddrt.

101) Select **ENGDESIGN-W-SOLIDWORKS\MY-TEMPLATES** for Save In File Folder.

102) Enter **CUSTOM-B** for File name.

103) Click **Save** from the Save Sheet Format dialog box.

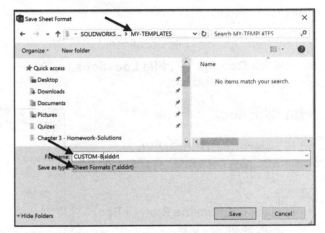

The book is designed to expose the new user to many tools, techniques and procedures. It may not always use the most direct tool or process.

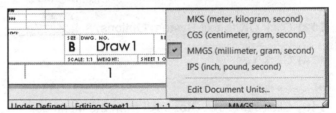

The Automatic Border tool lets you control every aspect of a sheet format's border, including zone layout and border size.

Save the Drawing Template.

104) Click **Save As** from the Menu bar.

105) Click **Drawing Templates (*.drwdot)** from the Save as type box.

106) Select **ENGDESIGN-W-SOLIDWORKS\MY-TEMPLATES** for Save In File Folder.

107) Enter **B-ANSI-MM** for File name.

108) Click **Save**.

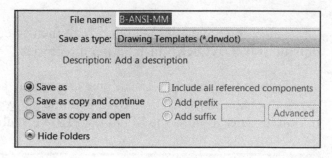

Set System Options - File Locations.

109) Click **Options** ⚙, **File Locations** from the Menu bar.

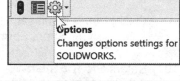

110) Click **Add**.

111) Select **ENGDESIGN-W-SOLIDWORKS\MY-TEMPLATES** for folder.

112) Click **OK** from the Browse For Folder dialog box.

113) Click **OK** to exit System Options.

114) Click **Yes**.

Close all files.

115) Click **Window**, **Close All** from the Menu bar.
SOLIDWORKS remains open, and no documents are displayed.

Utilize Drawing Template descriptive filenames that contain the size, dimension standard and units.

☀ Combine customized Drawing Templates and Sheet Formats to match your company's drawing standards. Save the empty Drawing Template and Sheet Format separately to reuse information.

Additional details on Drawing Templates, Sheet Format and Custom Properties are available in SOLIDWORKS Help Topics.

Keywords: Documents (templates, properties); Sheet Formats (new, new drawings, note text); Properties (drawing sheets); Customized Drawing Sheet Formats.

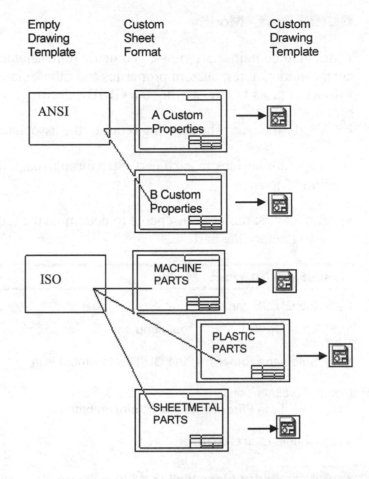

Review Drawing Templates

A custom Drawing Template was created from the default Drawing Template. You modified Sheet Properties and Document Properties to control the Sheet size, scale, annotations, dimensions and layers.

The Sheet Format contained Title block and Custom Property information. You inserted a Company Logo and modified the Title block.

The Save Sheet Format option was utilized to save the CUSTOM-B.SLDDRT Sheet Format. The File, Save As option was utilized to save the B-ANSI-MM.DRWDOT Template. The Sheet Format and Drawing Template were saved in the MY-TEMPLATES folder.

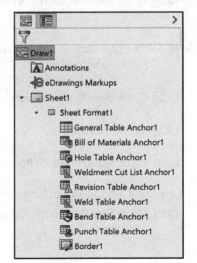

As an exercise, explore the sub-folders and their options under Sheet Format1.

GUIDE Part - Modify

A drawing contains part views, geometric dimensioning and tolerances, centerlines, center marks, notes, custom properties and other related information. Perform the following tasks before starting the GUIDE drawing:

- Verify the part. The drawing requires the associated part.

- View dimensions in each part. Step through each feature of the part and review all dimensions.

- Review the dimension scheme to determine the required dimensions and notes to manufacture the part.

Activity: GUIDE Part - Modify

Open the GUIDE part.

116) Click **Open** 📂 from the Menu bar.

117) Select the **folder** that the GUIDE document is in.

Modify the dimensions.
118) Select the **Filter Parts (*prt; *sldprt)** button.

119) Double-click **GUIDE**.

120) Click **Hidden Lines Visible** 📦 from the Heads-up View toolbar.

121) Double-click **Base-Extrude** from the GUIDE FeatureManager.

122) Click the **80** dimension.

123) Enter **100**mm.

124) Click **inside** the Graphics window.

Display a Shaded With Edges view. Save the GUIDE part.
125) Click **Shaded With Edges** 🔷 from the Heads-up View toolbar.

126) Click **Save** 💾.

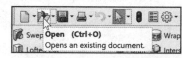

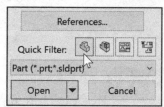

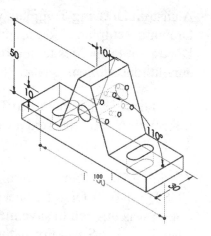

💡 Review part history with the Rollback bar to understand how the part was created. Position the Rollback bar at the top of the FeatureManager. Drag the Rollback bar below each feature. When working between features, right-click in the FeatureManager. Select Roll to Previous\Roll to End.

GUIDE Part - Drawing

The GUIDE drawing consists of multiple views, dimensions and annotations. The GUIDE part was designed for symmetry. Add or redefine dimensions in the drawing to adhere to a Drawing Standard. Add dimensions and notes to the drawing in order to correctly manufacture the part.

Address the dimensions for three features:

- The right Slot Cut is not dimensioned to an ASME Y14 standard.

- No dimensions exist for the left Mirror Slot Cut.

- The Guide Hole and Linear Pattern of Tapped Holes require notes.

The GUIDE part remains open. Create a new GUIDE drawing. Utilize the B-ANSI-MM Drawing Template. Utilize the Model View 🔯 tool to insert the Front view into Sheet1. Utilize the Auto-start Projected View option to project the Top, Right and Isometric views from the Front view. Note: You can also use the View Palette in the Task Pane.

Activity: GUIDE Part - Drawing

Create the GUIDE Drawing.

127) Click **New** 🗋 from the Menu bar.

128) Double-click **B-ANSI-MM** from the MY-TEMPLATES tab. The Model View PropertyManager is displayed.

129) Click **Cancel** ✖ from the Model View PropertyManager. The Draw# FeatureManager is displayed.

Review the Draw# FeatureManager.
130) Expand Sheet1 from the FeatureManager.

131) Expand Sheet Format1 from the FeatureManager. View your options.

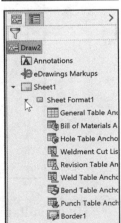

🔆 The current drawing name is Draw2 if the second new drawing is created in the same session of SOLIDWORKS. The current sheet name is Sheet1. Sheet1 is the current Sheet.

Insert four Drawing Views.

132) Click **Model View**  from the View Layout tab in the CommandManager. The Model View PropertyManager is displayed.

133) Double-click **GUIDE** from the Part/Assembly to Insert box.

Insert four views.
134) Check the **Create multiple views** box.

135) Click **Front, Top** and **Right** from the Orientation box. Note: All four views are selected. *Isometric should be selected by default.

136) Click **OK** ✔ from the Model View PropertyManager. Four views are displayed on Sheet1.

137) Click inside the **Isometric view boundary**. Drawing View4 PropertyManager is displayed.

138) Click **Shaded With Edges** 🔲 from the Display Style box.

139) Click **OK** ✔ from the Drawing View4 PropertyManager. If required, hide any annotations, origins or dimensions as illustrated.

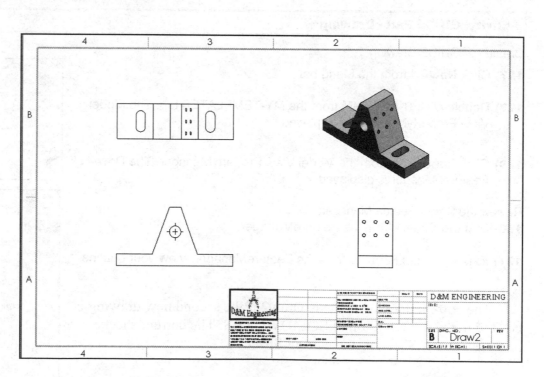

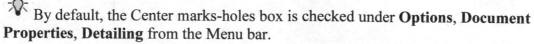

💡 By default, the Center marks-holes box is checked under **Options, Document Properties, Detailing** from the Menu bar.

Save the GUIDE drawing. Enter name.
140) Click **Save As** from the Menu bar.

141) Select **PROJECTS** for Save in folder. Enter **GUIDE** for file name. Drawing is the default Save as type.

142) Click **Save**. Note: The correct display modes need to be selected, dimensions to be added, along with Center Marks, Centerlines, additional views, Custom Properties, etc.

The DWG. NO. box in the Title block displays the part File name, GUIDE. The TITLE: box in the Title block displays GUIDE SUPPORT.

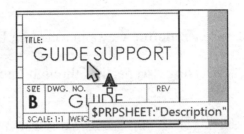

Predefined text in the CUSTOM-B Sheet Format links the Properties: $PRPSHEET:"Description" and $PRP:"SW-FileName." The Properties were defined in the GUIDE part utilizing File, Save As. Properties in the Title block are passed from the part to the drawing.

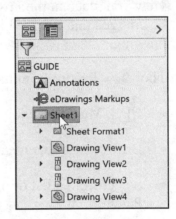

Always confirm the File name and Save in folder. Projects deal with multiple File names and folders. Select Save as type from the drop down list. Do not enter the extension. The file extension is entered automatically.

Each drawing has a unique file name. Drawing file names end with a .SLDDRW suffix. Part file names end with a .SLDPRT suffix. A drawing or part file can have the same prefix. A drawing or part file **can't have the same suffix**.

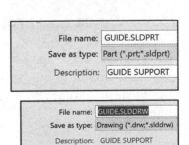

Example: Drawing file name GUIDE.SLDDRW. Part file name GUIDE.SLDPRT. The current file name is GUIDE.SLDDRW.

The GUIDE drawing contains three Principle views (Standard Orthographic views): Front, Top, Right and an Isometric view. You created the views with the Model View option.

Drawing views can be are inserted as follows:

- Utilize the **Model View** tool from the View Layout tab in the CommandManager.

- **Drag** and **drop** an active part view from the View Palette located in the Task Pane as illustrated. With an open part, drag and drop the selected view into the active drawing sheet.

- **Drag** and **drop** a part into the drawing to create three Standard Views.

- **Predefine** views in a custom Drawing Template.

- Drag a **hyperlink** through Internet Explorer.

The Top view and Right view are projected off the view you place in the Front view location. Any view can be dragged and dropped into the Front view location of a drawing.

The View Palette from the Task Pane populates when you:

- Click Make Drawing from Part/Assembly.

- Browse to a document from the View Palette.

- Select from a list of open documents in the View Palette.

Move Views and Properties of the Sheet

The GUIDE drawing contains four views. Reposition the view on a drawing. Provide approximately 1in. - 2in. (25mm - 50mm) between each view for dimension placement.

Move Views on Sheet1 to create space for additional Drawing View placement. The mouse pointer provides feedback in both the Drawing Sheet and Drawing View modes. The mouse pointer displays the Drawing Sheet icon when the Sheet properties and commands are executed.

The mouse pointer displays the Drawing View  icon when the View properties and commands are executed.

☀ View the mouse pointer for feedback to select Sheet, View, and Component and Edge properties in the Drawing.

Sheet Properties

- Sheet Properties display properties of the selected sheet. Right-click in the sheet boundary to view the available commands.

View Properties

- View Properties display properties of the selected view. Right-click inside the view boundary. Modify the View Properties in the Display Style box or the View Toolbar.

Component Properties

- Component Properties display properties of the selected component. Right-click on the face of the component. View the available options.

Edge Properties

- Edge Properties display properties of the selected geometry. Right-click on an edge inside the view boundary. View the available options.

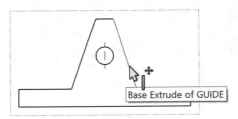

Base Extrude of GUIDE

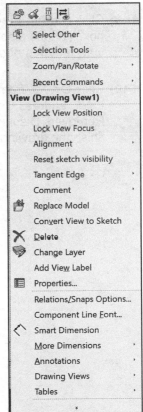

Activity: Move Views and Properties of the Sheet

Modify and move the front view.

143) Click inside the **Drawing View1** (Front) view boundary.

The mouse pointer displays the Drawing View 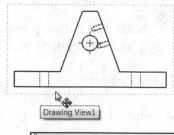 icon.
The view boundary is displayed in blue.

144) Click **Hidden Lines Visible** ⬡ from the Display Style box.

145) Position the **mouse pointer** on the edge of the Front view

until the Drawing Move View ✛ icon is displayed.

146) Click and drag **Drawing View1** in an upward vertical
direction. The Top and Right view move aligned to Drawing
View1 (Front).

Modify the top view.

147) Click inside the **Top view boundary** (Drawing View3). The
Drawing View3 PropertyManager is displayed.

148) Click **Hidden Lines Removed** ⬜ from the Display Style
box. Various display styles provide the ability to select and
view features of a part.

149) Click **OK** ✔ from the Drawing View3 PropertyManager.
Later, address Centerlines, Center Marks, Tangent Edges
Removed, Display styles, Customer properties, etc. to
finish the drawing.

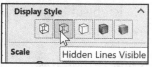

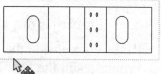

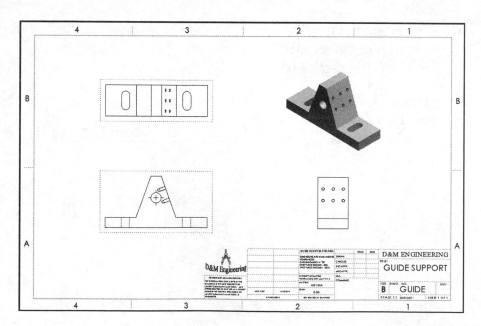

💡 Select the dashed view boundary to move a view in the drawing.

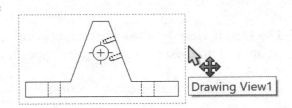

Auxiliary View, Section View, and Detail View

The GUIDE drawing requires additional views to document the part. Insert an Auxiliary view, Section view and Detail view from the View Layout tab in the CommandManager. Review the following view terminology before you begin the next activity.

Auxiliary View:

The Auxiliary view ⌖ drawing tool provides the ability to display a plane parallel to an angled plane with true dimensions.

A primary Auxiliary view is hinged to one of the six Principle Orthographic views. Create a Primary Auxiliary view that references the angled edge in the Front view.

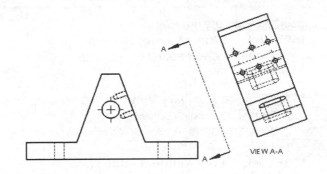

Section View:

The Section view ⇄ drawing tool provides the ability to display the interior features of a part. Define a cutting plane with a sketched line in a view perpendicular to the Section view.

Use the Section view sketch mode in conjunction with the Section tool user interface to create a Section view in the Top view.

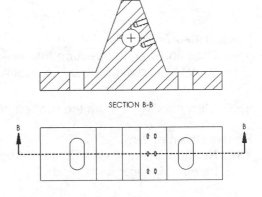

Detail View:

The Detail view 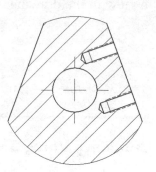 drawing tool provides the ability to enlarge an area of an existing view. Specify location, shape and scale.

Create a Detail view from a Section view with a 3:2 scale.

🔆 The book is designed to expose the new user to many tools, techniques and procedures. It does not always use the most direct tool or process.

🔆 You can add multiple breaks to models using the Model Break View tool. The model breaks are saved as configurations. Model break views are helpful when you need to shorten components especially for technical and marketing purposes. Additionally, the Model Break View tool lets you display breaks on drawing views per ASME Y14.3.

DETAIL C
SCALE 3 : 2

Activity: Auxiliary Drawing View

Insert an Auxiliary drawing view.
150) Click the **View Layout** tab from the CommandManager.

151) Click the **Auxiliary View** drawing tool. The Auxiliary View PropertyManager is displayed.

152) Click the **right angled edge** of the GUIDE in the Front view as illustrated.

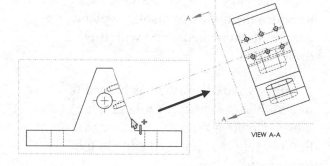

VIEW A-A

153) Click a **position** to the right and up from the Front view as illustrated.

Position the Auxiliary View.
154) Click and drag the **section line A-A midpoint** toward Drawing View1. The default label A is displayed in the Arrow box.

155) Click the **OK** from the PropertyManager.

Rename the new view.
156) **Rename** the new view (Drawing View#) to Auxiliary in the FeatureManager as illustrated.

Activity: Section Drawing View

Insert a Section drawing view.

157) Click the **Section View** ⇄ drawing tool. The Section View PropertyManager is displayed. The Section tab is selected by default.

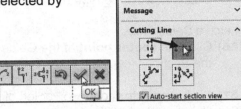

158) Click the **Horizontal** Cutting Line button as illustrated.

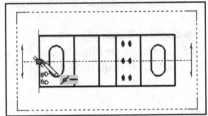

159) Locate the **midpoint of the left vertical line** as illustrated.

160) Click the **midpoint**. If needed, click **OK** ✓ from the Pop-up menu.

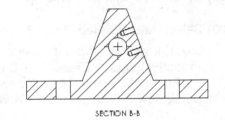

Position the Section drawing View.
161) Click a **position** above the Top view. The section arrows point downward.

162) If needed, check **Flip direction** from the Section Line box. The section arrows point upward. If required, enter **B** for Section View Name in the Label box.

163) Click **OK** ✓ from the Section View B-B PropertyManager. Section View B-B is displayed in the FeatureManager. If required, hide any annotations.

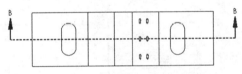

SECTION B-B

Save the drawing.
164) Click **Save** 💾. Note: If needed, insert a CenterMark in the Section view. Utilize the CenterMark tool under the Annotations tab.

🔅 The material in the GUIDE part determines the hatch pattern in the GUIDE drawing.

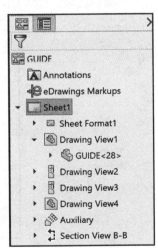

🔅 Use the Pack and Go tool to save and gather all related files for a model design (parts, assemblies, drawings, references, design tables, Design Binder content, decals, appearances, and scenes and SOLIDWORKS Simulation results) into a single folder or zip file.

Activity: Detail Drawing View

Insert a Detail drawing view.

165) Click the **Detail View** 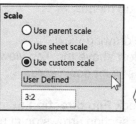 drawing tool. The Detail View PropertyManager is displayed.

Sketch a Detail circle.

166) Click the **center point** of the Guide Hole in the Section view.

167) Click a **position** to the lower left of the Guide Hole to complete the circle.

Position the Detail View.

168) Click a **position** to the right of the Section View. If required, enter **C** for Detail View Name in the Label box.

Modify the drawing view scale.

169) Check **Use custom scale** option.

170) Select **User Defined**. Enter **3:2** in the Scale box. Click **OK** ✔ from the Detail View C PropertyManager. Detail View C is displayed in the FeatureManager. Note: If needed, insert a CenterMark in the Detail view. Utilize the CenterMark tool under the Annotations tab.

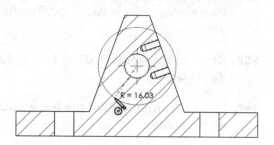

SECTION B-B

Scale
○ Use parent scale
○ Use sheet scale
● Use custom scale

User Defined

3:2

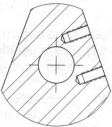

DETAIL C
SCALE 3 : 2

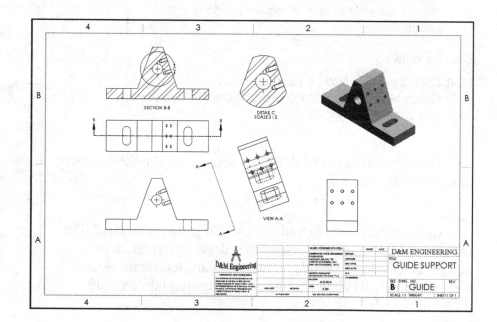

Save the GUIDE drawing.
171) Click **Save** 🖫.

Partial Auxiliary Drawing View - Crop Drawing View

Create a Partial Auxiliary view from the Full Auxiliary view. Sketch a closed profile in the active Auxiliary view. Create the Profile with a closed Spline. Create a Partial Auxiliary view.

Crop the view. The 6mm dimension references the centerline from the Guide Hole. For Quality Assurance and Inspection of the GUIDE part, add a dimension that references the Temporary Axis of the Guide Hole. Sketch a centerline collinear with the Temporary Axis.

Activity: Partial Auxiliary Drawing View - Crop Drawing View

Select the view.
172) Click **Zoom to Area** 🔍 from the Heads-up View toolbar.

173) Zoom in on the Auxiliary view.

174) Click **Zoom to Area** 🔍 to deactivate.

175) Click inside the **Auxiliary view boundary**. The Auxiliary PropertyManager is displayed.

Display Hidden Lines Removed.
176) Click **Hidden Lines Removed** ⬦.

Sketch a closed Spline profile.
177) Click the **Sketch** tab from the CommandManager.

178) Click the **Spline** ∿ Sketch tool.

179) Click **five or more positions** clockwise to create the closed Spline as illustrated. The first point is Coincident with the last point. The Sketch profile is closed.

Insert a Partial Auxiliary Drawing View.
180) Click the **Crop View** 🗐 drawing tool from the View Layout tab in the CommandManager.

181) Click **OK** ✔ from the Spline PropertyManager. The Crop View is displayed on Sheet1.

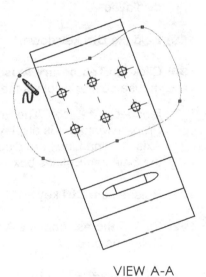

VIEW A-A

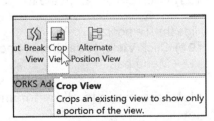

Display the Temporary Axes and insert a sketched centerline.

182) Click inside the **Auxiliary view boundary**. The Auxiliary PropertyManager is displayed.

183) Click **View**, **Hide/Show**, check **Temporary Axes** from the Menu bar. The Temporary Axis for the Guide Hole is displayed.

184) Click **Sketch** tab from the CommandManager.

185) Click the **Centerline** Sketch tool. The Insert Line PropertyManager is displayed.

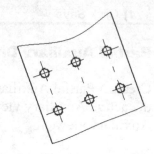

VIEW A-A

186) Sketch a **centerline** parallel above the Temporary axis. The centerline extends approximately 5mm to the left and right of the profile lines. If needed insert a Parallel relation to the temporary axis.

187) Right-click **Select** to deselect the Centerline Sketch tool.

Add a Collinear relation between the centerline and temporary axis.

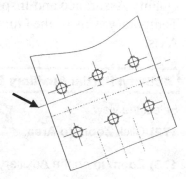

188) Click the **centerline**. The Line Properties PropertyManager is displayed.

189) Hold the **Ctrl** key down.

190) Click the **Temporary Axis**. The mouse pointer displays the Axis feedback icon. The Properties PropertyManager is displayed. Axis<1> and Line1 are displayed in the Selected Entities box.

191) Release the **Ctrl** key.

192) Click **Collinear** from the Add Relations box.

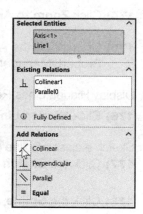

193) Click **OK** from the Properties PropertyManager.

Hide the Temporary Axis.

194) Click **View**, **Hide/Show**, uncheck **Temporary Axes** from the Menu bar.

Move the views to allow for ample spacing for dimensions and notes.

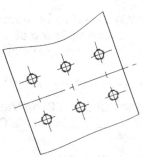

VIEW A-A

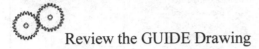 Additional information on creating a New Drawing, Model View, Move View, Auxiliary View, Section View, and Detail View are located in the SOLIDWORKS Help Topics section. Keywords: New (drawing document), Auxiliary View, Detail View, Section View and Crop View.

Review the GUIDE Drawing

You created a new drawing, GUIDE with the B-ANSI-MM Drawing Template. The GUIDE drawing utilized the GUIDE part in the Model View PropertyManager. The Model View PropertyManager allowed new views to be inserted with a View Orientation. You selected Front, Top, Right and Isometric to position the GUIDE views.

Additional views were required to fully detail the GUIDE. You inserted the Auxiliary Section, Detail, Partial Auxiliary and Crop view. You moved the views by dragging the view boundary. The next step is to insert the dimensions and annotations to detail the GUIDE drawing.

Display Modes and Performance

Display modes for a Drawing view are similar to a part. When applying Shaded With Edges, select either Tangent edges removed or Tangent edges As phantom from the System Options section.

Mechanical details require either the Hidden Lines Visible mode or the Hidden Lines Removed display mode. Select Shaded/Hidden Lines Removed to display Auxiliary views to avoid confusion.

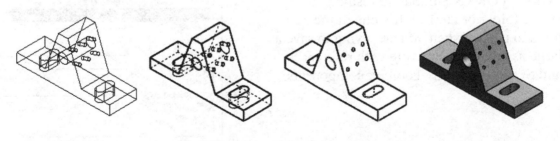

Wireframe   Hidden Line Visible   Hidden Line Removed   Shaded With Edges

Tangent Edges Visible provides clarity for feature edges. To address the ASME Y14.5 standard, use Tangent Edges With Font (Phantom lines) or Tangent Edges Removed. Right-click in the view boundary to access the Tangent Edge options.

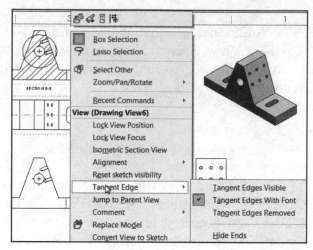

Drawing views can be displayed in High quality and Draft quality. In High quality, all model information is loaded into memory. By default, drawing views are displayed in High quality.

In Draft quality, only minimum model information is loaded into memory. Utilize Draft quality for large assemblies to increase performance.

🔅 Utilize **Options**, **System Options**, **Drawings**, **Display Style** to control the quality of a view.

By default, SOLIDWORKS will populate Section and Detail views before other views on your drawing.

Use the Pack and Go tool to save and gather all related files for a model design (parts, assemblies, drawings, references, design tables, Design Binder content, decals, appearances, and scenes, and SOLIDWORKS Simulation results) into a single folder or zip file. It's one of the best tools to utilize when you are trying to save a large assembly or drawing with references and SOLIDWORKS Toolbox components.

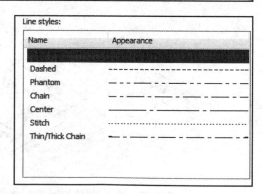

Detail Drawing

The design intent of this project is to work with dimensions inserted from parts and to incorporate them into the drawings. Explore methods to move, hide and recreate dimensions to adhere to a drawing standard.

There are other solutions to the dimensioning schemes illustrated in this project. Detail drawings require dimensions, annotations, tolerance, materials, Engineering Change Orders, authorization, etc. to release the part to manufacturing and other notes prior to production.

Review a hypothetical "worse case" drawing situation. You just inserted dimensions from a part into a drawing. The dimensions, extension lines and arrows are not in the correct locations. How can you address the position of these details? Answer: Dimension to an ASME Y14.5M standard.

No.	Situation:
1	Extension line crosses dimension line. Dimensions not evenly spaced.
2	Largest dimension placed closest to profile.
3	Leader lines overlapping.
4	Extension line crossing arrowhead.
5	Arrow gap too large.
6	Dimension pointing to feature in another view. Missing dimension – inserted into Detail view (not shown).
7	Dimension text over centerline, too close to profile.
8	Dimension from other view – leader line too long.
9	Dimension inside section lines.
10	No visible gap.
11	Arrows overlapping text.
12	Incorrect decimal display with whole number (millimeter), no specified tolerance.

Worst Case Drawing Situation

The ASME Y14.5 2009 standard defines an engineering drawing standard. Review the twelve changes made to the drawing to meet the standard.

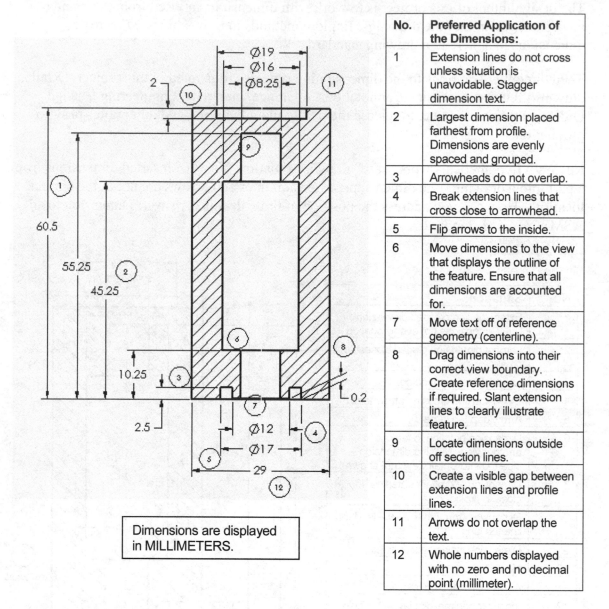

Dimensions are displayed in MILLIMETERS.

No.	Preferred Application of the Dimensions:
1	Extension lines do not cross unless situation is unavoidable. Stagger dimension text.
2	Largest dimension placed farthest from profile. Dimensions are evenly spaced and grouped.
3	Arrowheads do not overlap.
4	Break extension lines that cross close to arrowhead.
5	Flip arrows to the inside.
6	Move dimensions to the view that displays the outline of the feature. Ensure that all dimensions are accounted for.
7	Move text off of reference geometry (centerline).
8	Drag dimensions into their correct view boundary. Create reference dimensions if required. Slant extension lines to clearly illustrate feature.
9	Locate dimensions outside off section lines.
10	Create a visible gap between extension lines and profile lines.
11	Arrows do not overlap the text.
12	Whole numbers displayed with no zero and no decimal point (millimeter).

Apply these dimension practices to the GUIDE drawing. Manufacturing utilizes detailed drawings. A mistake on a drawing can cost your company a substantial loss in revenue. The mistake could result in a customer liability lawsuit.

As the designer, dimension and annotate your parts clearly to avoid common problems and mistakes.

Insert Dimensions from the part.
Dimensions you created for each part feature
are inserted into the drawing.

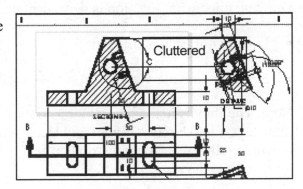

Select the first dimensions to display for the
Front view. Do not select the Import Items
into all Views option for complex drawings.
Dimension text is cluttered and difficult to
locate.

Follow a systematic, "one view at a time"
approach for complex drawings. Insert part
feature dimensions onto the Dim Layer for
this project.

Activity: Detail Drawing - Insert Model Items

Set the Dimension Layer.
195) Right-click in the drawing sheet.

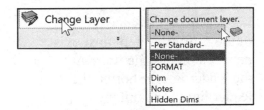

196) Click **Change Layer**. The Change Layer dialog
box is displayed.

197) Select **Dim** from the drop-down menu.

Insert dimensions into Drawing View1 (Front).
198) Click inside the **Drawing View1** boundary. The Drawing View1
PropertyManager is displayed.

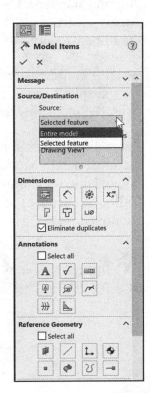

199) Click the **Model Items** tool from the Annotation tab in the
CommandManager. The Model Items PropertyManager is
displayed.

200) Select **Entire model** from the Source box. Drawing View1 is
displayed in the Destination box. At this time, do not click the Hole
callout button to import the Hole Wizard information into the
drawing. Accept the default settings as illustrated.

201) Click **OK** from the Model Items PropertyManager. Dimensions
are displayed in the Front view.

In SOLIDWORKS, inserted dimensions in the drawing are
displayed in gray. Imported dimensions from the part are displayed
in black.

Drawing dimension location is dependent on:

- Feature dimension creation.

- Selected drawing views.

Note: The Import items into all views option, first inserts dimensions into Section Views and Detail Views. The remaining dimensions are distributed among the visible views on the drawing.

Move Dimensions in the Same View

Move dimensions within the same view. Use the mouse pointer to drag dimensions and leader lines to a new location.

Leader lines reference the size of the profile. A gap must exist between the profile lines and the leader lines. Shorten the leader lines to maintain a drawing standard. Use the blue Arrow control buttons to flip the dimension arrows.

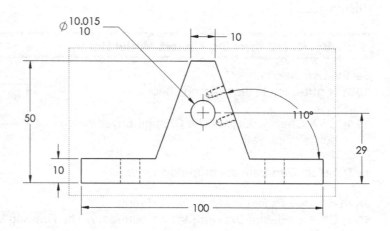

Insert part dimensions into the Top view. The Top view displays crowded dimensions. Move the overall dimensions. Move the Slot Cut dimensions.

Place dimensions in the view where they display the most detail. Move dimensions to the Auxiliary View. Hide the diameter dimensions and add Hole Callouts. Display the view with Hidden Lines Removed.

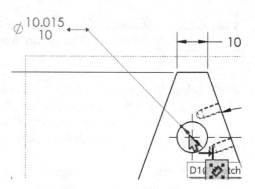

Illustrations may vary depending on your SOLIDWORKS release version.

Activity: Detail Drawing - Move Dimensions

Move the linear dimensions in Drawing View1.

202) Zoom to area 🔍 on Drawing View1.

203) Click and drag the vertical dimension text **10**, **29**, and **50** to the right as illustrated.

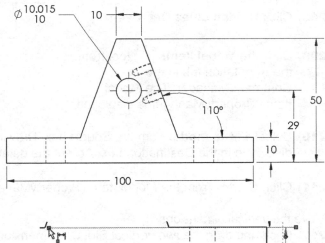

Create a gap between the extension lines and the profile lines for the 10mm vertical dimension.

204) Click the vertical dimension text **10**. The vertical dimension text, extension lines and the profile lines are displayed in blue.

205) Click and drag the **square blue endpoints** approximately 10mms from the right vertex as illustrated. A gap is created between the extension line and the profile.

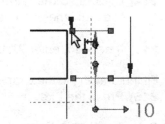

A gap exists between the profile line and the leader lines. Drag the blue endpoints to a vertex, to create a gap.

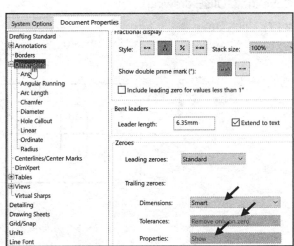

💡 The smallest linear dimension should be placed closest to the profile.

💡 Note the default settings for a drawing document under Dimensions. Trailing zeroes Dimensions default: **Smart**. Trailing zeroes Tolerances default: **Remove only on zero**. Trailing zeroes Properties default: **Show**.

Fit the drawing to the Graphics window.
206) Press the **f** key.

Insert dimensions into Drawing View3.
207) Click inside the **Drawing View3** boundary. The Drawing View3 PropertyManager is displayed.

208) Click **Hidden Lines Removed**.

209) Click the **Model Items** 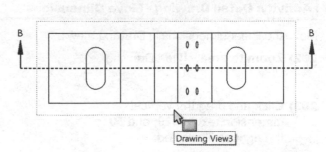 tool from the Annotation tab in the CommandManager. The Model Items PropertyManager is displayed.

210) Select **Entire model** from the Source box. Drawing View3 is displayed in the Destination box. Accept the default settings.

211) Click **OK** ✔ from the Model Items PropertyManager.

Move the vertical dimensions.
212) Click and drag the two vertical Slot Cut dimensions, **10** to the right of the Section arrow as illustrated.

213) Flip the **arrows** to the inside.

214) Click the **dimension text** and drag the text outside the leader lines. Hide all other dimensions and annotations as illustrated.

215) If needed, enter **2X** for the Radius as illustrated.

Fit the drawing to the Graphics window.
216) Press the **f** key.

Save the GUIDE drawing.
217) Click **Save** 💾.

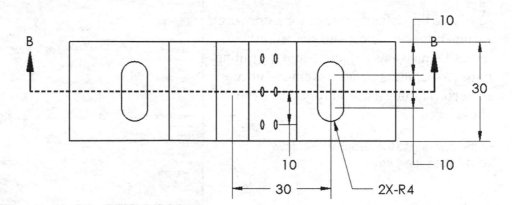

Insert dimensions into the Auxiliary View.

218) Click **inside** the Auxiliary view boundary.

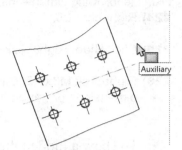

219) Click the **Model Items** Drawing tool. The Model Items PropertyManager is displayed.

220) Select **Entire Model** from the Source box. Auxiliary is displayed in the Destination box. View your options from the Dimensions box.

221) Click the **Hole Wizard Locations** button from the Dimensions box. Note: Do not click the Hole Callout button at this time. You will manually create the Hole Wizard Callout in the drawing.

222) Click **OK** from the Model Items PropertyManager.

223) **Move** the dimensions off the view as illustrated. If needed zoom in on the dimensions to move them.

The dimensions for the Linear Pattern of Holes are determined from the initial Hole Wizard position dimensions and the Linear Pattern dimensions. Your dimensioning standard requires the distance between the holes in a pattern.

Do not over dimension. In the next steps, hide the existing dimensions and add a new dimension.

In the next section, if needed, click **Options**, **Document Properties**, **Dimensions** from the Menu bar. Uncheck the **Add parentheses by default** box. Click **OK**.

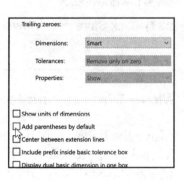

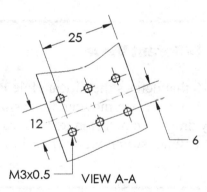

VIEW A-A

Hide the following dimensions.
224) Right-click **25**.

225) Click **Hide**. Right-click **6**. Click **Hide**.

226) Right-click **M3x0.5**. Click **Hide**. Note: if required, hide any other
dimensions or annotations. Do not hide the 12 dimension.

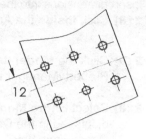

 To show a hidden dimension, click **View**, **Hide/Show**,
Annotations from the Menu bar.

VIEW A-A

Add a dimension.

227) Click **Smart Dimension** from the Annotation toolbar. The
Dimension PropertyManager is displayed.

228) Click **Smart dimensioning** in the Dimension Assist Tools box. Click
the **center point** of the bottom left hole. Click the **center point**
of the bottom right hole.

229) Click a **position** below the profile as illustrated.

230) Check **OK** from the Dimension PropertyManager.

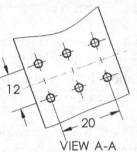

VIEW A-A

Save the drawing.
231) Click **Save**. View the results.

Move Dimensions to a Different View

Move the linear dimension 10 that defines the Linear Hole Pattern feature from Drawing
View3 (Top) to the Auxiliary view. When moving dimensions from one view to another,
utilize the Shift key and only drag the dimension text. Release the dimension text inside
the view boundary. The text will not switch views if positioned outside the view
boundary.

Activity: Move Dimensions to a Different View

Move dimensions from the Top view to the Auxiliary view.
232) Press the **z key** approximately 4 times to view the dimensions in the Top view.

233) Hold the **Shift** key down. Click and drag the vertical dimension **10** between the 2 holes from
the Top view to the Auxiliary view.

234) Release the **mouse button** and the **Shift** key when the mouse pointer is inside the
Auxiliary view boundary.

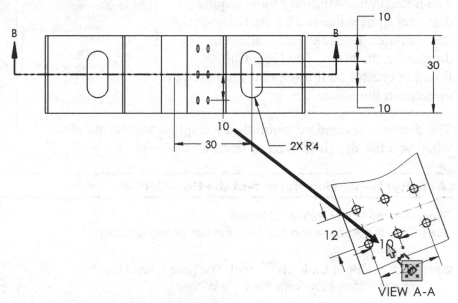

235) Click and drag the **dimensions** off the Auxiliary view.

236) Click and drag the **VIEW A-A** text off the view boundary. View the results.

Save the drawing.

237) Click **Save** 💾 . View the results.

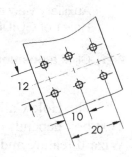

Dimension Holes and the Hole Callout

Simple holes and other circular geometry are dimensioned in various ways: Diameter, Radius and Linear (between two straight lines).

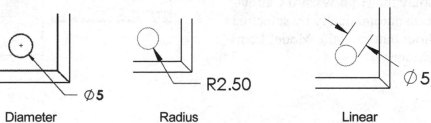

Diameter   Radius   Linear

The holes in the Auxiliary view require a diameter dimension and a note to represent the six holes. Use the Hole Callout to dimension the holes. The Hole Callout function creates additional notes required to dimension the holes.

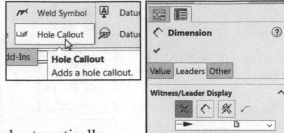

The dimension standard symbols are displayed automatically when you use the Hole Wizard feature.

Activity: Dimension Holes and the Hole Callout

Dimension the Linear Pattern of Holes.
238) Click the **Annotation** tab from the CommandManager.

239) Click the **Hole Callout** ⊔ø tool. The Hole Callout tool inserts information from the Hole Wizard.

240) Click the **circumference** of the lower left circle in the Auxiliary view as illustrated. The tool tip M3x0.5 Tapped Hole1 of GUIDE is displayed.

241) Click a **position** to the bottom left of the Auxiliary view.

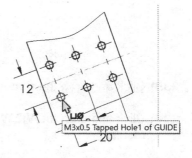

💡 The Hole Callout text displayed in the Dimension Text box depends on the options utilized in the Hole Wizard feature and the Linear Pattern feature.

Remove trailing zeroes for ASME Y14 millimeter display.
242) Select **.1** from the Primary Unit Tolerance/Precision box.

243) Click **OK** ✔ from the Dimension PropertyManager. The Hole Callout is deactivated.

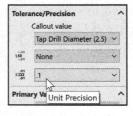

VIEW A-A

💡 Display the Hole Wizard Callout information automatically by selecting the Hole callout button in the Model Items PropertyManager.

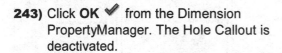

Know inch/mm decimal display. The ASME Y14.5-2009 standard states:

- For millimeter dimensions <1, display the leading zero. Remove trailing zeroes.

- For inch dimensions <1, delete the leading zero. The dimension is displayed with the same number of decimal places as its tolerance.

Symbols are located on the bottom of the Dimension Text dialog box. The current text is displayed in the text box. Example:

- <NUM_INST>: Number of Instances in a Pattern.

- <MOD-DIAM>: Diameter symbol $\emptyset$.

- <HOLE-DEPTH>: Deep symbol ↓.

- <HOLE-SPOT>: Counterbore symbol ⨆.

- <DIM>: Dimension value 3.

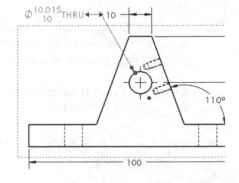

Fit the Drawing to the Graphics window.
244) Press the **f** key.

If needed, insert dimension text.
245) Click inside the **Drawing View1** boundary. The Drawing View1 PropertyManager is displayed.

246) Click the Guide Hole dimension ⌀**10.015/10** in Drawing View1.

247) Enter text **THRU** in the Dimension box. THRU is displayed on the drawing in blue.

248) Click **OK** ✔ from the Dimension PropertyManager.

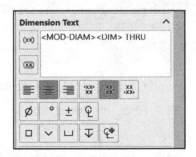

Save the drawing.
249) Click **Save** 💾.

 Access Notes, Hole Callouts and other Dimension Annotations through the Annotations toolbar menu. Access Annotations with the right mouse button, and click Annotations.

Center Marks and Centerlines

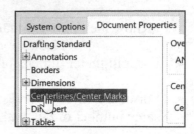

Hole centerlines are composed of alternating long and short dash lines. Centerlines indicate symmetry. Centerlines also identify the center of a circle, axes or cylindrical geometry.

Center Marks represents two perpendicular intersecting centerlines. The default Size and display of the Center Mark is set in Options, Document Properties, Dimensions, Centerlines/Center Marks as illustrated.

Center Marks are inserted on view creation by default. Modify the position and size of the Center Marks. The following three steps illustrate how to create a new Center Mark.

Activity: Center Marks and Centerlines

Insert a Center mark into the Front view. If a Center mark exists, skip the next few steps.

250) Click inside the **Drawing View1** boundary.

251) Click the **Center Mark** ⊕ tool from the Annotation tab in the CommandManager. The Center Mark PropertyManager is displayed.

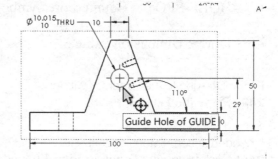

252) Click the **Guide Hole circumference**. The Center mark is inserted.

253) Click **OK** ✔ from the Center Mark PropertyManager.

Dimension standards require a gap between the Center Mark and the end points of the leader line. Currently, the leader line overlaps the Center Mark.

Insert Centerlines into the drawing views.

254) Click the **Centerline** ⊞ tool from the Annotation tab in the CommandManager. The Centerline PropertyManager is displayed.

255) Check the **Select View** box in the Auto Insert dialog box.

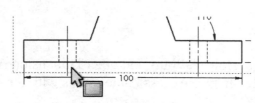

256) Click inside the Front view, **Drawing View1** boundary.

257) Click inside the **Detail view** boundary. Centerlines are displayed.

258) Click inside the **Section view** boundary. Centerlines are displayed.

259) Click **OK** ✔ from the Centerline PropertyManager.

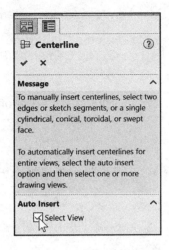

Save the drawing.
260) Click **Save** 💾.

Activate the Top view and insert Center marks.
261) Click inside the **Drawing View3** boundary.

262) Click the **Center Mark** ⊕ tool from the Annotation tab in the CommandManager. The Center Mark PropertyManager is displayed.

263) Uncheck the **Slot center mark** box.

264) Click the **four arcs**. The center marks are displayed.

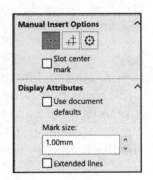

265) Uncheck the **Use document defaults** box.

266) Enter **1**mm for Mark size.

267) Un-check the **Extended lines** box. Accept the default settings.

268) Click **OK** ✔ from the Center Mark PropertyManager. View the results.

269) Click **Save** 💾. View the results.

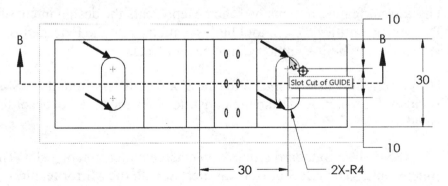

Insert an Annotation Centerline.

270) Click the **Centerline** 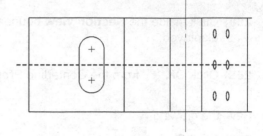 tool from the Annotation tab. The Centerline PropertyManager is displayed.

271) Click the **left edge** of the GUIDE hole as illustrated.

272) Click the **right edge** of the GUIDE hole. The centerline is displayed.

273) Click **OK** ✔ from the Centerline PropertyManager.

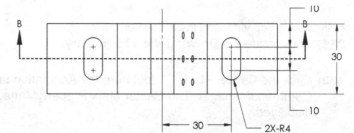

Save the drawing.

274) Click **Save** 💾. View the results.

💡 Click the dimension Palette rollover 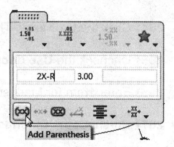 button to display the dimension palette. Use the dimension palette in the Graphics window to save mouse travel to the Dimension PropertyManager. Click on a dimension in a drawing view, and modify it directly from the dimension palette.

Modify the Dimension Scheme

The current feature dimension scheme represents the design intent of the GUIDE part. The Mirror Entities Sketch tool built symmetry into the Extruded Base sketch. The Mirror feature built symmetry into the Slot Cuts.

The current dimension scheme for the Slot Cut differs from the ASME 14.5M Dimension Standard for a slot. Redefine the dimensions for the Slot Cut according to the ASME 14.5 Standard.

The ASME 14.5 Standard requires an outside dimension of a slot. The radius value is not dimensioned. The left Slot Cut was created with the Mirror feature.

Create a centerline and dimension to complete the detailing of the Slot Cut. Sketch the vertical dimension, 10. The default arc conditions are measured from arc center point to arc center point. The dimension extension lines are tangent to the top arc and bottom arc.

Activity: Modify the Dimension Scheme

Modify the Slot Cut dimension scheme.

275) Right-click **Change Layer** in the Graphics window. The Change document Layers dialog box is displayed.

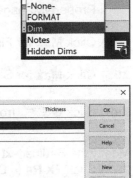

276) Click the **Layer Properties** icon from the Change document layer dialog box. The Layers dialog box is displayed.

Set the Show/Hide icon to modify the status of the Layers.

277) Click the **Dims** Layer **Show**. Dims is the current layer.

278) Click the **Notes** Layer Show.

279) Click the **Hidden Dims** Layer Hide.

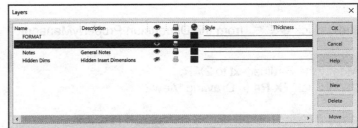

280) Click the **Format** Layer Show.

281) Click **OK** from the Layers dialog box.

282) Click **Zoom to Area** 🔍 from the Heads-up View toolbar.

283) **Zoom in** on the right slot of Drawing View3 as illustrated.

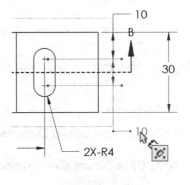

Hide the dimension between the two arc center points.

284) Click the **10** dimension on the right bottom side of Drawing View3.

285) Click the **Others** tab in the Dimension PropertyManager.

286) Select **Hidden Dims** from the Layer box in the Dimension PropertyManager.

287) Click **OK** ✔ from the Dimension PropertyManager. The 10 dimension is not displayed.

Create a vertical dimension.

288) Click the **Smart Dimension** Sketch tool. The Dimension PropertyManager is displayed.

289) Click **Smart dimensioning** from the Dimension Assist Tools box.

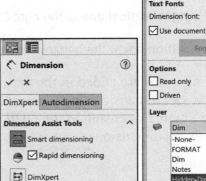

290) Click the top of the **top right arc**. Do not select the Center Mark of the Slot Cut.

291) Click the bottom of the **bottom right arc**. Do not select the Center Mark of the Slot Cut.

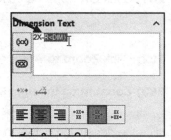

292) Click a **position** to the right of Drawing View3.

293) Click the **Leaders** tab in the Dimension PropertyManager.

294) Click **Max** for First arc condition.

295) Click **Max** for Second arc condition.

296) Click **OK** ✔ from the Dimension PropertyManager.

Modify the Radius text to 2X R.
297) Click **2X R4** in Drawing View2.

298) Delete **R<DIM>** in the Dimension text box.

299) Click **Yes** to the Confirm dimension value text override message.

300) Enter **R** for Dimension text.

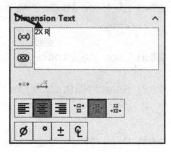

301) Click **OK** ✔ from the Dimension PropertyManager.

Add dimensions.
302) Click the **Smart Dimension** ✎ tool.

303) Click **Smart dimensioning** from the Dimension Assist Tools box.

304) Click the **left vertical line** of the right Slot Cut.

305) Click the **right vertical line** of the right Slot Cut.

306) Click a **position** below the horizontal profile line.

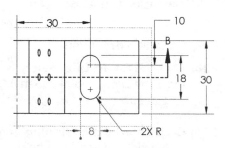

Drawing dimensions are added in the drawing document. Model dimensions are inserted from the part. Utilize Smart Dimension in the GUIDE drawing to create drawing dimensions.

307) Click the **centerline**.

308) Click the left **Slot Cut arc top center point**.

309) Click a **position** above the top horizontal line as illustrated. 30 is the dimension.

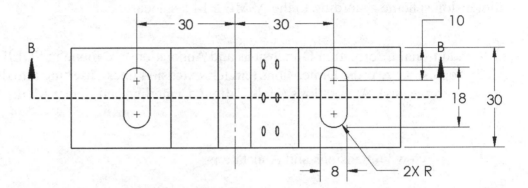

Insert a Reference dimension in the Section view.

310) Click the **left centerline** in the Section view. Click the **right centerline** in the Section view.

311) Click a **position** below the bottom horizontal line. Click the **Add Parenthesis** box in the Dimension Text.

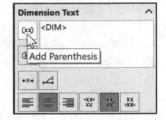

312) Click **OK** ✔ from the Dimension PropertyManager. View the results.

Save the drawing.

313) Click **Save** 💾.

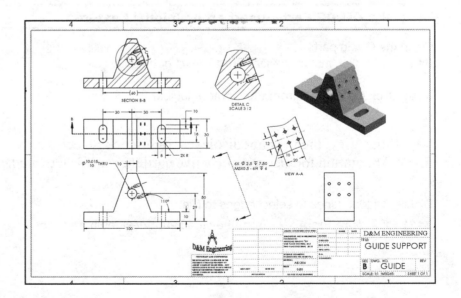

Design parts to maximize symmetry and geometric relations to fully define sketches. Minimize dimensions in the part. Insert dimensions in the drawing to adhere to machining requirements and Dimension Standards.

Examples: Insert reference dimensions by adding parentheses and redefine a slot dimension scheme according to the ASME Y14.5 standard.

Additional Information Dimensions and Annotations are found in SOLIDWORKS Help Topics. Keywords: Dimensions (circles, extension lines, inserting into drawings, move, parenthesis), Annotations (Hole Callout, Centerline and Center Mark).

 Review Dimensions and Annotations

You inserted part dimensions and annotations into the drawing. Dimensions were moved to new positions. Leader lines and dimension text were repositioned. Annotations were edited to reflect the drawing standard.

Centerlines and Center Marks were inserted into each view and were modified in the PropertyManager. You modified Hole Callouts, dimensions, annotations and referenced dimensions to conform to the drawing standard.

GUIDE Part - Insert an Additional Feature

The design process is dynamic. We do not live in a static world. Create and add an edge Fillet feature to the GUIDE part. Insert dimensions into Drawing View1; (Front).

Activity: GUIDE Part - Insert an Additional Feature

Open the Guide part.
314) Right-click inside the **Drawing View1** boundary.

315) Click **Open Part** from the Content toolbar.

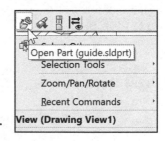

If the View menu is not displayed, right-click Select. Tools in the Annotation toolbar remain active until they are deactivated.

Display Hidden Lines to select edges to fillet.
316) Click **Wire Frame** from the Heads-up View toolbar.

Display an Isometric view.

317) Click **Trimetric view** 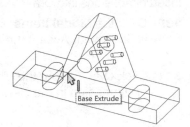 from the Heads-up View toolbar or from the Orientation dialog box.

Create a Fillet feature.

318) Click the **Fillet** feature tool. The Fillet PropertyManager is displayed. Click the **Manual** tab.

319) Click the **Constant Size** Fillet type.

320) Enter **1**mm in the Radius box. Symmetric is selected by default.

321) Click the **hidden edge** as illustrated.

322) Click the other **3 edges** as illustrated. Each edge is added to the Items to Fillet list. Accept the default settings.

323) Click **OK** ✔ from the Fillet PropertyManager. Fillet1 is displayed in the FeatureManager.

Save the GUIDE part.

324) Click **Save** .

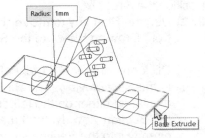

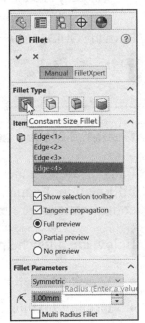

Open the Drawing.

325) Right-click the **GUIDE Part** GUIDE (Default‹ icon in the FeatureManager.

326) Click **Open Drawing**. The GUIDE drawing is displayed.

Fit the drawing to the Graphics window.

327) Press the **f** key. Note: Hide any unwanted annotations.

Display the GUIDE features.

328) **Expand** Drawing View1 from the FeatureManager. **Expand** the GUIDE part from the FeatureManager.

329) Click **Fillet1** from the Drawing FeatureManager. Insert the dimensions for the Fillet1 feature on the Dims Layer. The current layer should be Dims. If required, select Dims.

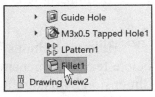

The Model Items, Import into drawing view, Import items into all views option insert dimensions into the Section view and Detail view before other types of views on the drawing. The Fillet feature dimension in the part is displayed in the Section view.

Insert dimensions into Drawing View1.

330) Click the **Model Items** 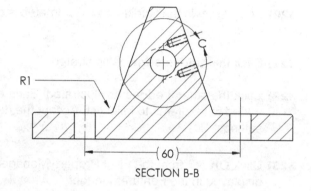 tool from the Annotation tab. The Model Items PropertyManager is displayed. Accept the default conditions.

331) Click **OK** ✔ from the Model Items PropertyManager. Note: Hide any unwanted annotation or dimension in the Detail view.

Position the Fillet text.

332) Use the Shift key to **move the R1 text** from Drawing View1 to the Section view.

333) Click and drag the **R1** text off the profile in the Section view. The R1 text is displayed either on the left or right side of the Section view. Note: It may come in directly into the Section view.

Save the GUIDE drawing.

334) Click **Save** 💾.

SECTION B-B

General Notes and Parametric Notes

Plan ahead for general drawing notes. Notes provide relative part or assembly information. Example: Material type, material finish, special manufacturing procedure or considerations, preferred supplier, etc.

Below are a few helpful guidelines to create general drawing notes:

- Use capital letters.

- Use left text justification.

- Font size should be the same size as the dimension text.

Create Parametric notes by selecting dimensions in the drawing. Example: Specify the Fillet radius of the GUIDE as a note in the drawing. If the radius is modified, the corresponding note is also modified.

💡 Hide superfluous feature dimensions. Do not delete feature dimensions. Recall a hidden dimension by using the View, Hide/Show, Hide/Show Annotations from the Menu bar. Utilize a Layer to Hide/Show superfluous feature dimensions with the Layer on/off icon.

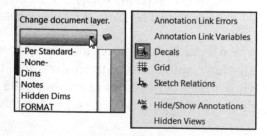

Activity: General Notes and Parametric Notes

Locate the R1 text used to create the edge Fillet.

335) Position the **mouse pointer** over the R1 text in the Section view. The text displays the dimension name: "D1@Fillet1 of Guide."

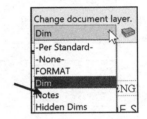

Insert a Parametric drawing note.

336) Click the **Notes** layer from the Layer toolbar.

337) Click the **Note** A tool from the Annotation tab in the CommandManager. The Note PropertyManager is displayed.

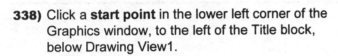

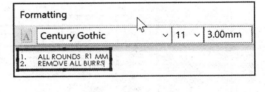

338) Click a **start point** in the lower left corner of the Graphics window, to the left of the Title block, below Drawing View1.

Type two lines of notes in the Note text box.

339) Line 1: Enter **1. ALL ROUNDS**.

340) Press the **Space** key.

341) Click **R1**. The radius value R1 is added to the text box.

342) Press the **Space** key.

343) Enter **MM**.

344) Press the **Enter** key.

345) Line 2: Enter **2. REMOVE ALL BURRS**.

346) Click **OK** ✔ from the Note PropertyManager.

Do not double dimension a drawing with a note and the corresponding dimension. Hide the radial dimension.

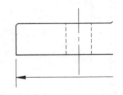

1.   ALL ROUNDS R1 MM
2.   REMOVE ALL BURRS

Hide the R1 dimension in the drawing.
347) Click the **R1** text in the Section view.

348) Click **Hidden Dims** from the Layer drop-down menu.

349) Click **OK** ✔ from the Dimension PropertyManager. R1 is hidden.

350) View the updated drawing.

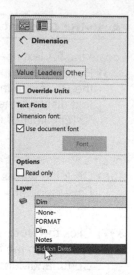

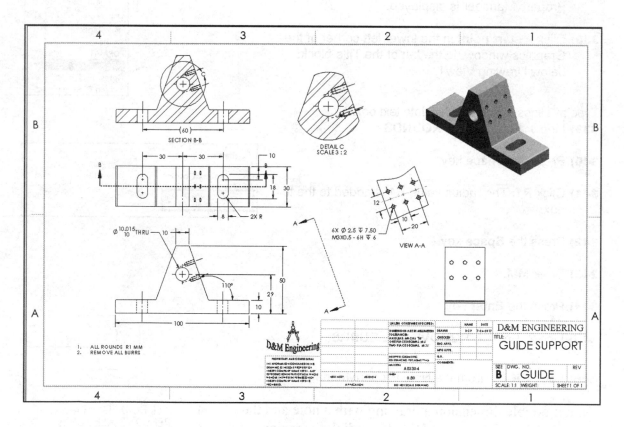

An Engineering Change Order (ECO) or Engineering Change Notice (ECN) is written for each modification in the drawing. Modifications to the drawing are listed in the Revision Table.

 Obtain the ECO number before you institute the change to manufacturing.

Revision Table

The Revision Table provides a drawing history. Changes to the document are recorded systematically in the Revision Table. Insert a Revision Table into the GUIDE drawing. The default columns are as follows: Zone, Rev, Description, Date and Approved.

Zone utilizes the row letter and column number contained in the drawing border. Position the Rev letter in the Zone area. Enter the Zone letter/number.

Enter a Description that corresponds to the ECO number. Modify the date if required. Enter the initials/name of the engineering manager who approved the revision.

Activity: Revision Table

Insert a Revision table.

351) Click **None** from the Layer drop-down menu.

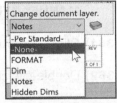

352) Click the **Revision Table** tool from the Consolidated toolbar in the CommandManager. The Revision Table PropertyManager is displayed. Accept the defaults.

353) Click **OK** ✔ from the Revision Table PropertyManager. The Revision table is displayed in the top right corner of the Sheet.

354) Zoom in on the first row in the table.

Insert a row. Create the first revision.
355) Right-click the **Revision Table**.

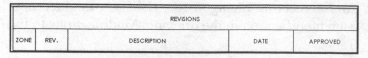

356) Click **Revisions, Add Revision**. The Revision letter A and the current date are displayed. The (A) Revision Symbol is displayed on the mouse pointer.

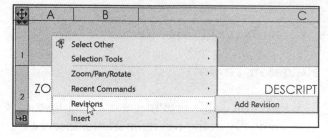

357) Click the **top face** of the GUIDE in the Isometric view to place the (A) icon as illustrated.

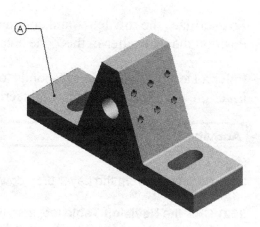

358) Click **OK** ✔ from the Revision Symbol PropertyManager.

359) Double-click the **text box** under the Description column.

360) Enter **ECO 32510 RELEASE TO MANUFACTURING** for DESCRIPTION.

361) Click a **position** outside of the text box.

362) Double-click the **text box** under APPROVED.

363) Enter Documentation Control Manager's Initials, **DCP**.

364) Click a **position** outside of the text box.

365) **View** the results.

		REVISIONS		
ZONE	REV.	DESCRIPTION	DATE	APPROVED
B2	A	ECO 32510 RELEASE TO MANUFACTURING	7/24/2017	DCP

The Revision Property, $PRP:"Revision," in the Title block is linked to the REV. Property in the REVISIONS table. The latest REV. letter is displayed in the Title block.

Revision Table Zone Numbers, Dates and Approved names are inserted as an exercise. The Revision Table and Engineering Change Order documentation are integrated in the drawing and the design process.

To save time, incorporate the Revision Table into your custom Sheet Formats.

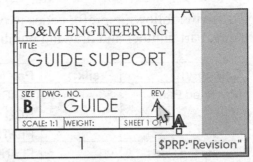

Part Number and Document Properties

Engineers manage the parts they create and modify. Each part requires a Part Number and Part Name. A part number is a numeric representation of the part. Each part has a unique number. Each drawing has a unique number. Drawings incorporate numerous part numbers or assembly numbers.

There are software applications that incorporate unique part numbers to create and perform:

- Bill of Materials

- Manufacturing procedures

- Cost analysis

- Inventory control/Just in Time (JIT)

You are required to procure the part and drawing numbers from the documentation control manager.

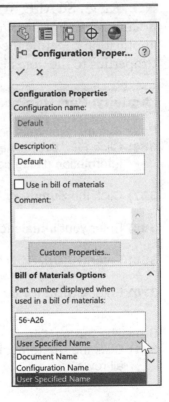

Utilize the following prefix codes to categories, created parts and drawings. The part name, part number and drawing numbers are as follows:

Category:	Prefix:	Part Name:	Part Number:	Drawing Number:
Machined Parts	56-	GUIDE	56-A26	56-22222
		ROD	56-A27	56-22223
		PLATE	56-A28	56-22224
Purchased Parts	99-	FLANGE BOLT	99-FBM8x1.25	999-551-8
Assemblies	10-	GUIDE-ROD	10-A123	10-50123

Link notes in the Title block to SOLIDWORKS Properties. The title of the drawing is linked to the GUIDE Part Description, GUIDE-SUPPORT. Create additional notes in the Title block that complete the drawing.

Additional notes are required in the Title block. The text box headings SIZE B, DWG. NO., REV., SCALE, WEIGHT and SHEET OF are entered in the SOLIDWORKS default Sheet Format. Properties are variables shared between documents and applications. Define the Document Properties in the GUIDE drawing. Link the Document Properties to the notes in the Title block.

Activity: Part Number and Document Properties

Enter Summary information for the GUIDE drawing.

366) Click **File**, **Properties** from the Menu bar. The Summary Information dialog box is displayed.

367) Click the **Summary** tab.

368) Enter your **initials** for Author. Example: DCP.

369) Enter **GUIDE, ROD** for Keywords.

370) Enter **GUIDE for customer ABC** for Comments.

Select the Custom tab.
371) Click the **Custom** tab. View the information in the rows. Fill in all rows.

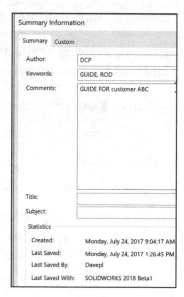

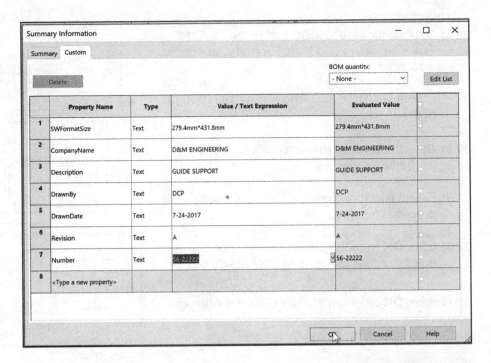

372) Add **DrawnBy** and **DrawnDate** in Custom Properties. Use your own initials and date.

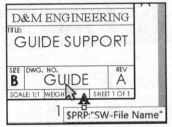

373) Click **OK** from the Summary Information dialog box.

The DrawnBy Property and DrawnDate Property are updated in the Title block. The current DWG NO. is the SOLIDWORKS File Name. Assign the Number Property to the DWG NO. in the Title block.

Link the Properties to the Drawing Notes.
374) Click inside **Sheet1**.

375) Right-click **Edit Sheet Format**.

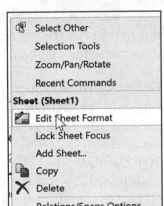

Edit the DWG NO.

376) Double-click the **GUIDE** text in the DWG NO. box. The Formatting and Note PropertyManager is displayed.

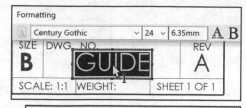

377) Delete **GUIDE**. Note: The Formatting dialog is still active.

378) Click the **Link to Property** 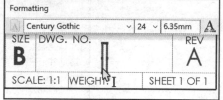 icon in the Properties dialog box.

379) Select **Number** from the Link to Property drop-down menu.

380) Click **OK** from the Link to Property box. The number is displayed in the DWG. NO. box.

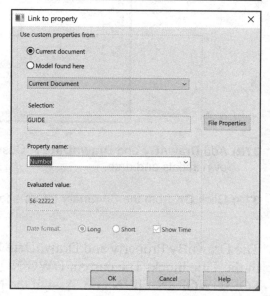

381) Click **OK** ✔ from the Note PropertyManager.

382) Right-click **Edit Sheet**.

Fit the drawing to the Graphics window.

383) Press the **f** key. The drawing remains in the Edit Sheet Format mode.

Save the Drawing.

384) Click **Save** 💾.

Note: Remain in the Edit Sheet Format mode through the next few steps. The Title block information is displayed in blue.

Define Custom Properties for Material and Surface Finish in the GUIDE part. Return to the GUIDE drawing to complete the Title block. The design team decided that the GUIDE part would be fabricated from 304 Stainless Steel. There are numerous types of Stainless steels for various applications. Select the correct material for the application. This is critical!

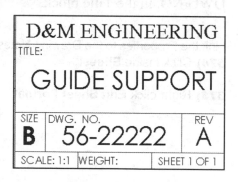

AISI 304 was set in the part. Open the GUIDE part and create the Material Custom Property. The AISI 304 value propagates to the drawing. Surface Finish is another important property in machining. The GUIDE utilizes an all around 0.80μm high grade machined finish. The numerical value refers to the roughness height of the machined material. Surface finish adds cost and time to the part. Work with the manufacturer to specify the correct part finish for your application.

Create the Material Custom Property.
385) Open the GUIDE part.

386) Click the **Configuration PropertyManager** icon.

387) Right-click **Default[GUIDE]**.

388) Click **Properties**. The Configuration Properties PropertyManager is displayed.

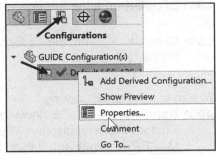

389) Click the **Custom Properties** button.

390) Click the **Configuration Specific** tab.

Specify the Material.
391) Click inside the **Property Name** box.

392) Select **Material** from the drop-down menu.

393) Click inside the **Value/Text Expression** box.

394) Select **Material** from the drop-down menu.

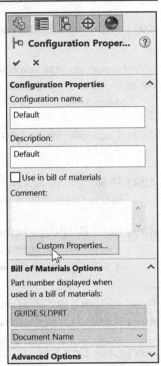

💡 Determine the correct units before you create Custom Properties linked to Mass, Density, Volume, Center of Mass, Moment of Inertia, etc. Recall Weight = Mass *Gravity. Modify the Title block text to include "g" for gravity, SI units in the Weight box.

Specify the Finish.
395) Click inside the **second row** under the Property Name.

396) Select **Finish** from the drop-down menu.

397) Enter **0.80** for Value/Text Expression as illustrated.

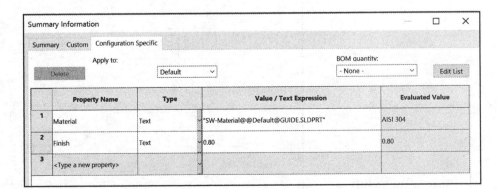

398) Click **OK** from the Summary Information box.

399) Click **OK** ✔ from the Configuration Properties PropertyManager.

Return to the GUIDE drawing.
400) **Return** to the GUIDE drawing. The Material Custom Property displays AISI 304 in the MATERIAL block and 0.80 in the FINISH block.

401) Right-click **Edit Sheet Format**.

Insert a micrometer symbol.
402) Double-click **0.80**. The Formatting toolbar is displayed.

403) Click **SWGrekc** from the Formatting toolbar for Font.

404) Press the **m** key to display the Greek letter μ. μ is the Greek letter for micro.

405) Click **Century Gothic** for Font. Enter **m** for meter. The text reads μm for micrometer. Click a **position** outside the text box.

Return to the Edit Sheet mode.
406) Right-click in the **Graphics window**.

407) Click **Edit Sheet**. The Title block is displayed in black.

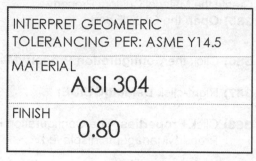

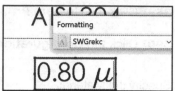

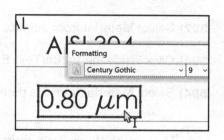

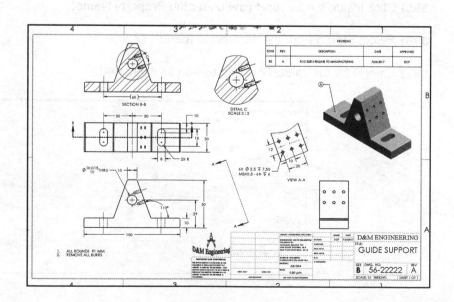

Fit the drawing to the Graphics window.
408) Press the **f** key.

409) Drag the **views** and **annotations** if required for spacing.

Save the GUIDE drawing.
410) Click **Save** 💾.

💡 Establish Custom Properties early in the design process. The MATERIAL and FINISH Custom Properties are established in the part and propagate to the Title block in the drawing.

💡 Create engineering procedures to define the location and values of Custom Properties. The REVISION Custom Property was created in the drawing in the REVISION Table. The REVISION Custom Property can also be established in the part and propagate to the drawing. Does the part or the drawing control the revision? Your company's engineering procedures determine the answer.

Additional annotations are required for drawings. Utilize the Surface Finish ✓ tool to apply symbols to individual faces/edges in the part or in the drawing. When an assembly contains mating parts, document their relationship. Add part numbers in the Used On section. Additional annotations are left as an exercise.

Exploded View

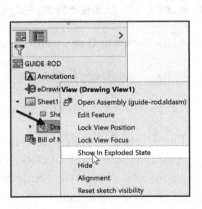

Add an Exploded view and Bill of Materials to the drawing. Add the GUIDE-ROD assembly Exploded view. The Bill of Materials reflects the components of the GUIDE-ROD assembly. Create a drawing with a Bill of Materials.

Perform the following steps:

- Create a new drawing from the B-ANSI-MM Drawing Template.

- Display the Exploded view of the assembly.

- Insert the Exploded view of the assembly into the drawing.

- Label each component with Balloon text.

- Create a Bill of Materials.

Activity: Exploded View

Close all parts and drawings.
411) Click **Windows**, **Close All** from the Menu bar.

Open the GUIDE-ROD assembly.
412) Open the **GUIDE-ROD** assembly from the PROJECTS folder. The GUIDE-ROD assembly is displayed in the Graphics window with the updates to the GUIDE.

Create a new drawing.
413) Click **Make Drawing from Part/Assembly** from the Menu bar.

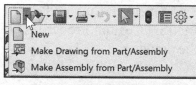

414) Double-click **B-ANSI-MM** from the MY-TEMPLATES tab. The Model View PropertyManager is displayed.

415) Click the **View Palette** tab in the Task Pane.

416) Click the **drop-down** arrow and click **GUIDE-ROD**.

Insert the GUIDE-ROD assembly using the View Palette tool.
417) Click and drag the ***Isometric** view from the View Palette onto Sheet1. The Drawing View1 PropertyManager is displayed. Note: You have the option to click and drag the *Isometric Exploded view directly into the sheet.

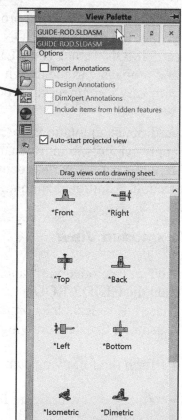

Display a Shaded With Edges, 1:1 scale view.
418) Click **Shaded With Edges** in the Display Style box.

419) Click the **Use custom scale** box.

420) Select **User Defined**.

421) Select **1:1** from the drop-down menu.

The View Palette contains images of Standard views, Annotation views, Section views, Flat patterns (sheet metal parts) and Isometric Exploded view (if an Exploded configuration is created) of the selected model.

Click the Browse button in the View Palette to locate additional models.

A goal of this book is to expose the new user to various tools, techniques and procedures. It may not always use the most direct tool or process.

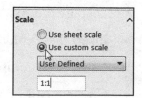

Display an Exploded view.
422) Click **inside** the Isometric view boundary.

423) Right-click **Show in Exploded State**. Note: You can also click **Properties**, check **Show in Exploded State**, and click **OK** from the Drawing View Properties dialog box.

424) Click **OK** ✔ from the Drawing View1 PropertyManager.

Note: The Explode view was created in Project 3. The Show in exploded state option is visible only if an Exploded configuration exists in the assembly.

Fit the drawing to the Graphics window.
425) Press the **f** key.

426) Click **OK** ✔ from the Drawing View1 PropertyManager.

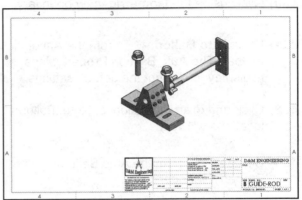

The drawing filename is the same as the assembly filename.

Save the GUIDE-ROD drawing.
427) Click **Save As** from the Menu bar.

428) Select **ENGDESIGN-W-SOLIDWORKS\PROJECTS** for folder. GUIDE-ROD is displayed for the File name.

Save the drawing.
429) Click **Save**.

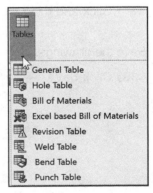

Balloons

Label each component with a unique item number. The item number is placed inside a circle. The circle is called Balloon text. List each item in a Bill of Materials table. Utilize the Auto Balloon tool to apply Balloon text to all components. Utilize the Bill of Materials tool to apply a BOM to the drawing.

The Circle Split Line option contains the Item Number and Quantity. Item number is determined by the order listed in the assembly FeatureManager. Quantity lists the number of instances in the assembly.

Magnetic lines are a convenient way to align balloons along a line at any angle. You attach balloons to magnetic lines, choose to space the balloons equally or not, and move the lines freely, at any angle in the drawing.

Activity: Auto Balloons with Magentic Lines

Insert Automatic Balloons.

430) Click inside the **Isometric view** boundary.

431) Click **Auto Balloon** from the Annotation toolbar. The Auto Balloon PropertyManager is displayed. Accept the default settings.

432) Click and **drag** the balloons. Note: Balloon locations will vary.

433) Click **OK** from the Auto Balloon PropertyManager.

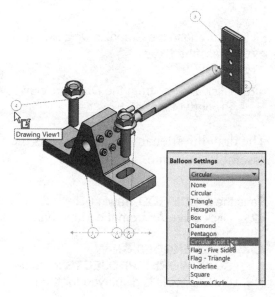

Display Item Number and Quantity.

434) Shift-Select the six **Balloon text**. The Balloon PropertyManager is displayed.

435) Select **Circular Split Line** for Settings from the drop-down menu. View the results in the Graphics window.

436) Click **OK** from the Balloon PropertyManager.

437) Click and **drag** the circular split line balloons.

Save the drawings.

438) Click **Save** .

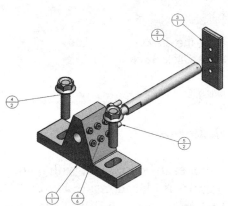

The symbol "?" is displayed when the Balloon attachment is not coincident with an edge or face. ASME Y14.2 defines the attachment display as an arrowhead for Edge and a dot for Face.

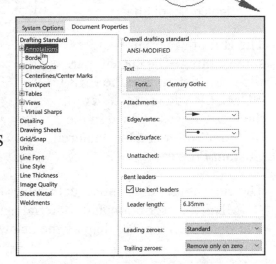

Arrowhead display is located under Options, Document Properties, Annotations or under the Leaders tab in the Dimension PropertyManager. Note: Edge/vertex attachment is selected by default.

Each Balloon references a component with a SOLIDWORKS File Name. The SOLIDWORKS File Name is linked to the Part Name in the Bill of Materials by default. Customize both the Balloon text and the Bill of Materials according to your company's requirements.

Bill of Materials

The Bill of Materials reflects the components of the GUIDE-ROD assembly. The Bill of Materials is a table in the assembly drawing that contains Item Number, Quantity, Part Number and Description by default. Insert additional columns to customize the Bill of Materials.

Part Number and Description are Custom Properties entered in the individual part documents. Insert additional Custom Properties to complete the Bill of Materials.

Activity: Bill of Materials

Create a Bill of Materials.

439) Click inside the **Isometric view** in the GUIDE-ROD drawing.

440) Click the **Bill of Materials** 🗔 tool from the Annotation tab. The Bill of Materials PropertyManager is displayed.

441) Click the **Open table template for Bill of Materials** ⭐ icon.

442) Double-click **bom-material.sldbomtbt** from the Select BOM Table dialog box.

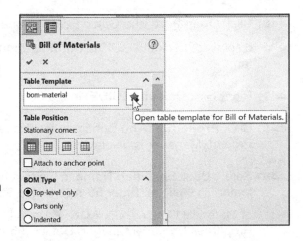

443) Click **OK** ✔ from the Bill of Materials PropertyManager.

444) Click a **position** in the upper left corner of Sheet1.

The Bill of Materials requires some editing. The current part file name determines the PART NUMBER values. Material for the GUIDE was defined in the Material Editor and the Material Custom Property.

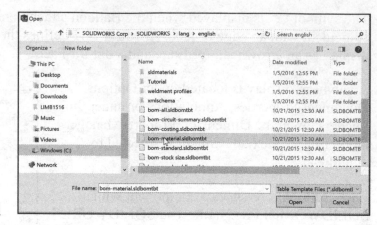

The current part description determines the DESCRIPTION values. Redefine the PART NUMBER, Description and Material in the BOM. Your BOM values and rows *may vary*. The final BOM should display the below PART NUMBERS, DESCRIPTION, and MATERIAL.

Modify the GUIDE part number in the BOM.

ITEM NO.	PART NUMBER	DESCRIPTION	MATERIAL	QTY.
1	56-A26	GUIDE SUPPORT	AISI 304	1
2	56-A27	ROD	AISI 304	1
3	56-A28	PLATE 56MM x 22MM	AISI 304	1
4	M8-1.25 x 30	FLANGE BOLT M8x1.25x30	AISI 304	2
5	56-333	CAP SCREW, 4MM	AISI 304	2
6	44-4434	CAP SCREW, 3MM	AISI 304	6

445) Right-click on the **GUIDE** in the Isometric view.

446) Click **Open Part**. The Guide FeatureManager is displayed. Click the GUIDE **ConfigurationManager** 🔧 tab.

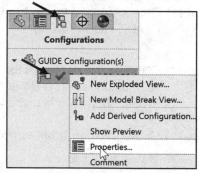

447) Right-click **Default [GUIDE]** in the ConfigurationManager.

448) Click **Properties**. The Configuration Properties PropertyManager is displayed.

449) Select **User Specified Name** in the spin box from the Bill of Materials Options. Enter **56-A26** for the Part Number in the Bill of Materials Options. Click **OK** ✔ from the Configuration Properties PropertyManager. Default [56-A26] is displayed in the ConfigurationManager.

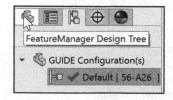

450) **Return** to the FeatureManager.

Return to the GUIDE-ROD drawing.
451) Click **Window**, **GUIDE-ROD - Sheet1** from the Menu bar.

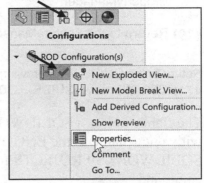

Modify the ROD part number.
452) Right-click on the **ROD** in the Isometric view.

453) Click **Open Part**. The ROD FeatureManager is displayed.

454) Click the ROD **ConfigurationManager** tab.

455) Right-click **Default [ROD]** from the ConfigurationManager.

456) Click **Properties**.

457) Select **User Specified Name** from the drop-down menu in the Bill of Materials Options.

458) Enter **56-A27** for the Part Number in the Bill of Materials Options.

459) Click **OK** ✔ from the Configuration Properties PropertyManager.

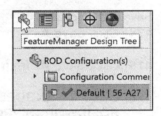

460) Return to the FeatureManager.

Return to the GUIDE-ROD drawing.
461) Click **Window**, **GUIDE-ROD - Sheet1** from the Menu bar.

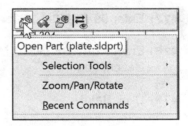

Modify the PLATE part number.
462) Right-click on the **PLATE** in the Isometric view.

463) Click **Open Part**.

464) Click the PLATE **ConfigurationManager** tab.

465) Right-click **Default [PLATE]** from the ConfigurationManager.

466) Click **Properties**.

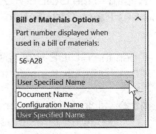

467) Select **User Specified Name** in the spin box from the Bill of Materials Options.

468) Enter **56-A28** for the Part Number in the Bill of Materials Options.

469) Click **OK** ✔ from the Configuration Properties PropertyManager.

470) Return to the FeatureManager.

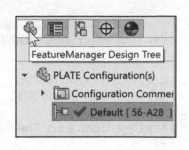

Return to the GUIDE-ROD drawing.
471) Click **Window**, **GUIDE-ROD - Sheet1** from the Menu bar.

The flange bolt is a SOLIDWORKS library part. Copy the part with a new name to the ENGDESIGN-W-SOLIDWORKS\PROJECTS folder with the Save As command.

Utilize the flange bolt Part Number for File name.

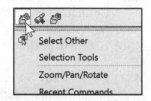

Modify the flange bolt.
472) Right-click a **flange bolt** in the Isometric view.

473) Click **Open Part**.

Save the flange bolt.
474) Click **Save As** from the Menu bar.

475) Click **Save As** from the dialog box.

476) Select the **ENGDESIGN-W-SOLIDWORKS\PROJECTS** folder.

477) Enter **99-FBM8-1-25** for the File name.

478) Enter **FLANGE BOLT M8x1.25x30** for Description.

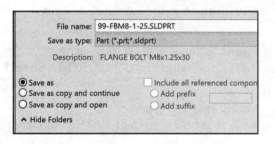

479) Click **Save**.

Return to the GUIDE-ROD drawing.
480) Press **Ctrl+Tab** to return to the GUIDE-ROD drawing.

Open the GUIDE-ROD assembly.
481) Right-click inside the **Isometric view** boundary on Sheet1.

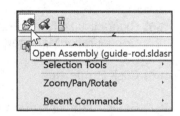

482) Click **Open Assembly**. The GUIDE-ROD assembly FeatureManager is displayed.

483) Expand the Hardware folder from the FeatureManager. The flange bolt displays the new name.

Return to the GUIDE-ROD drawing.
484) Press **Ctrl+Tab** to return to the GUIDE-ROD drawing.

Update the Bill of Materials.
485) Rebuild 🔘 the model.

Fit the drawing to the Graphics window.
486) Press the **f** key.

🔆 If the Bill of Materials does not update when changes to a part have been made, then open the assembly. Return to the drawing. Issue a Rebuild.

As an exercise, update the Bill of Materials and fill in all of the columns and address the Title block as illustrated below.

Save the GUIDE-ROD drawing.
487) Click **Save** 💾.

488) Click **Yes** to update.

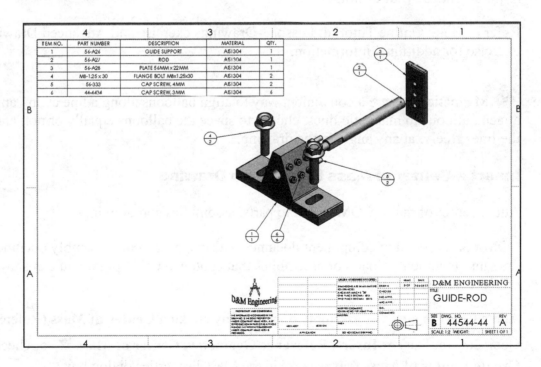

ITEM NO.	PART NUMBER	DESCRIPTION	MATERIAL	QTY.
1	56-A26	GUIDE SUPPORT	AISI 304	1
2	56-A27	ROD	AISI 304	1
3	56-A28	PLATE 56MM x 22MM	AISI 304	1
4	M8-1.25 x 30	FLANGE BOLT M8x1.25x30	AISI 304	2
5	56-333	CAP SCREW, 4MM	AISI 304	2
6	44-4434	CAP SCREW, 3MM	AISI 304	6

D&M ENGINEERING

TITLE
GUIDE-ROD

SIZE **B** DWG. NO. 44544-44 REV **A**

SCALE: 1:2 WEIGHT: SHEET 1 OF 1

Review of the Parametric Notes, Revision Table and Bill of Materials

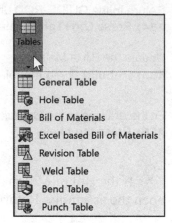

You created a Parametric note in the drawing by inserting a part dimension into the Note text box. The Revision Table was inserted into the drawing to maintain drawing history.

The Bill of Materials listed the Item Number, Part Number, Description, Material and Quantity of components in the assembly.

You developed Custom Properties in the part and utilized the Properties in the drawing and Bill of Materials.

Drawings and Custom Properties are an integral part of the design process. Part, assemblies and drawings all work together.

From your initial design concepts, you created parts and drawings that fulfill the design requirements of your customer.

Refer to Help, Online Tutorial, Lesson3-Drawings exercise and Advanced Drawings exercise for additional information.

Magnetic lines are a convenient way to align balloons along a line at any angle. You attach balloons to magnetic lines, choose to space the balloons equally or not, and move the lines freely, at any angle in the drawing.

Insert a Center of Mass Point into a Drawing

Add a center of mass (COM) point to parts, assemblies and drawings.

COM points added in component documents also appear in the assembly document. In a drawing document of parts or assemblies that contain a COM point you can show and reference the COM point.

Add a COM to a part or assembly document by clicking **Center of Mass** (Reference Geometry toolbar) or **Insert, Reference Geometry, Center of Mass** ⊕ or check the **Create Center of Mass feature** box in the Mass Properties dialog box.

The center of mass of the model is displayed in the graphics window and in the FeatureManager design tree just below the origin.

The position of the **COM** point 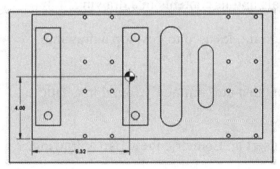 updates when the model's center of mass changes. The COM point can be suppressed and unsuppressed for configurations.

You can measure distances and add reference dimensions between the COM point and entities such as vertices, edges and faces.

If you want to display a reference point where the CG was located at some particular point in the FeatureManager, you can insert a Center of Mass Reference Point. See SOLIDWORKS Help for additional information.

Add a center of mass (COM) point to a drawing view. The center of mass is a selectable entity in drawings and you can reference it to create dimensions.

In a drawing document, click **Insert, Model Items**. The Model Items PropertyManager is displayed. Under Reference Geometry, click the **Center of Mass** icon. Enter any needed additional information. Click **OK** from the Model Items PropertyManager. View the results in the drawing.

The part or assembly **needs to have a COM before** you can view the COM in the drawing. To view the center of mass in a drawing, click **View**, **Hide/Show**, **Center of mass**.

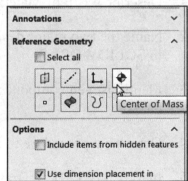

Project Summary

In this Project you developed two drawings: GUIDE drawing and the GUIDE-ROD assembly drawing. The drawings contained three standard views, (principle views) and an Isometric view. The drawings utilized a custom Sheet Format containing a Company logo, Title block and Custom Properties. You incorporated the GUIDE part dimensions into the drawing.

The Drawing toolbar contained the Model View tool and the Projected View tool to develop standard views. Additional views were required and utilized the Auxiliary, Detail and Section view tools. Dimensions were inserted from the part and added to the drawing.

You used two major design modes in the drawings: Edit Sheet Format and Edit Sheet.

The detailed GUIDE drawing included annotations and Custom Properties. The GUIDE-ROD assembly drawing incorporated a Bill of Materials and additional Custom Properties.

To show a hidden dimension, click **View**, **Hide/Show**, **Annotations** from the Menu bar.

Questions

1. Describe a Bill of Materials and its contents in a drawing.

2. Name the two major design modes used to develop a drawing in SOLIDWORKS.

3. Identify seven components that are commonly found in a Title block.

4. Describe a procedure to insert an Isometric view into a drawing.

5. In SOLIDWORKS, Drawing file names end with a _____ suffix.

6. In SOLIDWORKS, Part file names end with a _____ suffix.

7. Can a part and drawing have the same name?

8. True or False. In SOLIDWORKS, if a part is modified, the drawing is automatically updated.

9. True or False. In SOLIDWORKS, when a dimension in the drawing is modified, the part is automatically updated.

10. Name three guidelines to create General Notes in a drawing.

11. True or False. Most engineering drawings use the following font: Times New Roman - All small letters.

12. What are Leader lines? Provide an example.

13. Name the three ways that Holes and other circular geometry can be dimensioned.

14. Describe Center Marks. Provide an example.

15. How do you calculate the maximum and minimum variation?

16. Describe the differences between a Drawing Template and a Sheet Format.

17. Describe the key differences between a Detail view and a Section view.

18. Describe a Revision table and its contents.

Exercises

Exercise 4.1: L - BRACKET Drawing

Create the A (ANSI) Landscape - IPS - Third Angle L-BRACKET drawing as illustrated below.

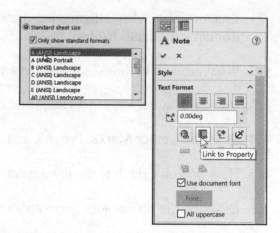

- First create the part from the drawing, then create the drawing.

- Insert the Front, Top, Right and Shaded Isometric view using the View Palette tool from the Task Pane. Hide the Top view.

- Insert dimensions into the Sheet. Think about the required extension line gaps needed between the Feature lines. Think about the proper view and position for your dimensions. Use the default A (ANSI) Landscape Sheet Format/Size.

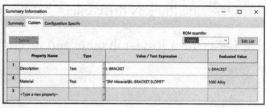

- Insert Custom Properties: Material, Revision, Number, Description, DrawnBy, DrawnDate, and CompanyName.

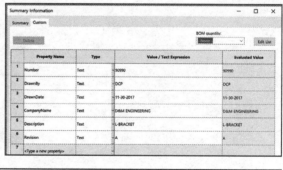

- Material is 1060 Alloy.

- Insert Company and Third Angle projection icons. The icons are available in the homework folder.

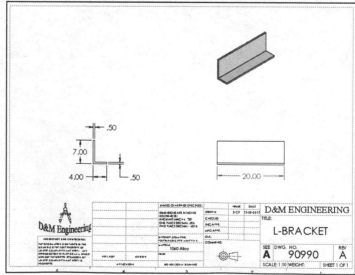

Exercise 4.2: T - SECTION Drawing

Create the A (ANSI) Landscape - IPS - Third Angle
T-SECTION drawing as illustrated below.

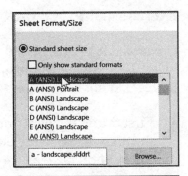

- First create the part from the drawing, then create the
  drawing. Use the default A (ANSI) Landscape Sheet
  Format/Size.

- Insert the Front, Top, Right and
  Shaded Isometric view using the
  View Palette tool from the Task Pane.
  Hide the Top view.

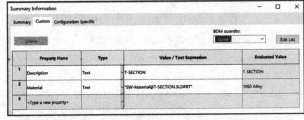

- Insert dimensions into the Sheet.
  Think about the proper view for your
  dimensions.

- Think about the needed Extension
  line gaps between the Feature lines.

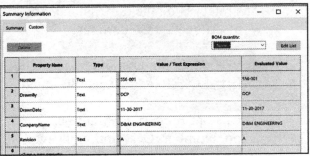

- Insert Custom Properties: Material,
  Revision, Number, Description,
  DrawnBy, DrawnDate, and
  CompanyName.

- Material is 1060 Alloy.

- Insert Company and
  Third Angle projection
  icons. The icons are
  available in the
  homework folder.

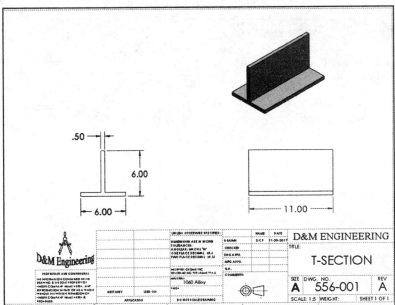

Exercise 4.3: FLAT BAR - 3HOLE Drawing

Create the A (ANSI) Landscape - IPS - Third Angle 3HOLES drawing as illustrated. Do not display Tangent Edges. Do not dimension to hidden lines.

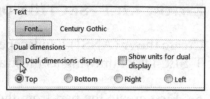

- First create the part from the drawing, then create the drawing. Use the default A (ANSI) Landscape Sheet Format/Size.

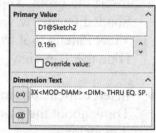

- Insert the Front, Top and Shaded Isometric view as illustrated. Insert dimensions. Address needed gaps. Address display modes.

- Add a Smart (Linked) Parametric note for MATERIAL THICKNESS in the drawing as illustrated. Hide the dimension in the Top view. Insert needed Centerlines.

- Modify the Hole dimension text to include 3X THRU EQ. SP. and 2X as illustrated.

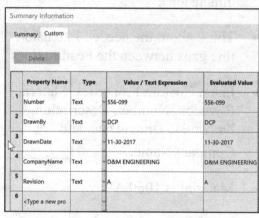

- Insert Custom Properties: Material, Revision, Number, Description, DrawnBy, DrawnDate, and CompanyName.

- Material is 1060 Alloy.

- Insert Company and Third Angle projection icons. The icons are available in the homework folder.

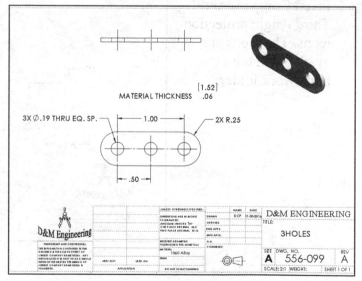

Exercise 4.4: CYLINDER Drawing

Create the A (ANSI) Landscape - IPS - Third Angle CYLINDER drawing as illustrated. Do not display Tangent Edges. Do not dimension to hidden lines.

- First create the part from the drawing, then create the drawing. Use the default A (ANSI) Landscape Sheet Format/Size.

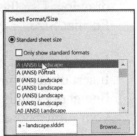

- Insert views as illustrated. Insert dimensions. Think about the proper view for your dimensions. Address needed extension line gaps. Address proper display modes.

- Insert Company and Third Angle projection icons. The icons are available in the homework folder.

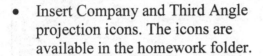

- Insert needed Centerlines, Center Marks and Annotations.

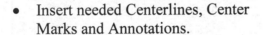

- Insert Custom Properties: Material, Revision, Number, Description, DrawnBy, DrawnDate, and CompanyName.

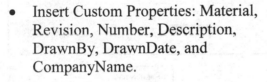

- Material is AISI 1020.

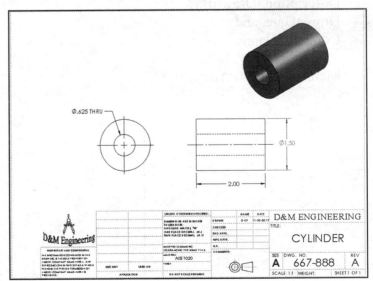

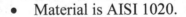

Exercise 4.5: PRESSURE PLATE Drawing

Create the A (ANSI) Landscape - IPS - Third Angle
PRESSURE PLATE drawing. Do not display Tangent edges.
Do not dimension to hidden lines.

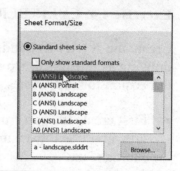

- First create the part from the drawing, then create the
  drawing. Use the default A (ANSI) Landscape Sheet
  Format/Size.

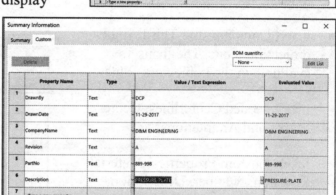

- Insert the Front and Right view as illustrated.
  Insert dimensions. Address needed extension
  line gaps. Think about the proper view for
  your dimensions. Address proper display
  modes.

- Insert Company and Third
  Angle projection icons. The
  icons are available in the
  homework folder.

- Insert Centerlines, Center
  Marks and Annotations.

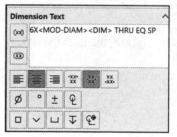

- Insert Custom Properties:
  Material, Revision, Number,
  Description, DrawnBy,
  DrawnDate, and
  CompanyName.

- Material is 1060 Alloy.

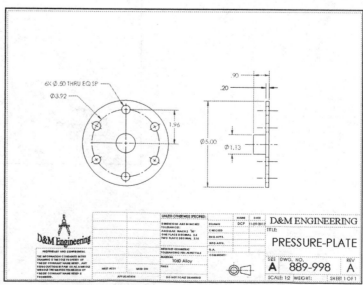

Exercise 4.6: PLATE-1 Drawing

Create the A (ANSI) Landscape - MMGS - Third Angle PLATE-1 drawing as illustrated below. Do not display Tangent edges. Do not dimension to Hidden lines.

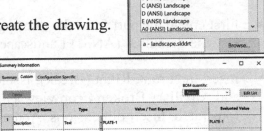

- First create the part from the drawing, then create the drawing. Use the default A (ANSI) Landscape Sheet Format/Size.

- Insert the Front and Right view as illustrated. Insert dimensions. Address extension line gaps. Think about the **proper view** for your dimensions. Address proper display modes.

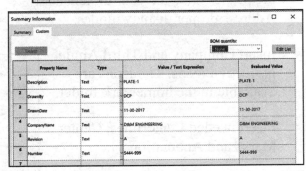

- Insert Company and Third Angle projection icons. The icons are available in the homework folder.

- Insert needed Centerlines, Center Marks and annotations.

- Insert Custom Properties: Material, Revision, Number, Description, DrawnBy, DrawnDate, and CompanyName.

- Material is 1060 Alloy.

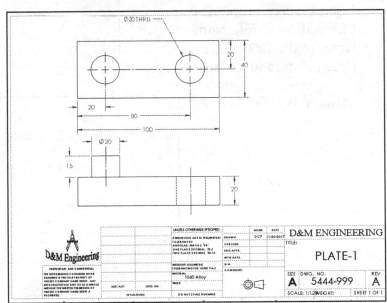

Exercise 4.7: SHAFT-1 Drawing

Create the A (ANSI) Landscape - IPS - Third Angle PLATE-1 drawing as illustrated below. Do not display Tangent edges. Phantom lines are fine.

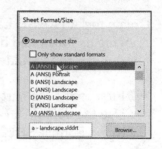

- First create the part from the drawing, then create the drawing. Use the default A (ANSI) Landscape Sheet Format/Size.

- Insert the Front, Right (Break), Isometric and Auxiliary Broken Crop view as illustrated. Insert dimensions. Address needed extension line gaps. Think about the proper view for your dimensions. Address proper display modes.

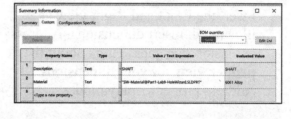

- Insert Company and Third Angle projection icons. The icons are available in the homework folder.

- Insert needed Centerlines, Center Marks and Annotations.

- Insert Custom Properties: Material, Revision, Number, Description, DrawnBy, DrawnDate, and CompanyName.

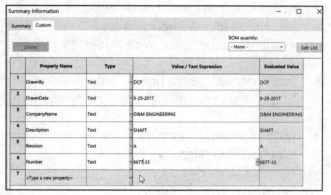

- Material is 1060 Alloy.

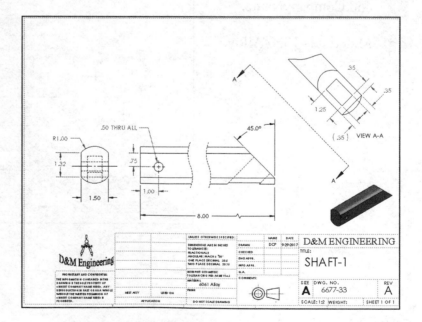

Exercise 4.8: TUBE-2 Drawing

Create the A (ANSI) Landscape - IPS - Third Angle TUBE-2 drawing as illustrated below. Do not display Tangent edges. Phantom lines are fine.

- First create the part from the drawing, then create the drawing. Use the default A (ANSI) Landscape Sheet Format/Size.

- Insert views as illustrated. Insert dimensions. Address extension line gaps. Think about the proper view for your dimensions.

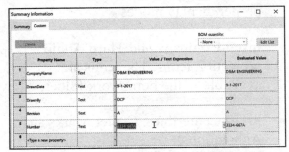

- Insert Company and Third Angle projection icons. Icons are available in the homework folder.

- Insert Centerlines, Center Marks and Annotations.

- Address proper display modes.

- Insert Custom Properties: Material, Revision, Number, Description, DrawnBy, DrawnDate, and CompanyName.

- Material is 6061 Alloy.

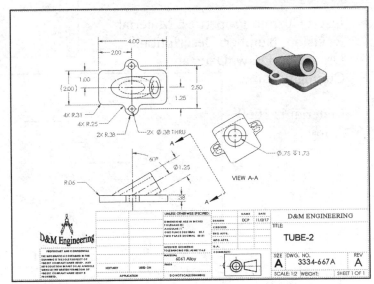

Exercise 4.9: PLATE Drawing

Create the A (ANSI) Landscape - IPS - Third Angle FLAT-PLATE drawing as illustrated below. Do not display Tangent edges. Phantom lines are fine.

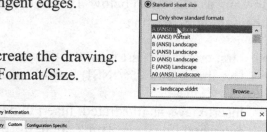

- First create the part from the drawing, then create the drawing. Use the default A (ANSI) Landscape Sheet Format/Size.

- Insert the Front, Top, Right and Isometric views as illustrated. Insert dimensions. Think about the proper view for your dimensions. Address proper display modes.

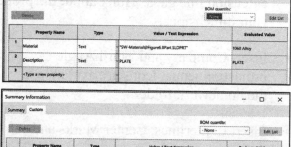

- Insert Company and Third Angle projection icons. The icons are available in the homework folder.

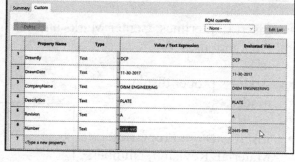

- Insert Centerlines, Center Marks and Annotations.

- Insert Custom Properties: Material, Revision, Number, Description, DrawnBy, DrawnDate, and CompanyName.

- Material is 1060 Alloy.

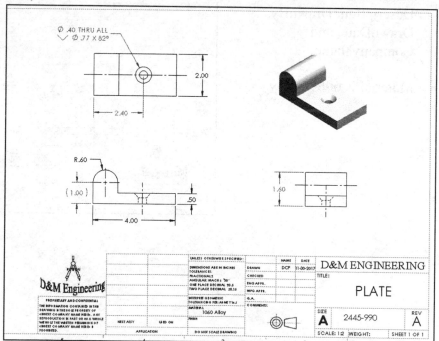

Exercise 4.10: FRONT-SUPPORT Assembly Drawing

Create the A (ANSI) Landscape - Third Angle FRONT-SUPPORT Assembly drawing with a Bill of Materials and balloons.

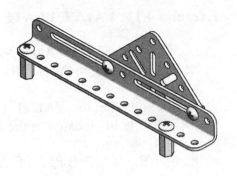

- Copy all needed files. Open the FRONT-SUPPORT assembly on your local hard drive. The FRONT-SUPPORT assembly is located in the Chapter 4 Homework folder.

- Use the default A (ANSI) Landscape Sheet Format/Size. Insert an Isometric Shaded With Edges view.

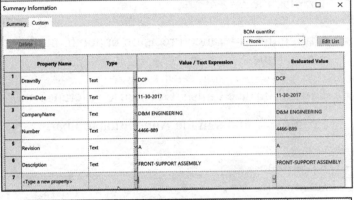

- Insert a Bill of Materials. Display the top level sub-assemblies and parts only in the Bill of Materials.

- Select bom-material.sldbomtbt from the Select BOM Table dialog box.

- Resize the text in the Title box and DWG. NO. box. Insert Custom Properties: Description, DrawnBy, DrawnDate, CompanyName etc.

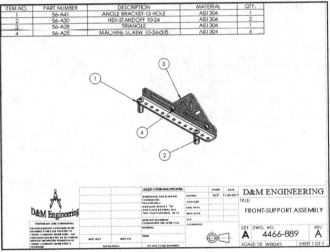

- Insert Circular Auto-Balloons as illustrated.

- Insert Custom Properties: Revision, Number, Description, DrawnBy, DrawnDate, and CompanyName.

- Insert Company and Third Angle projection icons. The icons are available in the homework folder.

Exercise 4.11: VALVE PLATE Drawing.

Create the A-ANSI Third Angle VALVE PLATE Drawing
document according to the ASME 14.5 standard.

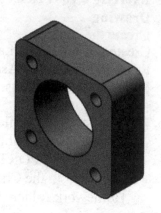

- Copy and open the VALVE PLATE part. The VALVE
  PLATE part is located in the Chapter 4 Homework folder.

- Create the VALVE PLATE drawing.

- Insert three views: Front, Top and Right. Insert the proper
  Display modes.

- Utilize the Tolerance/Precision option to modify the decimal
  place value.

- Utilize Surface Finish, Geometric Tolerance and Datum

  Feature ⒜ tools in the Annotation toolbar.

- Create the
  following
  drawing.

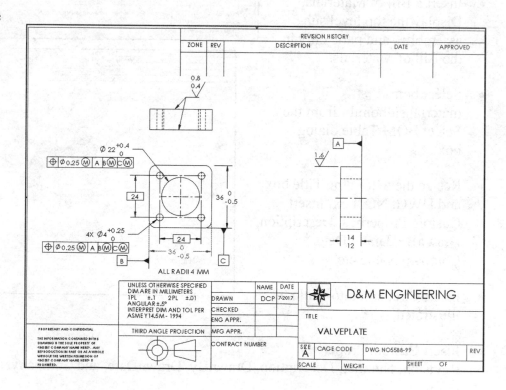

Exercise 4.12: GUIDE eDrawing

Create the GUIDE eDrawing. A SOLIDWORKS eDrawing is a compressed document that does not require the corresponding part or assembly. A SOLIDWORKS eDrawing is animated to display multiple views and dimensions. Review the eDrawing SOLIDWORKS Help Topics for additional functionality. The eDrawings Professional version contains additional options to mark up a drawing.

- Open the GUIDE drawing. Note: The Finished GUIDE drawing is located in the Chapter 4 Homework folder.

- Click File, Publish to eDrawing from the Main menu.

- Click the Animate button. Click Play. View the drawing.

- Click Stop. View the Menu features. Return to the initial drawing. Click the Reset button.

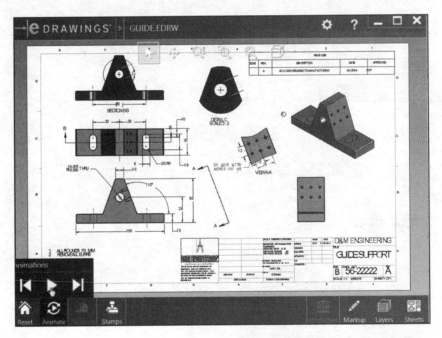

- Save the GUIDE eDrawing.

- Close the eDrawing.

Refer to Help, Online Tutorial, eDrawings exercise for additional information.

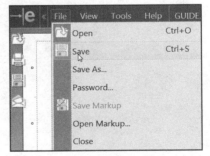

Notes:

Project 5

Extrude and Revolve Features

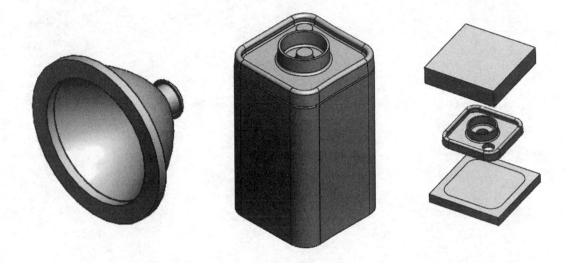

Below are the desired outcomes and usage competencies based on the completion of Project 5.

Project Desired Outcomes:	Usage Competencies:
• Obtain your customer's requirements for the FLASHLIGHT assembly.	• Ability to incorporate Design Intent into sketches, features, parts and assemblies.
• Two Part Templates: 　○ PART-IN-ANSI 　○ PART-MM-ISO	• Aptitude to apply Document Properties in a custom Part Template.
• Four key parts: 　○ BATTERY 　○ BATTERYPLATE 　○ LENS 　○ BULB	• Specific knowledge of the following features: Extruded Boss/Base, Instant3D, Extruded Cut, Revolved Boss/Base, Revolved Cut, Dome, Shell, Circular Pattern and Fillet.
• Core and Cavity Tooling for the BATTERYPLATE.	• Understanding of the Mold tools: Scale, Parting Lines, Parting Surfaces, Shut-off Surfaces, Tooling Split and Draft.

Notes:

Project 5 - Extrude and Revolve Features

Project Objective

Design a FLASHLIGHT assembly according to the customer's requirements. The FLASHLIGHT assembly will be cost effective, serviceable and flexible for future manufacturing revisions.

Design intent is the process in which the model is developed to accept future changes. Build design intent into the FLASHLIGHT sketches, features, parts and assemblies. Create two custom Part Templates. The Part Template is the foundation for the FLASHLIGHT parts.

Create the following parts:

- BATTERY

- BATTERYPLATE

- LENS

- BULB

The other parts for the FLASHLIGHT assembly are addressed in Project 5. Create the Core and Cavity mold tooling required for the BATTERYPLATE.

On the completion of this project, you will be able to:

- Apply design intent to sketches, features, parts, and assemblies.

- Select the best profile for a sketch.

- Select the proper Sketch plane.

- Create a template: English and Metric units.

- Set Document Properties.

- Customize the SOLIDWORKS CommandManager toolbar.

- Insert/Edit dimensions.

- Insert/Edit relations.

- Use the following SOLIDWORKS features:
    - Instant3D
    - Extruded Boss/Base
    - Extruded Cut

o Revolved Boss/Base

o Revolved Boss Thin

o Revolved Cut Thin

o Dome

o Shell

o Circular Pattern

o Fillet

- Use the following Mold tools:

  o Draft, Scale, Parting Lines, Shut-off Surfaces, Parting Surfaces, and Tooling Split

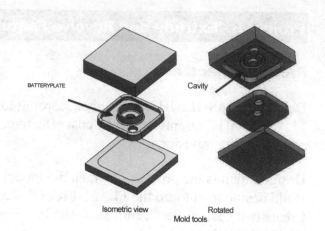

Project Overview

Start the design of the FLASHLIGHT assembly according to the customer's requirements. The FLASHLIGHT assembly will be cost effective, serviceable and flexible for future manufacturing revisions.

A template is the foundation for a SOLIDWORKS document. A template contains document settings for units, dimensioning standards and other properties.

Create two part templates for the FLASHLIGHT Project:

- PART-IN-ANSI

- PART-MM-ISO

Create four parts for the FLASHLIGHT assembly in this Project:

- BATTERY

- BATTERYPLATE

- LENS

- BULB

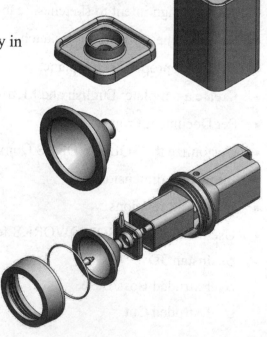

Parts models consist of 3D features. Features are the building blocks of a part.

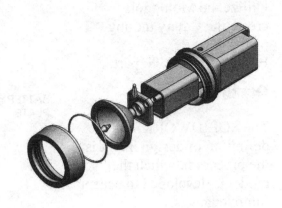

A 2D Sketch Plane is required to create an Extruded feature. Utilize the sketch geometry and sketch tools to create the following features:

- Extruded Boss/Base

- Extruded Cut

Utilize existing faces and edges to create the following features:

- Fillet

- Chamfer

This project introduces you to the Revolved feature.

Create two parts for the FLASHLIGHT assembly in this section:

- LENS

- BULB

A Revolved feature requires a 2D sketch profile and a centerline. Utilize sketch geometry and sketch tools to create the following features:

- Revolved Boss/Base

- Revolved Boss-Thin

- Revolved Cut

Utilize existing faces to create the following features:

- Shell

- Dome

- Hole Wizard

Utilize the Extruded Cut feature to create a Circular Pattern.

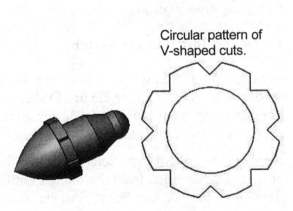

Circular pattern of V-shaped cuts.

Utilize the Mold tools to create the Cavity tooling plates for the BATTERYPLATE part.

Design Intent

The SOLIDWORKS definition of design intent is the process in which the model is developed to accept future changes.

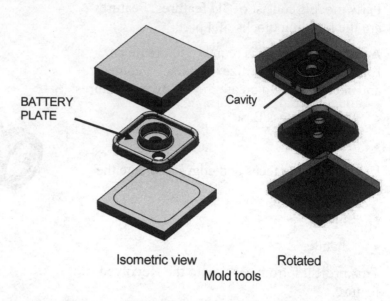

BATTERY PLATE

Cavity

Isometric view   Rotated

Mold tools

Models behave differently when design changes occur. Design for change. Utilize geometry for symmetry, reuse common features and reuse common parts.

Build change into the following areas:

1. Sketch

2. Feature

3. Part

4. Assembly

5. Drawing

See Project 9 (Intelligent modeling techniques) for additional information.

1. Design Intent in the Sketch

In SOLIDWORKS, relations between sketch entities and model geometry, in either 2D or 3D sketches, are an important means of building in design intent. In this chapter we will only address 2D sketches.

Apply design intent in a sketch as the profile is created. A profile is determined from the Sketch Entities. Example: Rectangle, Circle, Arc, Point, Slot etc.

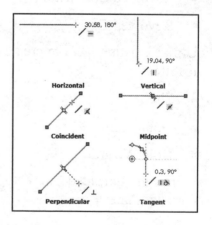

Develop design intent as you sketch with Geometric relations. Sketch relations are geometric constraints between sketch entities or between a sketch entity and a plane, axis, edge, or vertex. Relations can be added automatically or manually.

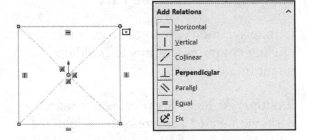

A rectangle contains Horizontal, Vertical, and Perpendicular automatic Geometric relations. Apply design intent using added Geometric relations. Example: Horizontal, Vertical, Collinear, Perpendicular, Parallel etc.

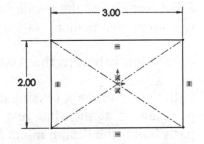

Example A: Apply design intent to create a square profile. Sketch a rectangle. Apply the Center Rectangle tool. Note: No construction reference centerline or Midpoint relation is required with the Center Rectangle tool. Insert dimensions to define the square.

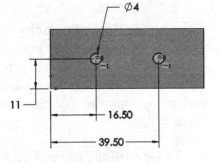

Example B: Develop a rectangular profile. Apply the Corner Rectangle tool. The bottom horizontal midpoint of the rectangular profile is located at the Origin. Add a Midpoint relation between the horizontal edge of the rectangle and the Origin. Insert two dimensions to define the width and height of the rectangle as illustrated.

2. Design Intent in the Feature

Build design intent into a feature by addressing symmetry, feature selection, and the order of feature creation.

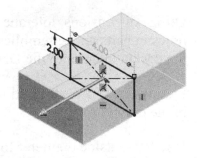

Example A: The Boss-Extrude1 feature (Base feature) remains symmetric about the Front Plane. Utilize the Mid Plane End Condition option in Direction 1. Modify the depth and the feature remains symmetric about the Front Plane.

Example B: Do you create each tooth separate using the Extruded Cut feature? No. Create a single tooth and then apply the Circular Pattern feature. Create 34 teeth for a Circular Pattern feature. Modify the number of teeth from 32 to 24.

3. Design Intent in the Part

Utilize symmetry, feature order and reusing common features to build design intent into the part.

Example A: Feature order. Is the entire part symmetric? Feature order affects the part. Apply the Shell feature before the Fillet feature and the inside corners remain perpendicular.

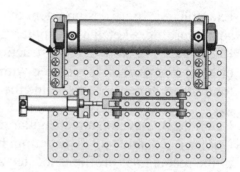

4. Design Intent in the Assembly

Utilizing symmetry, reusing common parts and using the Mate relation between parts builds the design intent into an assembly.

Example A: Reuse geometry in an assembly. The assembly contains a linear pattern of holes. Insert one screw into the first hole. Utilize the Component Pattern feature to copy the machine screw to the other holes.

5. Design Intent in the Drawing

Utilize dimensions, tolerance and notes in parts and assemblies to build the design intent into the Drawing.

Example A: Tolerance and material in the drawing.

Insert an outside diameter tolerance +.000/-.002 into the TUBE part. The tolerance propagates to the drawing.

Define the Custom Property MATERIAL in the part. The MATERIAL Custom Property propagates to the drawing.

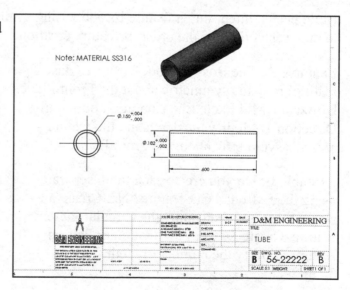

Project Situation

You work for a company that specializes in providing promotional tradeshow products. The company is expecting a sales order for 100,000 flashlights with a potential for 500,000 units next year. Prototype drawings of the flashlight are required in three weeks.

You are the design engineer responsible for the project. You contact the customer to discuss design options and product specifications. The customer informs you that the flashlights will be used in an international marketing promotional campaign. Key customer requirements:

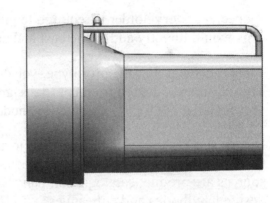

- Inexpensive reliable flashlight.

- Available advertising space of 10 square inches, 64.5 square centimeters.

- Lightweight semi-indestructible body.

- Self-standing with a handle.

Your company's standard product line does not address the above key customer requirements. The customer made it clear that there is no room for negotiation on the key product requirements.

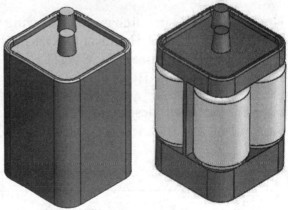

You contact the salesperson and obtain additional information on the customer and product. This is a very valuable customer with a long history of last minute product changes. The job has high visibility with great future potential.

In a design review meeting, you present a conceptual sketch. Your colleagues review the sketch. The team's consensus is to proceed with the conceptual design.

The first key design decision is the battery. The battery type directly affects the flashlight body size, bulb intensity, case structure integrity, weight, manufacturing complexity and cost.

Review two potential battery options:

- A single 6-volt lantern battery.

- Four 1.5-volt D cell batteries.

The two options affect the product design and specification. Think about it.

A single 6-volt lantern battery is approximately 25% higher in cost and 35% more in weight. The 6-volt lantern battery does provide higher current capabilities and longer battery life.

A special battery holder is required to incorporate the four 1.5 volt D cell configuration. This would directly add to the cost and design time of the FLASHLIGHT assembly.

Time is critical. For the prototype, you decide to use a standard 6-volt lantern battery. This eliminates the requirement to design and procure a special battery holder. However, you envision the four D cell battery model for the next product revision.

Design the FLASHLIGHT assembly to accommodate both battery design options. Battery dimensional information is required for the design. Where do you go? Potential sources are product catalogs, company web sites, professional standards organizations, design handbooks and colleagues.

The team decides to purchase the following parts: 6-volt BATTERY, LENS ASSEMBLY, SWITCH and an O-RING. Model the following purchased parts: BATTERY, LENS assembly, SWITCH and the O-RING. The LENS assembly consists of the LENS and the BULB.

Your company will design, model and manufacture the following parts: BATTERYPLATE, LENSCAP and HOUSING.

Purchased Parts:	Designed Parts:
BATTERY	BATTERYPLATE
LENS assembly	MOLD TOOLING
*SWITCH	*LENSCAP
*O-RING	*HOUSING

*Parts addressed in Project 6.

The BATTERYPLATE, LENSCAP and HOUSING are plastic parts. Review the injection molded manufacturing process and the SOLIDWORKS Mold tools. Modify the part features to eject the part from the mold. Create the MOLD TOOLING for the BATTERYPLATE.

Part Template

Units are the measurement of physical quantities. Millimeter dimensioning and decimal inch dimensioning are the two most common unit types specified for engineering parts and drawings. The FLASHLIGHT project is designed in inch units and manufactured in millimeter units. Inch units are the primary unit and Millimeter units are the secondary unit.

Create two Part templates:

- PART-IN-ANSI

- PART-MM-ISO

Save the Part templates in the MY-TEMPLATES folder. System Options, File Locations option controls the file folder location of SOLIDWORKS documents. Utilize the File Locations option to reference your Part templates in the MY-TEMPLATES folder. Add the MY-TEMPLATES folder path name to the Document Templates File Locations list.

Activity: Create Two Part Templates

Create a PART-IN-ANSI Template.

1)   Click **New** 🗋 from the Menu bar.

2)   Double-click **Part** from the default Templates tab from the Menu bar.

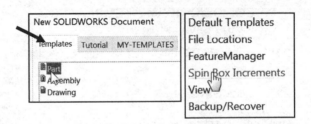

Set Dimensioning Standard.

3)   Click **Options** ⚙ from the Menu bar.

4)   Click the **System Options** tab.

5)   Click **Spin Box Increments**. View the default settings.

6)   Click **inside** the English units box.

7)   Enter **.10**in.

8)   Click **inside** the Metric units box.

9)   Enter **2.50**mm.

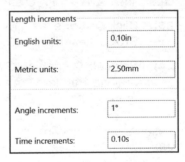

Set Document Properties. Set drafting standard, units, and precision.

10)   Click the **Document Properties** tab.

11)   Select **ANSI** from the Overall drafting standard drop-down menu.

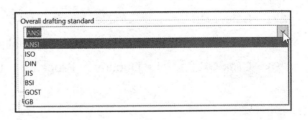

Set units and precision.

12) Click **Units**.

13) Select **IPS** for Unit system.

14) Select **.123** for Basic unit length decimal place.

15) Select **millimeters** for Dual dimension length unit.

16) Select **None** for Basic unit angle decimal place.

17) Click **OK** from the Document Properties - Units dialog box.

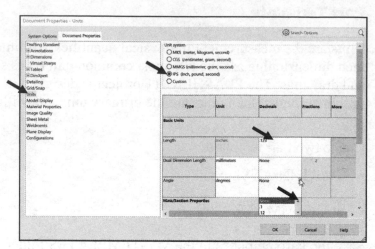

Save the template. Enter name.

18) Click **Save As** from the drop-down Menu bar.

19) Select **Part Templates (*.prtdot)** from the Save as type box.

20) Select **ENGDESIGN-W-SOLIDWORKS\MY-TEMPLATES** for the Save in folder.

21) Enter **PART-IN-ANSI** for File name.

22) Click **Save**.

Utilize the PART-IN-ANSI template to create the PART-MM-ISO template.

23) Click **Options** ⚙, **Document Properties** tab.

24) Select **ISO** from the Overall drafting standard drop-down menu.

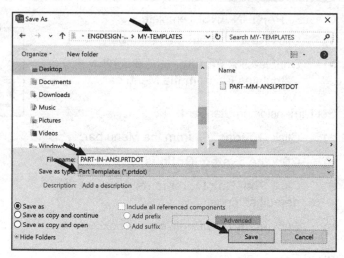

Set units and precision.

25) Click **Units**.

26) Select **MMGS** for Unit system.

27) Select **.12** for Basic unit length decimal place.

28) Select **None** for Basic unit angle decimal place.

29) Click **OK** from the Document Properties - Units dialog box.

Save the template. Enter name.

30) Click **Save As** from the drop-down Menu bar.

31) Select **Part Templates (*.prtdot)** from the Save as type box.

32) Select **ENGDESIGN-W-SOLIDWORKS\MY-TEMPLATES** for the Save in folder.

33) Enter **PART-MM-ISO** for File name.

34) Click **Save**.

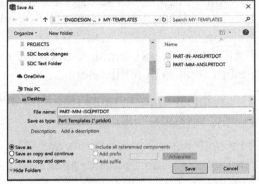

If needed, set the System Options for File Locations to display in the New dialog box.

35) Click **Options** ⚙ from the Menu bar.

36) Click **File Locations** from the System Options tab.

37) Select **Document Templates** from Show folders for.

38) Click the **Add** button.

39) Select the **MY-TEMPLATES** folder.

40) Click **OK** from the Browse for Folder dialog box.

41) Click **OK** from the System Options dialog box.

Close all documents.

42) Click **Windows**, **Close All** from the Menu bar.

Each folder listed in the System Options, File Locations, Document Templates, Show Folders For option produces a corresponding tab in the New SOLIDWORKS Document dialog box. The order in the Document Templates box corresponds to the tab order in the New dialog box.

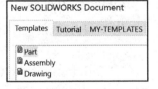

The MY-TEMPLATES tab is visible when the folder contains SOLIDWORKS Template documents. Create the PART-MM-ANSI template as an exercise.

To remove Tangent edges on a model, click **Display/Selections** from the Options menu, and check the **Removed** box.

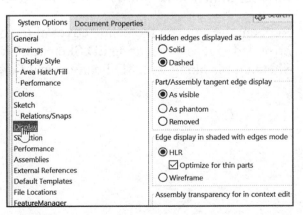

The PART-IN-ANSI template contains Document Properties settings for the parts contained in the FLASHLIGHT assembly. Substitute the PART-MM-ISO or PART-MM-ANSI template to create the identical parts in millimeters.

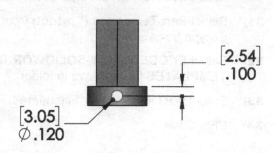

The primary units in this Project are IPS (inch, pound, second).

The optional secondary units are MMGS (millimeter, gram, second) and are indicated in brackets [ ].

Illustrations are provided in both inch and millimeter. Utilize inch, millimeter or both.

To set dual dimensions, select Options, Document Properties, Dimensions. Check the Dual dimensions display box as illustrated.

To set dual dimensions for an active document, check the Dual Dimension box in the Dimension PropertyManager.

Enter toolbars, features in SOLIDWORKS Help Search category to review the function of each Features toolbar.

Additional information on System Options, Document Properties, File Locations and Templates is found in SOLIDWORKS Help. Keywords: Options (detailing, units); templates; Files (locations); menus and toolbars (features, sketch).

Model about the Origin; this provides a point of reference for your dimensions to fully define the sketch.

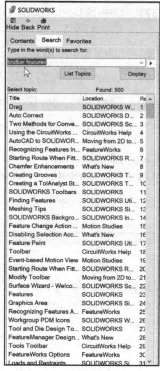

Redeem the code on the inside cover of the book. View the provided videos on creating 2D Sketches, Sketch Planes and Sketch tools along with 3D Features and Design Intent.

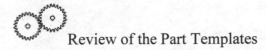 Review of the Part Templates

You created two Part templates: PART-IN-ANSI and PART-MM-ISO. Note: Other templates were created in the previous project. The Document Properties Dimensioning Standard, units and decimal places are stored in the Part Templates.

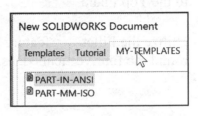

The File Locations System Option, Document Templates option controls the reference to the MY-TEMPLATES folder.

Note: In some network locations and school environments, the File Locations option must be set to MY-TEMPLATES for each session of SOLIDWORKS.

You can exit SOLIDWORKS at any time during this project. Save your document. Select File, Exit from the Menu bar.

BATTERY Part

The BATTERY is a simplified representation of a purchased OEM part. Represent the battery terminals as cylindrical extrusions. The battery dimensions are obtained from the ANSI standard 908D.

A 6-volt lantern battery weighs approximately 1.38 pounds (0.62kg). Locate the center of gravity closest to the center of the battery.

Create the BATTERY part. Use features to create parts. Features are building blocks that add or remove material.

Utilize the Instant3D tool to create the Extruded Boss/Base feature vs. using the Boss-Extrude PropertyManager.

The Extrude Boss/Base features add material. The Base feature (Boss-Extrude1) is the first feature of the part. Note: The default End Condition for Instant3D is Blind.

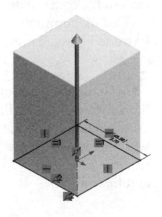

Apply symmetry. Use the Center Rectangle Sketch tool on the Top Plane. The 2D Sketch profile is centered at the Origin.

Extend the profile perpendicular (⊥) to the Top Plane.

Utilize the Fillet feature to round the four vertical edges.

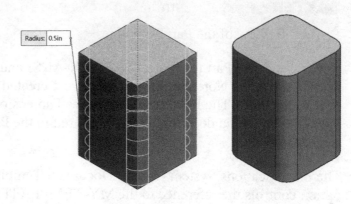

The Extruded Cut feature removes material from the top face. Utilize the top face for the Sketch plane. Utilize the Offset Entity Sketch tool to create the profile.

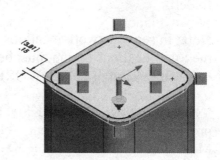

Utilize the Fillet feature to round the top narrow face.

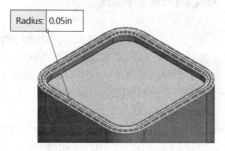

The Extruded Boss/Base feature adds material. Conserve design time. Represent each of the terminals as a cylindrical Extruded Boss feature.

Think design intent. When do you use the various End Conditions and Geometric sketch relations? What are you trying to do with the design? How does the component fit into the assembly? Design for change and flexibility.

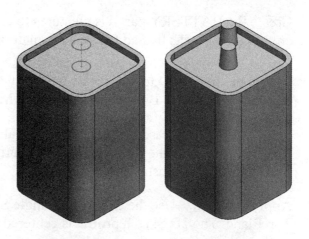

BATTERY Part - Extruded Boss/Base Feature

The Extruded Boss/Base feature requires:

- Sketch plane (Top).

- Sketch profile (Rectangle).
  - Geometric relations and dimensions.

- End Condition Depth (Blind) in Direction 1.

Create a new part named BATTERY. Insert an Extruded Boss/Base feature. Extruded features require a Sketch plane. The Sketch plane determines the orientation of the Extruded Base feature. The Sketch plane locates the Sketch profile on any plane or face.

The Top Plane is the Sketch plane. The Sketch profile is a rectangle. The rectangle consists of two horizontal lines and two vertical lines.

Geometric relations and dimensions constrain the sketch in 3D space. The Blind End Condition in Direction 1 requires a depth value to extrude the 2D Sketch profile and to complete the 3D feature.

🔅 Alternate between the Features tab and the Sketch tab in the CommandManager to display the available Feature and Sketch tools for the Part document.

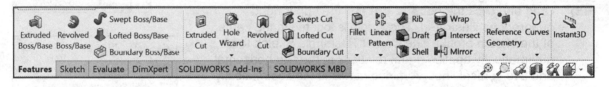

Activity: BATTERY Part - Create the Extruded Base Feature

Create a New part.

43) Click **New** ⬜ from the Menu bar.

44) Click the **MY-TEMPLATES** tab. The MY-TEMPLATES tab was created earlier in the book.

45) Double-click **PART-IN-ANSI**, [**PART-MM-ISO**].

Save the part. Enter name and description.

46) Click **Save** 💾.

47) Select **PROJECTS** for Save in folder.

48) Enter **BATTERY** for File name.

49) Enter **BATTERY**, **6-VOLT** for Description.

50) Click **Save**. The Battery FeatureManager is displayed.

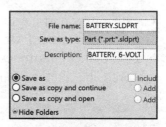

Select the Sketch plane.

51) Right-click **Top Plane** from the FeatureManager. This is your Sketch plane.

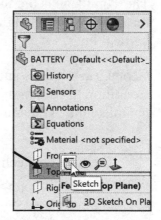

Sketch the 2D Sketch profile centered at the Origin.

52) Click **Sketch** ✏️ from the Context toolbar. The Sketch toolbar is displayed.

53) Click the **Center Rectangle** ⬜ Sketch tool. The Center Rectangle icon is displayed.

54) Click the **Origin**. This is your first point.

55) Drag and click the **second point** in the upper right quadrant as illustrated. The Origin is located in the center of the sketch profile. The Center Rectangle Sketch tool automatically applies equal relations to the two horizontal and two vertical lines. A midpoint relation is automatically applied to the Origin.

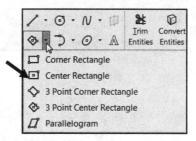

A goal of this book is to expose the new SOLIDWORKS user to various tools, techniques and procedures. The text may not always use the most direct tool or process.

💡 Click **View**, **Hide/Show**, **Sketch Relations** from the Main menu to view sketch relations in the Graphics area.

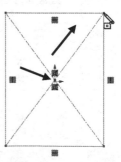

Dimension the sketch.

56) Click the **Smart Dimension** ✎ Sketch tool.

57) Click the **top horizontal line**.

58) Click a **position** above the horizontal line.

59) Enter **2.700**in, [**68.58**] for width.

60) Click the **Green Check mark** ✔ in the Modify dialog box.

61) Enter **2.700**in, [68.58] for height as illustrated.

62) Click the **Green Check mark** ✔ in the Modify dialog box. The black Sketch status is fully defined. Click **OK** ✔ from the Dimension PropertyManager.

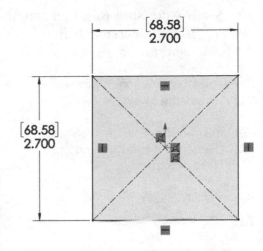

Exit the Sketch.
63) Click **Exit Sketch**.

Insert an Extruded Boss/Base feature. Apply the Instant3D tool. The Instant3D tool provides the ability to drag geometry and dimension manipulator points to resize or to create features directly in the Graphics window.

Display an Isometric view. Use the on-screen ruler.
64) Press the **space bar** to display the Orientation dialog box. Click **Isometric view** ▣.

65) Click the **front horizontal line** as illustrated. A green arrow is displayed.

66) Click and drag the **arrow** upward.

67) Click the on-screen ruler at **4.1**in, [104.14] as illustrated. This is the depth in direction 1. The extrude direction is upwards. Boss-Extrude1 is displayed In the FeatureManager.

Check the Boss-Extrude1 feature depth dimension.
68) Right-click **Boss-Extrude1** from the FeatureManager.

69) Click **Edit Feature** 🗔 from the Context toolbar. 4.100in is displayed for depth. Blind is the default End Condition. Note: If you did not select the correct depth, input the depth in the Boss-Extrude1 PropertyManager.

70) Click **OK** ✔ from the Boss-Extrude1 PropertyManager.

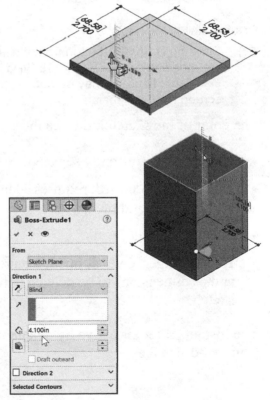

Modify the **Spin Box Increments** in System Options to
display different increments in
the on-screen ruler.

Fit the part to the Graphics window.
71) Press the **f** key.

Rename the Boss-Extrude1 feature.
72) Rename **Boss-Extrude1** to
Base Extrude.

Save the BATTERY.
73) Click **Save** 💾.

Modify the BATTERY.
74) Click **Base Extrude** from the
FeatureManager. Instant3D
is activated by default.

75) Drag the **manipulator point**
upward and click the on-
screen ruler to create a
5.000in, [127] depth as
illustrated. Blind is the default End Condition.

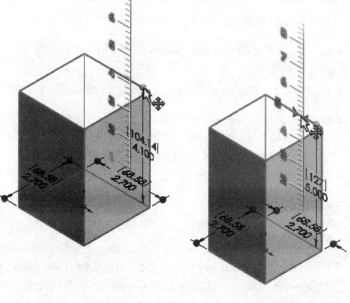

Return to the 4.100 depth.
76) Click the **Undo** ↶ button from the Menu bar. The depth of
the model is 4.100in, [104.14]. Blind is the default End
Condition. Practice may be needed to select the correct on-
screen ruler dimension.

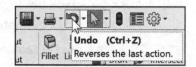

The color of the sketch indicates the sketch status.

- **Light Blue** - Currently selected.

- **Blue** - Under defined, requires additional geometric
  relations and or dimensions.

- **Black** - Fully defined.

- **Red/Orange** - Over defined, requires geometric relations
  and or dimensions to be deleted or redefined to solve the
  sketch.

The Instant3D tool is active by default in the Features toolbar located in the
CommandManager.

BATTERY Part - Fillet Feature Edge

Fillet 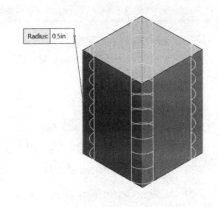 features remove sharp edges. Utilize Hidden Lines Visible from the Heads-up View toolbar to display hidden edges.

An edge Fillet feature requires:

- A selected edge

- Fillet radius

Select a vertical edge. Select the Fillet feature from the Features toolbar. Enter the Fillet radius. Add the other vertical edges to the Items To Fillet option.

The order of selection for the Fillet feature is not predetermined. Select edges to produce the correct result. The Fillet feature uses the Fillet PropertyManager. The Fillet PropertyManager provides the ability to select either the Manual or FilletXpert tab.

Each tab has a separate menu and PropertyManager. The Fillet PropertyManager and FilletXpert PropertyManager display the appropriate selections based on the type of fillet you create.

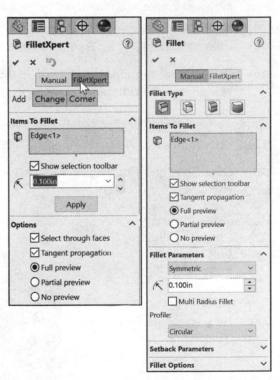

The FilletXpert automatically manages, organizes and reorders your fillets in the FeatureManager design tree. The FilletXpert PropertyManager provides the ability to add, change or corner fillets in your model. The PropertyManager remembers its last used state. View the SOLIDWORKS tutorials for additional information on fillets.

Use the Fillet tool to create symmetrical conic shaped fillets for parts, assemblies, and surfaces. You can apply conic shapes to *Constant Size*, *Variable Size*, and *Face* fillets.

Activity: BATTERY Part - Fillet Feature Edge

Display Hidden edges.

77) Click **Hidden Lines Visible** from the Heads-up View toolbar.

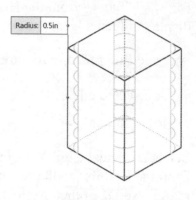

Insert a Fillet feature.

78) Click the **Fillet** feature tool. The Fillet PropertyManager is displayed.

79) Click the **Manual** tab.

80) Click **Constant Size Fillet** for type.

81) Click the **left front vertical edge** as illustrated to show the mouse pointer edge icon. Edge<1> is displayed in the Items To Fillet box. The fillet option pop-up toolbar is displayed. Options are model dependent.

82) Select the **Connected to start face, 3 Edges icon**. The four selected edges are displayed in the Edges, Faces, Features, and Loop box.

83) Enter **.500**in, **[12.7]** for Radius. Accept the default settings.

84) Click **OK** from the Fillet PropertyManager. Fillet1 is displayed in the FeatureManager.

Display an Isometric, Shaded with Edges view.

85) Click **Isometric view**.

86) Click **Shaded With Edges** from the Heads-up View toolbar.

Rename the feature.

87) Rename **Fillet1** to **Side Fillets** in the FeatureManager.

Save the BATTERY.

88) Click **Save**.

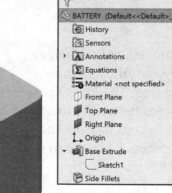

BATTERY Part - Extruded Cut Feature

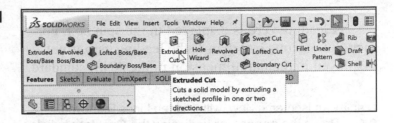

An Extruded Cut feature removes material. An Extruded Cut feature requires:

- Sketch plane (Top face)

- Sketch profile (Offset Entities)

- End Condition depth (Blind) in Direction 1

The Offset Entity Sketch tool uses existing geometry, extracts an edge or face and locates the geometry on the current Sketch plane.

Offset the existing Top face for the 2D sketch. Utilize the default Blind End Condition in Direction 1.

Activity: BATTERY Part - Extruded Cut Feature

Select the Sketch plane.

89) Right-click the **Top face** of the BATTERY in the Graphics window. Base Extruded is highlighted in the FeatureManager.

Create a sketch.

90) Click **Sketch** from the Context toolbar. The Sketch toolbar is displayed.

Display the Top face.

91) Press the **space bar** to display the Orientation dialog box.

92) Click **Top view**.

Offset the existing geometry from the boundary of the Sketch plane.

93) Click the **Offset Entities** Sketch tool. The Offset Entities PropertyManager is displayed.

94) Enter **.150**in, [**3.81**] for the Offset Distance.

95) If needed check the **Reverse** box. The new Offset profile displays inside the original profile.

96) Click **OK** from the Offset Entities PropertyManager.

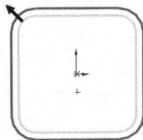

💡 A leading zero is displayed in the spin box. For inch dimensions less than 1, the leading zero is not displayed in the part dimension in the ANSI standard.

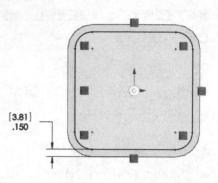

Display an Isometric view, with Hidden Lines Removed.
97) Press the **space bar** to display the Orientation dialog box.

98) Click **Isometric view** 🔲.

99) Click **Hidden Lines Removed** ⬡ from the Heads-up View toolbar.

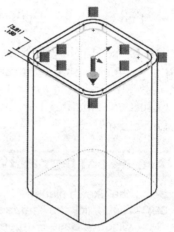

Insert an Extruded Cut feature. As an exercise, use the Instant3D tool to create the Extruded Cut feature. In this section, the Extruded-Cut PropertyManager is used.

100) Click the **Extruded Cut** 🔲 feature tool. The Cut-Extrude PropertyManager is displayed.

101) Enter **.200**in, **[5.08]** for Depth in Direction 1. Accept the default settings.

102) Click **OK** ✔ from the Cut-Extrude PropertyManager. Cut-Extrude1 is displayed in the FeatureManager.

Rename the feature.
103) Rename **Cut-Extrude1** to **Top Cut** in the FeatureManager.

Save the BATTERY
104) Click **Save** 💾.

The Cut-Extrude PropertyManager contains numerous options. The Reverse Direction option determines the direction of the Extrude. The Extruded Cut feature is valid only when the direction arrow points into material to be removed.

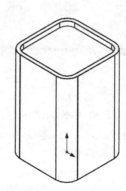

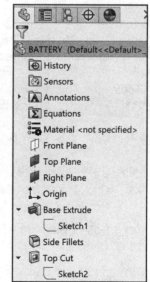

The Flip side to cut option determines if the cut is to the inside or outside of the Sketch profile.

The Flip side to cut arrow points outward. The Extruded Cut feature occurs on the outside of the BATTERY.

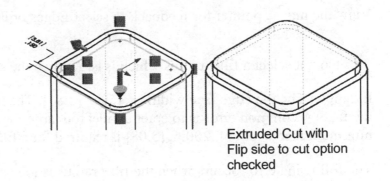

Extruded Cut with
Flip side to cut option
checked

BATTERY Part - Fillet Feature

The Fillet ⬚ feature tool rounds sharp edges with a constant radius by selecting a face. A Fillet requires a:

- Selected face

- Fillet radius

Activity: BATTERY Part - Fillet Feature Face

Insert a Fillet feature on the top face.

105) Click the **Fillet** ⬚ feature tool. The Fillet PropertyManager is displayed.

106) Click the **Manual** tab.

107) Click the **top thin face** as illustrated. Note the face icon feedback symbol. Face<1> is displayed in the Items To Fillet box.

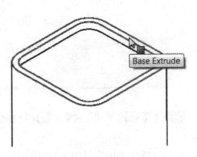

108) Click the **Constant Size Fillet** for Fillet Type.

109) Enter .050in, [1.27] for Radius.

110) Click **OK** ✔ from the Fillet PropertyManager. Fillet1 is displayed in the FeatureManager.

Rename the feature.
111) Rename **Fillet1** to **Top Face Fillet**.

Fit the model to the Graphics window. Display Hidden Lines Visible.
112) Press the **f** key.

113) Click **Hidden Lines Visible** ⬚.

Save the BATTERY.
114) Click **Save** 💾.

View the mouse pointer for feedback to select Edges or Faces for the fillet.

 Do not select a fillet radius which is larger than the surrounding geometry.

Example: The top edge face width is .150in, [3.81]. The fillet is created on both sides of the face. A common error is to enter a Fillet too large for the existing geometry. A minimum face width of .200in, [5.08] is required for a fillet radius of .100in, [2.54].

The following error occurs when the fillet radius is too large for the existing geometry.

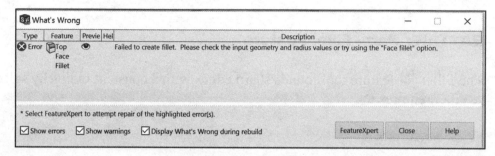

Avoid the fillet rebuild error. Use the FeatureXpert to address a constant radius fillet build error or manually enter a smaller fillet radius size.

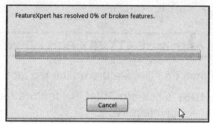

BATTERY Part - Extruded Boss/Base Feature

The Extruded Boss feature requires a truncated cone shape to represent the geometry of the BATTERY terminals. The Draft Angle option creates the tapered shape.

Sketch the first circle on the Top face. Utilize the Ctrl key to copy the first circle.

The dimension between the center points is critical. Dimension the distance between the two center points with an aligned dimension. The dimension text toggles between linear and aligned. An aligned dimension is created when the dimension is positioned between the two circles.

An angular dimension is required between the Right Plane and the centerline. Acute angles are less than 90°. Acute angles are the preferred dimension standard. The overall BATTERY height is a critical dimension. The BATTERY height is 4.500in, [114.3].

Calculate the depth of the extrusion: For inches: 4.500in - (4.100in Base-Extrude height - .200in Offset cut depth) = .600in. The depth of the extrusion is .600in.

For millimeters: 114.3mm - (104.14mm Base-Extrude height - 5.08mm Offset cut depth) = 15.24mm. The depth of the extrusion is 15.24mm.

Activity: BATTERY Part - Extruded Boss Feature

Select the Sketch plane.

115) Right-click the **top face** of the Top Cut feature in the Graphics window. This is your Sketch plane.

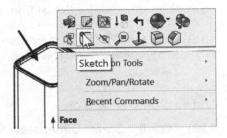

Create the sketch.

116) Click **Sketch** from the Context toolbar. The Sketch toolbar is displayed.

117) Click **Top view** .

Sketch the Close profile.

118) Click the **Circle** Sketch tool. The Circle PropertyManager is displayed.

119) Click the **center point** of the circle coincident to the Origin .

120) Drag and click the **mouse pointer** to the right of the Origin as illustrated.

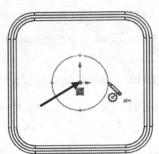

Add a dimension.

121) Click the **Smart Dimension** Sketch tool.

122) Click the **circumference** of the circle.

123) Click a **position** diagonally to the right.

124) Enter **.500**in, [**12.7**].

125) Click the **Green Check mark** in the Modify dialog box. The black sketch is fully defined.

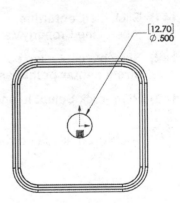

Copy the sketched circle.

126) Right-click **Select** to de-select the Smart Dimension Sketch tool.

127) Hold the **Ctrl** key down.

128) Click and drag the **circumference** of the circle to the upper left quadrant as illustrated.

129) Release the **mouse button**.

130) Release the **Ctrl** key. The second circle is selected. If needed click **OK** ✔ from the Circle PropertyManager.

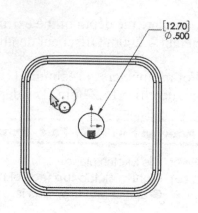

Add an Equal relation.

131) Hold the **Ctrl** key down.

132) Click the **circumference of the first circle**. The Properties PropertyManager is displayed. Both circles are selected and are displayed in green.

133) Release the **Ctrl** key.

134) Right-click **Make Equal** from the Context toolbar.

135) Click **OK** ✔ from the Properties PropertyManager. The second circle remains selected.

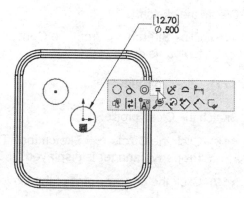

Add an aligned dimension.

136) Click the **Smart Dimension** ↖ Sketch tool.

137) Click the **two center points** of the two circles.

138) Click a **position** off the profile in the upper left corner.

139) Enter **1.000**in, **[25.4]** for the aligned dimension.

140) Click the **Green Check mark** ✔ in the Modify dialog box.

Insert a centerline.

141) Click the **Centerline** ✐ Sketch tool. The Insert Line PropertyManager is displayed.

142) Sketch a centerline between the **two circle center points** as illustrated.

143) Right-click **Select** to end the line.

💡 Double-click to end the centerline.

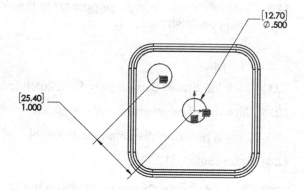

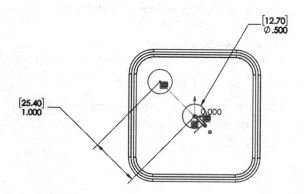

💡 Press the Enter key to accept the value in the Modify dialog box. The Enter key replaces the Green Check mark.

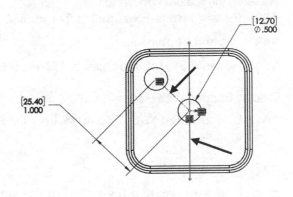

Add an angular dimension.

144) Click the **Smart Dimension** ✎ Sketch tool.

145) Click the **centerline** between the two circles.

146) Click the **Right Plane** (vertical line) from the FeatureManager.

147) Click a **position** between the centerline and the Right Plane, off the profile.

148) Enter **45**. Click **OK** ✔ from the Dimension PropertyManager.

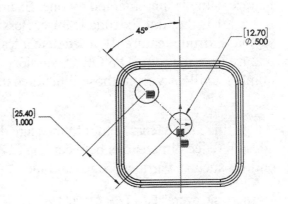

Fit the model to the Graphics window.
149) Press the **f** key.

Hide the Right Plane if needed. Save the model.
150) Right-click **Right Plane** in the FeatureManager. Click **Hide** from the Context toolbar.

151) Click **Save** 💾.

💡 Create an angular dimension between three points or two lines. Sketch a centerline/construction line when an additional point or line is required.

Display an Isometric view. Insert an Extruded Boss feature.
152) Click **Isometric view** 🧊.

153) Click the **Extruded Boss/Base** 🗍 feature tool. The Boss-Extrude PropertyManager is displayed. Blind is the default End Condition Type.

154) Enter **.600**in, **[15.24]** for Depth in Direction 1.

155) Click the **Draft ON/OFF** button.

156) Enter **5**deg in the Draft Angle box.

157) Click **OK** ✔ from the Boss-Extrude PropertyManager. The Boss-Extrude2 feature is displayed in the FeatureManager.

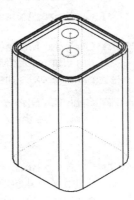

Rename the feature and sketch.

158) Rename **Boss-Extrude2** to **Terminals**.

159) **Expand** Terminals.

160) Rename **Sketch3** to **Sketch-TERMINALS**.

Display Shaded With Edges. Save the model.

161) Click **Shaded With Edges** 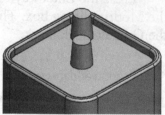 from the Heads-up View toolbar.

162) Click **Save** .

Each time you create a feature of the same feature type, the feature name is incremented by one. Example: Boss-Extrude1 is the first Extrude feature. Boss-Extrude2 is the second Extrude feature. If you delete a feature, rename a feature or exit a SOLIDWORKS session, the feature numbers will vary from those illustrated in the text.

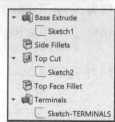

Utilize the Measure tool to measure distances and add reference dimensions between the COM point and entities such as vertices, edges and faces.

Measure the overall BATTERY height.

163) Click **Front view** .

164) Click the **Measure** tool from the Evaluate tab in the CommandManager. The Measure - BATTERY dialog box is displayed.

165) Click the **Show XYZ Measurements** option. This should be the *only active* option.

166) Click the **top edge** of the battery terminal as illustrated.

167) Click the **bottom edge** of the battery. The overall height, Delta Y is 4.500, [114.3]. Apply the Measure tool to ensure a proper design.

168) **Close** the Measure - BATTERY dialog box.

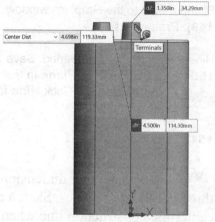

The Measure tool provides the ability to display custom settings. Click **Units/Precision** from the Measure dialog box. View your options. Click **OK**.

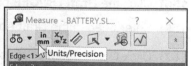

The Selection Filter  option toggles the Selection Filter toolbar. When Selection Filters are activated, the mouse pointer displays the Filter icon . The Clear All Filters tool removes the current Selection Filters. The Help icon displays the SOLIDWORKS Online Users Guide.

Display a Trimetric view.
169) Click **Trimetric view** from the Heads-up View toolbar.

Save the BATTERY.
170) Click **Save** .

 View the provided videos on creating 2D Sketching, Sketch Planes and Sketch tools along with 3D Features and Design Intent to enhance your experience in this section.

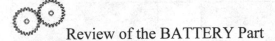 Review of the BATTERY Part

The BATTERY utilized a 2D Sketch profile located on the Top Plane. The 2D Sketch profile utilized the Center Rectangle Sketch tool. The Center Rectangle Sketch tool applied equal geometric relations to the two horizontal and two vertical lines. A midpoint relation was added to the Origin.

The Extruded Boss/Base feature was created using the Instant3D tool. Blind was the default End Condition. The Fillet feature rounded sharp edges. All four edges were selected to combine common geometry into the same Fillet feature. The Fillet feature also rounded the top face. The Sketch Offset Entity created the profile for the Extruded Cut feature.

The Terminals were created with an Extruded Boss feature. You sketched a circular profile and utilized the Ctrl key to copy the sketched geometry.

A centerline was required to locate the two holes with an angular dimension. The Draft Angle option tapered the Extruded Boss feature. All feature names were renamed.

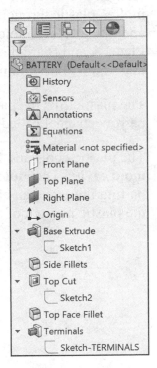

Injection Molded Process

Lee Plastics of Sterling, MA, is a precision injection molding company. Through the World Wide Web (www.leeplastics.com), review the injection molded manufacturing process.

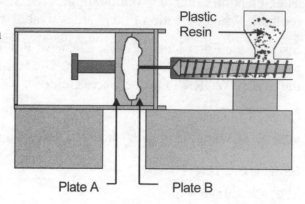

Plastic Resin

Plate A   Plate B

The injection molding process is as follows:

An operator pours the plastic resin, in the form of small dry pellets, into a hopper. The hopper feeds a large augur screw. The screw pushes the pellets forward into a heated chamber. The resin melts and accumulates into the front of the screw.

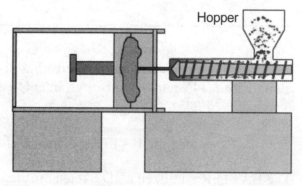

Hopper

At high pressure, the screw pushes the molten plastic through a nozzle, to the gate and into a closed mold (Plates A & B). Plates A and B are the machined plates that you will design in this project.

The plastic fills the part cavities through a narrow channel called a gate.

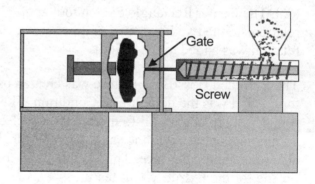

Gate

Screw

The plastic cools and forms a solid in the mold cavity. The mold opens (along the parting line), and an ejection pin pushes the plastic part out of the mold into a slide.

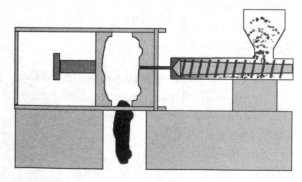

Injection Molded Process
(Courtesy of Lee Plastics, Inc.)

BATTERYPLATE Part

The BATTERYPLATE is a critical plastic part. The BATTERYPLATE:

- Aligns the LENS assembly.

- Creates an electrical connection between the BATTERY and LENS.

Design the BATTERYPLATE. Utilize features from the BATTERY to develop the BATTERYPLATE. The BATTERYPLATE is manufactured as an injection molded plastic part. Build Draft into the Extruded Boss/Base features.

Edit the BATTERY features. Create two holes from the original sketched circles. Apply the Instant3D tool to create an Extruded Cut feature.

Modify the dimensions of the Base feature. Add a 3° draft angle.

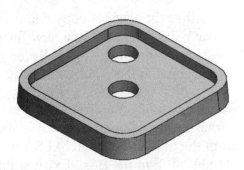

🔅 A sand pail contains a draft angle. The draft angle assists the sand to leave the pail when the pail is flipped upside down.

Insert an Extruded Boss/Base feature. Offset the center circular sketch.

The Extruded Boss/Base feature contains the LENS. Create an inside draft angle. The draft angle assists the LENS into the Holder.

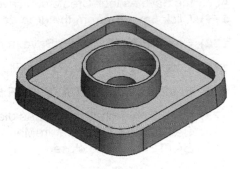

Insert a Face Fillet and a Multi-radius Edge Fillet to remove sharp edges. Plastic parts require smooth edges. Group Fillet features together into a folder.

Perform a Draft Analysis on the part and create the Core and Cavity mold tooling.

🔅 Group fillets together into a folder to locate them quickly. Features listed in the FeatureManager must be continuous in order to be placed as a group into a folder.

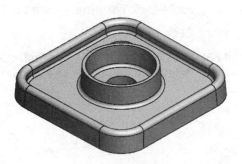

Save As, Delete, Edit Feature and Modify

Create the BATTERYPLATE part from the BATTERY part.
Utilize the Save As tool from the Menu bar to copy the
BATTERY part to the BATTERYPLATE part.

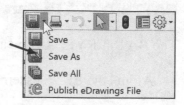

- Utilize the Save As/Save as command to save
  the file in another file format.

- Utilize the Save as copy and continue command
  to save the document to a new file name without
  replacing the active document.

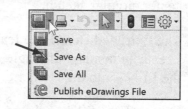

- Utilize the Save as copy and open command to
  save the document to a new file name that becomes the active document. The original
  document remains open. References to the original document are not automatically
  assigned to the copy.

Reuse existing geometry. Create two holes. Delete the Terminals feature and reuse the
circle sketch. Select the sketch in the FeatureManager. Create an Extruded Cut feature
from the Sketch-TERMINALS using the Instant3D tool. Blind is the default End
Condition. Edit the Boss-Extrude feature. Modify the overall depth. Rebuild the model.

Activity: BATTERYPLATE Part - Save As, Delete, Modify, and Edit Feature

Apply the Save As tool. Create and save a new part. Enter name and description.
171) Click **Save As** from the drop-down Menu bar.

172) Select **PROJECTS** for Save In folder.

173) Enter **BATTERYPLATE** for File name.

174) Enter **BATTERY PLATE, FOR 6-VOLT** for Description.

175) Click **Save** from the Save As dialog box. The
BATTERYPLATE FeatureManager is displayed. The
BATTERY part is closed.

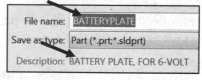

Delete the Terminals feature.
176) Right-click **Terminals** from the FeatureManager.

177) Click **Delete**.

178) Click **Yes** from the Confirm Delete dialog box. Do not
delete the two-circle sketch, Sketch-TERMINALS.

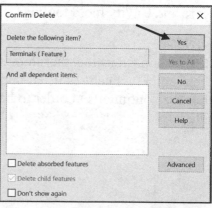

Create an Extruded Cut feature from the Sketch-TERMINALS using Instant3D.

179) Click **Sketch-TERMINALS** from the FeatureManager. If needed, Show Sketches.

180) Click the **circumference** of the center circle as illustrated. An arrow is displayed.

181) Hold the **Alt** key down.

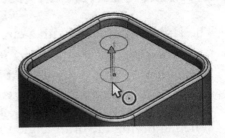

182) Drag the **arrow** downward below the model to create a hole in Direction 1.

183) Release the mouse button on the **vertex** as illustrated. This ensures a Through All End Condition with model dimension changes.

184) Release the **Alt** key. Boss-Extrude1 is displayed in the FeatureManager.

185) Rename the **Boss-Extrude** feature to **Holes** in the FeatureManager.

Edit the Base Extrude feature.

186) Right-click **Base Extrude** from the FeatureManager.

187) Click **Edit Feature** 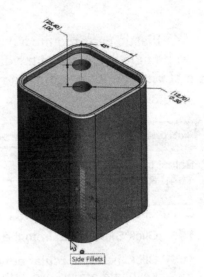 from the Context toolbar. The Base Extrude PropertyManager is displayed.

Modify the overall depth.

188) Enter .400in, [10.16] for Depth in Direction 1.

189) Click the **Draft ON/OFF** button.

190) Enter 3.00deg in the Angle box.

191) Click **OK** ✔ from the Base Extrude PropertyManager.

Fit the model to the Graphics window.

192) Press the **f** key.

Save the BATTERYPLATE.

193) Click **Save** 💾 .

🔅 Modify the **Spin Box Increments** in System Options to display different increments for the Instant3D on-screen ruler.

To delete both the feature and the sketch at the same time, select the Also delete absorbed features check box from the Confirm Delete dialog box.

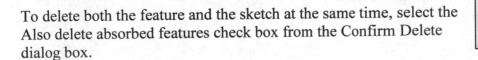

BATTERYPLATE Part - Extruded Boss Feature

The Holder is created with a circular Extruded Boss/Base feature. Utilize the Offset Entities ⌐ Sketch tool to create the second circle. Apply a draft angle of 3° in the Extruded Boss feature.

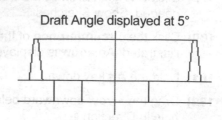

Draft Angle displayed at 5°

When applying the draft angle to the two concentric circles, the outside face tapers inwards and the inside face tapers outwards.

☀ Plastic parts require a draft angle. Rule of thumb: 1° to 5° is the draft angle. The draft angle is created in the direction of pull from the mold. This is defined by geometry, material selection, mold production and cosmetics. Always verify the draft with the mold designer and manufacturer.

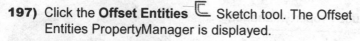

Activity BATTERYPLATE Part - Extruded Boss Feature

Select the Sketch plane.
194) Right-click the **top face** of Top Cut. This is your Sketch plane.

Create the sketch.
195) Click **Sketch** ⌐ from the Context toolbar.

196) Click the **top circular edge** of the center hole. Note: Use the keyboard arrow keys or the middle mouse button to rotate the sketch if needed.

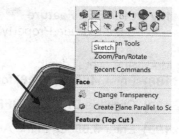

197) Click the **Offset Entities** ⌐ Sketch tool. The Offset Entities PropertyManager is displayed.

198) Enter **.300**in, **[7.62]** for Offset Distance. Accept the default settings.

199) Click **OK** ✓ from the Offset Entities PropertyManager.

200) Drag the **dimension** off the model.

Create the second offset circle.
201) Click the **offset circle** in the Graphics window.

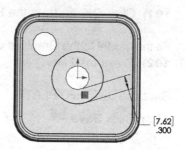

202) Click the **Offset Entities** ⌐ Sketch tool. The Offset Entities PropertyManager is displayed.

203) Enter **.100**in, **[2.54]** for Offset Distance.

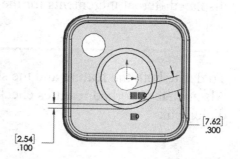

204) Click **OK** ✔ from the Offset Entities PropertyManager. Drag the dimension off the model. Two offset concentric circles define the sketch.

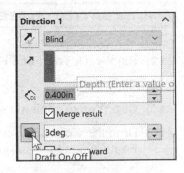

Insert an Extruded Boss/Base feature.

205) Click the **Extruded Boss/Base** 🗔 feature tool. The Boss-Extrude PropertyManager is displayed.

206) Enter **.400**in, **[10.16]** for Depth in Direction 1.

207) Click the **Draft ON/OFF** button.

208) Enter **3**deg in the Angle box.

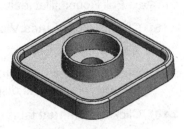

209) Click **OK** ✔ from the Boss-Extrude PropertyManager. The Boss-Extrude1 feature is displayed in the FeatureManager.

Rename the feature.

210) Rename the **Boss-Extrude1** feature to **Holder** in the FeatureManager.

Save the model.

211) Click **Save** 💾.

BATTERYPLATE Part - Fillet Features: Full Round and Multiple Radius Options

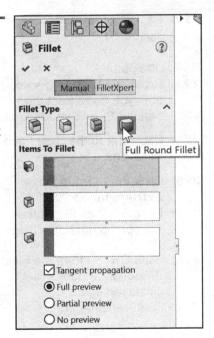

Use the Fillet 🗔 feature tool to smooth rough edges in a model. Plastic parts require fillet features on sharp edges. Create two Fillets. Utilize different techniques. The current Top Face Fillet produced a flat face. Delete the Top Face Fillet. The first Fillet feature is a Full round fillet. Insert a Full round fillet feature on the top face for a smooth rounded transition.

The second Fillet feature is a Multiple Size fillet. Select a different radius value for each edge in the set. Select the inside and outside edge of the Holder. Select all inside tangent edges of the Top Cut. A Multiple Size fillet is utilized next as an exercise. There are machining instances where radius must be reduced or enlarged to accommodate tooling. Note: There are other ways to create Fillets.

💡 Group Fillet features into a Fillet folder. Placing Fillet features into a folder reduces the time spent for your mold designer or toolmaker to look for each Fillet feature in the FeatureManager.

Activity: BATTERYPLATE Part - Fillet Features: Full Round, Multiple Size Options

Delete the Top Edge Fillet.

212) Right-click **Top Face Fillet** from the FeatureManager.

213) Click **Delete**.

214) Click **Yes** to confirm delete.

215) Drag the **Rollback** bar below Top Cut in the FeatureManager as illustrated.

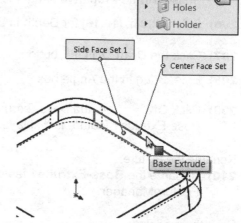

Create a Full Round fillet feature.

216) Click **Hidden Lines Visible** 🗇 from the Heads-up View toolbar.

217) Click the **Fillet** 🗇 feature tool. The Fillet PropertyManager is displayed.

218) Click the **Manual** tab.

219) Click the **Full Round Fillet button** for Fillet Type.

220) Click the **inside Top Cut face** for Side Face Set 1 as illustrated.

221) Click **inside** the Center Face Set box.

222) Click the **top face** for Center Face Set as illustrated.

Rotate the part.

223) Press the **Left Arrow** key until you can select the outside Base Extrude face.

224) Click **inside** the Side Face Set 2 box.

225) Click the **outside Base Extrude face** for Side Face Set 2 as illustrated. Accept the default settings.

226) Click **OK** ✔ from the Fillet PropertyManager. Fillet1 is displayed in the FeatureManager.

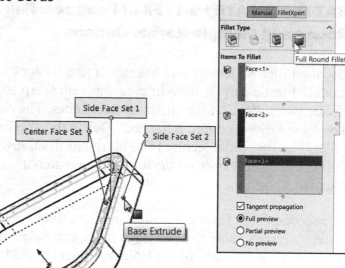

Rename Fillet1.

227) Rename **Fillet1** to **TopFillet**.

Display an Isometric view with Hidden Lines Removed. Save the
BATTERYPLATE.

228) Click **Isometric view** .

229) Click **Hidden Lines Removed** from the Heads-up View toolbar.

230) Drag the **Rollback bar** to the bottom of the FeatureManager.

231) Click **Save** .

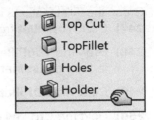

Create a Multiple Size fillet feature.

232) Click the **bottom outside circular edge** of the Holder as illustrated.

233) Click the **Fillet** feature tool. The Fillet PropertyManager is displayed.

234) Click the **Constant Size Fillet** button for Fillet Type.

235) Enter **.050**in, **[1.27]** for Radius.

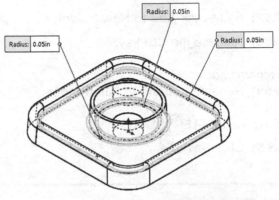

236) Click the **bottom inside circular edge** of the Top Cut as illustrated.

237) Click the **inside edge** of the Top Cut.

238) Check the **Tangent propagation** box.

239) Check the **Multiple Radius Fillet** box.

Modify the Fillet values.

240) Click the **Radius** box for the Holder outside edge.

241) Enter **0.060**in, **[1.52]**.

242) Click the **Radius** box for the Top Cut inside edge.

243) Enter **0.040**in, **[1.02]**.

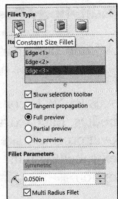

244) Click **OK** from the Fillet PropertyManager. Fillet is displayed in the FeatureManager.

Rename the Fillet2 folder.

245) Rename **Fillet#** to **HolderFillet**.

Display Shaded With Edges.

246) Click **Shaded With Edges** from the Heads-up View toolbar. View the results in the Graphics window.

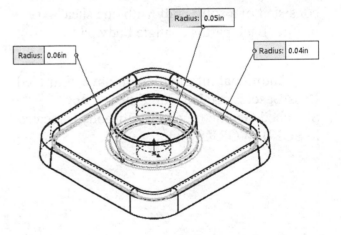

Group the Fillet features into a new folder.

247) Click **TopFillet** from the FeatureManager.

248) Drag the **TopFillet** feature directly above the HolderFillet feature in the FeatureManager.

249) Click **HolderFillet** in the FeatureManager.

250) Hold the **Ctrl** key down.

251) Click **TopFillet** in the FeatureManager.

252) Right-click **Add to New Folder**.

253) Release the **Ctrl** key.

Rename Folder1.

254) Rename **Folder1** to **FilletFolder**.

Save the BATTERYPLATE.

255) Click **Save** .

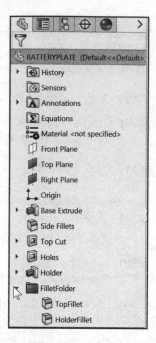

Multi-body Parts and the Extruded Boss/Base Feature

A Multi-body part has separate solid bodies within the same part document.

A WRENCH consists of two cylindrical bodies. Each extrusion is a separate body. The oval profile is sketched on the right plane and extruded with the Up to Body End Condition option.

The BATTERY and BATTERYPLATE parts consisted of a solid body with one sketched profile. Each part is a single body part.

Additional information on Save, Extruded Boss/Base, Extruded Cut, Fillets, Copy Sketched Geometry, and Multi-body is located in SOLIDWORKS Help.

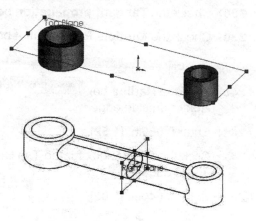

Multi-body part
Wrench

Refer to **Help, SOLIDWORKS Tutorials, Advanced Techniques, Multibody Parts** for additional information.

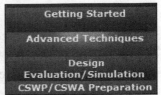

 Review of the BATTERYPLATE Part

The Save As option was utilized to copy the BATTERY part to the BATTERYPLATE part. You created a hole in the BATTERYPLATE using Instant3D and modified features using the PropertyManager.

The BATTERYPLATE is a plastic part. The Draft Angle option was added in the Extruded Base (Boss-Extrude1) feature.

The Holder Extruded Boss utilized a circular sketch and the Draft Angle option. The Sketch Offset tool created the circular ring profile. Multi radius Edge Fillets and Face Fillets removed sharp edges.

Similar Fillet features were grouped together into a folder. Features were renamed in the FeatureManager.

The BATTERY and BATTERYPLATE utilized an Extruded Boss/Base feature.

It is considered best practice to fully define all sketches in the model. However, there are times when this is not practical, generally when using the spline tool to create a complex freeform shape.

 Sketch dimensions are displayed in black.

 Feature dimensions are displayed in blue.

LENS Part

Create the LENS. The LENS is a purchased part. The LENS utilizes a Revolved Base feature.

Sketch a centerline and a closed profile on the Right Plane. Insert a Revolved Base feature. The Revolved Base feature requires an axis of revolution and an angle of revolution.

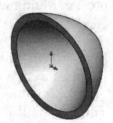

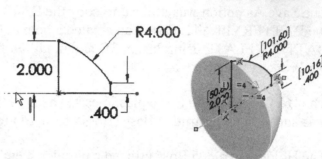

Insert the Shell feature. The Shell feature provides uniform wall thickness. Select the front face as the face to be removed.

Utilize the Convert Entities sketch tool to extract the back circular edge for the sketched profile. Insert an Extruded Boss feature from the back of the LENS.

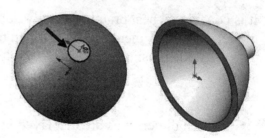

Sketch a single profile. Insert a Revolved Thin feature to connect the LENS to the BATTERYPLATE. The Revolved Thin feature requires thickness.

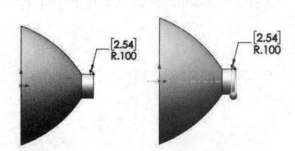

Insert a Counterbore Hole feature using the Hole Wizard tool. The BULB is located inside the Counterbore Hole.

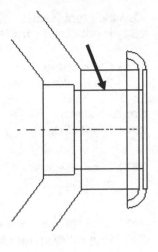

Insert the front Lens Cover with an Extruded Boss/Base feature. The Extruded Boss/Base feature is sketched on the Front Plane. Add a transparent Lens Shield with the Extruded Boss feature.

LENS Part - Revolved Base Feature

Create the LENS with a Revolved Base feature. The solid Revolved Base feature requires:

- Sketch plane (Right)
- Sketch profile
- Centerline
- Angle of Revolution (360°)

The profile lines reference the Top and Front Planes. Create the curve of the LENS with a 3-point arc.

Activity: LENS Part - Create a Revolved Base Feature

Create the New part.

256) Click **New** 📄 from the Menu bar.

257) Click the **MY-TEMPLATES** tab.

258) Double-click **PART-IN-ANSI**, **[PART-MM-ISO]** from the Template dialog box.

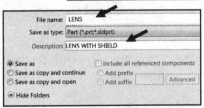

Save the part. Enter name and description.

259) Click **Save** 💾.

260) Select **PROJECTS** for Save in folder.

261) Enter **LENS** for File name.

262) Enter **LENS WITH SHIELD** for Description.

263) Click **Save**. The LENS FeatureManager is displayed.

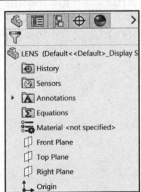

View the Front Plane.

264) Right click **Front Plane** from the FeatureManager.

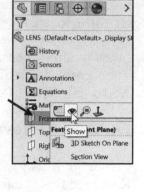

265) Click **Show** 👁 from the Context toolbar. Hide unwanted planes in the FeatureManager if needed.

Create a sketch.

266) Right-click **Right Plane** from the FeatureManager.

267) Click **Sketch** ⬜ from the Context toolbar. The Sketch toolbar is displayed.

268) Click the **Centerline** 〰 Sketch tool. The Insert Line PropertyManager is displayed.

269) Sketch a **horizontal centerline** collinear to the Top Plane, through the Origin ⌞ as illustrated, approximately 4 inches.

Sketch the profile. Create three lines.

270) Click the **Line** ╱ Sketch tool. The Insert Line PropertyManager is displayed.

271) Sketch a **vertical line** collinear to the Front Plane coincident with the Origin, approximately 2 inches.

272) Sketch a **horizontal line** coincident with the Top Plane, approximately 2 inches.

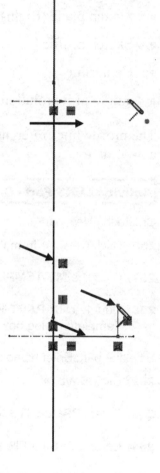

273) Sketch a **vertical line** approximately 1/3 the length of the first line. Right-click **End Chain**.

Create a 3 Point Arc. A 3 Point Arc requires three points.

274) Click the **3 Point Arc** ⌒ Sketch tool from the Consolidated Centerpoint Arc toolbar. Note the mouse pointer feedback icon.

275) Click the **top point** on the left vertical line. This is your first point.

276) Drag the **mouse pointer** to the right.

277) Click the **top point** on the right vertical line. This is your second point.

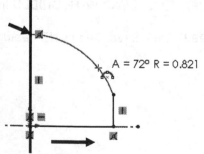

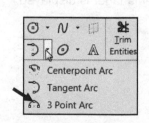

A = 72° R = 0.821

278) Drag the **mouse pointer** upward.

279) Click a **position** on the arc.

Add an Equal relation.
280) Right-click **Select** to deselect the sketch tool.

281) Click the **left vertical** line.

282) Hold the **Ctrl** key down.

283) Click the **horizontal** line. The Properties PropertyManager is displayed. The selected entities are displayed in the Selected Entities box.

284) Release the **Ctrl** key.

285) Right-click **Make Equal** from the Context toolbar.

286) Click **OK** ✔ from the Properties PropertyManager.

Add dimensions.
287) Click the **Smart Dimension** ↰ Sketch tool.

288) Click the **left vertical** line.

289) Click a **position** to the left of the profile.

290) Enter **2.000**in, **[50.8]**.

291) Click the **right vertical** line.

292) Click a **position** to the right of the profile.

293) Enter **.400**in, **[10.16]**. Click the **arc**.

294) Click a **position** to the right of the profile.

295) Enter **4.000**in, **[101.6]**. The black sketch is fully defined.

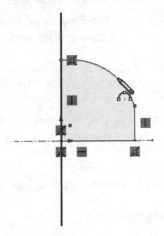

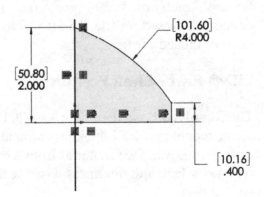

💡 Utilize **Tools, Sketch Tools, Check Sketch for Feature** option to determine if a sketch is valid for a specific feature and to understand what is wrong with a sketch.

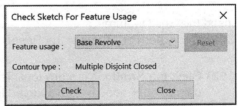

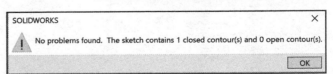

Insert the Revolved Base feature.

296) Click the **Revolved Boss/Base** 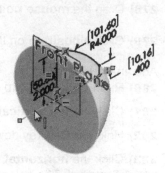 feature tool. The Revolve PropertyManager is displayed.

297) If needed, click the **horizontal centerline** for the axis of revolution. Note: The direction arrow points clockwise.

298) Click **OK** ✔ from the Revolve PropertyManager. Revolve1 is displayed in the FeatureManager.

Rename the feature.

299) Rename **Revolve1** to **BaseRevolve**.

Save the model.

300) Click **Save** 💾.

Display the axis of revolution.

301) Click **View**, **Hide/Show**, **Temporary Axes** from the Menu bar.

Revolve features contain an axis of revolution. The axis of revolution utilizes a sketched centerline, edge or an existing feature/sketch or a Temporary Axis. The solid Revolved feature contains a closed profile. The Revolved thin feature contains an open or closed profile.

LENS Part - Shell Feature

The Revolved Base feature is a solid. Utilize the Shell feature to create a constant wall thickness around the front face. The Shell feature removes face material from a solid. The Shell feature requires a face and thickness. Use the Shell feature to create thin-walled parts.

Activity: LENS Part - Shell Feature

Insert the Shell feature.

302) Click the **front face** of the BaseRevolve feature.

303) Click the **Shell** 🗔 feature tool. The Shell1 PropertyManager is displayed. Face<1> is displayed in the Faces to Remove box.

304) Enter **.250**in, **[6.35]** for Thickness.

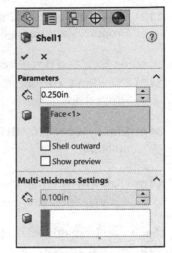

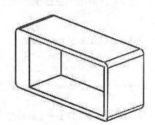

Display the Shell feature.

305) Click **OK** ✔ from the Shell1
PropertyManager. Shell1 is displayed in
the FeatureManager.

306) Right-click **Front Plane** from the
FeatureManager.

307) Click **Hide** 👁 from the Context toolbar.

Rename the feature.
308) Rename **Shell1** to **LensShell**.

Click **Save** 💾.

💡 To insert rounded corners inside a shelled part, apply the
Fillet feature before the Shell feature. Select the Multi-thickness
option to apply different thicknesses.

Extruded Boss/Base Feature and Convert Entities Sketch tool

Create the LensNeck. The LensNeck houses the BULB base and
is connected to the BATTERYPLATE. Use the Extruded
Boss/Base feature. The back face of the Revolved Base feature is
the Sketch plane.

Utilize the Convert Entities Sketch tool to extract the back circular face to the Sketch
plane. The new curve develops an On Edge relation. Modify the back face, and the
extracted curve updates to reflect the change. No sketch dimensions are required.

Activity: Extruded Boss Feature and Convert Entities Sketch tool

---

Rotate the LENS.
309) **Rotate** the LENS with the middle mouse button to display the back face as illustrated. The

Rotate ↻ icon is displayed.

Sketch the profile.
310) Right-click the **back face** for the Sketch plane.
BaseRevolve is highlighted in the FeatureManager.
This is your Sketch plane.

311) Click **Sketch** 📝 from the Context toolbar. The Sketch
toolbar is displayed.

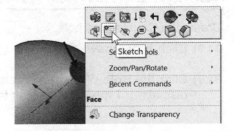

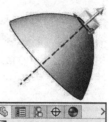

312) Click the **Convert Entities** 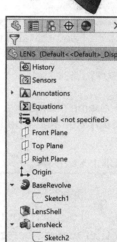 Sketch tool. The Convert Entities PropertyManager is displayed.

Insert an Extruded Boss/Base feature.

313) Click the **Extruded Boss/Base** feature tool. The Boss-Extrude PropertyManager is displayed.

314) Enter **.400**in, **[10.16]** for Depth in Direction 1. Accept the default settings.

315) Click **OK** ✔ from the Boss-Extrude PropertyManager. Boss-Extrude1 is displayed in the FeatureManager.

Display an Isometric view. Rename the feature and save the model.

316) Click **Isometric view** 🔲.

317) Rename **Boss-Extrude1** as **LensNeck**.

318) Click **Save** 💾.

LENS Part - Hole Wizard

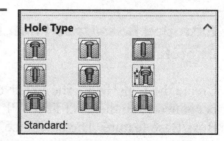

The LENS requires a Counterbore hole. Apply the Hole Wizard 🔘 feature. The Hole Wizard feature assists in creating complex and simple holes. The Hole Wizard Hole type categories are *Counterbore, Countersink, Hole, Straight Tap, Tapered Tap, Legacy Hole* (holes created before SOLIDWORKS 2000), *Counterbore Slot, Countersink Slot,* and *Slot.*

Specify the user parameters for the custom Counterbore hole. The parameters are *Description, Standard, Screw Type, Hole Size, Fit, Counterbore diameter, Counterbore depth* and *End Condition*. Select the face or plane to locate the hole profile.

Insert a Coincident relation to position the hole center point. Dimensions for the Counterbore hole are provided in both inches and millimeters.

Activity: LENS Part - Hole Wizard Counterbore Hole Feature

Create the Counterbore hole.

319) Click **Front view** 🔲.

320) Click the **Hole Wizard** 🔘 feature tool. The Hole Specification PropertyManager is displayed. Type is the default tab.

321) Click the **Counterbore** icon as illustrated.

Note: For a metric hole, skip the next few steps.

For inch Counterbore hole:
322) Select **ANSI Inch** for Standard. Select **Hex Bolt** for Type.

323) Select ½ for Size. Check the **Show custom sizing** box.

324) Click inside the **Counterbore Diameter** value box.

325) Enter **.600**in. Click inside the **Counterbore Depth** value box.

326) Enter **.200**in. Select **Through All** for End Condition. Click the **Position** tab.

327) Click the small **inside back face** of the LensShell feature as illustrated. Do not select the Origin. LensShell is highlighted in the FeatureManager.

The Point tool icon is displayed.

328) Click the **origin**. A Coincident icon relation is displayed.

Deselect the Point tool.
329) Right-click **Select** in the Graphics window.

Note: For an inch hole, skip the next few steps to address millimeter.

For millimeter Counterbore hole:
330) Select **Ansi Metric** for Standard.

331) Enter **Hex Bolt** for Type. Select **M5** for Size.

332) Click **Through All** for End Condition. Check the **Show custom sizing** box.

333) Click inside the **Counterbore Diameter** value box.

334) Enter **15.24**. Click inside the **Counterborebore Depth** value box.

335) Enter **5**. Click the **Position** tab.

336) Click the small **inside back face** of the LensShell feature as illustrated. Do not select the Origin. LensShell is highlighted in the FeatureManager. The Point tool icon is displayed.

337) Click the **origin**. A Coincident relation is displayed.

Deselect the Point tool.
338) Right-click **Select** in the Graphics window.

Return to the Type tab.
339) Click the **Type** tab.

Add the new hole type to your Favorites list.
340) Expand the Favorites box.

341) Click the **Add or Update Favorite**  icon.

342) Enter **CBORE for ½ Hex Head Bolt**.

343) Click **OK** from the Add or Update a Favorite
dialog box.

344) Click **OK** ✔ from the Hole Specification
PropertyManager.

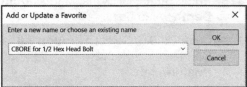

Expand the Hole feature.
345) Expand the CBORE feature in the
FeatureManager. Note: Sketch3 and Sketch4
created the CBORE feature.

Display the Section view.
346) Click **Right Plane** from the FeatureManager.

347) Click **Section view** 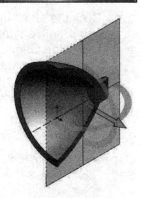 from the Heads-up View toolbar.

348) Click **OK** ✔ from the Section View PropertyManager.

349) Click **Isometric view** .

Display the Full view.
350) Click **Section view** from the Heads-up View toolbar.

Rename the feature and save the model.
351) Rename **CBORE for ½ Hex Head Bolt1** to **BulbHole**.

352) Click **Save** .

LENS Part - Revolved Boss Thin Feature

Create a Revolved Boss Thin feature. Rotate an open sketched profile around an axis.
The sketch profile must be open and cannot cross the axis. A Revolved Boss Thin feature
requires:

- Sketch plane (Right Plane)

- Sketch profile (Center point arc)

- Axis of Revolution (Temporary axis)

- Angle of Rotation (360)

- Thickness .100in [2.54]

A Revolved feature produces silhouette edges in 2D views. A silhouette edge represents the extent of a cylindrical or curved face.

Select the Temporary Axis for Axis of Revolution. Select the Revolved Boss feature. Enter .100in [2.54] for Thickness in the Revolve PropertyManager. Enter 360° for Angle of Revolution.

Activity: LENS Part - Revolved Boss Thin Feature

Create a sketch.
353) Right-click **Right Plane** from the FeatureManager.

354) Click **Sketch** from the Context toolbar.

355) Click **Right view**.

356) Zoom in on the LensNeck.

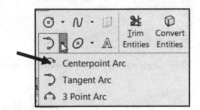

357) Click the **Centerpoint Arc** Sketch tool.

358) Click the **top horizontal edge** of the LensNeck. Do not select the midpoint of the silhouette edge.

359) Click the **top right corner** of the LensNeck.

360) Drag the **mouse pointer** counterclockwise to the left.

361) Click a **position** directly above the first point as illustrated. If needed, insert a Vertical relation between the end third and first point.

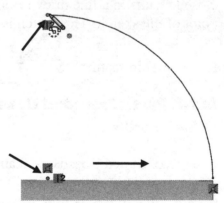

Add a dimension.
362) Click the **Smart Dimension** Sketch tool.

363) Click the **arc**.

364) Click a **position** to the right of the profile.

365) Enter **.100**in, **[2.54]**. The Sketch is fully defined.

Insert a Revolved Thin feature.

366) Click the **Revolved Boss/Base** feature tool. The Revolve PropertyManager is displayed.

367) Select **Mid-Plane** for Revolve Type in the Thin Feature box.

368) Enter **.050**in, **[1.27]** for Direction1 Thickness.

369) Click the **Temporary Axis** for Axis of Revolution.

370) Click **OK** ✔ from the Revolve PropertyManager.

Rename the feature.
371) Rename **Revolve-Thin1** to **LensConnector**.

Fit the model to the Graphics window and save.
372) Press the **f** key.

373) Click **Save** .

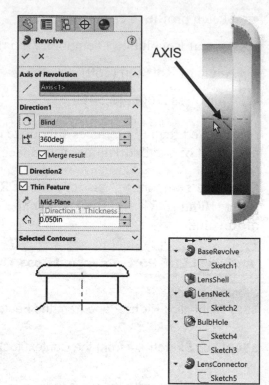

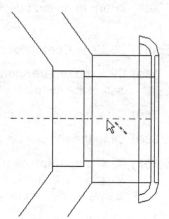

 A Revolved sketch that remains open results in a Revolve

Thin feature ——⌐ . A Revolved sketch that is automatically closed results in a line drawn from the start point to the end point of the sketch. The sketch is closed and results in a non-

Revolve Thin feature ——⌐ .

LENS Part - Extruded Boss/Base Feature and Offset Entities

Use the Extruded Boss/Base feature tool to create the front LensCover. Utilize the Offset Entities Sketch tool to offset the outside circular edge of the Revolved feature. The Sketch plane for the Extruded Boss feature is the front circular face.

The Offset Entities Sketch tool requires an Offset Distance and direction. Utilize the Bi-direction option to create a circular sketch in both directions. The extrude direction is away from the Front Plane.

Activity: LENS Part - Extruded Boss Feature and Offset Entities

Create the Sketch.

374) Click **Isometric view** 🔷 from the Heads-up View toolbar.

375) Click **Hidden Lines Removed** ⬜ from the Heads-up View toolbar.

376) Right-click the **front circular face** for the Sketch plane.

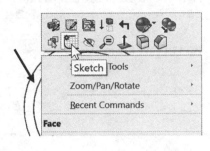

377) Click **Sketch** ⬛ from the Context toolbar.

378) Click **Front view** ⬜.

Offset the selected edge.

379) Click the **outside circular edge** of the LENS in the Graphics window.

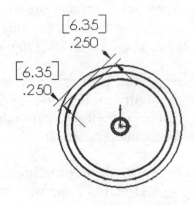

380) Click the **Offset Entities** ⬛ Sketch tool. The Offset Entities PropertyManager is displayed.

381) Check the **Bi-directional** box.

382) Enter **.250**in, **[6.35]** for Offset Distance. Accept the default settings.

383) Click **OK** ✔ from the Offset Entities PropertyManager.

Display an Isometric view with Shaded With edges.

384) Click **Isometric view** 🔷 from the Heads-up View toolbar.

385) Click **Shaded With Edges** ⬜ from the Heads-up View toolbar.

Insert an Extruded Boss/Base feature.

386) Click the **Extruded Boss/Base** ⬛ feature tool. The Boss-Extrude PropertyManager is displayed.

387) Enter **.250**in, **[6.35]** for Depth in Direction 1. Accept the default settings.

388) Click **OK** ✔ from the Boss-Extrude PropertyManager. Boss-Extrude2 is displayed in the PropertyManager.

Verify the position of the extruded feature.
389) Click the **Top view** ⬚ . View the extruded feature.

Rename the feature. Display an Isometric view. Save the model.
390) Rename **Boss-Extrude2** to **LensCover**.

391) Click **Isometric view** ◈ .

392) Click **Save** 💾 .

LENS Part - Extruded Boss Feature and Transparency

Apply the Extruded Boss/Base feature to create the LensShield. Utilize the Convert Entities Sketch tool to extract the inside circular edge of the LensCover and place it on the Front plane.

Apply the Transparent Optical property to the LensShield to control the ability for light to pass through the surface. Transparency is an Optical Property found in the Color PropertyManager. Control the following properties:

- **Diffuse amount, Specular amount, Specular spread, Reflection amount, Transparent amount and Luminous intensity.**

Activity: LENS Part - Extruded Boss Feature and Transparency

Create the sketch.
393) Right-click **Front Plane** from the FeatureManager. This is your Sketch plane.

394) Click **Sketch** ⬚ from the Context toolbar. The Sketch toolbar is displayed.

395) Click **Isometric view** ◈ .

396) Click the **front inner circular edge** of the LensCover (Boss-Extrude2) as illustrated.

397) Click the **Convert Entities** 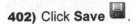 Sketch tool. The circle is projected onto the Front Plane.

Insert an Extruded Boss feature.

398) Click the **Extruded Boss/Base** feature tool. The Boss-Extrude FeatureManager is displayed.

399) Enter **.100**in, **[2.54]** for Depth in Direction 1.

400) Click **OK** from the Boss-Extrude PropertyManager. Boss-Extrude3 is displayed in the FeatureManager.

Rename the feature. Save the model.
401) Rename **Boss-Extrude3** to **LensShield**.

402) Click **Save**.

Change Transparency of the LensShield.
403) Right-click **LensShield** in the FeatureManager.

404) Click **Change Transparency**. **View** the results.

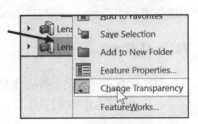

Hide the axis of revolution. Display Shaded with Edges.
405) Click **View**, **Hide/Show**, and uncheck **Temporary Axes** from the Menu bar.

406) Click **Shaded with Edges** from the Heads-up toolbar.

Save the model.
407) Click **Save**.

🔍 Additional information on Revolved Boss/Base, Shell, Hole Wizard and Appearance is located in SOLIDWORKS Help Topics. Keywords: Revolved (features), Shells, Hole Wizard (Counterbore) and Appearances.

Refer to Help, SOLIDWORKS Tutorials, Revolve and Swept for additional information.

Review of the LENS Part

The LENS feature utilized a Revolved Base feature. A Revolved feature required an axis, profile and an angle of revolution. The Shell feature created a uniform wall thickness.

You utilized the Convert Entities Sketch tool to create the Extruded Boss feature for the LensNeck. The Counterbore hole was created using the Hole Wizard feature.

The Revolved Thin feature utilized a single 3 Point Arc. Geometric relations were added to the silhouette edge to define the arc. The LensCover and LensShield utilized existing geometry to Offset and Convert the geometry to the sketch. The Color and Optics PropertyManager determined the LensShield transparency.

BULB Part

The BULB fits inside the LENS. Use the Revolved feature as the Base feature for the BULB.

Insert the Revolved Base  feature from a sketched profile on the Right Plane.

Insert a Revolved Boss 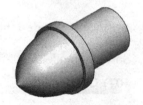 feature using a Spline sketched profile. The profile utilizes a complex curve called a Spline (Non-Uniform Rational B-Spline or NURB). Draw Splines with control points.

Insert a Revolved Cut Thin  feature at the base of the BULB. A Revolved Cut Thin feature removes material by rotating an open sketch profile about an axis.

Insert a Dome  feature at the base of the BULB. A Dome feature creates spherical or elliptical shaped geometry. Use the Dome feature to create the Connector feature of the BULB. The Dome feature requires a face and a height value.

Insert a Circular Pattern  feature from an Extruded Cut feature.

BULB Part - Revolved Base Feature

Create the new part, BULB. The BULB utilizes a solid Revolved Base feature.

The solid Revolved Base feature requires a:

- Sketch plane (Right Plane)
- Sketch profile (Lines)
- Axis of Revolution (Centerline)
- Angle of Rotation (360°)

Utilize the centerline to create a diameter dimension for the profile. The flange of the BULB is located inside the Counterbore hole of the LENS. Align the bottom of the flange with the Front Plane. The Front Plane mates against the Counterbore face.

Activity: BULB Part - Revolved Base Feature

Create a New part.

408) Click **New** ⬜ from the Menu bar.

409) Click the **MY-TEMPLATES** tab.

410) Double-click **PART-IN-ANSI**, **[PART-MM-ISO]**.

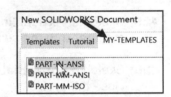

Save the part. Enter name and description.

411) Click **Save** 💾.

412) Select **PROJECTS** for Save in folder.

413) Enter **BULB** for File name.

414) Enter **BULB FOR LENS** for Description.

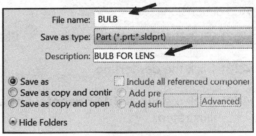

415) Click **Save**. The BULB FeatureManager is displayed.

Select the Sketch plane.

416) Right-click **Right Plane** from the FeatureManager. This is your Sketch plane.

Create the sketch.

417) Click **Sketch** 📝 from the Context toolbar. The Sketch toolbar is displayed.

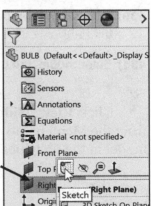

Sketch a centerline.

418) Click the **Centerline** ✎ Sketch tool. The Insert Line PropertyManager is displayed.

419) Sketch a **horizontal centerline** through the Origin ⌊.

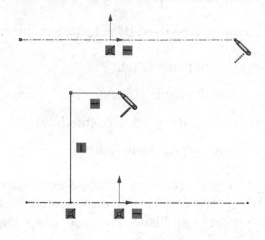

Create six profile lines.

420) Click the **Line** ✎ Sketch tool. The Insert Line PropertyManager is displayed.

421) Sketch a **vertical line** to the left of the Front Plane.

422) Sketch a **horizontal line** with the endpoint coincident to the Front Plane.

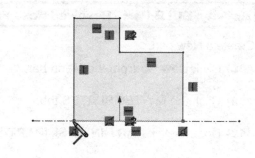

423) Sketch a short **vertical line** towards the centerline, collinear with the Front Plane.

424) Sketch a **horizontal line** to the right.

425) Sketch a **vertical line** with the endpoint collinear with the centerline.

426) Sketch a **horizontal line** to the first point to close the profile.

Add dimensions.

427) Click the **Smart Dimension** ✎ Sketch tool.

428) Click the **centerline**.

429) Click the **top right horizontal line** as illustrated.

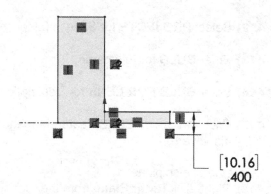

430) Click a **position** below the centerline and to the right.

431) Enter **.400**in, **[10.016]**.

[10.16]
.400

💡 Click **View, Hide/Show, Sketch Relations** from the Menu bar to display the relations of the model in the Graphics window.

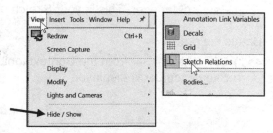

432) Click the **centerline**.

433) Click the **top left horizontal line**.

434) Click a **position** below the centerline and to the left.

435) Enter .590in, [**14.99**]. Click the **top left horizontal line**.

436) Click a **position** above the profile.

437) Enter .100in, [**2.54**]. Click the **top right horizontal line**.

438) Click a **position** above the profile.

439) Enter .500in, [**12.7**].

Fit the model to the Graphics window.
440) Press the **f** key.

Insert a Revolved Base feature.

441) Click the **Revolved Boss/Base** feature tool. The Revolve PropertyManager is displayed. Accept the default settings.

442) Click **OK** from the Revolve PropertyManager. Revolve1 is displayed in the FeatureManager.

Display an Isometric view. Save the model.
443) Click **Isometric view**.

444) Click **Save**.

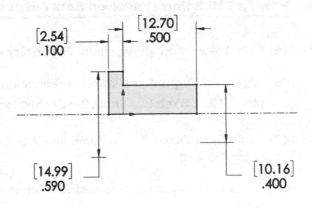

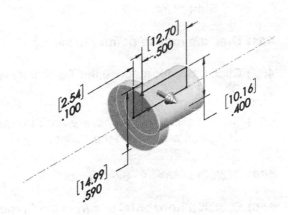

BULB Part - Revolved Boss Feature and Spline Sketch Tool

The BULB requires a second solid Revolved feature. The profile utilizes a complex curve called a Spline (Non-Uniform Rational B-Spline or NURB).
Draw Splines with control points. Adjust the shape of the curve by dragging the control points.

For additional flexibility, deactivate the Snaps option in Document Properties for this model.

Activity: BULB Part - Revolved Boss Feature and Spline Sketch Tool

Create the sketch.

445) Click **View**, **Hide/Show**, check **Temporary Axes** from the Menu bar.

446) Right-click **Right Plane** from the FeatureManager for the Sketch plane. Click **Sketch** from the Context toolbar.

447) Click **Right view**. The Temporary Axis is displayed as a horizontal line.

448) Press the **z** key approximately five times to view the left vertical edge.

Sketch the profile.

449) Click the **Spline** Sketch tool.

450) Click the **left vertical edge** of the Base feature for the Start point.

451) Drag the **mouse pointer** to the left.

452) Click a **position** above the Temporary Axis for the Control point.

453) Double-click the **Temporary Axis** to create the End point and to end the Spline.

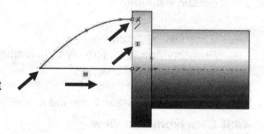

454) Click the **Line** Sketch tool.

455) Sketch a **horizontal line** from the Spline endpoint to the left edge of the Revolved feature.

456) Sketch a **vertical line** to the Spline start point, collinear with the left edge of the Revolved feature. Dimensions are not required to create a feature.

Insert a Revolved Boss feature.

457) De-select the Line Sketch tool. Right-click **Select**.

458) Click the **Temporary Axis** from the Graphics window as illustrated.

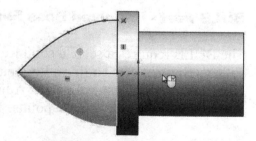

459) Click the **Revolved Boss/Base** feature tool. The Revolve PropertyManager is displayed. Accept the default settings.

460) Click **OK** from the Revolve PropertyManager.

461) Click **Isometric view** .

462) Click **Save** .

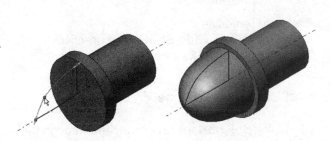

The points of the Spline dictate the shape of the Spline. Edit the control points in the sketch to produce different shapes for the Revolved Boss feature.

BULB Part - Revolved Cut Thin Feature

A Revolved Cut Thin feature removes material by rotating an open sketch profile around an axis. Sketch an open profile on the Right Plane. Add a Coincident relation to the silhouette and vertical edge. Insert dimensions.

Sketch a Centerline to create a diameter dimension for a revolved profile. The Temporary axis does not produce a diameter dimension.

Note: If lines snap to grid intersections, uncheck Tools, Sketch Settings, Enable Snapping for the next activity.

Activity: BULB Part - Revolved Cut Thin Feature

Create the sketch.
463) Right-click **Right Plane** from the FeatureManager.

464) Click **Sketch** from the Context toolbar.

465) Click **Right view** from the Heads-up View toolbar.

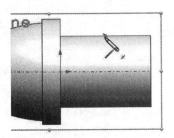

466) Click the **Line** Sketch tool.

467) Click the **midpoint** of the top silhouette edge.

468) Sketch a **line** downward and to the right as illustrated.

469) Sketch a **horizontal line** to the right vertical edge.

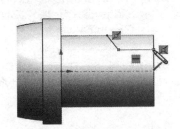

470) De-select the Line Sketch tool. Right-click **Select**.

If needed, add a Coincident relation.
471) Click the **endpoint** of the line.

472) Hold the **Ctrl** key down.

473) Click the right **vertical edge**.

474) Release the **Ctrl** key.

475) Click **Coincident** ⊀ .

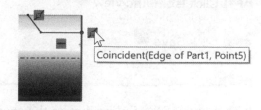

476) Click **OK** ✔ from the Properties PropertyManager.

Sketch a Centerline.
477) Click **View**, **Hide/Show**, and un-check **Temporary Axes** from the Menu bar.

478) Click the **Centerline** ✎ Sketch tool.

479) Sketch a **horizontal centerline** through the Origin.

Add dimensions.
480) Click the **Smart Dimension** ✎ Sketch tool.

481) Click the **horizontal centerline**.

482) Click the **short horizontal line**.

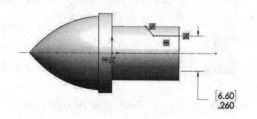

483) Click a **position** below the profile to create a diameter dimension.

484) Enter .260in, [6.6].

485) Click the **short horizontal line**.

486) Click a **position** above the profile to create a horizontal dimension. Enter **.070**in, [**1.78**]. The Sketch is fully defined and is displayed in black.

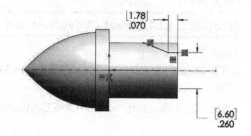

💡 For Revolved features, the ∅ symbol is not displayed in the part. The ∅ symbol is displayed when inserted into the drawing.

Insert the Revolved Cut Thin feature.
487) De-select the Smart Dimension Sketch tool. Right-click **Select**. Click the **centerline** in the Graphics window.

488) Click the **Revolved Cut** 🏛 feature tool. The Cut-Revolve PropertyManager is displayed.

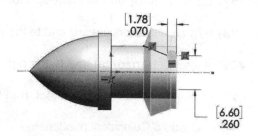

489) Click **No** to the Warning Message, "Would you like the sketch to be automatically closed?" The Cut-Revolve PropertyManager is displayed.

490) Check the **Thin Feature** box.

491) Enter **.150**in, **[3.81]** for Thickness.

492) Click the **Reverse Direction** box.

493) Click **OK** ✓ from the Cut-Revolve PropertyManager. Cut-Revolve-Thin1 is displayed in the FeatureManager.

Save the model.
494) Click **Save** 💾.

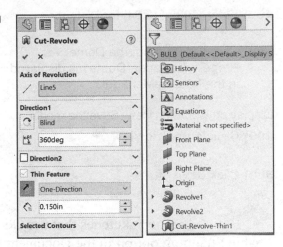

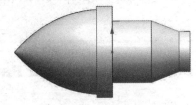

BULB Part - Dome Feature

The Dome 🔵 feature creates spherical or elliptical shaped geometry. Use the Dome feature to create the Connector feature of the BULB. The Dome feature requires a face and a height/distance value.

Activity: BULB Part - Dome Feature

Insert the Dome feature.
495) Click the **back circular face** of Revolve1. Revolve1 is highlighted in the FeatureManager.

496) Click **Insert**, **Features**, **Dome** 🔵 from the Main menu. The Dome PropertyManager is displayed. Face1 is displayed in the Parameters box.

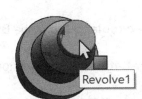

497) Enter **.100**in, **[2.54]** for Distance.

498) Click **OK** ✔ from the Dome
PropertyManager. Dome1 is displayed
in the FeatureManager.

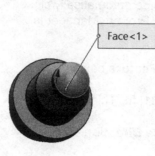

Display an Isometric view. Save the model.

499) Click **Isometric view** 🧊.

500) Click **Save** 💾.

🔆 Before creating sketches that use Geometric relations, check the
Enable Snapping option in the Document Properties dialog box.

BULB Part - Circular Pattern Feature

A Pattern feature creates one or more instances of a feature or a
group of features. The Circular Pattern feature places the
instances around an axis of revolution.

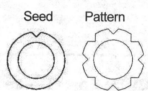

Seed   Pattern

The Circular Pattern 🔘 feature requires a seed feature. The
seed feature is the first feature in the pattern. The seed feature in
this section is a V-shaped Extruded Cut feature.

Activity: BULB Part - Circular Pattern Feature

Create the Seed Cut feature.

501) Right-click the **front face** of the Base feature, Revolve1 in the
Graphics window for the Sketch plane as illustrated. Revolve1
is highlighted in the FeatureManager.

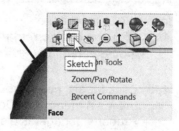

502) Click **Sketch** 📝 from the Context toolbar. The Sketch toolbar
is displayed.

503) Click the **outside circular edge** of the BULB.

504) Click the **Convert Entities** 🔲 Sketch tool.

505) Click **Front view** 🔲 from the Heads-up View toolbar.

506) Zoom in on the top half of the BULB.

Sketch a centerline.

507) Click the **Centerline** 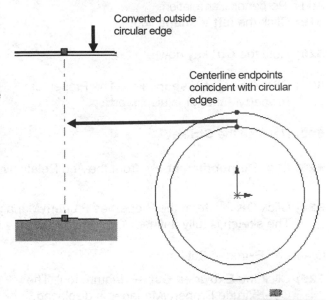 Sketch tool.

508) Sketch a **vertical centerline** coincident with the top and bottom circular circles and coincident with the Right plane.

Converted outside circular edge

Centerline endpoints coincident with circular edges

Sketch a V-shaped line.

509) Click **Tools**, **Sketch Tools**, **Dynamic Mirror** from the Menu bar. The Mirror PropertyManager is displayed.

510) Click the **centerline** from the Graphics window.

511) Click the **Line** Sketch tool.

512) Click the **midpoint** of the centerline.

513) Click the coincident **outside circle edge to the left** of the centerline.

Deactivate the Dynamic Mirror tool.
514) Click **Tools**, **Sketch Tools**, **Dynamic Mirror** from the Menu bar.

Trim unwanted geometry.

515) Click the **Trim Entities** Sketch tool. The Trim PropertyManager is displayed.

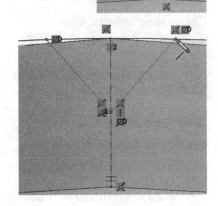

516) Click **Power trim** from the Options box.

517) Click a **position** in the Graphics window and drag the mouse pointer until it intersects the circle circumference.

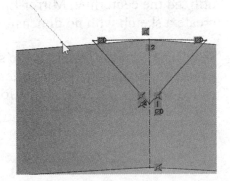

518) Click **OK** from the Trim PropertyManager.

Add a Perpendicular relation.
519) Click the **left V shape** line.

520) Hold the **Ctrl** key down.

521) Click the **right V shape** line. The Properties PropertyManager is displayed.

522) Release the **Ctrl** key.

523) Click **Perpendicular** ⊥ from the Add Relations box.

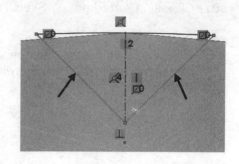

524) Click **OK** ✓ from the Properties PropertyManager. The sketch is fully defined.

Create an Extruded Cut feature.
525) Click the **Extruded Cut**  feature tool. The Cut-Extrude PropertyManager is displayed.

526) Click **Through All** for End Condition in Direction 1. Accept the default settings.

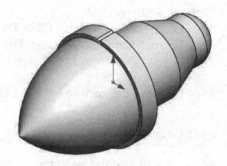

527) Click **OK** ✓ from the Cut-Extrude PropertyManager. The Cut-Extrude1 feature is displayed in the FeatureManager.

Display an Isometric view.
528) Click **Isometric view** .

Fit the drawing to the Graphics window.
529) Press the **f** key.

Save the model.
530) Click **Save** 💾.

💡 Reuse Geometry in the feature. The Cut-Extrude1 feature utilized the centerline, Mirror Entity, and geometric relations to create a sketch with no dimensions.

The Cut-Extrude1 feature is the seed feature for the pattern. Create four copies of the seed feature. A copy of a feature is called an Instance. Modify the four copies Instances to eight.

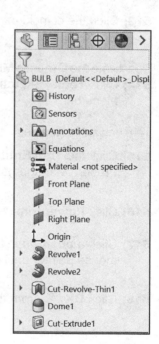

Insert the Circular Pattern feature.

531) Click the **Cut-Extrude1** feature from the FeatureManager.

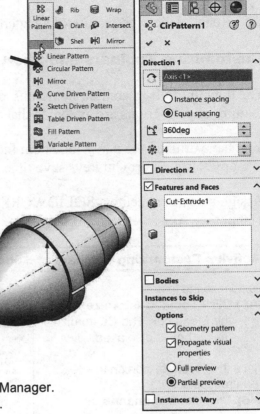

532) Click the **Circular Pattern** feature tool. The Circular Pattern PropertyManager is displayed. Cut-Extrude1 is displayed in the Features to Pattern box.

533) Click **View, Hide/Show,** and check **Temporary Axes** from the Menu bar. Click the **Temporary Axis** from the Graphics window. Axis<1> is displayed in the Pattern Axis box.

534) Enter **4** in the Number of Instances box. Check the **Equal spacing** box.

535) Enter **360deg**.

536) Check the **Geometry pattern** box. Accept the default settings.

537) Click **OK** from the Circular Pattern PropertyManager. CirPattern1 is displayed in the FeatureManager.

Edit the Circular Pattern feature.

538) Right-click **CirPattern1** from the FeatureManager.

539) Click **Edit Feature** from the Context toolbar. The CirPattern1 PropertyManager is displayed.

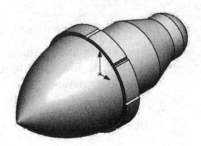

540) Enter **8** in the Number of Instances box.

541) Click **OK** from the CirPattern1 PropertyManager.

Rename the feature.

542) Rename **Cut-Extrude1** to **SeedCut**.

Hide the reference geometry.

543) Click **View, Hide/Show,** and uncheck **Temporary Axes** from the Menu bar.

Save the model.

544) Click **Save** .

Rename the seed feature of a pattern to locate it quickly for future assembly.

Customizing Toolbars and Short Cut Keys

The default toolbars contain numerous icons that represent basic functions. Additional features and functions are available that are not displayed on the default toolbars.

You have utilized the z key for Zoom In/Out, the f key for Zoom to Fit and Ctrl-C/Ctrl-V to Copy/Paste. Short Cut keys save time.

Assign a key to execute a SOLIDWORKS function. Create a Short Cut key for the Temporary Axis.

Activity: Customizing Toolbars and Short Cut Keys

Customize the toolbar.

545) Click **Tools**, **Customize** from the Menu bar. The Customize dialog box is displayed.

Place the Freeform icon on the Features toolbar.

546) Click the **Commands** tab.

547) Click **Features** from the category text box.

548) Drag the **Freeform** feature icon into the Features toolbar. The Freeform feature option is displayed.

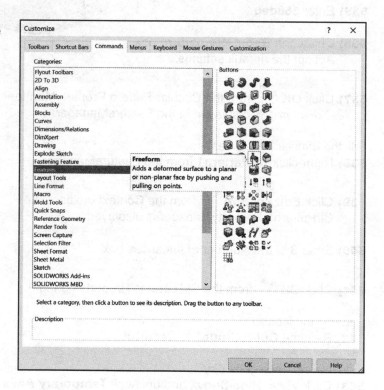

Customize the keyboard for the Temporary Axes.

549) Click the **Keyboard** tab from the Customize dialog box.

550) Select **View** for Categories.

551) Select **Temporary Axes** for Commands.

552) Enter **R** for Shortcut(s) key.

553) Click **OK**.

554) Press the **R** key to toggle the display of the Temporary Axes.

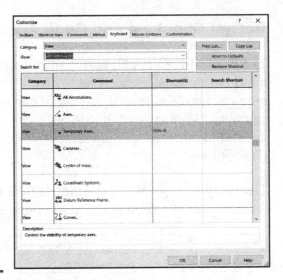

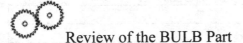 Test the proposed Short cut key before you customize your keyboard. Refer to the default Keyboard Short cut table in the Appendix.

Refer to Help, SOLIDWORKS Tutorials, Pattern Features for additional information.

Review of the BULB Part

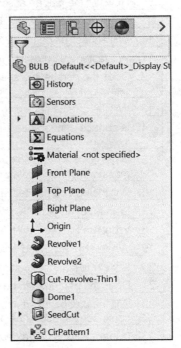

The Revolved Base feature utilized a sketched profile on the Right plane and a centerline. The Revolved Boss feature utilized a Spline sketched profile. A Spline is a complex curve.

You created the Revolved Cut Thin feature at the base of the BULB to remove material. A centerline was inserted to add a diameter dimension. The Dome feature was inserted on the back face of the BULB. The Circular Pattern feature was created from an Extruded Cut feature. The Extruded Cut feature utilized existing geometry and required no dimensions.

Toolbars and keyboards were customized to save time. Always verify that a Short cut key is not predefined in SOLIDWORKS.

Design Checklist and Goals before Plastic Manufacturing

The BATTERYPLATE part is manufactured from a plastic resin. A plastic part requires mold base plates, named Plate A and Plate B, to create the mold. A plastic part also requires knowledge of the material and manufacturing process. Critical decisions need to be made. What type of plastic resin do you use? The answer is derived from the development team: designer, resin supplier, mold maker and mold base, (Plate A & Plate B) supplier.

Ticona (www.ticona.com) supplies their customers with a designer's checklist to assist the designer in selecting the correct material.

TICONA-Celanese Typical Design Checklist
1.   What is the function of the part?
2.   What is the expected lifetime of the part?
3.   What agency approvals are required? (UL, FDA, USDA, NSF, USP, SAE, MIL spec)
4.   What electrical characteristics are required and at what temperatures?
5.   What temperature will the part see and for how long?
6.   What chemicals will the part be exposed to?
7.   Is moisture resistance necessary?
8.   How will the part be assembled? Can parts be combined into one plastic part?
9.   Is the assembly going to be permanent or one time only?
10. Will adhesives be used? Some resins require special adhesives.
11. Will fasteners be used? Will threads be molded in?
12. Does the part have a snap fit? Glass filled materials will require more force to close the snap fit but will deflect less.
13. Will the part be subjected to impact? If so, radius the corners.
14. Is surface appearance important? If so, beware of weld lines, parting lines, ejector location and gate location.
15. What color is required for the part? Is a specific match required or will the part be color-coded? Some glass or mineral filled materials do not color as well as unfilled materials.
16. Will the part be painted? Is primer required? Will the part go through a high temperature paint oven?
17. Is weathering or UV exposure a factor?
18. What are the required tolerances? Can they be relaxed to make molding more economical?
19. What is the expected weight of the part? Will it be too light (or too heavy)?
20. Is wear resistance required?
21. Does the part need to be sterilized? With what methods (chemical, steam, radiation)?
22. Will the part be insert molded or have a metal piece press fit in the plastic part? Both methods result in continuous stress in the part.
23. Is there a living hinge designed in the part? Be careful with living hinges designed for crystalline materials such as acetal.
24. What loading and resulting stress will the part see? And, at what temperature and environment.
25. Will the part be loaded continuously or intermittently? Will permanent deformation or creep be an issue?
26. What deflections are acceptable?
27. Is the part moldable? Are there undercuts? Are there sections that are too thick or thin?
28. Will the part be machined?
29. What is the worst possible situation the part will be in? Worst-case environment.

In "Designing With Plastic - The Fundamentals," Ticona lists three design goals for creating injection molded parts:

Goal 1: Maximize Functionality

Mold bases are costly. Design functionality into the part. A single plastic chassis replaces several sheet metal components.

Reduce assembly time and part weight whenever it is appropriate.

Sheet Metal
Assembly

Injection Molded
Thermoplastic

Goal 2: Optimize Material Selection

For material selection, consider the following elements: part functionality, operating environment, cost/price constraint and any special cosmetic requirements.

Several materials should be selected with a developed list of advantages and disadvantages for review.

For maximum product performance at a competitive cost in this project, use Celanese® (Nylon 6/6) (P/A 6/6), where:

- Celanese® is the registered trademark name.

- Nylon 6/6 is the common plastic name.

- P/A 6/6 (Polyhexamethyleneadipamide) is the chemical name.

Goal 3: Minimize Material Use

Optimize part wall thickness for cost and performance.

The minimum volume of plastic that satisfies the structural, functional, appearance and molding requirements of the application is usually the best choice.

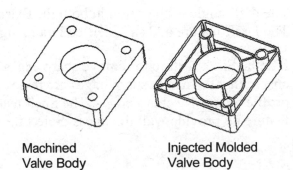

Machined
Valve Body

Injected Molded
Valve Body

Mold Base

Designing a custom mold base is expensive
and time consuming. An example of a mold
base supplier is Progressive Components
(www.procomps.com).

The mold base positions the mold cavities.
The mold base plates are machined to create
the mold cavities. The mold base is designed
to withstand high pressure, heating and
cooling sequences during the fabrication
process.

The mold base assembly is composed of the
following two plates and a variety of support
plates, ejector plates, pins and bushings:

- PLATE A

- PLATE B

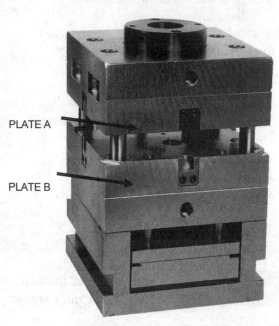

PLATE A

PLATE B

Courtesy of Progressive Components
Wauconda, IL USA

Applying SOLIDWORKS Features
for Mold Tooling Design

SOLIDWORKS features such as Draft, Fillet and Shell assist in the design of plastic
parts. Utilize the Draft tool to analyze the part before creating the Mold Tooling, PLATE
A and PLATE B.

Most molded parts require Draft. Plastic molded injection parts require a draft angle to
eject the part from the mold. To properly eject the part, design parts with a draft angle in
the direction of the mold movement. A draft angle of 1° - 3° is the design rule for most
injection molded parts. There are exceptions based on material type and draw depth. For
illustration purposes, a draft angle of 1° is used in the BATTERYPLATE.

The draft angle is an option in both the Extruded Boss/Base and Extruded Cut features.
The Draft feature adds a specified Draft Angle to faces of an existing feature.

Use the Fillet feature to remove sharp edges from plastic parts. For thin walled parts,
ensure that the inside sharp edges are removed. Sketch Arcs and 2D Fillets in the 2D
profile to remove sharp edges. The Shell feature, Extruded Cut feature or Extruded Thin
feature all provide wall thickness. Select the correct wall thickness for a successful part.

If the wall thickness is too thin, the part will structurally fail. If the wall thickness is too thick, the part will be overweight and too costly to manufacture. This will increase cycle time and decrease profits.

There are numerous types of plastic molds. SOLIDWORKS contains a variety of Mold tools to assist the designer, mold maker and mold manufacturer. For a simple Cavity Mold follow the following steps:

- *Draft analysis*. A draft analysis examines the faces of the model for sufficient draft, to ensure that the part ejects properly from the tooling.

- *Undercut detection*. Identifies trapped areas that prevent the part from ejecting.

- *Parting Lines*. This tool has two functions:

   o Verifies that you have draft on your model, based on the angle you specify.

   o Creates a parting line from which you create a parting surface. The Parting Lines tool includes the option to select an edge and have the system Propagate to all edges.

- *Shut-off Surfaces*. Creates surface patches to close up through holes in the molded part.

- *Parting Surfaces*. Extrude from the parting line to separate mold cavity from core. You can also use a parting surface to create an interlock surface.

- *Ruled Surface*. Adds draft to surfaces on imported models. You can also use the Ruled Surface tool to create an interlock surface.

- *Tooling Split*. Creates the core and cavity bodies, based on the steps above.

Activity: SOLIDWORKS Features for Mold Tooling Design

Create the New part.

555) Click **New** 📄 from the Menu bar.

556) Click the **MY-TEMPLATES** tab.

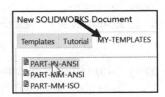

557) Double-click **PART-IN-ANSI**, [**PART-MM-ISO**].

Save the part. Enter name and description.

558) Click **Save** 💾. Select **PROJECTS** for Save in folder.

559) Enter **MOLD-TOOLING** for File name.

560) Enter **MOLD FEATURES FOR TOOLING** for Description. Click **Save**. The MOLD-TOOLING FeatureManager is displayed.

Insert a part.
561) Click **Insert**, **Part** from the Menu bar.

562) Double-click **BATTERYPLATE** from the PROJECTS folder. The Insert Part PropertyManager is displayed.

563) Click **OK** ✔ from the Insert Part PropertyManager. The BATTERYPLATE part is the first feature in the MOLD-TOOLING part.

564) Click **OK**.

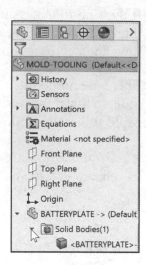

Utilize the Insert Part option to maintain the original part and a simplified FeatureManager. The Insert Part maintains an External reference to the original part ->. Modify the original part and the Insert Part updates on a Rebuild. Apply the Insert Part option to SOLIDWORKS parts and imported parts.

Display the Mold toolbar.
565) Right-click in the **gray area** of the Feature toolbar. Click **Mold Tools**. The Molds toolbar is displayed.

Perform a Draft Analysis.
566) Click the **Draft** 🔲 tool from the Mold toolbar. The Draft PropertyManager is displayed. Note: This is the same Draft tool as displayed in the Features toolbar.

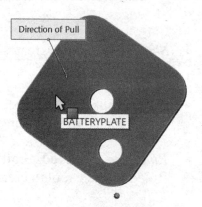

567) Click the **DraftXpert** tab. The DraftXpert PropertyManager is displayed. Click the **Add** tab.

Create the Direction of Pull.
568) **Rotate** the part and click the **bottom face**. Face<1> is displayed in the Neutral Plane box.

569) Click the **Reverse Direction** arrow. The direction of pull arrow points upwards. Display an **Isometric view**.

Select a face to draft.
570) **Zoom in** and click the **inside hole face** of the center hole.

571) **Zoom in** and click the **inside hole face** of the small hole. The selected faces are displayed.

572) Enter **1deg** for Draft angle.

573) Check the **Auto Paint** box.

574) Click the **Apply** button.

575) Rotate the **part** to display the faces that require a draft. The faces are displayed in different colors. View the results in the Graphics window.

576) Click **OK** ✔ from the DraftXpert PropertyManager. Draft1 is displayed.

By default, the faces that require draft are displayed in red. The Top Cut, Extruded Cut, and the Holes Extruded Cut requires a draft. The bottom flat surface will become the Parting surface.

Modify the BATTERYPLATE part In-Context of the MOLD-TOOLING part. Both parts are open during the modification session.

Modify the BATTERYPLATE In-Context.
577) Right-click **BATTERYPLATE** in the FeatureManager.

578) Click **Edit In Context**. The BATTERYPLATE FeatureManager is displayed.

Insert Draft into the Top Cut.
579) Right-click **Top Cut** in the BATTERYPLATE FeatureManager.

580) Click **Edit Feature** 📝. The Top Cut PropertyManager is displayed.

581) If needed, click the **Draft on/off** button. Enter **1**deg for Draft Angle. Click **OK** ✔ from the Top Cut PropertyManager.

Insert a draft into the holes.
582) Right-click **Holes** in the BATTERYPLATE FeatureManager. Click **Edit Feature** 📝. The Holes PropertyManager is displayed. If needed, click the **Draft on/off** button. Enter **1**deg for Draft Angle. Click **OK** ✔ from the Holes PropertyManager.

Return to the MOLD-TOOLING part.
583) Press **Ctrl-Tab**. The MOLD TOOLING FeatureManager is displayed.

584) Rebuild 🔵 the model.

Scale:

The Scale feature increases or decreases the part's volume by a specified factor. The hot thermoplastic utilized in this part cools during the molding process and hardens and shrinks. Increase the part with the Scale feature to compensate for shrinkage.

Insert the Scale feature.

585) Click **Scale** 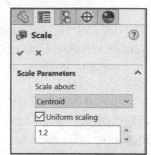 from the Mold toolbar. The Scale PropertyManager is displayed.

586) Enter **1.2** for Scale Factor.

587) Click **OK** ✔ from the Scale PropertyManager. Scale1 is displayed in the FeatureManager.

Parting Lines:

The Parting Lines are the boundary between the core and cavity surfaces. The Parting Lines create the edges of the molded part. The Parting Lines consist of the bottom eight edges.

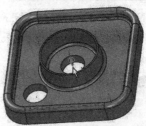

Insert the Parting Lines feature.

588) Click **Parting Lines** ⊕ from the Mold toolbar. The Parting Line PropertyManager is displayed.

Display the Temporary Axis.
589) Click **View**, **Hide/Show**, **Temporary Axes** from the Menu bar.

590) Click the **Temporary Axis** displayed through the center hole for the Direction of Pull.

591) Click **Reverse Direction** if required. The direction arrow points upward.

592) Enter **1deg** for Draft Angle.

593) Click the **Draft Analysis** button. The Parting Lines box contains the eight bottom edges of the BATTERYPLATE.

594) Click **OK** ✔ from the Parting Line PropertyManager. Parting Line1 is displayed in the FeatureManager.

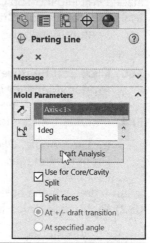

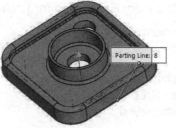

Hide the Temporary Axes.
595) Click **View**, **Hide/Show**, uncheck **Temporary Axes** from the
Menu bar.

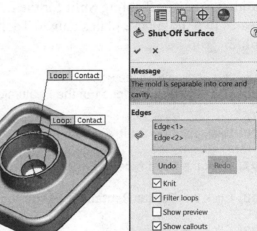

Shut-off Surfaces:

Holes and windows that penetrate through an entire part require
separate mold tooling. A shut-off area in a mold is where two
pieces of the mold tooling contact each other. These areas require
a Shut-off Surface. There are two Thru Holes in the
BATTERYPLATE.

Insert the Shut-off Surfaces feature.

596) Click **Shut-off Surfaces** from
the Mold toolbar. The Shut-Off
Surface PropertyManager is
displayed. Accept the defaults
settings.

597) Click **OK** from the Shut-Off
Surface PropertyManager. A
surface closes off the opening of
each hole. Shut-Off Surface1 is
displayed in the FeatureManager.

598) **Rotate** the model and view the closed surface.

Parting Surface:

A Parting Surface is utilized to separate the mold tooling components. A Parting Surface
extends radially, a specified distance from the Parting Lines. Usually, the surface is
normal to the direction of pull for simple molds. The boundary of the Parting Surface
must be larger than the profile for the Tooling Split.

Insert the Parting Suface feature.

599) Click **Parting Surfaces** from the Mold toolbar. The
Parting Surface PropertyManager is displayed.

600) Click **Top view**.

601) Enter **1.00**in, **[25.4]** for Distance. Accept the default
settings.

602) Click **OK** from the Parting Surface PropertyManager.
A Parting Surface (Radiate Surface) is created from the
Parting Lines and the Distance value.

Fit the model to the Graphics window.
603) Press the **f** key.

Save the model.
604) Click **Save** 💾.

Tooling Split:

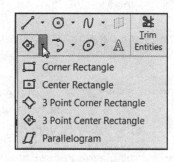

The Tooling Split separates the solid bodies from the parting surfaces. The solid bodies become the mold tooling and the molded part. The Tooling Split for the BATTERYPLATE creates two mold base plates named Tooling Split1 and Tooling Split2.

Insert the Tooling Split feature.
605) Click **Parting Surface1** in the FeatureManager.

606) Click **Tooling Split** 📐 from the Mold toolbar.

607) Click the **front face** of Parting Surface1.

608) Click the **Center Rectangle** Sketch tool.

609) Click the **Origin**.

610) Sketch a **rectangle** as illustrated about the Origin.

De-select the Center Rectangle Sketch tool.
611) Right-click **Select**.

Add an Equal relation.
612) Click the **top horizontal line** of the rectangle.

613) Hold the **Ctrl key** down.

614) Click the **left vertical line, right vertical line,** and **bottom horizontal line** of the rectangle.

615) Release the **Ctrl key**.

616) Click **Equal** = .

617) Click **OK** ✔ from the Properties PropertyManager.

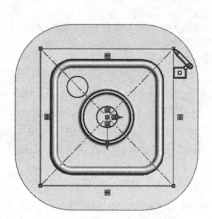

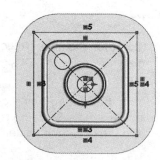

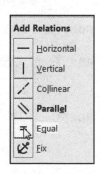

Add a dimension.

618) Click the **Smart Dimension** Sketch tool.

619) Click the **top horizontal** line.

620) Enter **4.000**in, **[101.60]**.

621) Click **Exit Sketch**. The Tooling Split PropertyManager is displayed.

Enter values for the Block size.
622) Click **Front view**.

623) Enter **1.00**in, **[25.4]** for Depth in Direction 1.

624) Enter **.500**in, **[12.7]** for Depth in Direction 2.

625) Click **Isometric view** .

626) Click **OK** from the Tooling Split PropertyManager. Tooling Split1 is displayed in the FeatureManager.

The Mold tools produce Surface Bodies and Solid Bodies. The Solid Bodies are utilized in an assembly for the actual mold base. The Surface Bodies were utilized as interim features to create the mold tooling parts. Hide the Surface Bodies to display the solid mold tooling parts.

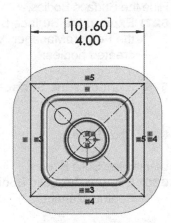

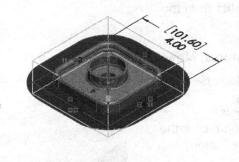

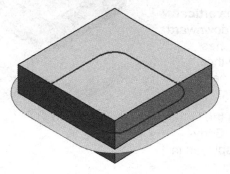

Hide the Surface Bodies.
627) Expand the Surface Bodies folder in the FeatureManager. View the created bodies.

628) Right-click the **Surface Bodies** folder.

629) Click **Hide**.

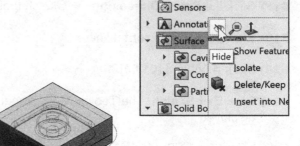

Display the Solid Bodies.
630) Expand the Solid Bodies folder in the FeatureManager.

View each entry.
631) Click each **entry** in the Solid Bodies folder.

Separate Tooling Split1.
632) Click **Tooling Split1** from the FeatureManager.

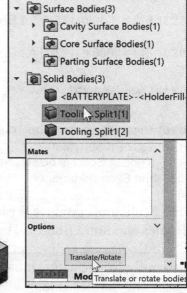

633) Click **Insert**, **Features**, **Move/Copy** from the Menu bar. The Move/Copy Body PropertyManager is displayed. If needed, click the **Translate/Rotate** button.

634) Click and drag the **vertical axis of the Triad** downward approximately 3inches (75mm) to display the BATTERYPLATE part.

635) Click **OK** from the PropertyManager. Body-Move/Copy1 is displayed in the FeatureManager.

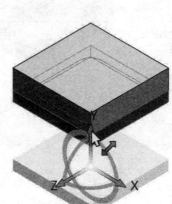

Separate Tooling Split1[2].

636) Click **Tooling Split1[2]** from the FeatureManager.

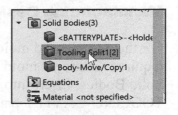

637) Click **Insert**, **Features**, **Move/Copy**  from the Menu bar.

638) Click and drag the **vertical axis** of the Triad upward approximately 4in, [100] to display the BATTERYPLATE part.

639) Click **OK**  from the PropertyManager. Body-Move/Copy2 is displayed in the FeatureManager.

Rotate the view.

640) Click the **Up Arrow key** to display the model as illustrated.

Save the MOLD-TOOLING part.

641) Click **Save** .

642) Click **Yes** to save the referenced part (BATTERYPLATE).

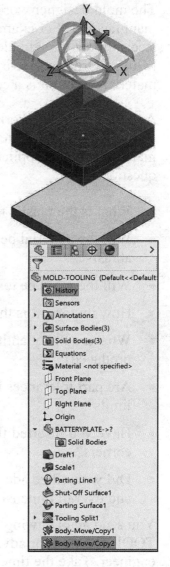

Manufacturing Design Issues

For the experienced mold designer, the complete mold base requires additional mold tooling, lines and fittings. Utilize the Solid Bodies, Insert into New Part option to create the MOLD-TOOLING-CAVITY part from Body2 in the Solid Bodies folder.

The mold designer works with a material supplier and mold manufacturer. For example, Bohler-Uddeholm provides mold steels in rod, bar and block form. Cutting operations include Milling, Drilling, Sawing, Turning and Grinding. A wrong material selection is a costly mistake.

As a designer, you work with the material supplier to determine the best material for the project. For example, a material supplier application engineer asks the following questions:

- What is the plastic material utilized in this application?

- Will the material be corrosive or abrasive?

- Will the mold be textured?

- How important is the surface finish?

- What are the quantities of the part to be produced?

- Are part tolerances held within close limits?

- Have you inserted fillets on all sharp corners?

- Did you utilize adequate wall thickness and overall dimensions?

You send an eDrawing of your MOLD-TOOLING part for advice to the application engineer. Take the time early in the design process to determine the tooling material.

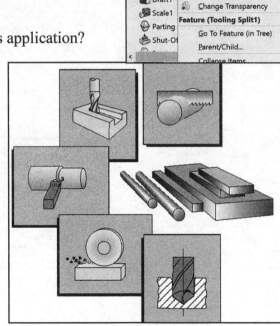

Cutting Operations for Tool Steels
Courtesy of Bohler-Uddeholm
www.bucorp.com

Project Summary

You are designing a FLASHLIGHT assembly that is cost effective, serviceable and flexible for future design revisions. The FLASHLIGHT assembly consists of various parts. The BATTERY, BATTERYPLATE, LENS and BULB parts were modeled in this project.

This Project concentrated on the Extruded Boss/Base feature and the Revolved Boss/Base feature. The Extruded Boss/Base feature required a Sketch plane, Sketch profile and End Condition (Depth). The BATTERY and BATTERYPLATE parts incorporated an Extruded Boss/Base (Boss-Extrude1) feature. A Revolved feature requires a Sketch plane, Sketch profile, Axis of Revolution and an Angle of Rotation.

You also utilized the Extruded Boss/Base, Extruded Cut, Fillet, Chamfer, Revolved Cut, Revolved Thin-Cut, Shell, Hole Wizard, Dome and Circular Pattern features.

You addressed the following Sketch tools in this project: Sketch, Smart Dimension, Line, Rectangle, Circle, Tangent Arc and Centerline. You addressed additional Sketch tools that utilized existing geometry: Add Relations, Display/Delete Relations, Mirror Entities, Convert Entities and Offset Entities. Geometric relations were utilized to build symmetry into the sketches.

The BATTERYPLATE utilized various Mold tools to develop the Core and Cavity tooling plates required in the manufacturing process. Plastic parts require Draft. The Draft Analysis tool determined areas of positive and negative draft. Features were modified to accommodate the manufacturing process.

Practice these concepts with the project exercises. The other parts for the FLASHLIGHT assembly are addressed in Project 6.

View the provided videos in the book to enhance the user experience.

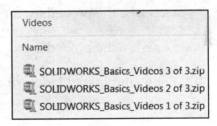

Questions

1. Provide a short explanation of the following features:

Fillet	Extruded Cut	Draft	Scale
Extruded Boss	Revolved Base	Shell	Move/Copy
Revolved Cut Thin	Circular Pattern	Hole Wizard	Parting Surface

2. True or False. Design intent exists only in the sketch.

3. Describe a Symmetric relation.

4. Describe an Angular dimension.

5. Describe a draft angle. When do you apply a draft angle in a model?

6. When do you apply the Mirror feature? Utilize the SOLIDWORKS Online Users Guide to determine the difference between the Sketch Mirror and the Dynamic Mirror Sketch tool.

7. Describe the Spline Sketch tool.

7.a Identify the required information for a Circular Pattern Feature.

8. Describe the procedure for adding a feature icon to the default Feature toolbar.

9. Describe the procedure to create a Short Cut key to Show/Hide Planes.

10. For a simple mold, describe the procedure to create the core and cavity mold tooling.

Exercise 5.1: FLAT PLATE

- Create the ANSI FLAT PLATE part on the Top Plane. The FLAT PLATE is machined from 0.090in, [2.3mm] 6061 Alloy stock. Primary Unit system: IPS.

- The 8.688in, [220.68mm] x 5.688in, [144.48mm] FLAT PLATE contains a Linear Pattern of ⌀.190in, [4.83mm] Thru Holes.

- The holes are equally spaced, .500in, [12.70mm] apart.

- Utilize the Geometric Pattern option for the Linear Pattern feature.

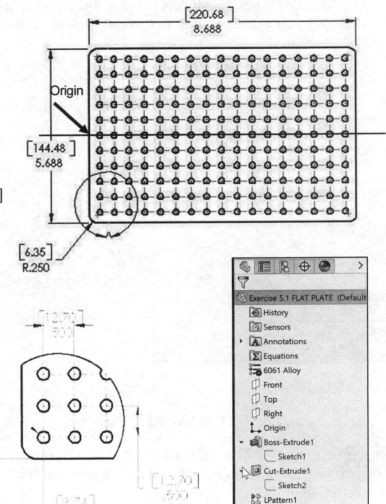

Exercise 5.2: IM15-MOUNT Part

Create the ANSI IM15-MOUNT part as illustrated on the Right plane. The IM15-MOUNT part is machined from 0.060in, [1.5mm] 304 Stainless Steel flat stock. Primary Unit system: IPS.

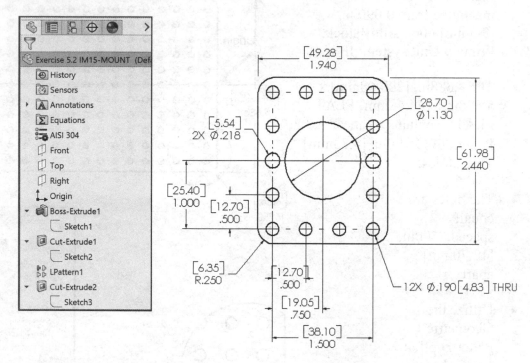

Exercise 5.3: ANGLE BRACKET Part

Create the ANSI ANGLE BRACKET part. The Boss-Extrude1 feature is sketched with an L-Shaped profile on the Right Plane.

The ANGLE BRACKET part is machined from 0.060in, [1.5mm] 304 Stainless Steel flat stock. Primary Unit system: IPS.

- Create the Boss-Extrude1 feature.

- Apply the Mid Plane option for Direction 1 End Condition.

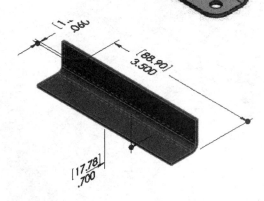

- Create the second Extrude feature on the Top view for the hole.

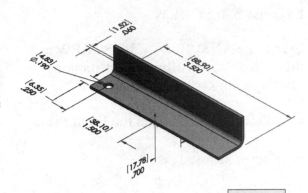

- Use the Linear Pattern feature to create six additional holes along the ANGLE BRACKET.

- Add Fillets to the four edges, .250in.

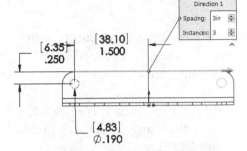

- Create the third Extrude feature.

- Apply the Linear Pattern feature.

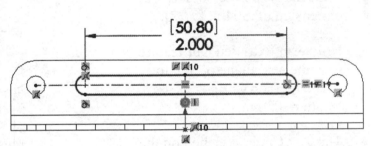

- Create the fourth Extrude feature for the center slot.

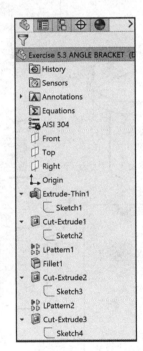

Exercise 5.4: SCREW

Create the ANSI 10-24 x 3/8 SCREW as illustrated. Note: A Simplified version. Primary Unit system: IPS.

- Sketch a centerline on the Front Plane.

- Sketch a closed profile.

- Utilize a Revolved Feature. For metric size, utilize an M4x10 machine screw.

- Edit the Revolved Base Sketch. Apply the Tangent Arc tool with Trim Entities. Enter an Arc dimension of .304in, [7.72].

- Utilize an Extruded Cut feature with the Mid Plane option to create the Top Cut. Depth = .050in.

- The Top Cut is sketched on the Front Plane.

- Utilize the Convert Entities Sketch tool to extract the left edge of the profile.

- Utilize the Circular Pattern feature and the Temporary Axis to create four Top Cuts.

- Use the Fillet Feature on the top circular edge, .01in to finish the simplified version of the SCREW part.

- Material - AISI 304.

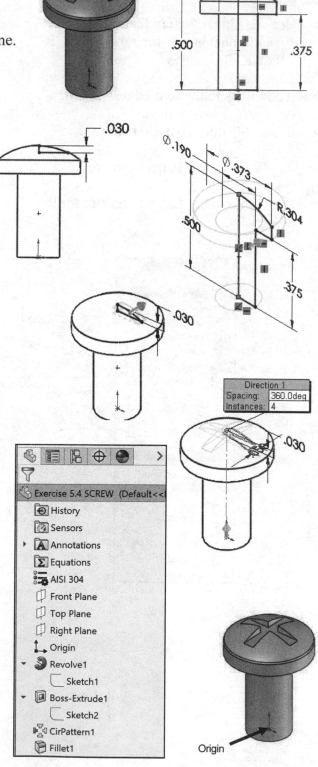

Exercise 5.5

Create the illustrated ANSI part. All edges of the model are not located on perpendicular planes. Think about the steps required to build the model. Insert two features: Extruded Boss/Base and Extruded Cut.

Note the location of the Origin. Select the Right Plane as the Sketch plane. Apply construction geometry. Insert the required geometric relations and dimensions for Sketch1. Calculate the overall mass of the part.

Given:
A = 3.00, B = 1.00
Material: 6061 Alloy
Density = .097 lb/in^3
Units: IPS
Decimal places = 2

🔅 There are numerous ways to build the models in this section.

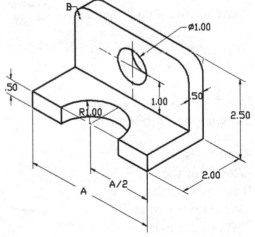

Origin

Exercise 5.6

Create the illustrated ANSI part. Think about the required steps to build this part. Insert four features: Boss-Extrude1, Cut-Extrude1, Cut-Extrude2 and Fillet. Note the location of the Origin.

Select the Right Plane as the Sketch plane. The part Origin is located in the lower left corner of the sketch. Calculate the overall mass of the part.

Given:
A = 4.00
B = R.50
Material: 6061 Alloy
Density = .0975 lb/in^3
Units: IPS
Decimal places = 2

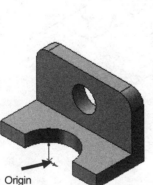

Origin

Exercise 5.7

Create the illustrated ANSI part. Think about the required steps to build this part. Insert three features: Boss-Extrude1, Boss-Extrude2, and Mirror. Three holes are displayed with a 1.00in diameter.

Note the location of the Origin.

Create Sketch1. Select the Top Plane as the Sketch plane. Apply the Tangent Arc and Line Sketch tool. Insert the required Geometric relations and dimensions. Calculate the overall mass of the part.

💡 There are numerous ways to build the models in this section.

Given:
A = Ø1.00
All Thru Holes
Material: Brass
Density = .307 lb/in^3
Units: IPS
Decimal places = 2

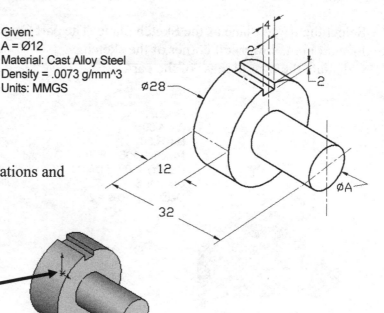

Origin

Exercise 5.8

Create the illustrated ANSI part. Think about the required steps to build this part. Insert a Revolved Base feature and an Extruded Cut feature to build this part.

Note the location of the Origin. Select the Front Plane as the Sketch plane.

Apply the Centerline Sketch tool for the Revolve1 feature.

Insert the required geometric relations and dimensions for Sketch1.

Sketch1 is the profile for the Revolve1 feature. Calculate the overall mass of the part.

Given:
A = Ø12
Material: Cast Alloy Steel
Density = .0073 g/mm^3
Units: MMGS

Ø28

12

32

ØA

Origin

Exercise 5.9

Create the illustrated ANSI part. Think about the required steps to build this part.

Insert two features: Boss-Extrude1 and Revolve1.

Note the location of the Origin.

Select the Top Plane as the Sketch plane.

Apply construction geometry.

Apply the Tangent Arc and Line Sketch tool.

Insert the required geometric relations and dimensions. Calculate the overall mass of the part.

Given:
A = 60, B = 40, C = 8
Material: Cast Alloy Steel
Density = .0073 g/mm^3
Units: MMGS

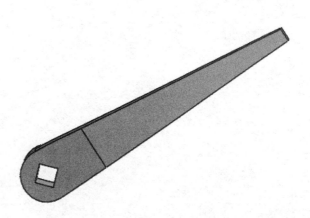

Origin

Exercise 5.10

Create the illustrated ANSI part. Think about the required steps to build this part. Apply the Extruded Boss/Base, Extruded Cut and Shell features. View the provided drawing to obtain a few dimensions (next page). As an engineer, you may not have all of the needed information. You are the designer. Fill in the gaps.

Units: MMGS.

Decimal Places: 2

Part Origin is arbitrary in this example.

Material: Plain Carbon Steel.

Calculate the overall mass of the part in grams.

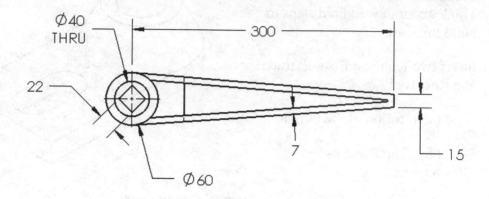

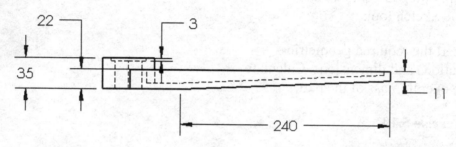

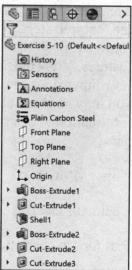

Exercise 5.11: Ice Cream Cone

Create a traditional Ice Cream cone as illustrated. Think about where you would start. Think about the design features that create the model. Why does the cone use ribs? Ribs are used for structural integrity.

Use a standard Cake Ice Cream cone and measure all dimensions (approximately).

View the sample FeatureManager. Your FeatureManager can (should) be different. This is just one way to create this part.

Be creative. Use the Wrap feature and/or the Text Sketch entities tool.

Below are sample models from my Freshman Engineering class.

Below are sample models from my Freshman Engineering (Cont:).

Exercise 5.12: Gem® Paper clip

Create a simple paper clip. Create an ANSI - IPS model.

Apply material.

What are the dimensions? Measure or estimate needed dimensions from a small gem paper clip.

Apply (lines and arcs) as the path and a circular diameter profile (0.010in).

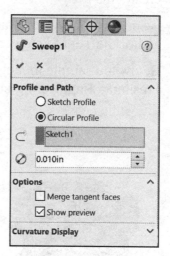

Exercise 5.13: Anvil Spring

Create a Variable pitch spring (ANSI - IPS) as illustrated.

Sketch a circle, Coincident to the Origin on the Top plane with a .235in dimension.

Create the Helix/Spiral feature.

Create a Region parameters table.

Enter the following information as illustrated. Coils 1, 2, 5, 6 & 7 are the closed ends of the spring. The pitch needs to be slightly larger than the wire.

Enter .021in for the Pitch.

Enter .080in for the free state of the two active coils.

Enter Start angle of 0deg.

Create the Swept Boss feature.

Enter .015in for Depth (Circular Profile).

Add material to the model.

View the results.

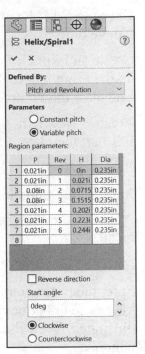

	P	Rev	H	Dia
1	0.021in	0	0in	0.235in
2	0.021in	1	0.021i	0.235in
3	0.08in	2	0.0715	0.235in
4	0.08in	3	0.1515	0.235in
5	0.021in	4	0.202i	0.235in
6	0.021in	5	0.223i	0.235in
7	0.021in	6	0.244i	0.235in
8				

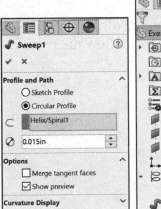

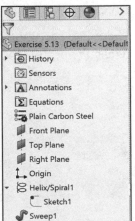

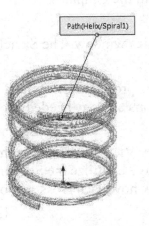

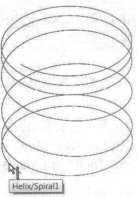

Exercise 5.14: Explicit Equation Driven Curve tool

Create an Explicit Equation Driven Curve on the Front plane. Revolve the curve. Calculate the volume of the solid.

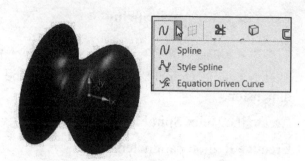

Create a New part. Use the default ANSI, IPS Part template.

Create a 2D Sketch on the Front Plane.

Activate the Equation Driven Curve Sketch f_x tool from the Consolidated drop-down menu.

Enter the Equation yx as illustrated.

Enter the parameters x1, x2 that define the lower and upper bounds of the equation as illustrated. View the curve in the Graphics window.

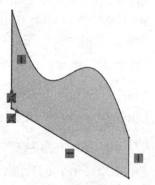

Size the curve in the Graphics window. The Sketch is under defined.

Insert three lines to close the profile as illustrated. Fully define your sketch. Enter dimensions and any needed geometric relation.

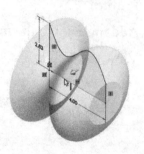

Create the Revolved feature. View the results in the Graphics window. Revolve1 is displayed. Utilize the Section tool parallel with the Right plane to view how each cross section is a circle.

Apply Brass for material.

Precision = 2.

Calculate the volume of the part using the Mass Properties tool. View the results.

Add Relations
— Horizontal
| Vertical
⩑ Coincident

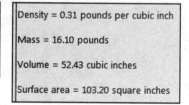

Density = 0.31 pounds per cubic inch

Mass = 16.10 pounds

Volume = 52.43 cubic inches

Surface area = 103.20 square inches

Exercise 5.15: Advanced Part

Build the illustrated model below.

Calculate the overall mass of the model. Units - IPS. Decimal places: 2. Material: AISI 304.

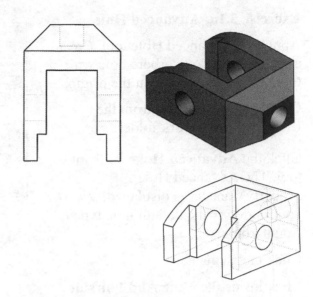

Think about the steps that you would take to build the illustrated part.

Insert the required geometric relations and dimensions.

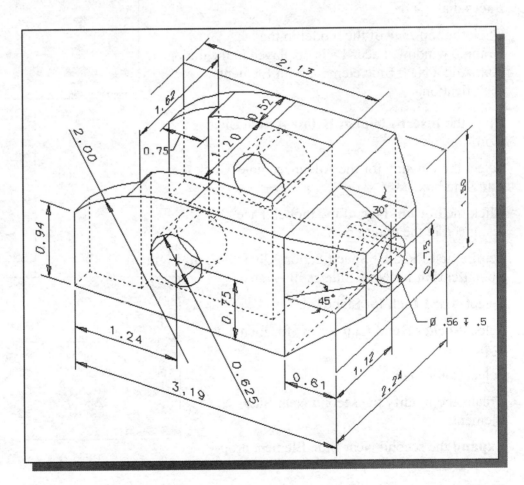

Exercise 5.16: Advanced Hole

Apply the Advanced Hole tool. Create and modify a Counterbore hole with three elements position at the origin.

Open **Advanced Hole** from the Chapter 5 Homework folder.

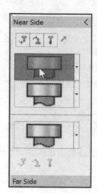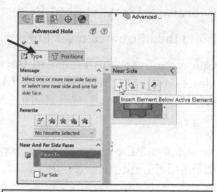

Click the **Advanced Hole** 🔲 feature tool. The Advanced Hole PropertyManager is displayed. View your options. The default hole type is Counterbore.

Click the **Type** tab.

Click inside the **Near And Far side Faces** dialog box.

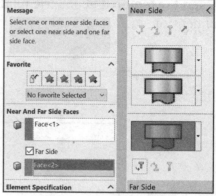

Click the **top face** of the model in the Graphic window. Face<1> is displayed in the dialog box. Add a Near Side element with Element specification.

Click the **Insert Element Below Active Element** ⇊ icon.

Select the Far side for the Advanced hole. Click the **Far Side** box.

Click the **bottom face** of the model. Face<2> is displayed in the dialog box.

Set the first Element Specification. Click the **Top Element Specification** for Near Side as illustrated.

Select **ANSI Inch** for Standard.

Select **Socket Head Cap Screw** for Element Type.

Select ¼ for Size.

Create and modify the second Near Side Element.

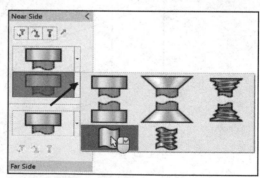

Expand the second Near Side Element drop-down menu as illustrated.

Click the **Straight** icon as illustrated.

Set the second Element Specification.

Select **ANSI Inch** for Standard.

Select **Screw Clearances** for Type.

Select ¼ for Size.

Select **Blind** for End Condition.

Enter **1.0in** for Depth. View the updated model in the Graphics area.

The Far Side element is currently set to Counterbore. Modify the Counterbore to a Straight tapped thread element.

Expand the Far Side Element drop-down menu as illustrated.

Click the **Straight Tapped Thread** icon as illustrated.

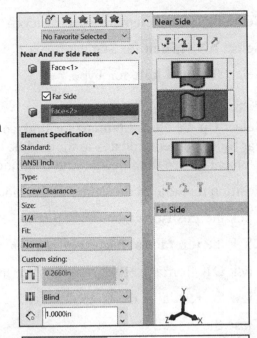

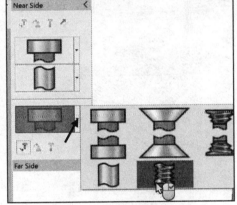

Set the Far Side Element Specification.

Select **ANSI Inch** for Standard.

Select **Tapped hole** for Type.

Select **¼-20** for Size.

Select **Up To Next Element** for End Condition. The Far Side Element updates to the middle Near Side Element due to the selected End Condition.

Position the hole in the center of the block.

Click the **Positions** tab.

Click the **top face** at the origin.

Click **OK** from the Hole Position PropertyManager.

View the results in the Graphics area.

Close the model.

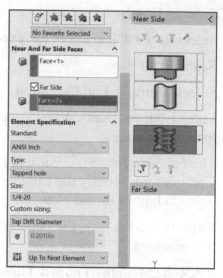

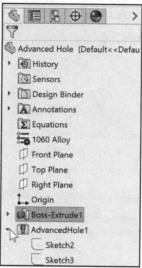

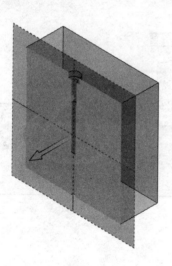

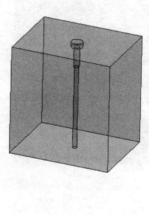

Thread Feature

The Thread ⬚ Feature provides the ability to create threads for standard and non-standard threads for shafts and holes. For prototypes and digital simulation use the Thread feature. Select the start thread location, specify an offset, set end conditions, specify the type, size, diameter, pitch and rotation angle and choose options such as right-hand or left-hand thread.

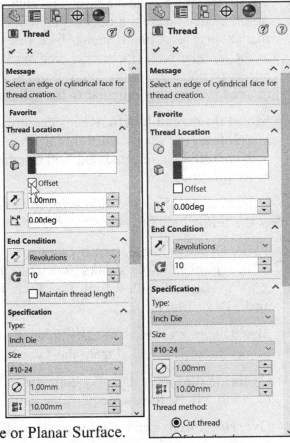

The Thread feature uses the Thread PropertyManager. The Thread PropertyManager provides the following selections:

Thread Location. The Thread Location box provides the ability to select the following options:

- **Edge of Cylinder**. Select a circular edge.

- **Optional Start Location**. Provides the ability to select a Vertex, Edge, Plane or Planar Surface.

- **Offset**. The offset value must be positive. You can also enter an equation by starting with = (equal sign).

- **Offset Distance**. Distance of the Offset.

- **Start Angle**. Defines the starting location of the helix. The start angle must be positive. You can also enter an equation by starting with = (equal sign).

End Condition. The End Condition box provides the ability to select the following options:

- **Blind**. Specify a value for Depth. Terminates the thread at the desired distance from the starting location, taking into account any input for offset.

- **Revolutions**. Terminates the helix at the desired number of revolutions from the starting location, taking into account any input for offset. The value must be positive and greater than 0.00. You can also enter an equation by starting with = (equal sign).

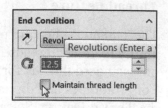

- **Up to Selection**. Select a vertex (sketch, model, or reference points), edge (sketch, model or reference axis), plane, or planar surface. A plane, planar face, or edge must be parallel to the circular edge (i.e. perpendicular to the thread axis). A vertex or point acts as the point through which a plane is created that is perpendicular to the cylindrical axis.

- **Maintain thread length**. Keeps the thread at a constant length from the start surface. It only displays if the Start Surface Offset is selected and End Condition is set to Blind or Revolutions.

Specification. The Specification box provides the ability to select the following options:

- **Type**. Displays library part files initially found in C:\ProgramData\SolidWorks\SOLIDWORKS YYYY\Thread Profiles. **Inch Die, Inch Tap, Metric Die and SP4xx Bottle**.

- **Size**. Displays each of the configurations found in the library part file selected from the Type list.

- **Override diameter**. Display diameter of the cylindrical face and will be the diameter of the helix.

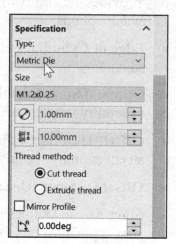

- **Override Pitch**. Enter a value or start with = to create an equation. The value in the dimension input field is determined from the selected configuration's profile sketch and the length of a vertical construction line drawn from the origin of the model.

- **Thread method**. Select one of the following:

- **Cut thread**.

- **Extruded thread**.

- **Mirror Profile**. Select one of the following:

- **Mirror horizontally**.

- **Mirror vertically**.

- **Locate Profile**. Select a new point on the thread profile sketch from the graphics area to attach to the helical path. You can select any vertex created by the entities and sketch points in the profile sketch. The button is disabled when no circular edge has been selected.

Thread Options. The Thread Options box provides the ability to select the following options:

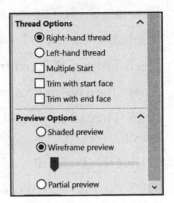

- **Right-hand thread**. Creates a right hand thread.

- **Left-hand thread**. Creates a left hand thread.

- **Multiple Start**. Set the number of starts to define the number of times the thread is created in an evenly-spaced circular pattern around the hole or shaft. The image shows a four-start thread with a different color per thread.

- **Trim with start face**. Provides the ability to align threads to start faces.

- **Trim with end face**. Provides the ability to align threads to end face.

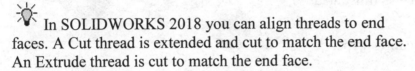 In SOLIDWORKS 2018 you can align threads to end faces. A Cut thread is extended and cut to match the end face. An Extrude thread is cut to match the end face.

Preview Options. The Preview Options box provides the ability to select the following options:

- **Shaded preview**. Displays a fully tessellated preview of the thread.

- **Wireframe preview**. Displays a wireframe preview of the thread.

- **Partial preview**. Adjusts the number of wires displayed in the wireframe.

Exercise 5.17: Thread Wizard

Apply the Thread Wizard to create a thread feature.

Open **Thread Wizard** from the Chapter 5 homework folder.

Apply the Thread Feature.

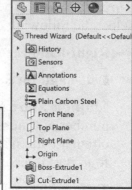

Click **Thread** from the Hole Wizard drop-down menu.

Click **OK** if needed.

Click the **font circular edge** as illustrated. Edge1 is displayed in the Thread Location box.

Select **Up To Selection** for End Condition. Click inside the **End Condition** box.

Click the **front circular face** as illustrated. Face<1> is displayed.

Select **Inch Die** for Specification Type.

Select **#8-36** for Size.

Select **Cut thread** for Thread method.

Select **Right-hand** thread for Thread options.

Select **Trim with start face**.

Select **Trim with end face**.

Select **Shaded preview**.

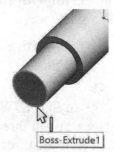

Click **OK** ✔ from the Thread PropertyManager. Thread1 is created and is displayed.

Zoom in on the thread section. View the Trim with start face option.

Close the part.

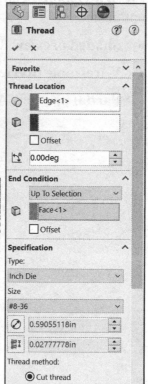

☼ You can align threads to end faces. A Cut thread is extended and cut to match the end face. An Extrude thread is cut to match the end face. In the PropertyManager, under Thread Options, select Trim with start face and Trim with end face.

Project 6

Swept, Lofted and Additional Features

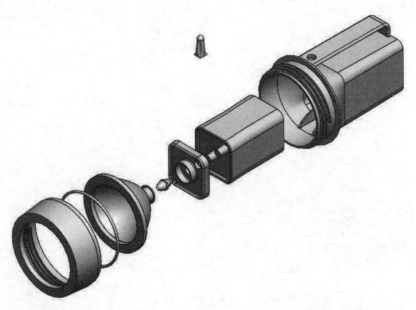

Below are the desired outcomes and usage competencies based on the completion of Project 6.

Project Desired Outcomes:	Usage Competencies:

• Four FLASHLIGHT parts:
   o O-RING
   o SWITCH
   o LENSCAP
   o HOUSING
	• Ability to apply the following 3D features: Extruded Boss/Base, Extruded Cut, Swept Boss/Base, Shell, Lofted Boss/Base, Draft, Rib, Linear Pattern, Circular Pattern, Mirror, Dome, and Revolved Cut.

• Three O-RING configurations: o Small, Medium, and Large	• Create and edit a Design Table for various configurations.

• Four assemblies:
   o LENSANDBULB assembly
   o CAPANDLENS assembly
   o BATTERYANDPLATE assembly
   o FLASHLIGHT assembly
	• Skill to combine multiple features to create components.

	• Knowledge of the Bottom-up assembly modeling method using Standard mates and the Quick mate procedure.

Notes:

Project 6 - Swept, Lofted, and Additional Features

Project Objective

Create four parts: O-RING, SWITCH, LENSCAP and HOUSING of the FLASHLIGHT assembly. Create four assemblies: LENSANDBULB, CAPANDLENS, BATTERYANDPLATE and the final FLASHLIGHT assembly.

On the completion of this project, you will be able to:

- Select the best Sketch profile.

- Choose the proper Sketch plane.

- Develop three O-RING configurations using a Design Table.

- Create an Assembly Template with Document Properties.

- Apply Bottom-up assembly modeling techniques.

- Calculate assembly interference.

- Export SOLIDWORKS files.

- Utilize the following SOLIDWORKS features:

  o Swept Boss/Base.

  o Lofted Boss/Base.

  o Extruded Boss/Base.

  o Revolved Boss/Base.

  o Extruded Cut.

  o Revolved Cut.

  o Draft.

  o Dome.

  o Rib.

  o Circular Pattern and Linear Pattern.

  o Mirror.

Project Overview

Project 6 introduces the Swept
Boss/Base and Lofted Boss/Base
features. The O-RING utilizes a simple
Swept Base feature. The SWITCH
utilizes the Lofted Base feature. The
LENSCAP and HOUSING utilize the
Swept Boss and Lofted Boss feature.

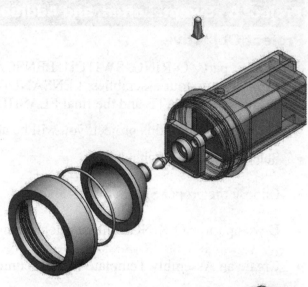

A simple Swept Base feature requires a
path and *profile*. Sketch the path and
enter the circular profile diameter. The
profile follows the path to create the
following Swept features:

- Swept Base

- Swept Boss

The Lofted feature requires a minimum of two profiles
sketched on different planes. The profiles are blended
together to create the following Lofted features:

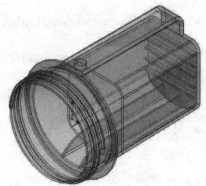

- Lofted Base

- Lofted Boss

Utilize existing features to create the Rib, Linear
Pattern and Mirror features.

Utilize existing faces to create the Draft and Dome
features.

The LENSCAP and HOUSING combines the Extruded
Boss/Base, Extruded Cut, Revolved Cut Thin, Shell
and Circular Pattern with the Swept and Lofted
features.

Features in the SWITCH, LENSCAP and HOUSING
have been simplified for educational purposes.

Create four assemblies in this project:

1. LENSANDBULB assembly

2. CAPANDLENS assembly

3. BATTERYANDPLATE assembly

4. FLASHLIGHT assembly

Create an inch and metric Assembly template.

- ASM-IN-ANSI

- ASM-MM-ISO

Develop an understanding of assembly modeling techniques.

Combine the LENSANDBULB assembly, CAPANDLENS assembly, BATTERYANDPLATE assembly, HOUSING part and SWITCH part to create the FLASHLIGHT assembly.

Create the following Standard Mate types: Coincident, Concentric and Distance.

Utilize the following tools: Insert Component 🗗 , Hide/Show 🕶 , Suppress ↓🗗 , Mate 🖉 , Move Component 🗗 , Rotate Component 🖉 , Exploded View 🖉 and Interference Detection 🖉 .

Project Situation

Communication is a major component to a successful development program. Provide frequent status reports to the customer and to your team members.

Communicate with suppliers. Ask questions. Check on details. What is the delivery time for the BATTERY, LENS and SWITCH parts?

Talk to colleagues to obtain useful manufacturing suggestions and ideas. Your team decided that plastic injection molding is the most cost effective method to produce large quantities of the desired part.

Investigate surface finishes that support the customer's advertising requirement. You have two fundamental choices:

- Adhesive label

- Silk-screen

There are time, quantity and cost design constraints. For the prototype, use an adhesive label. In the final product, create a silk screen.

Investigate the options on O-Ring material. Common O-Ring materials for this application are Buna-N (Nitrile®) and Viton®. Be cognizant of compatibility issues between O-RING materials and lubricants.

The LENSCAP encloses the LENS. The HOUSING protects the BATTERY. How do you design the LENSCAP and HOUSING to ease the transition from development to manufacturing? Answer: Review the fundamental design rules behind the plastic injection manufacturing process:

- Maintain a constant wall thickness. Inconsistent wall thickness creates stress.

- Create a radius on all corners. No sharp edges. Sharp edges create vacuum issues when removing the mold. Filleting inside corners on plastic parts will strengthen the part and improve plastic flow characteristics around corners in the mold.

- Understand the Draft feature. The Draft feature is commonly referred to as plus draft or minus draft. Plus draft adds material. Minus draft removes material.

- Allow a minimum draft angle of 1°. Draft sides and internal ribs. Draft angles assist in removing the part from the mold.

- Ribs are often added to improve strength. Ribs must also be added to improve plastic flow to entirely fill the cavity and to avoid a short shot—not enough material to fill the mold.

- Generally use a 2:3 ratio for rib thickness compared to overall part thickness.

- Shrinkage of the cooling plastic requires that the cavity is larger than the finished part. Shrinkage around the core will cause the part to bind onto the core. The Draft feature helps eject the part from the mold. Shrinkage on the cavity side will cause the part to pull from the mold. Less draft is needed on the cavity faces than on the core faces.

- Sink marks occur where there is an area with thicker material than the rest of the part wall thickness. A problem may occur where a tall rib with draft meets a face of the part.

- Mold cavities should be vented to allow trapped gases to escape through the channels.

Obtain additional information on material and manufacturing from material suppliers or mold manufacturers. Example: GE Plastics of Pittsfield, MA, provides design guidelines for selecting raw materials and creating plastic parts and components.

O-RING Part - Swept Base Feature

The O-RING is positioned between the LENSCAP and the LENS.

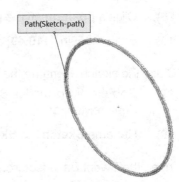

Path(Sketch-path)

A Swept Boss/Base feature can be simple or complex. The O-RING is classified as a simple Swept Base feature. A complex Swept Boss feature utilizes 3D curves and Guide Curves.

Create the O-RING with the Swept Base 🍃 feature. The Swept Base feature requires two sketches or a sketch path and a circular profile.

Utilize the PART-IN-ANSI template created in Project 2 for inch units. Utilize the PART-MM-ISO template for millimeter units. Millimeter dimensions are provided in brackets [x].

💡 For non-circular sketch profiles, create the sketch profile on a perpendicular plane to the path and use a pierce relation to locate the profile on the path.

Activity: O-RING Part - Swept Base Feature

Create a New part. Enter name and description.

1) Click **New** ⬜ from the Menu bar.

2) Click the **MY-TEMPLATES** tab.

3) Double-click **PART-IN-ANSI**, **[PART-MM-ISO]**.

4) Click **Save**.

5) Select **PROJECTS** for the Save in folder.

6) Enter **O-RING** for File name.

7) Enter **O-RING FOR LENS** for Description.

8) Click **Save**. The O-RING FeatureManager is displayed.

Create the Swept path.

9) Right-click **Front Plane** from the FeatureManager for the Sketch plane. This is your Sketch plane.

10) Click **Sketch** 🖉 from the Context toolbar. The Sketch toolbar is displayed.

11) Click the **Circle** ⊙ Sketch tool. The Circle PropertyManager is displayed.

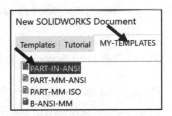

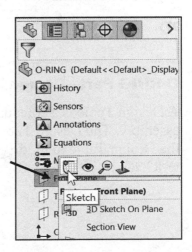

12)   Sketch a circle centered at the **Origin** ⌊↑→.

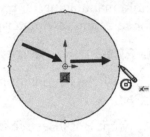

Add a dimension.

13)   Click the **Smart Dimension** ✎ Sketch tool.

14)   Click the **circumference** of the circle.

15)   Click a **position** off the profile.

16)   Enter **4.350**in, **[110.49]** as illustrated.

Close the sketch. Rename the sketch.

17)   **Rebuild** 🛢 the model. Sketch1 is displayed in the FeatureManager.

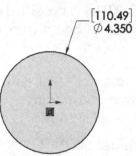

18)   Rename **Sketch1** to **Sketch-path**.

Insert the Swept Base feature.

19)   Click the **Swept Boss/Base** 🛠 feature tool. The Sweep PropertyManager is displayed.

20)   **Expand** O-RING from the fly-out FeatureManager.

21)   Click the **Circular Profile** box.

22)   Click **Sketch-path** from the fly-out FeatureManager. Sketch-path is displayed.

23)   Enter **0.125**in for Diameter.

24)   Click **OK** ✔ from the Sweep PropertyManager. Sweep1 is displayed in the FeatureManager.

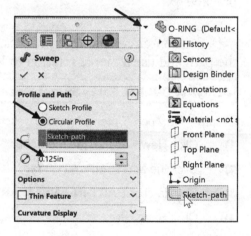

Rename the feature.

25)   Rename **Sweep1** to **Base-Sweep**.

Display an Isometric view. Save the model.

26)   Click **Isometric view** 🧊.

27)   Click **Save** 💾.

O-RING Part - Design Table

A Design Table is a spreadsheet used to create multiple configurations in a part or assembly. The Design Table controls the dimensions and parameters in the part. Utilize the Design Table to modify the overall path diameter and profile diameter of the O-RING.

Create three configurations of the O-RING:

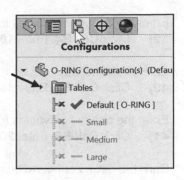

- Small

- Medium

- Large

The O-RING contains two dimension names in the Design Tables. Parts contain hundreds of dimensions and values. Rename dimension names for clarity.

The part was initially designed in inches. You are required to manufacture the configurations in millimeters. Modify the part units to millimeters.

Activity: O-RING Part - Design Table

Modify the Primary Units.

28) Click **Options** ⚙, **Document Properties** tab from the Menu bar. Select **Units**.

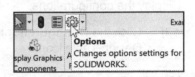

29) Select **MMGS**. Select **.12** for Basic unit length decimal place.

30) Click **OK** from the Documents Properties - Units box.

Insert a Design Table.
31) Click **Insert**, **Tables**, **Design Table** from the Menu bar. The Auto-create option is selected. Accept the default settings.

32) Click **OK** ✔ from the Design Table PropertyManager.

33) Click **D1@Sketch-path**. Hold the **Ctrl** key down.

34) Click **D3@Sweep1**. Release the **Ctrl** key.

35) Click **OK** from the Dimensions dialog box.

Note: The dimension variable name will be different if sketches or features were deleted.

The input dimension names and default values are automatically entered into the Design Table. The value Default is entered in Cell A3.

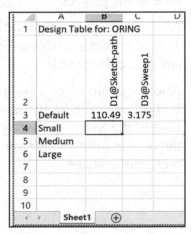

The values for the O-RING are entered in Cells B3 through C6. The sketch-path diameter is controlled in Column B. The sketch-profile diameter is controlled in Column C.

Enter the three configuration names.
36) Click **Cell A4**. Enter **Small**.

37) Click **Cell A5**. Enter **Medium**.

38) Click **Cell A6**. Enter **Large**.

Enter the dimension values for the Small configuration.
39) Click **Cell B4**. Enter **100**. Click **Cell C4**. Enter **3**.

Enter the dimension values for the Medium configuration.
40) Click **Cell B5**. Enter **150**. Click **Cell C5**. Enter **4**.

Enter the dimension values for the Large configuration.
41) Click **Cell B6**. Enter **200**. Click **Cell C6**. Enter **10**.

Build the three configurations.
42) Click a **position** inside the Graphics window.

43) Click **OK** to generate the three configurations. The Design Table icon is displayed in the FeatureManager.

44) **Rebuild** 🔵 the model.

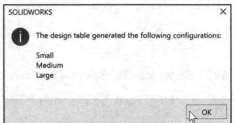

	A	B	C	D
1	Design Table for: ORING			
2		D1@Sketch-path	D3@Sweep1	
3	Default	110.49	3.175	
4	Small	100	3	
5	Medium	150	4	
6	Large	200	5	
7				
8				
9				
10				

Sheet1 ⊕

Display the configurations.

45) Click the **ConfigurationManager** 🔲 tab. View the Design Table icon.

View the three configurations.
46) Double-click **Small**. Double-click **Medium**.

47) Double-click **Large**. Double-click **Default**.

Return to the FeatureManager.
48) Click the **FeatureManager** 🔷 tab.

Save the model.
49) Click **Save** 💾.

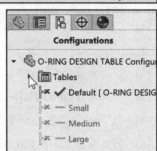

SOLIDWORKS ✕

ⓘ The design table generated the following configurations:

Small
Medium
Large

OK

Configurations

⯆ 🔶 O-RING DESIGN TABLE Configu
   📋 Tables
   ✖ ✓ Default [O-RING DESIG
   ✖ — Small
   ✖ — Medium
   ✖ ⌇ Large

The O-RING is classified as a simple Swept feature. A complex Swept feature utilizes 3D curves and Guide Curves. Investigate additional Swept features later in the project.

An example of a complex Swept is a Thread. A Thread requires a Helical/Spiral curve for the path and a circular profile.

Another example of a complex Swept is a violin body. The violin body requires Guide Curves to control the Swept geometry. Without Guide Curves the profile follows the straight path to produce a rectangular shape.

Thread

With Guide Curves, the profile follows the path and the Guide Curve geometry to produce the violin body.

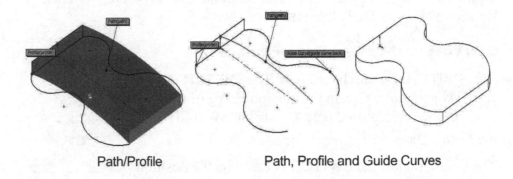

Path/Profile   Path, Profile and Guide Curves

Additional information on Swept and Pierce are found in SOLIDWORKS Help Topics. Keywords: Swept (overview, simple sweep) and Pierce (relations). Refer to Help, SOLIDWORKS Tutorials, Revolves and Sweeps for additional information.

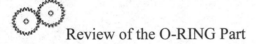 Review of the O-RING Part

The O-RING utilized the Swept feature. The Swept feature required a sketch path and a sketch profile. The path was a circle sketched on the Front Plane. The profile was the diameter of the circular profile.

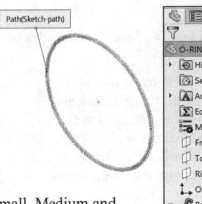

For non-circular sketch profiles, create the sketch profile on a perpendicular plane to the path and use a pierce relation to locate the profile on the path.

Swept features can be simple or complex. Small, Medium and Large configurations of the O-RING were developed with a Design Table. The Design Table icon displayed in the ConfigurationManager is an Excel based spread sheet utilized to create variations of dimensions and features.

💡 The Sketch toolbar contains two areas: Sketch Entities and Sketch Tools. There are numerous ways to access these areas: Select the Sketch 🔲 icon from the Sketch toolbar, right-click and select the Sketch icon from the Context toolbar or click Tools, Sketch Entities or Sketch Tools from the Menu bar.

SWITCH Part-Lofted Base Feature

The SWITCH is a purchased part. The SWITCH is a complex assembly. Create the outside casing of the SWITCH as a simplified part. Create the SWITCH with the Lofted Base ⬇ feature.

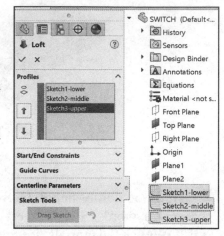

The orientation of the SWITCH is based on the position in the assembly. The SWITCH is comprised of three cross section profiles. Sketch each profile on a different plane.

The first plane is the Top Plane. Create two reference planes parallel to the Top Plane.

Sketch one profile on each plane. The design intent of the sketch is to reduce the number of dimensions.

Utilize symmetry, construction geometry and Geometric relations to control three sketches with one dimension.

Insert a Lofted feature. Select the profiles to create the Loft feature.

Insert the Dome feature to the top face of the Loft and modify the Loft Base feature. If the Dome feature is not displayed in the Feature CommandManager, click Insert, Features, Dome from the Main menu.

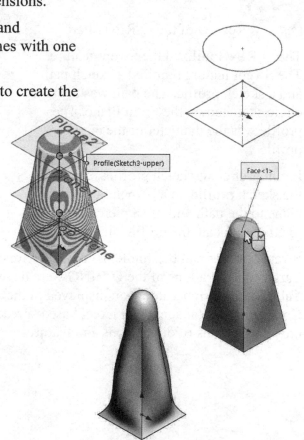

💡 Specify an elliptical dome for cylindrical or conical models or a continuous dome for polygonal models.

Activity: Switch Part - Loft Base Feature

Create a New part. Enter name and description.

50) Click **New** from the Menu bar.

51) Click the **MY-TEMPLATES** tab.

52) Double-click **PART-IN-ANSI**, **[PART-MM-ISO]**.

53) Click **Save**.

54) Select **PROJECTS** for the Save in folder.

55) Enter **SWITCH** for File name.

56) Enter **BUTTON STYLE** for Description.

57) Click **Save**. The SWITCH FeatureManager is displayed.

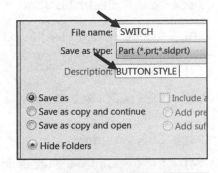

If you upgrade from an older version, your Planes may be displayed by default. Click View, Hide/Show, Planes from the Menu bar to display all reference planes. Hide unwanted planes in the FeatureManager.

Display the Top Plane.

58) Right-click **Top Plane** from the FeatureManager.

59) Click **Show**.

Display an Isometric view.

60) Click **Isometric view** from the Heads-up View toolbar.

Insert two Reference planes.

61) Hold the **Ctrl** key down.

62) Click and drag the **Top Plane** upward. The Plane PropertyManager is displayed.

63) Release the **mouse button**.

64) Release the **Ctrl** key.

65) Enter **.500**in, **[12.7]** for Offset Distance.

66) Enter **2** for # of Planes to Create.

67) Click **OK** from the Plane PropertyManager. Plane1 and Plane2 are displayed in the FeatureManager.

68) Click **Front view** to display Plane1 and Plane2 offset from the Top plane.

> Hold the Ctrl key down. Drag the Top Plane upward. Pick an edge, not the handles.

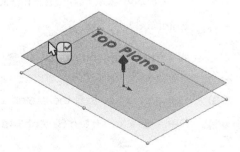

Display a Top view.

69)   Click **Top view**

Insert Sketch1. Sketch1 is a square on the Top Plane centered about the
Origin.

70)   Right-click **Top Plane** from the FeatureManager.

71)   Click **Sketch** from the Context toolbar.

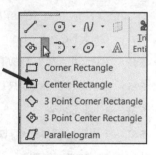

72)   Click the **Center Rectangle** tool from the Consolidated Sketch
toolbar.

73)   Click the **Origin**.

74)   Click a **point** in the upper right hand of the
window as illustrated. The Center Rectangle
Sketch tool automatically inserts a midpoint
relation about the Origin and Vertical and
Horizontal relations.

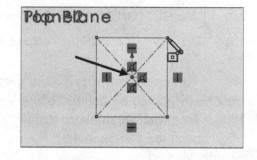

75)   Click **OK** from the Rectangle
PropertyManager.

Add an Equal relation between the four edges.

76)   Click the **left vertical line**.

77)   Hold the **Ctrl** key down.

78)   Click the **top horizontal** line.

79)   Click the **right vertical** line.

80)   Click the **bottom horizontal** line.

81)   Release the **Ctrl** key.

82)   Click **Equal** = .

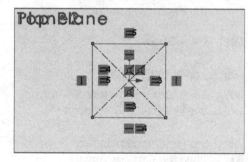

Add a dimension.

83)   Click the **Smart Dimension** Sketch tool.

84)   Click the **top horizontal** line.

85)   Click a **position** above the profile.

86)   Enter **.500**in, [**12.7**].

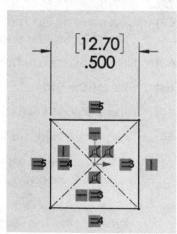

Close, fit and rename the sketch.

87)   **Rebuild** the model.

88)   Press the **f** key to fit the sketch to the Graphics area.

89)   Rename **Sketch1** to **Sketch1-lower**.

Save the model.

90)   Click **Save**

Display a Top view. Insert Sketch2 on Plane1.

91) Click **Top view** .

92) Right-click **Plane1** from the FeatureManager.

93) Click **Sketch** from the Context toolbar.

94) Click the **Circle** ⊙ Sketch tool. The Circle PropertyManager is displayed.

95) Sketch a **Circle** centered at the Origin as illustrated.

Add a Tangent relation.

96) Right-click **Select** to deselect the Circle Sketch tool.

97) Click the **circumference** of the circle.

98) Hold the **Ctrl** key down.

99) Click the **top horizontal Sketch1-lower line**. The Properties PropertyManager is displayed.

100) Release the **Ctrl** key.

101) Click **Tangent** ⌀.

102) Click **OK** ✔ from the Properties PropertyManager.

Close and rename the sketch.

103) Click **Exit Sketch**.

104) Rename **Sketch2** to **Sketch2-middle**.

Display an Isometric view.

105) Click **Isometric view** 🔲. View the results.

Display a Top view. Insert Sketch3 on Plane2.

106) Click **Top view** .

107) Right-click **Plane2** from the FeatureManager.

108) Click **Sketch** from the Context toolbar. The Sketch toolbar is displayed.

109) Click the **Centerline** Sketch tool.

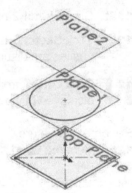

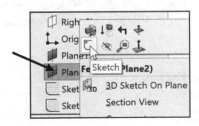

110) Sketch a **centerline** coincident with the Origin and the upper right corner point as illustrated.

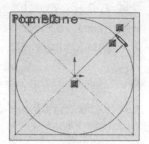

111) Click the **Point** ▪ Sketch tool.

112) Click the **midpoint** of the right diagonal centerline. The Point PropertyManager is displayed.

113) Click **Circle** ⊙ from the Sketch toolbar.

114) Sketch a **Circle** centered at the Origin to the midpoint of the diagonal centerline.

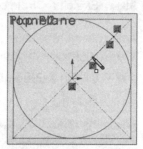

Close and rename the sketch.
115) Click **Exit Sketch**.

116) Rename **Sketch3** to **Sketch3-upper**.

Display an Isometric view. Insert the Lofted Base feature.

117) Click the **Lofted Boss/Base** 🔫 feature tool. The Loft PropertyManager is displayed.

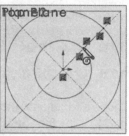

118) Click **Isometric view** 🔳 from the Heads-up View toolbar.

119) Right-click in the **Profiles box**.

120) Click **Clear Selections**.

121) Click **Sketch1-lower** from the fly-out FeatureManager.

122) Click **Sketch2-middle** as illustrated.

123) Click **Sketch3-upper** as illustrated. The selected sketch entities are displayed in the Profiles box.

124) Click **OK** ✔ from the Loft PropertyManager. Loft1 is displayed in the FeatureManager.

Rename the feature.
125) Rename **Loft1** to **Base Loft**.

Hide the planes.
126) Click **View**, **Hide/Show,** uncheck **Planes** from the Menu bar.

Save the part.
127) Click **Save** 💾.

The system displays a preview curve and loft as you select the profiles. Use the Up button and Down button in the Loft PropertyManager to rearrange the order of the profiles.

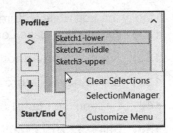

🔆 Redefine incorrect selections efficiently. Right-click in the Graphics window, and click Clear Selections to remove selected profiles. Select the correct profiles.

SWITCH Part - Dome Feature

Insert the Dome feature on the top face of the Lofted Base feature. The Dome feature forms the top surface of the SWITCH. Note: You can specify an elliptical dome for cylindrical or conical models and a continuous dome for polygonal models. A continuous dome's shape slopes upwards, evenly on all sides.

Activity: SWITCH Part - Dome Feature

Insert the Dome feature.
128) Click the **top face** of the Base Loft feature in the Graphics window.

129) Click the **Dome** ⏣ feature tool. The Dome PropertyManager is displayed. Face<1> is displayed in the Faces to Dome box. Note: If the Dome feature is not displayed in the Feature CommandManager, click Insert, Features, Dome from the Main menu.

Enter distance.
130) Enter .20in, [**5.08**] for Distance. View the dome feature in the Graphics window.

131) Click **OK** ✔ from the Dome PropertyManager.

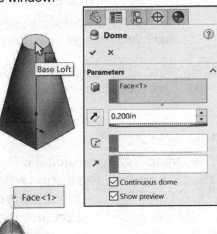

Experiment with the Dome feature to display different results. Click Insert, Features,

Freeform ⏣ to view the Freeform PropertyManager. As an exercise replace the Dome feature with a Freeform feature.

In the next section, modify the offset distance between the Top Plane and Plane1.

Modify the Loft Base feature.

132) **Expand** the Base Loft feature.

133) Right-click on **Annotations** in the FeatureManager.

134) Click **Show Feature Dimensions**.

135) Click on the Plane1 offset dimension, **.500**in, **[12.700]**.

136) Enter **.125**in, **[3.180]**. Click **Rebuild**.

137) Click **OK** ✔ from the Dimension PropertyManager.

Hide Feature dimensions.

138) Right-click on **Annotations** in the FeatureManager.

139) Un-check **Show Feature Dimensions**.

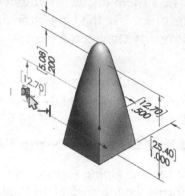

Display Performance Evaluation.

140) Click **Performance Evaluation** from the Evaluate tab in the CommandManager. View the results.

Close SWITCH Performance Evaluation.

141) Click **Close** from the Performance Evaluation dialog box.

142) Click **Save** 💾.

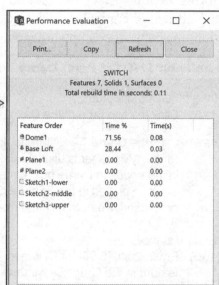

The Dome feature created the top for the SWITCH. The Performance Evaluation report displays the rebuild time for the Dome feature and the other SWITCH features. As feature geometry becomes more complex, the rebuild time increases.

 Review of the SWITCH Part

The SWITCH utilized the Lofted Base and Dome feature. The Lofted Base feature required three planes: Sketch1-lower, Sketch2-middle and Sketch3-upper. A profile was sketched on each plane. The three profiles were combined to create the Lofted Base feature.

The Dome feature created the final feature for the SWITCH.

🔅 The SWITCH utilized a simple Lofted Base feature. Lofts become more complex with additional Guide Curves. Complex Lofts can contain hundreds of profiles.

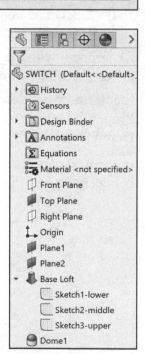

Four Major Categories of Solid Features

The LENSCAP and HOUSING combine the four major categories of solid features:

- Extrude: Requires one profile.

- Revolve: Requires one profile and axis of revolution.

- Swept: Requires one profile and one path.

- Lofted: Requires two or more profiles sketched on different planes.

Identify the simple features of the LENSCAP and HOUSING. Extrude and Revolve are simple features. Only a single sketch profile is required. Swept and Loft are more complex features. Two or more sketches are required.

Example: The O-RING was created as a Swept.

Could the O-RING utilize an Extruded feature?

Answer: No. Extruding a circular profile produces a cylinder.

Can the O-RING utilize a Revolved feature?
Answer: Yes. Revolving a circular profile about a centerline creates the O-RING.

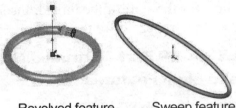

Revolved feature   Sweep feature

A Swept feature is required if the O-RING contained a non-circular path. Example: A Revolved feature does not work with an elliptical path or a more complex curve as in a paper clip. Combine the four major features and additional features to create the LENSCAP and HOUSING.

LENSCAP Part

The LENSCAP is a plastic part used to position the LENS to the HOUSING. The LENSCAP utilizes an Extruded Boss/Base, Extruded Cut, Extruded Thin Cut, Shell, Revolved Cut and Swept features.

The design intent for the LENSCAP requires that the Draft Angle be incorporated into the Extruded Boss/Base and Revolved Cut feature. Create the Revolved Cut feature by referencing the Extrude Base feature geometry. If the Draft angle changes, the Revolved Cut also changes.

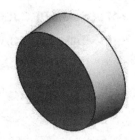

Insert an Extruded Boss/Base 🗐 feature with a circular profile on the Front Plane. Use a Draft option in the Boss-Extrude PropertyManager. Enter 5deg for Draft angle.

Insert an Extruded Cut  feature. The Extruded Cut feature should be equal to the diameter of the LENS Revolved Base feature.

Insert a Shell feature. Use the Shell feature for a constant wall thickness.

Insert a Revolved Cut feature on the back face. Sketch a single line on the Silhouette edge of the Extruded Base.

Utilize the Thin Feature option in the Cut-Revolve PropertyManager.

Utilize a Swept feature for the thread. Insert a new reference plane for the start of the thread. Insert a Helical Curve for the path. Sketch a trapezoid for the profile.

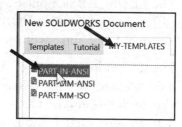

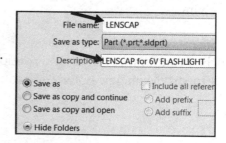

LENSCAP Part - Extruded Boss/Base, Extruded Cut and Shell Features

Create the LENSCAP. Review the Extruded Boss/Base, Extruded Cut and Shell features introduced in Project 5. The first feature is a Boss-Extrude feature. Select the Front Plane for the Sketch plane. Sketch a circle centered at the Origin for the profile. Utilize a Draft angle of 5deg.

Create an Extruded Cut feature on the front face of the Base feature. The diameter of the Extruded Cut equals the diameter of the Revolved Base feature of the LENS.

The Shell feature removes the front and back face from the solid LENSCAP.

Activity: LENSCAP Part - Extruded Base, Extruded Cut, and Shell Features

Create a New part. Enter name and description.

143) Click **New** from the Menu bar.

144) Click the **MY-TEMPLATES** tab.

145) Double-click **PART-IN-ANSI**, [**PART-MM-ISO**].

146) Click **Save**.

147) Select **PROJECTS** for the Save in folder.

148) Enter **LENSCAP** for File name.

149) Enter **LENSCAP for 6V FLASHLIGHT** for Description.

150) Click **Save**. The LENSCAP FeatureManager is displayed.

Create the sketch for the Extruded Base feature.

151) Right-click **Front Plane** from the FeatureManager.

152) Click **Sketch** from the Context toolbar.

153) Click the **Circle** Sketch tool. The Circle PropertyManager is displayed.

154) Sketch a **circle** centered at the Origin.

Add a dimension.

155) Click the **Smart Dimension** Sketch tool.

156) Click the **circumference** of the circle.

157) Click a **position** off the profile. Enter **4.900**in, **[124.46]**.

Insert an Extruded Boss/Base feature.

158) Click the **Extruded Boss/Base** feature tool. The Boss-Extrude PropertyManager is displayed. Blind is the default End Condition in Direction 1.

159) Click the **Reverse Direction** box.

160) Enter **1.725**in, **[43.82]** for Depth in Direction 1.

161) Click the **Draft On/Off** button.

162) Enter **5deg** for Angle. Click the **Draft outward** box.

163) Click **OK** from the Boss-Extrude PropertyManager. Boss-Extrude1 is displayed in the FeatureManager.

164) Rename **Boss-Extrude1** to **Base Extrude**.

165) Click **Save**.

Create the sketch for the Extruded Cut feature.

166) Right-click the **front face** for the Sketch plane.

167) Click **Sketch** from the Context toolbar. The Sketch toolbar is displayed.

168) Click the **Circle** Sketch tool. The Circle PropertyManager is displayed.

169) Sketch a **circle** centered at the Origin.

Add a dimension.

170) Click the **Smart Dimension** Sketch tool.

171) Click the **circumference** of the circle.

172) Click a **position** off the profile.

173) Enter **3.875**in, **[98.43]**.

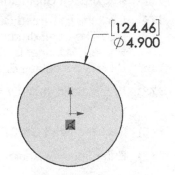

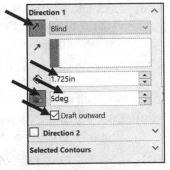

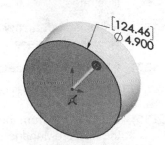

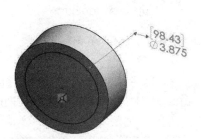

Insert an Extruded Cut feature.

174) Click the **Extruded Cut**  feature tool. The Cut-Extrude PropertyManager is displayed. Blind is the default End Condition.

175) Enter **.275**in, **[6.99]** for Depth in Direction 1.

176) Click the **Draft On/Off** button.

177) Enter **5**deg for Angle. Accept the default settings.

178) Click **OK** ✔ from the Cut-Extrude PropertyManager. Cut-Extrude1 is displayed in the FeatureManager.

179) Rename **Cut-Extrude1** to **Front-Cut**.

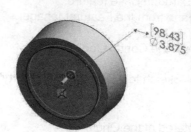

Insert the Shell feature.

180) Click the **Shell** feature tool. The Shell1 PropertyManager is displayed.

181) Click the **front face** of the Front-Cut as illustrated.

182) Press the **left arrow** approximately 8 times to view the back face.

183) Click the **back face** of the Base Extrude.

184) Enter **.150**in, **[3.81]** for Thickness.

185) Click **OK** ✔ from the Shell1 PropertyManager. Shell1 is displayed in the FeatureManager.

Display an Isometric view.

186) Press the **space bar** to display the Orientation dialog box.

187) Click **Isometric view** ⬛.

Display the inside of the Shell. Save the model.

188) Click **Right view** ⬜.

189) Click **Hidden Lines Visible** ⬛ from the Heads-up View toolbar.

190) Click **Save** 💾.

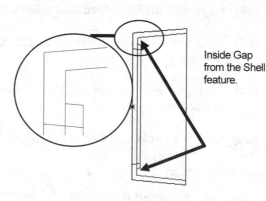

Inside Gap from the Shell feature.

💡 Use the inside gap created by the Shell feature to seat the O-RING in the assembly.

LENSCAP Part - Revolved Thin Cut Feature

The Revolved Thin Cut  feature removes material by rotating a sketched profile around a centerline.

The Right Plane is the Sketch plane. The design intent requires that the Revolved Cut maintains the same Draft angle as the Extruded Base feature.

Utilize the Convert Entities Sketch tool to create the profile. Small thin cuts are utilized in plastic parts. Utilize the Revolved Thin Cut feature for cylindrical geometry in the next activity.

Utilize a Swept Cut for non-cylindrical geometry. The semi-circular Swept Cut profile is explored in the project exercises.

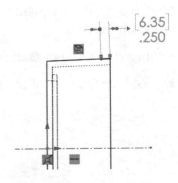

Sweep Cut Example

Activity: LENSCAP Part - Revolved Thin Cut Feature

Create the sketch.

191) Right-click **Right Plane** from the FeatureManager. This is your Sketch plane.

192) Click **Sketch** from the Context toolbar. The Sketch toolbar is displayed.

Sketch a centerline.

193) Click the **Centerline** Sketch tool. The Insert Line PropertyManager is displayed.

194) Sketch a **horizontal centerline** through the Origin as illustrated.

Create the profile.

195) Right-click **Select** to deselect the Centerline Sketch tool.

196) Click the **top silhouette outside edge** of Base Extrude as illustrated.

197) Click the **Convert Entities** Sketch tool.

198) Click and drag the **left endpoint 2/3** towards the right endpoint.

199) Release the **mouse button**.

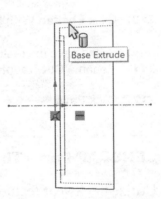

Add a dimension.

200) Click the **Smart Dimension** Sketch tool.

201) Click the **line**. Do not click the midpoint. The aligned dimension arrows are parallel to the profile line.

202) Drag the **text upward** and to the right.

203) Enter **.250**in, **[6.35]**.

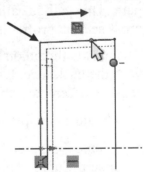

Insert a Revolved Cut feature.

204) Click **Revolved Cut**  from the Features toolbar. Do not close the Sketch. The warning message states, "The sketch is currently open."

205) Click **No**. The Cut-Revolve PropertyManager is displayed.

206) Click the **Reverse Direction** box in the Thin Feature box. The arrow points counterclockwise. Enter **.050in**, **[1.27]** for Direction 1 Thickness.

207) Click **OK** ✔ from the Cut-Revolve PropertyManager. Cut-Revolve-Thin1 is displayed in the FeatureManager.

Display the Revolved Thin Cut feature.
208) **Rotate** the part to view the back face.

209) Click **Isometric view** 🔲.

210) Click **Shaded With Edges** 🔲.

Rename and save the feature.
211) Rename **Cut-Revolve-Thin1** to **BackCut**.

212) Click **Save** 💾.

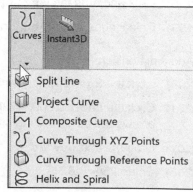

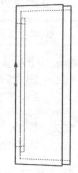

LENSCAP Part - Thread, Swept Feature and Helix/Spiral Curve

Utilize the Swept feature 🪱 to create the required threads. The thread requires a spiral path. This path is called the ThreadPath. The thread requires a Sketched profile. This cross section profile is called the ThreadSection.

The plastic thread on the LENSCAP requires a smooth lead in. The thread is not flush with the back face. Use an Offset plane to start the thread. There are numerous steps required to create a thread:

- Create a new plane for the start of the thread.

- Create the Thread path. Utilize Convert Entities and Insert, Curve, Helix/Spiral.

- Create a large thread cross section profile for improved visibility.

- Insert the Swept feature.

- Reduce the size of the thread cross section.

Activity: LENSCAP Part - Thread, Swept Feature, and Helix/Spiral Curve

Create the offset plane.

213) **Rotate** ↻ and **Zoom to Area** 🔍 on the back face of the LENSCAP.

214) Click the **narrow back face** of the Base Extrude feature. Note the mouse feedback icon.

215) Click **Insert**, **Reference Geometry, Plane** from the Menu bar. The Plane PropertyManager is displayed.

216) Enter **.450**in, [**11.43**] for Distance.

217) Click the **Flip offset** box.

218) Click **OK** ✔ from the Plane PropertyManager. Plane1 is displayed in the FeatureManager.

219) Rename **Plane1** to **ThreadPlane**.

Display the Isometric view with Hidden Lines Removed.

220) Click **Isometric view** 🧊.

221) Click **Hidden Lines Removed** ⬜.

Save the model.

222) Click **Save** 💾.

Utilize the Convert Entities Sketch tool to extract the back circular edge of the LENSCAP to the ThreadPlane.

Create the Thread path.

223) Right-click **ThreadPlane** from the FeatureManager.

224) Click **Sketch** from the Context toolbar.

225) Click the **back inside circular edge** of the Shell as illustrated.

226) Click the **Convert Entities** 🗗 Sketch tool.

227) Click **Top view** 🗗. The circular edge is displayed on the ThreadPlane.

228) Click **Hidden Lines Visible** 🗗 from the Heads-up View toolbar. View the results.

💡 Access the Plane tool from the Consolidated Reference Geometry toolbar.

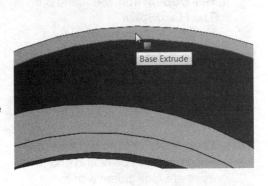

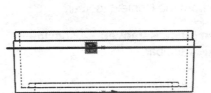

Insert the Helix/Spiral curve path.

229) Click **Insert, Curve, Helix/Spiral** from the Menu bar. The Helix/Spiral PropertyManager is displayed.

230) Enter **.250**in, **[6.35]** for Pitch.

231) Check the **Reverse** direction box.

232) Enter **2.5** for Revolutions.

233) Enter **0**deg for Starting angle. The Helix start point and end point are Coincident with the Top Plane.

234) Click the **Clockwise** box.

235) Click the **Taper Helix** box.

236) Enter **5**deg for Angle.

237) Uncheck the **Taper outward** box.

238) Click **OK** ✔ from the Helix/Spiral PropertyManager.

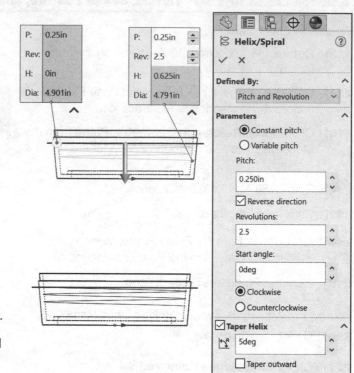

Rename the feature.

239) Rename **Helix/Spiral1** to **ThreadPath**.

Save the model.

240) Click **Save** 💾.

The Helix tapers with the inside wall of the LENSCAP. Position the Helix within the wall thickness to prevent errors in the Swept.

Sketch the profile on the Top Plane. Position the profile to the Top right of the LENSCAP in order to pierce to the ThreadPath in the correct location.

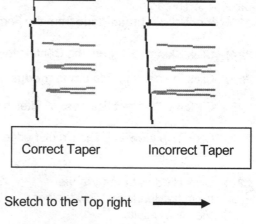

Correct Taper   Incorrect Taper

If required, Show the ThreadPlane.

241) Right-click **ThreadPlane** from the FeatureManager.

242) Click **Show** from the Context toolbar.

243) Click **Hidden Lines Removed** ⬡ from the Heads-up View toolbar.

Select the Plane for the Thread.

244) Right-click **Top Plane** from the FeatureManager.

Sketch to the Top right ⟶

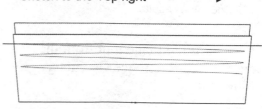

Sketch the profile.

245) Click **Sketch** 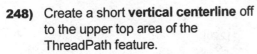 from the Context toolbar.

246) Click **Top view** .

247) Click the **Centerline** Sketch tool.

248) Create a short **vertical centerline** off to the upper top area of the ThreadPath feature.

249) Create a second **centerline** horizontal from the Midpoint to the left of the vertical line.

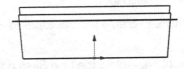

250) Create a third centerline coincident with the left horizontal endpoint. Drag the **centerline upward** until it is approximately the same size as the right vertical line.

251) Create a fourth **centerline** coincident with the left horizontal endpoint. Drag the **centerline** downward until it is approximately the same size as the left vertical line as illustrated.

Add an Equal relation.

252) Right-click **Select** to de-select the Centerline Sketch tool.

253) Click the **right vertical** centerline.

254) Hold the **Ctrl** key down.

255) Click the **two left vertical** centerlines.

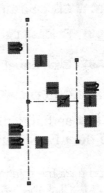

256) Release the **Ctrl** key. The selected sketch entities are displayed in the Selected Entities box.

257) Click **Equal** =

258) Click **OK** from the Properties PropertyManager.

Utilize centerlines and construction geometry with geometric relations to maintain relationships with minimal dimensions.

Check **View, Hide/Show, Sketch Relations** from the Menu bar to show/hide sketch relation symbols.

Add a dimension.

259) Click the **Smart Dimension** Sketch tool.

260) Click the two **left vertical endpoints**. Click a **position** to the left.

261) Enter **.500**in, **[12.7]**.

Sketch the profile. The profile is a trapezoid.

262) Click the **Line** Sketch tool.

263) Click the **endpoints** to create the trapezoid as illustrated.

264) Right-click **Select** to de-select the Line Sketch tool.

Add an Equal relation.
265) Click the **left vertical** line.

266) Hold the **Ctrl** key down.

267) Click the **top** and **bottom lines** of the trapezoid.

268) Release the **Ctrl** key.

269) Click **Equal** .

270) Click **OK** from the Properties PropertyManager.

Click and drag the sketch to a position above the top right corner of the LENSCAP.

Add a Pierce relation.
271) Click the **left midpoint** of the trapezoid.

272) Hold the **Ctrl** key down.

273) Click the **starting left back edge** of the ThreadPath.

274) Release the **Ctrl** key.

275) Click **Pierce** from the Add Relations box. The sketch is fully defined.

276) Click **OK** from the Properties PropertyManager.

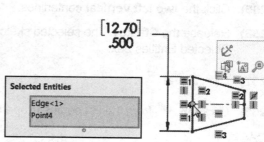

Selected Entities

Edge<1>
Point4

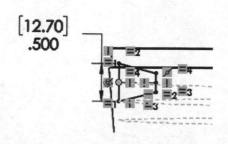

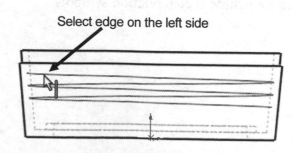

Select edge on the left side

Display an Isometric view.

277) Click **Isometric view** .

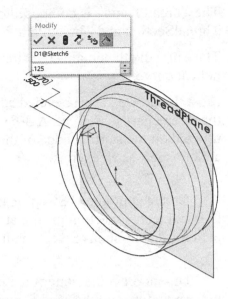

Modify the dimension.

278) Double-click the **.500** dimension text.

279) Enter **.125**in, **[3.18]**.

Close the sketch.

280) **Rebuild** the model.

Rename the sketch.

281) Rename **Sketch#** to **ThreadSection**.

Save the model.

282) Click **Save** .

Insert the Swept feature.

283) Click the **Swept Boss/Base** feature tool. The Sweep PropertyManager is displayed.

284) Click **Sketch Profile**.

285) Click **ThreadSection** for the fly-out FeatureManager.

286) Click inside the **Path** box.

287) Click **ThreadPath** from the fly-out FeatureManager. ThreadPath is displayed.

288) Click **OK** from the Sweep PropertyManager. Sweep1 is displayed in the FeatureManager.

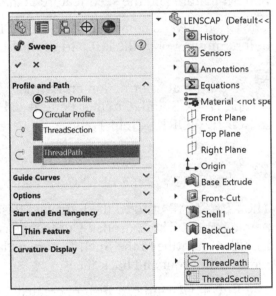

Rename the sketch. Display Shaded With Edges. Hide all planes.

289) Rename **Sweep1** to **Thread**.

290) Click **Shaded With Edges** from the Heads-up View toolbar. If needed, hide the ThreadPath and all planes.

Save the model.

291) Click **Save** .

Swept geometry cannot intersect itself. If the ThreadSection geometry intersects itself, the cross section is too large. Reduce the cross section size and recreate the Swept feature.

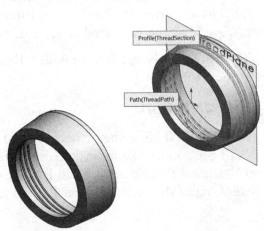

The Thread feature is composed of the following: ThreadSection and ThreadPath.

The ThreadPath contains the circular sketch and the helical curve.

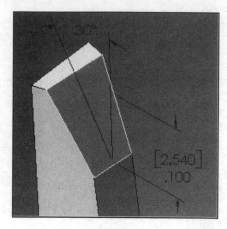

Most threads require a beveled edge or smooth edge for the thread part start point. A 30º Chamfer feature can be utilized on the starting edge of the trapezoid face. This action is left as an exercise.

 Create continuous Swept features in a single step. Pierce the cross section profile at the start of the swept path for a continuous Swept feature.

Un-suppress the Pattern feature to resolve both the Pattern feature and the seed feature at the same time.

The LENSCAP is complete. Review the LENSCAP before moving onto the last part of the FLASHLIGHT.

Additional information on Extruded Base/Boss, Extruded Cut, Swept, Helix/Spiral, Circular Pattern and Reference Planes are found in SOLIDWORKS Help Topics.

Review of the LENSCAP Part

The LENSCAP utilized the Extruded Base feature with the Draft Angle option. The Extruded Cut feature created an opening for the LENS. You utilized the Shell feature with constant wall thickness to remove the front and back faces.

The Revolved Thin Cut feature created the back cut with a single line. The line utilized the Convert Entities tool to maintain the same draft angle as the Extruded Base feature.

You utilized a Swept feature with a Helical Curve and Thread profile to create the thread.

HOUSING Part

The HOUSING is a plastic part utilized to contain the BATTERY and to support the LENS. The HOUSING utilizes an Extruded Boss/Base, Lofted Boss, Extruded Cut, Draft, Swept, Rib, Mirror and Linear Pattern features.

Insert an Extruded Boss/Base (Boss-Extrude1)
feature centered at the Origin.

Insert a Lofted Boss ⬇ feature. The first profile is the
converted circular edge of the Extruded Base. The
second profile is a sketch on the BatteryLoftPlane.

Insert the second Extruded Boss/Base (Boss-Extrude2)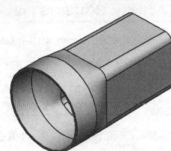
feature. The sketch is a converted edge from the Loft Boss.
The depth is determined from the height of the BATTERY.

Insert a Shell 🗖 feature to create a thin walled part.

Insert the third Extruded Boss (Boss-Extrude3) 🗖 feature.
Create a solid circular ring on the back circular face of the
Boss-Extrude1 feature. Insert the Draft feature to add a draft
angle to the circular face of the HOUSING. The design intent
for the Boss-Extrude1 feature requires you to maintain the
same LENSCAP draft angle.

Insert a Swept 🌀 feature for the Thread. Insert a Swept
feature for the Handle. Reuse the Thread profile from the
LENSCAP part. Insert an Extruded Cut to create the hole for
the SWITCH.

Insert the Rib ◈ feature on the back face of the HOUSING.

Insert a Linear Pattern ▦ feature to create a row of Ribs.

Insert a Rib ◈ feature along the bottom of the HOUSING.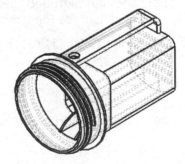

Utilize the Mirror ▥ feature to create the second Rib.

💡 Reuse geometry between parts. The LENSCAP thread is the same as the HOUSING thread. Copy the ThreadSection from the LENSCAP to the HOUSING.

💡 Reuse geometry between features. The Linear Pattern and Mirror Pattern utilized existing features.

Activity: HOUSING Part - Extruded Base Feature

Create the New part. Enter name and description.

292) Click **New** 🗋 from the Menu bar.

293) Click the **MY-TEMPLATES** tab.

294) Double-click **PART-IN-ANSI**, [**PART-MM-ISO**].

295) Click **Save**. Select **PROJECTS** for the Save in folder.

296) Enter **HOUSING** for File name. Enter **HOUSING FOR 6VOLT FLASHLIGHT** for Description.

297) Click **Save**. The HOUSING FeatureManager is displayed.

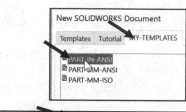

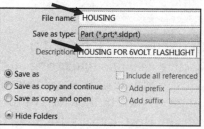

Create the Base sketch.

298) Right-click **Front Plane** from the FeatureManager. This is your Sketch plane. Click **Sketch** 🖉 from the Context toolbar.

299) Click the **Circle** ⊙ Sketch tool. The Circle PropertyManager is displayed.

300) Sketch a circle centered at the **Origin** ↳ as illustrated.

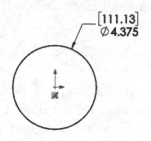

Add a dimension.

301) Click the **Smart Dimension** 🖉 Sketch tool.

302) Click the **circumference**. Enter **4.375**in, [**111.13**].

Insert an Extruded Boss/Base feature.

303) Click the **Extruded Boss/Base** 🗊 feature tool. The Boss-Extrude PropertyManager is displayed.

304) Enter **1.300**, [**33.02**] for Depth in Direction 1. Accept the default settings.

305) Click **OK** ✔ from the Boss-Extrude1 PropertyManager.

306) Click **Isometric view** 🔲. Note the location of the Origin.

Rename the feature and save the model.

307)   Rename **Boss-Extrude1** to **Base Extrude**.

308)   Click **Save** 🖫 .

HOUSING Part - Lofted Boss Feature

The Lofted Boss feature is composed of two profiles. The first sketch is named Sketch-Circle. The second sketch is named Sketch-Square.

Create the first profile from the back face of the Extruded feature. Utilize the Convert Entities sketch tool to extract the circular geometry to the back face.

Create the second profile on an Offset Plane. The FLASHLIGHT components must remain aligned to a common centerline. Insert dimensions that reference the Origin and build symmetry into the sketch. Utilize the Mirror Entities Sketch tool.

> **Activity: HOUSING Part - Lofted Boss Feature**

Create the first profile.

309)   Right-click the **back face** of the Base Extrude feature. This is your Sketch plane.

310)   Click **Sketch** ✎ from the Context toolbar. The Sketch toolbar is displayed.

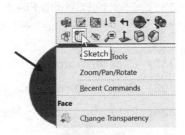

311)   Click the **Convert Entities** 🗇 Sketch tool to extract the face to the Sketch plane.

Close and rename the sketch.

312)   Right-click **Exit Sketch**.

313)   Rename **Sketch2** to **SketchCircle**.

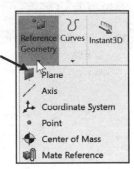

Create an offset plane.

314)   Click the **back face** of the Base Extrude feature.

315)   Click **Plane** from the Consolidated Reference Geometry Features toolbar. The Plane PropertyManager is displayed.

316)   Enter **1.300**in, **[33.02]** for Distance.

Verify the Plane position.

317)   Click **Top view** ▱. View the Graphics window. Click **OK** ✔ from the Plane PropertyManager. Plane1 is displayed in the FeatureManager.

Rename Plane1. Rebuild and save the model.

318)   Rename **Plane1** to **BatteryLoftPlane**.

319)   **Rebuild** 🛢 the model.

320)   Click **Save** 🖫 .

Create the second profile.

321) Right-click **BatteryLoftPlane** in the FeatureManager.

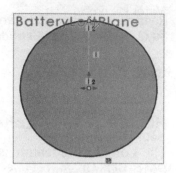

322) Click **Sketch** 🗐 from the Context toolbar. The Sketch toolbar is displayed.

323) Click **Back view** 📦.

324) Click the **circumference** of the circle.

325) Click the **Convert Entities** 🗐 Sketch tool.

326) Click the **Centerline** ⟋ Sketch tool.

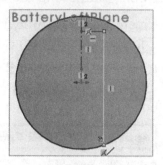

327) Sketch a **vertical centerline** coincident to the Origin and to the top edge of the circle as illustrated.

328) Click the **Line** ⟋ Sketch tool.

329) Sketch a **horizontal line** to the right side of the centerline as illustrated.

330) Sketch a **vertical line** down to the circumference as illustrated.

331) Click the **Sketch Fillet** ⌐ Sketch tool. The Sketch Fillet PropertyManager is displayed. If needed clear the Entities to Fillet box.

332) Click the **horizontal line** in the Graphics window.

333) Click the **vertical line** in the Graphics window.

334) Enter .1in, [**2.54**] for Radius.

335) Click **OK** ✔ from the Sketch Fillet PropertyManager.

336) Click **OK** ✔ from the Sketch Fillet PropertyManager. View the Sketch Fillet in the Graphics window.

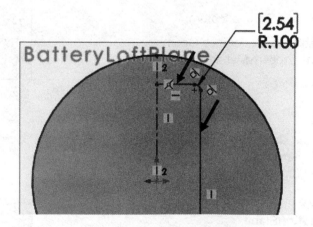

Mirror the profile.

337) Click the **Mirror Entities** Sketch tool. The Mirror PropertyManager is displayed.

338) Click the **horizontal line**, **fillet** and **vertical line**. The selected entities are displayed in the Entities to mirror box.

339) Click inside the **Mirror about** box.

340) Click the **centerline** from the Graphics window.

341) Click **OK** from the Mirror PropertyManager.

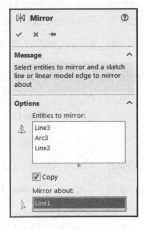

Trim unwanted geometry.

342) Click the **Trim Entities** Sketch tool. The Trim PropertyManager is displayed.

343) Click **PowerTrim** from the Options box.

344) Click a **position** to the far right of the circle.

345) Drag the **mouse pointer** to intersect the circle.

346) Perform the same **actions** on the left side of the circle.

347) Click **OK** from the Trim PropertyManager.

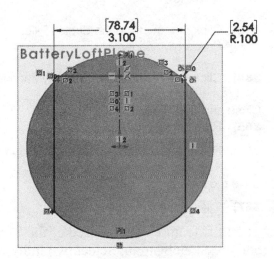

Add dimensions.

348) Click the **Smart Dimension** Sketch tool.

Create a horizontal dimension.
349) Click the **left vertical** line.

350) Click the **right vertical** line.

351) Click a **position** above the profile.

352) Enter **3.100**in, **[78.74]**. View the results.

Create a vertical dimension.

353) Click the **Origin**.

354) Click the **top horizontal** line.

355) Click a **position** to the right of the profile.

356) Enter **1.600**in, **[40.64]**.

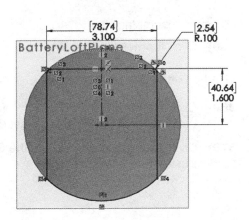

Modify the fillet dimension.

357) Double-click the **.100** fillet dimension.

358) Enter **.500**in, **[12.7]**.

Remove all sharp edges.

359) Click the **Sketch Fillet**  Sketch tool. The Sketch Fillet PropertyManager is displayed.

360) Enter **.500**in, **[12.7]** for Radius.

361) Click the **lower left corner point** as illustrated.

362) Click the **lower right corner point** as illustrated.

363) Click **OK** ✔ from the Sketch Fillet PropertyManager.

364) Click **OK** ✔ from the PropertyManager.

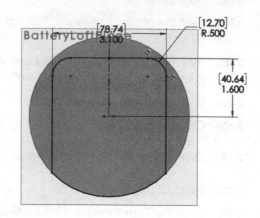

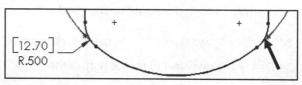

Close and rename the sketch.

365) Right-click **Exit Sketch**.

366) Rename the **Sketch3** to **SketchSquare**.

Save the model.

367) Click **Save** 💾.

The Loft feature is composed of the SketchSquare and the SketchCircle. Select two individual profiles to create the Loft. The Isometric view provides clarity when selecting Loft profiles.

Display an Isometric view.

368) Click **Isometric view** 🧊.

Insert a Lofted Boss feature.

369) Click the **Lofted Boss/Base** ⬇ feature tool. The Loft PropertyManager is displayed.

370) Click the **upper right side** of the SketchCircle as illustrated.

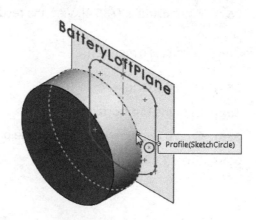

371) Click the **upper right side** of the SketchSquare as illustrated or from the Fly-out FeatureManager. The selected entities are displayed in the Profiles box.

372) Click **OK** ✔ from the Loft PropertyManager.

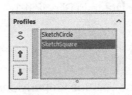

Rename Loft1. Save the model.

373) Rename **Loft1** to **Boss-Loft1**.

374) Click **Save** 💾.

💡 Organize the FeatureManager to locate Loft profiles and planes. Insert the Loft reference planes directly before the Loft feature. Rename the planes, profiles and guide curves with clear descriptive names.

HOUSING Part - Second Extruded Boss/Base Feature

Create the second Extruded Boss/Base feature from the square face of the Loft. How do you estimate the depth of the second Extruded Boss/Base feature? Answer: The Boss-Extrude1 feature of the BATTERY is 4.100in, [104.14mm].

Ribs are required to support the BATTERY. Design for Rib construction. Ribs add strength to the HOUSING and support the BATTERY. Use a 4.400in, [111.76mm] depth as the first estimate. Adjust the estimated depth dimension later if required in the FLASHLIGHT assembly.

The Extruded Boss/Base feature is symmetric about the Right Plane. Utilize Convert Entities to extract the back face of the Loft Base feature. No sketch dimensions are required.

Activity: HOUSING Part - Second Extruded Boss/Base Feature

Select the Sketch plane.

375) **Rotate** the model to view the back.

376) Right-click the **back face** of Boss-Loft1. This is your Sketch plane.

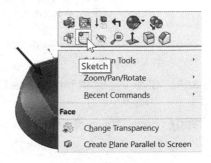

Create the sketch.

377) Click **Sketch** 🖊 from the Context toolbar. The Sketch toolbar is displayed.

378) Click the **Convert Entities** 🗗 Sketch tool.

Insert the second Extruded Boss/Base feature.

379) Click the **Extruded Boss/Base** feature tool. The Boss-Extrude PropertyManager is displayed.

380) Enter **4.400**in, **[111.76]** for Depth in Direction 1.

381) Click the **Draft On/Off** box.

382) Enter **1**deg for Draft Angle.

383) Click **OK** ✓ from the Boss-Extrude PropertyManager.

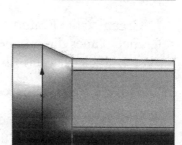

Display a Right view.

384) Click **Right view** ⬚.

Rename Boss-Extrude2. Save the model.

385) Rename **Boss-Extrude2** to **Boss-Battery**.

386) Click **Save** 💾.

HOUSING Part - Shell Feature

The Shell feature removes material. Use the Shell feature to remove the front face of the HOUSING. In the injection-molded process, the body wall thickness remains constant.

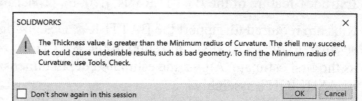

🔆 A dialog box is displayed if the thickness value is greater than the Minimum radius of Curvature.

SOLIDWORKS   ✕

⚠️ The Thickness value is greater than the Minimum radius of Curvature. The shell may succeed, but could cause undesirable results, such as bad geometry. To find the Minimum radius of Curvature, use Tools, Check.

☐ Don't show again in this session   OK   Cancel

Activity: HOUSING Part - Shell Feature

Insert the Shell feature.

387) Click **Isometric view** 🔲. If needed **Show** the BatteryLoftPlane from the FeatureManager as illustrated.

388) Click the **Shell** 🔲 feature tool. The Shell1 PropertyManager is displayed.

389) Click the **front face** of the Base Extrude feature as illustrated.

390) Enter **.100**in, **[2.54]** for Thickness.

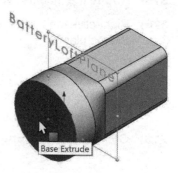

391) Click **OK**  from the Shell1
PropertyManager. Shell1 is displayed in the
FeatureManager.

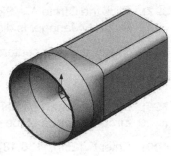

💡 The Shell feature position in the FeatureManager determines the geometry of
additional features. Features created before the Shell contained the wall thickness
specified in the Thickness option. Position features of different thickness such as the Rib
feature and Thread Swept feature after the Shell. Features inserted after the Shell remain
solid.

HOUSING Part - Third Extruded Boss/Base Feature

The third Extruded Boss/Base feature creates a solid circular ring on the back circular
face of the Boss-Extrude1 feature. The solid ring is a cosmetic stop for the LENSCAP
and provides rigidity at the transition of the HOUSING. Design for change. The Extruded
Boss/Base feature updates if the Shell thickness changes.

Utilize the Front plane for the sketch. Select the inside circular edge of the Shell. Utilize
Convert Entities to obtain the inside circle. Utilize the Circle Sketch tool to create the
outside circle. Extrude the feature towards the front face.

Activity: HOUSING Part - Third Extruded Boss Feature

Select the Sketch plane.
392) Right-click **Front Plane** from the FeatureManager. This is
your Sketch plane.

Create the sketch.
393) Click **Sketch** ✏ from the Context toolbar. The Sketch
toolbar is displayed.

394) Zoom-in and click the **front inside circular edge** of Shell1
as illustrated.

395) Click the **Convert Entities** ⬭ Sketch tool.

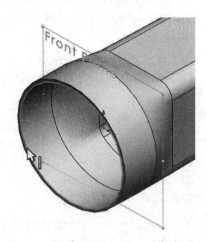

Create the outside circle. Display the Front view.
396) Click **Front view** ⬛.

397) Click the **Circle** ⊙ Sketch tool. The Circle
PropertyManager is displayed.

398) Sketch a **circle** centered at the Origin.

Add a dimension.

399) Click the **Smart Dimension** ⟨ Sketch tool. Click the
circumference of the circle.

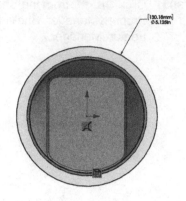

400) Enter **5.125**in, [**130.18**].

Insert the third Extruded Boss/Base feature.

401) Click the **Extruded Boss/Base** 🗐 feature tool. The
Boss-Extrude PropertyManager is displayed.

402) Enter **.100**in, [**2.54**] for Depth in Direction 1. The Extrude
arrow points to the front.

403) Click **OK** ✔ from the Boss-Extrude PropertyManager.

Display an Isometric view.

404) Click **Isometric view** 🔲.

Rename Boss-Extrude3. Hide the BatteryLoftPlane. Save the model.

405) Rename **Boss-Extrude3** to **Boss-Stop**.

406) If needed, **Hide** the BatteryLoftPlane as illustrated.

407) Click **Save** 💾.

HOUSING Part - Draft Feature

The Draft feature tapers selected model faces by a specified angle by utilizing a Neutral
Plane or Parting Line. The Neutral Plane option utilizes a plane or face to determine the
pull direction when creating a mold.

The Parting Line option drafts surfaces around a parting line of a mold. Utilize the
Parting Line option for non-planar surfaces. Apply the Draft feature to solid and surface
models.

A 5deg draft is required to ensure proper thread mating between the LENSCAP and the
HOUSING. The LENSCAP Extruded Boss/Base (Boss-Extrude1) feature has a 5deg
draft angle.

The outside face of the Extruded Boss/Base (Boss-Extrude1) feature HOUSING requires
a 5° draft angle. The inside HOUSING wall does not require a draft angle. The Extruded
Boss/Base (Boss-Extrude1) feature has a 5° draft angle. Use the Draft feature to create the
draft angle. The front circular face is the Neutral Plane. The outside cylindrical surface is
the face to draft.

You created the Extruded Boss/Base and Extruded Cut features with the Draft Angle
option. The Draft feature differs from the Extruded feature, Draft Angle option. The Draft
feature allows you to select multiple faces to taper.

In order for a model to eject from a mold, all faces must draft away from the parting line which divides the core from the cavity. Cavity side faces display a positive draft and core side faces display a negative draft. Design specifications include a minimum draft angle, usually less than 5deg.

For the model to eject successfully, all faces must display a draft angle greater than the minimum specified by the Draft Angle. The Draft feature, Draft Analysis Tools and DraftXpert utilize the draft angle to determine what faces require additional draft base on the direction of pull.

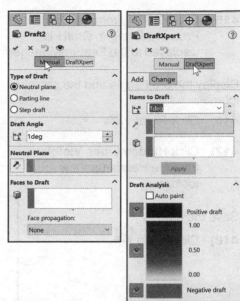

You can apply a draft angle as a part of an Extruded Boss/Base or Extruded Cut feature.

The Draft PropertyManager provides the ability to select either the *Manual* or *DraftXpert* tab. Each tab has a separate menu and option selections. The Draft PropertyManager displays the appropriate selections based on the type of draft you create.

The DraftXpert PropertyManager provides the ability to manage the creation and modification of all Neutral Plane drafts. Select the draft angle and the references to the draft. The DraftXpert manages the rest.

Activity: HOUSING Part - Draft Feature

Insert the Draft feature.

408) Click the **Draft** ◨ feature tool. The Draft PropertyManager is displayed.

409) Click the **Manual** tab.

410) Click **Neutral plane** for Type of Draft.

411) **Zoom-in** and click the thin **front circular face** of Base Extrude. The front circular face is displayed in the Neutral Plane box. Note: The face feedback icon. Face<1> is displayed.

412) Click inside the **Faces to draft** box.

413) Click the **outside cylindrical face** as illustrated. Note: The face feedback icon.

414) Enter **5**deg for Draft Angle.

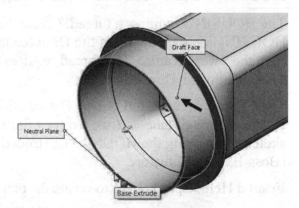

415) Click **OK** ✔ from the Draft PropertyManager. Draft1 is displayed in the FeatureManager.

Display the draft angle and the straight interior.

416) Click **Right view** 🔲.

417) Click **Hidden Lines Visible** 🔲 from the Heads-up View toolbar.

Save the model.

418) Click **Save** 💾.

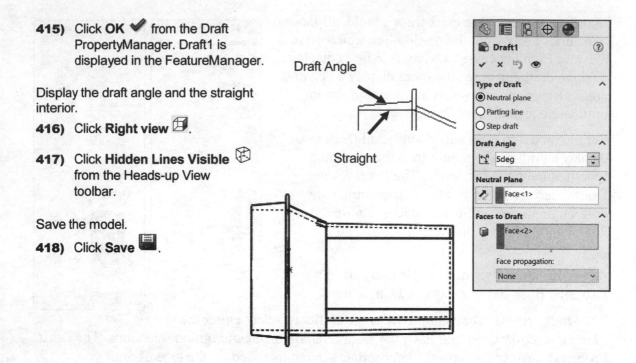

💡 The order of feature creation is important. Apply threads after the Draft feature for plastic parts to maintain a constant thread thickness.

HOUSING Part - Thread with Swept Feature

The HOUSING requires a thread. Create the threads for the HOUSING on the outside of the Draft feature. Create the thread with the Swept feature. The thread requires two sketches: ThreadPath and ThreadSection.

The LENSCAP and HOUSING Thread utilize the same technique. Create a ThreadPlane. Utilize Convert Entities to create a circular sketch referencing the HOUSING Extruded Boss/Base (Boss-Extrude1) feature.

Insert a Helix/Spiral curve to create the path.

Reuse geometry between parts. The ThreadSection is copied from the LENSCAP and is inserted into the HOUSING Top Plane.

Activity: HOUSING Part - Thread with Swept Feature

Insert the ThreadPlane.

419) Click **Isometric view** ⬡ from the Heads-up View toolbar.

420) Click **Hidden Lines Removed** ⬠ from the Heads-up View toolbar.

421) Click the **thin front circular face**, Base Extrude.

422) Click **Plane** from the Consolidated Reference Geometry toolbar. The Plane PropertyManager is displayed.

423) Check the **Flip offset** box.

424) Enter **.125**in, **[3.18]** for Distance. Accept the default settings.

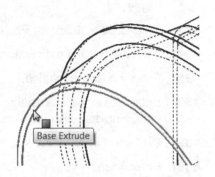

425) Click **OK** ✔ from the Plane PropertyManager. Plane2 is displayed in the FeatureManager.

Rename Plane2. Save the model.

426) Rename **Plane2** to **ThreadPlane**.

427) Click **Save** 💾.

Insert the ThreadPath.

428) Right-click **ThreadPlane** from the FeatureManager. This is your Sketch plane.

429) Click **Sketch** ✏ from the Context toolbar. The Sketch toolbar is displayed.

430) Click the **front outside circular edge** of the Base Extrude feature as illustrated.

431) Click the **Convert Entities** ⬠ Sketch tool. The circular edge is displayed on the ThreadPlane.

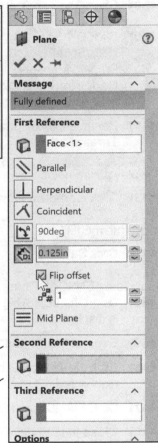

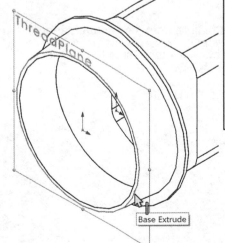

Insert the Helix/Spiral curve.

432) Click the **Helix and Spiral** tool from the Consolidated Curves toolbar as illustrated. The Helix/Spiral PropertyManager is displayed.

433) Enter **.250**in, **[6.35]** for Pitch.

434) Click the **Reverse direction** box.

435) Enter **2.5** for Revolution.

436) Enter **180** in the Start angle box. The Helix start point and end point are Coincident with the Top Plane.

437) Click the **Taper Helix** box.

438) Enter **5**deg for Angle.

439) Check the **Taper outward** box.

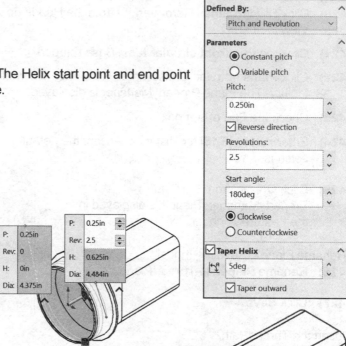

440) Click **OK** ✔ from the Helix/Spiral PropertyManager. HexliSpiral1 is displayed in the FeatureManager.

441) Rename **Helix/Spiral1** to **ThreadPath**.

Display an Isometric view.

442) Click **Isometric view** .

Save the model.

443) Click **Save** .

Copy the LENSCAP ThreadSection.

444) Open the LENSCAP part. The LENSCAP FeatureManager is displayed.

445) Expand the Thread feature from the FeatureManager.

446) Click the **ThreadSection** sketch. ThreadSection is highlighted.

447) Click **Edit**, **Copy** from the Menu bar.

448) Close the LENSCAP.

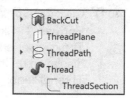

Open the HOUSING.
449) **Return** to the Housing.

Paste the LENSCAP ThreadSection.
450) Click **Top Plane** from the HOUSING FeatureManager.

451) Click **Edit**, **Paste** from the Menu bar. The ThreadSection is displayed on the Top Plane. The new Sketch7 name is added to the bottom of the FeatureManager.

452) **Hide** ThreadPlane.

Rename Sketch7. Save the model.
453) Rename **Sketch7** to **Thread Section**.

454) Click **Save** .

Add a Pierce relation.
455) Right-click **ThreadSection** from the FeatureManager.

456) Click **Edit Sketch**.

457) Click **ThreadSection** from the HOUSING FeatureManager.

458) **Zoom in** on the Midpoint of the ThreadSection.

459) Click the **Midpoint** of the ThreadSection.

460) Click **Isometric view** .

461) Hold the **Ctrl** key down.

462) Click the **right back edge of the ThreadPath**. Note: Do not click the end point. The Properties PropertyManager is displayed. The selected entities are displayed in the Selected Entities box.

463) Release the **Ctrl** key.

464) Click **Pierce** from the Add Relations box.

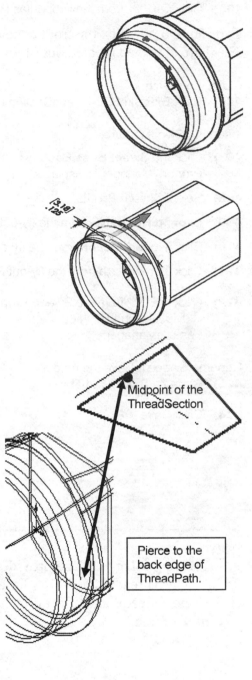

Midpoint of the ThreadSection

Pierce to the back edge of ThreadPath.

465) Click **OK** from the Properties PropertyManager.

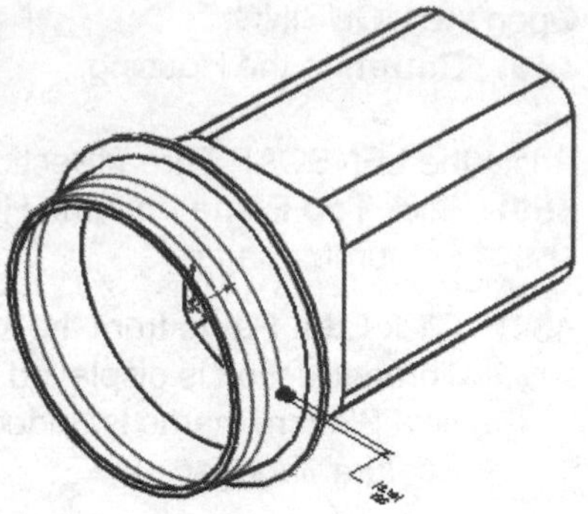

Caution: Do not click the front edge of the Thread path. The Thread is then created out of the HOUSING.

Close the sketch.

466) Click **Exit Sketch**. ThreadSection is fully defined.

Insert the Swept feature.

467) Click the **Swept Boss/Base** feature tool. The Swept PropertyManager is displayed.

468) Select **Sketch Profile**.

469) **Expand** HOUSING from the fly-out FeatureManager.

470) Click **ThreadSection** from the fly-out FeatureManager.

471) Click **ThreadPath** from the fly-out FeatureManager.

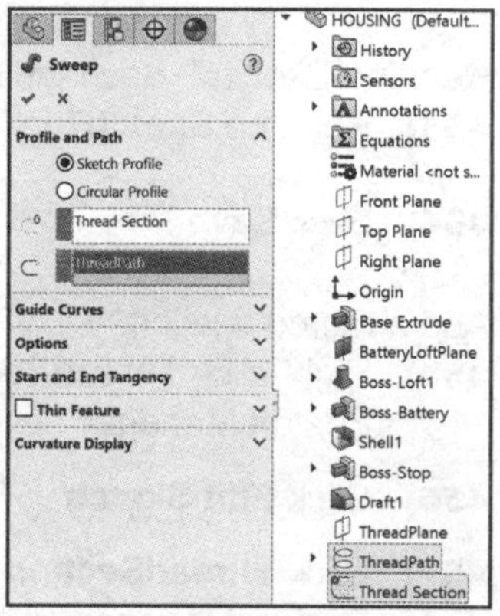

472) Click **OK** from the Sweep PropertyManager. Sweep1 is displayed in the FeatureManager.

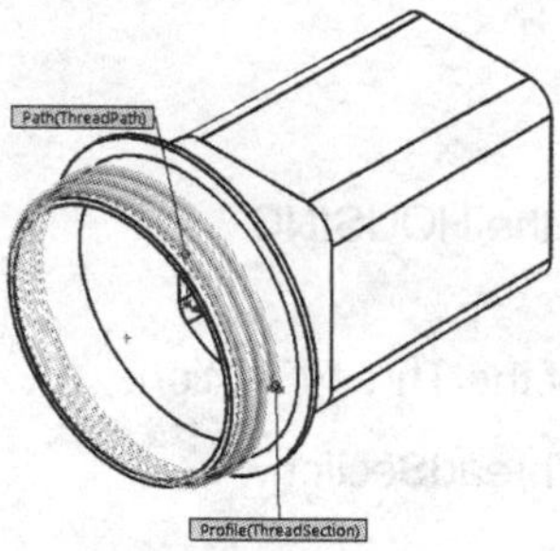

Rename Sweep1. Save the model.

473) Rename **Sweep1** to **Thread**.

474) Click **Save**.

Creating a ThreadPlane provides flexibility to the design. The ThreadPlane allows for a smoother lead. Utilize the ThreadPlane offset dimension to adjust the start of the thread.

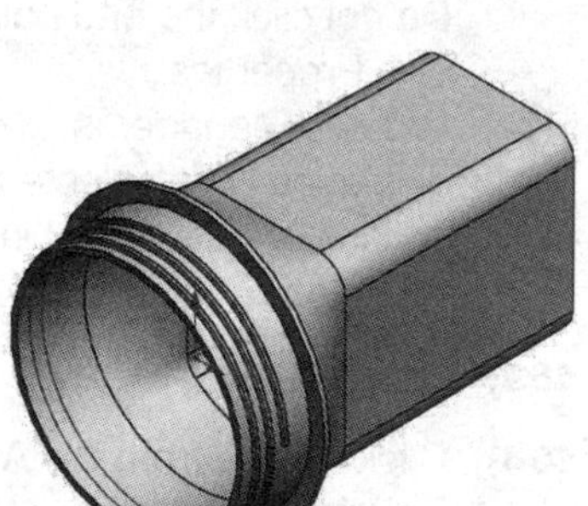

HOUSING Part - Handle with Swept Feature

Create the handle with the Swept feature. The Swept feature consists of a sketched path and cross section profile. Sketch the path on the Right Plane. The sketch uses edges from existing features. Sketch the profile on the back circular face of the Boss-Stop feature.

Activity: HOUSING Part - Handle with Swept Feature

Create the Swept path sketch.

475) Right-click **Right Plane** from the FeatureManager.

476) Select **Sketch** from the Context toolbar. The Sketch toolbar is displayed.

477) Click **Right view**.

478) Click **Hidden Lines Removed** from the Heads-up View toolbar.

479) Click the **Line** Sketch tool.

480) Sketch a **vertical line** up from the top corner as illustrated.

481) Sketch a **horizontal line** over to the left below the top of the Boss Stop as illustrated. Add any needed relations.

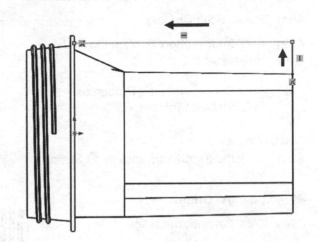

Insert a 2D Fillet.

482) Click the **Sketch Fillet** Sketch tool.

483) Click the **right top corner** of the sketch lines as illustrated.

484) Enter .500in, [**12.7**] for Radius.

485) Click **OK** from the Sketch Fillet PropertyManager.

486) Click **OK** from the PropertyManager.

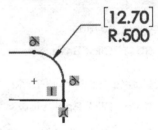

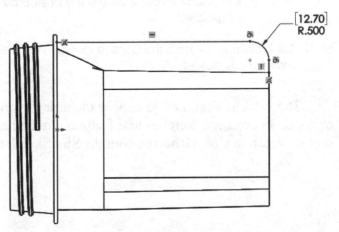

Add an Intersection relation.

487) Click the **bottom end point** of the vertical line.

488) Hold the **Ctrl** key down.

489) Click the **right vertical edge** of the Housing.

490) Click the **horizontal edge** of the Housing.

491) Release the **Ctrl** key.

492) Click **Intersection** ✕ from the Add Relations box.

493) Click **OK** ✔ from the Properties PropertyManager.

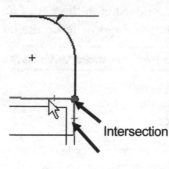
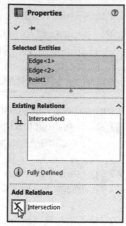

Intersection

Add a dimension.

494) Click the **Smart Dimension** ✛ Sketch tool.

495) Click the **Origin**.

496) Click the **horizontal** line.

497) Click a **position** to the right.

498) Enter **2.500**in, **[63.5]**.

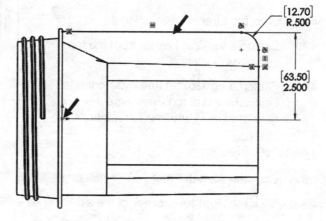

Close and rename the sketch.

499) Click **Exit Sketch**.

500) Rename **Sketch8** to **HandlePath**.

Save the model.

501) Click **Save** 💾.

Create the Swept Profile.

502) Click **Back view** 🔲.

503) Right-click the **back circular face** of the Boss-Stop feature as illustrated.

504) Click **Sketch** 📝 from the Context toolbar. The Sketch toolbar is displayed.

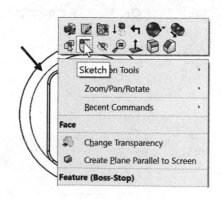

💡 The book is designed to expose the user to various methods in creating sketches and features. In the next section create a slot **without using** the Slot Sketch tool.

505) Click the **Centerline** Sketch tool.

506) Sketch a **vertical centerline** collinear with the Right Plane, coincident to the Origin.

507) Sketch a **horizontal centerline**. The left end point of the centerline is coincident with the vertical centerline on the Boss-Stop feature. Do not select existing feature geometry.

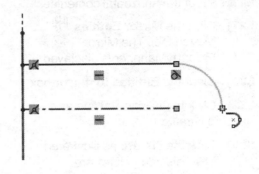

508) **Zoom in** on the top of the Boss-Stop.

509) Click the **Line** Sketch tool.

510) Sketch a **line** above the horizontal centerline as illustrated.

511) Click the **Tangent Arc** Sketch tool.

512) Sketch a **90° arc**.

513) Right-click **Select** to exit the Tangent Arc tool.

Add an Equal relation.

514) Click the **horizontal** centerline.

515) Hold the **Ctrl** key down.

516) Click the **horizontal** line.

517) Release the **Ctrl** key.

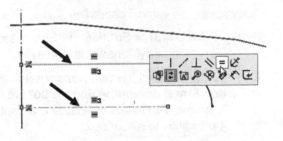

518) Right-click **Make Equal** from the Context toolbar.

519) Click **OK** from the Properties PropertyManager.

Add a Horizontal relation.

520) Click the **right end point** of the tangent arc.

521) Hold the **Ctrl** key down.

522) Click the **arc center point**.

523) Click the **left end point** of the centerline.

524) Release the **Ctrl** key.

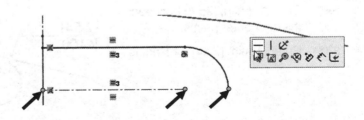

525) Right-click **Make Horizontal** from the Context toolbar.

526) Click **OK** from the Properties PropertyManager.

Mirror about the horizontal centerline.

527) Click the **Mirror Entities** ⊨⊧ Sketch tool. The Mirror PropertyManager is displayed.

528) Clear the **Entities to mirror** box.

529) Click the **horizontal** line as illustrated.

530) Click the **90° arc** as illustrated. The selected entities are displayed in the Entities to mirror box.

531) Click inside the **Mirror about** box.

532) Click the **horizontal** centerline.

533) Click **OK** ✔ from the Mirror PropertyManager.

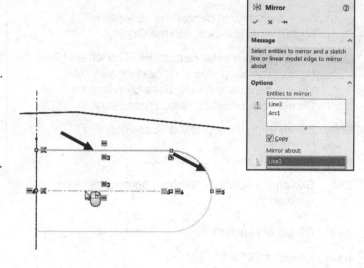

Mirror about the vertical centerline.

534) Click the **Mirror Entities** ⊨⊧ Sketch tool. The Mirror PropertyManager is displayed.

535) Window-Select the **two horizontal lines**, the **horizontal centerline** and the **90° arc** for Entities to mirror. The selected entities are displayed in the Entities to mirror box.

536) Click inside the **Mirror about** box.

537) Click the **vertical** centerline.

538) Click **OK** ✔ from the Mirror PropertyManager.

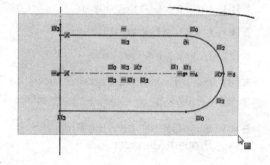

Add dimensions.

539) Click the **Smart Dimension** ⟋ Sketch tool.

540) Enter **1.000**in, [**25.4**] between the arc center points.

541) Enter **.100**in, [**2.54**] for Radius.

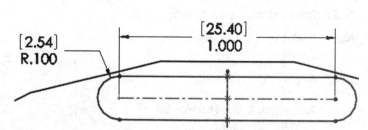

Add a Pierce relation.

542) Right-click **Select**.

543) Click **Isometric view** 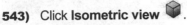.

544) Click the **top midpoint** of the Sketch profile.

545) Hold the **Ctrl** key down.

546) Click the **line** from the Handle Path.

547) Release the **Ctrl** key.

548) Click **Pierce** from the Add Relations box. The sketch is fully defined.

549) Click **OK** from the Properties PropertyManager.

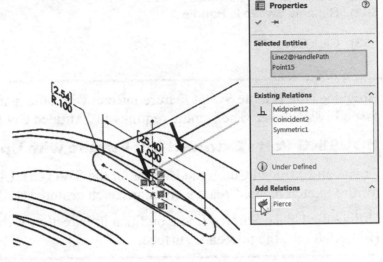

Close and rename the sketch.

550) Click **Exit Sketch**.

551) Rename **Sketch8** to **HandleProfile**.

552) **Hide** ThreadPlane and BatteryLoftPlane if needed.

Insert the Swept feature.

553) Click the **Swept Boss/Base** feature tool. The Sweep PropertyManager is displayed.

554) Select **Sketch Profile**.

555) **Expand** HOUSING from the fly-out FeatureManager.

556) Click inside the **Profile** box.

557) Click **HandleProfile** from the fly-out FeatureManager.

558) Click **HandlePath** from the fly-out FeatureManager.

559) Click **OK** from the Sweep PropertyManager. Sweep2 is displayed in the FeatureManager.

Fit the profile to the Graphics window.

560) Press the **f** key.

Display Shaded With Edges.

561) Click **Shaded With Edges** from the Heads-up View toolbar.

Rename Sweep. Save the model.
562) Rename **Sweep** to **Handle**.

563) Click **Save** 💾.

How does the Handle Swept feature interact with other parts in the FLASHLIGHT assembly? Answer: The Handle requires an Extruded Cut to insert the SWITCH.

HOUSING Part - Extruded Cut Feature with Up To Surface

Create an Extruded Cut in the Handle for the SWITCH. Utilize the top face of the Handle for the Sketch plane. Create a circular sketch centered on the Handle.

Utilize the Up To Surface End Condition in Direction 1. Select the inside surface of the HOUSING for the reference surface.

Activity: HOUSING Part - Extruded Cut Feature with Up To Surface

Select the Sketch plane.
564) Right-click the **top face** of the Handle. Handle is highlighted in the FeatureManager. This is your Sketch plane.

Create the sketch.
565) Click **Sketch** ✏ from the Context toolbar. The Sketch toolbar is displayed.

566) Click **Top view** 🔲.

567) Click **Circle** ⊙ from the Sketch toolbar. The Circle PropertyManager is displayed.

568) Sketch a **circle** on the Handle near the front as illustrated.

Deselect the circle sketch tool.
569) Right-click **Select**.

Add a Vertical relation.
570) Click the **Origin**.

571) Hold the **Ctrl** key down.

572) Click the **center point** of the circle. The Properties PropertyManager is displayed. The selected entities are displayed in the Selected Entities box.

573) Release the **Ctrl** key.

574) Click **Vertical** │.

575) Click **OK** ✔ from the Properties PropertyManager.

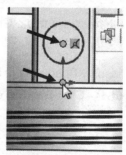

Add dimensions.

576) Click the **Smart Dimension** ⟨ Sketch tool.

577) Enter **.510**in, [**12.95**] for diameter.

578) Enter **.450**in, [**11.43**] for the distance from the Origin.

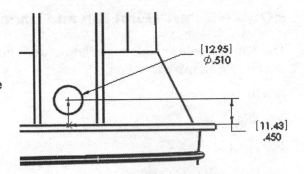

Insert an Extruded Cut feature.

579) **Rotate** the model to view the inside Shell1.

580) Click the **Extruded Cut** ⟨ feature tool. The Cut-Extrude PropertyManager is displayed.

581) Select the **Up To Surface** End Condition in Direction 1.

582) Click the **top inside face** of the Shell1 as illustrated.

583) Click **OK** ✔ from the Cut-Extrude PropertyManager. The Cut-Extrude1 feature is displayed in the FeatureManager.

Rename Cut-Extrude1. Display an Isometric view.

584) Rename the **Cut-Extrude1** feature to **SwitchHole**.

585) Click **Isometric view** ⬢ from the Heads-up View toolbar.

Save the model.

586) Click **Save** 💾.

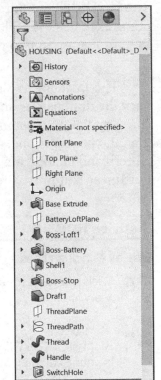

HOUSING Part - First Rib and Linear Pattern Feature

The Rib ⬗ feature adds material between contours of existing geometry. Use Ribs to add structural integrity to a part.

A Rib requires:

- A sketch

- Thickness

- Extrusion direction

The first Rib profile is sketched on the Top Plane. A 1° draft angle is required for manufacturing. Determine the Rib thickness by the manufacturing process and the material.

🔆 Rule of thumb states that the Rib thickness is ½ the part wall thickness. The Rib thickness dimension is .100 inches, [2.54mm] for illustration purposes.

The HOUSING requires multiple Ribs to support the BATTERY. A Linear Pattern feature creates multiple instances of a feature along a straight line. Create the Linear Pattern feature in two directions along the same vertical edge of the HOUSING.

🔆 The Instance to Vary option in the Linear Pattern PropertyManager allows you to vary the dimensions and locations of instances in a feature pattern *after it is created*. You can vary the dimensions of a series of instances, so that each instance is larger or smaller than the previous one. You can also change the dimensions of a single instance in a pattern and change the position of that instance relative to the seed feature of the pattern. For linear patterns, you can change the spacing between the columns and rows in the pattern.

Activity: HOUSING Part - First Rib and Linear Pattern Feature

Display all hidden lines.

587) Click **Hidden Lines Visible** ⬚ from the Heads-up View toolbar.

Create the sketch.

588) Right-click **Top Plane** from the FeatureManager. This is your Sketch plane.

589) Click **Sketch** ⬚ from the Context toolbar.

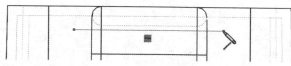

590) Click **Top view** ⬚.

591) Click the **Line** ✏ Sketch tool.

592) Sketch a **horizontal line** as illustrated. The endpoints are located on either side of the Handle.

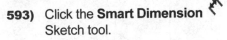 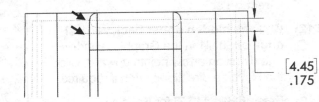

Add a dimension.

593) Click the **Smart Dimension** Sketch tool.

594) Click the **inner back** edge.

595) Click the **horizontal** line.

596) Click a **position** to the right off the profile.

597) Enter **.175**in, **[4.45]**.

Insert the Rib feature.

598) Click the **Rib** feature tool. The Rib PropertyManager is displayed.

599) Click the **Both Sides** button.

600) Click the **Draft On/Off** box.

601) Enter **1**deg for Draft Angle.

602) Enter **.100**in, **[2.54]** for Rib Thickness.

603) Click the **Parallel to Sketch** button. The Rib direction arrow points to the back. Flip the material side if required. Select the Flip material side check box if the direction arrow does not point towards the back.

604) Click **Front view**.

605) Click the **Rib** sketch as illustrated inside the HOUSING.

606) Click **OK** from the Rib PropertyManager. Rib1 is displayed in the FeatureManager.

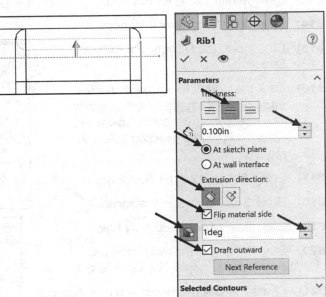
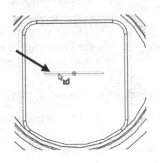

Display an Isometric view. Save the model.

607) Click Isometric view.

608) Click **Save**.

Existing geometry defines the Rib boundaries. The Rib does not penetrate through the wall.

Insert the Linear Pattern feature.

609) **Zoom to Area** on Rib1. Click **Rib1** from the FeatureManager.

610) Click the **Linear Pattern** feature tool. Rib1 is displayed in the Features to Pattern box.

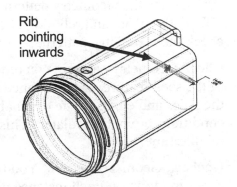

Rib pointing inwards

611) Click inside the **Direction 1 Pattern Direction** box.

612) Click the **hidden upper back vertical edge** of Shell1 in the Graphics window. The direction arrow points upward. Click the Reverse direction button if required.

613) Enter **.500**in, **[12.7]** for Spacing.

614) Enter **3** for Number of Instances.

615) Click inside the **Direction 2 Pattern** Direction box.

616) Click the hidden **lower back vertical edge** of Shell1 in the Graphics window. The direction arrow points downward. Click the Reverse direction button if required.

617) Enter **.500**in, **[12.7]** for Spacing.

618) Enter **3** for Number of Instances.

619) Click the **Pattern seed only** box.

620) Drag the Linear Pattern **Scroll bar** downward to display the Options box.

621) Check the **Geometry pattern** box. Accept the default values.

622) Click **OK** ✔ from the Linear Pattern PropertyManager. LPattern1 is displayed in the FeatureManager.

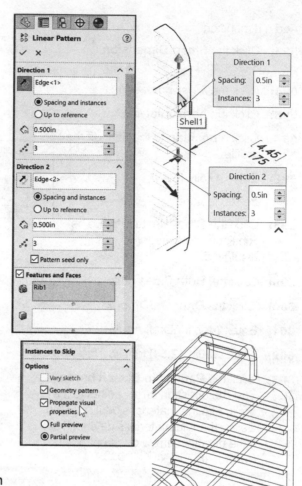

Display an Isometric view. Save the model.

623) Click **Isometric view** 📦 .

624) Click **Save** 💾 .

💡 Utilize the Geometry pattern option to efficiently create and rebuild patterns. Know when to check the Geometry pattern.

Check Geometry pattern. You require an exact copy of the seed feature. Each instance is an exact copy of the faces and edges of the original feature. End conditions are not calculated. This option saves rebuild time.

Uncheck Geometry pattern. You require the end condition to vary. Each instance will have a different end condition. Each instance is offset from the selected surface by the same amount.

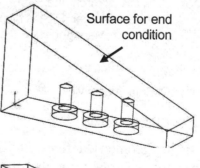

Surface for end condition

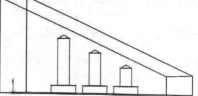

💡 Suppress Patterns when not required. Patterns contain repetitive geometry that takes time to rebuild. Pattern features also clutter the part during the model creation process. Suppress patterns as you continue to create more complex features in the part. Unsuppress a feature to restore the display and load into memory for future calculations. Hide features to improve clarity. Show feature to display hidden features.

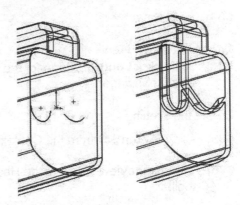

Rib sketches are not required to be fully defined. The Linear Rib option blends sketched geometry into existing contours of the model.

Example: Create an offset reference plane from the inside back face of the HOUSING.

Sketch two under defined arcs. Insert a Rib feature with the Linear option. The Rib extends to the Shell walls.

HOUSING Part - Second Rib Feature

The Second Rib feature supports and centers the BATTERY. The Rib is sketched on a reference plane created through a point on the Handle and parallel with the Right Plane. The Rib sketch references the Origin and existing geometry in the HOUSING. Utilize an Intersection and Coincident relation to define the sketch.

Activity: HOUSING Part - Second Rib Feature

Insert a Reference plane for the second Rib feature. Create a Parallel Plane at Point.

625) Click **Wireframe** ⊞ from the Heads-up View toolbar.

626) **Zoom to Area** 🔍 on the back right side of the Handle.

627) Click **Plane** from the Features toolbar. The Plane PropertyManager is displayed.

628) Click **Right Plane** from the fly-out FeatureManager.

629) Click the **vertex** (point) at the back right of the handle as illustrated. Plane1 is fully defined.

630) Click **OK** ✔ from the Plane PropertyManager. Plane1 is displayed in the FeatureManager.

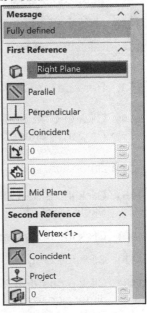

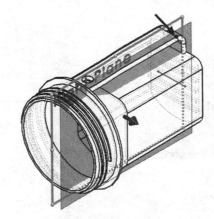

Rename Plane1.
631) Rename **Plane1** to **LongRibPlane**.

Fit to the Graphics window. Save the model.
632) Press the **f** key.

633) Click **Save** .

Create the second Rib.
634) Right-click **LongRibPlane** for the FeatureManager.

Create the sketch.

635) Click **Sketch** from the Context toolbar.

636) Click **Right view** from the Heads-up View toolbar.

637) Click the **Line** Sketch tool.

638) Sketch a **horizontal line**. Do not select the edges of the Shell1 feature.

Deselect the Line Sketch tool.
639) Right-click **Select**.

Add a Coincident relation.
640) Click **BatteryLoftPlane** from the FeatureManager.

641) Hold the **Ctrl** key down.

642) Click the **left end point** of the horizontal sketch line.

643) Release the **Ctrl** key. Click **Coincident**.

644) Click **OK** from the Properties PropertyManager.

Add a dimension.

645) Click the **Smart Dimension** Sketch tool.

646) Click the **horizontal** line.

647) Click the **Origin**.

648) Click a **position** for the vertical linear dimension text.

649) Enter **1.300**in, **[33.02]**.

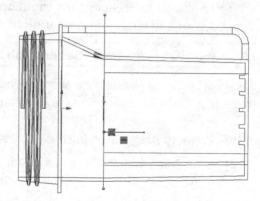

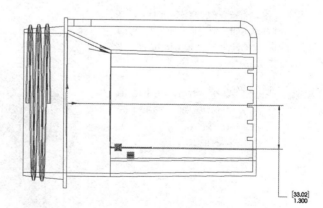

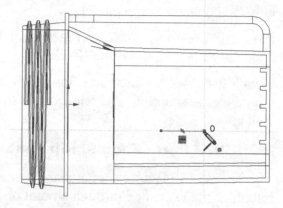

When the sketch and reference geometry become complex, create dimensions by selecting Reference planes and the Origin in the FeatureManager.

Dimension the Rib from the Origin, not from an edge or surface for design flexibility. The Origin remains constant. Modify edges and surfaces with the Fillet feature.

Sketch an arc.

650) **Zoom to Area** on the horizontal Sketch line.

651) Click the **Tangent Arc** Sketch tool.

652) Click the **left end** point of the horizontal line.

653) Click the **intersection** of the Shell1 and Boss Stop features. The sketch is displayed in black and is fully defined. If needed add an Intersection relation between the endpoint of the Tangent Arc, the left vertical Boss-Stop edge and the Shell1 Silhouette edge of the lower horizontal inside wall.

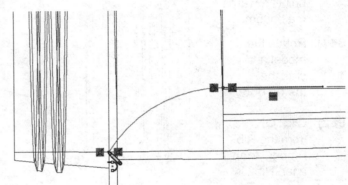

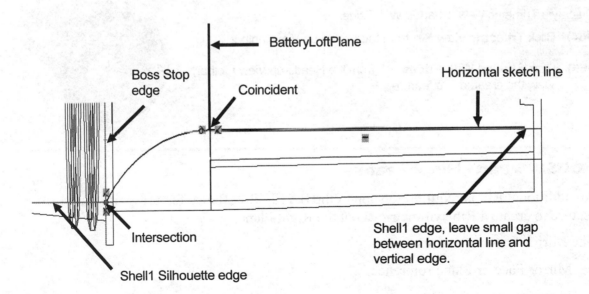

BatteryLoftPlane

Boss Stop edge

Coincident

Horizontal sketch line

Intersection

Shell1 Silhouette edge

Shell1 edge, leave small gap between horizontal line and vertical edge.

Insert the Rib feature.

654) Click the **Rib** 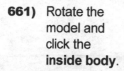 feature tool. The Rib PropertyManager is displayed.

655) Click the **Both Sides** box.

656) Click the **Draft On/Off** box.

657) Enter **1**deg for Angle.

658) Enter **.075**in, **[1.91]** for Rib Thickness.

659) Click the **Draft outward** box.

660) Click the **Flip material side** box if required. The direction arrow points towards the bottom.

661) Rotate the model and click the **inside body**.

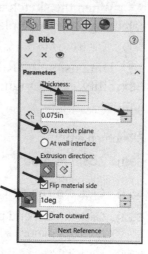

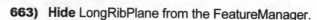

662) Click **OK** ✔ from the Rib PropertyManager. Rib2 is displayed in the FeatureManager.

663) **Hide** LongRibPlane from the FeatureManager.

Display a Trimetric view. Shaded With Edge.

664) Click **Trimetric view** 📦 from the Heads-up View toolbar.

665) Click **Shaded With Edges** 📦 from the Heads-up View toolbar. View the created Rib feature.

HOUSING Part - Mirror Feature

An additional Rib is required to support the BATTERY. Reuse features with the Mirror feature to create a Rib symmetric about the Right Plane.

The Mirror feature requires:

- Mirror Face or Plane reference.

- Features or Faces to Mirror.

Utilize the Mirror feature. Select the Right Plane for the Mirror Plane. Select the second Rib for the Features to Mirror.

Activity: HOUSING Part - Mirror Feature

Insert the Mirror feature.

666) Click the **Mirror** ⬚ feature tool. The Mirror PropertyManager is displayed.

667) Click inside the **Mirror Face/Plane** box.

668) Click **Right Plane** from the fly-out FeatureManager.

669) Click **Rib2** for Features to Mirror from the fly-out FeatureManager.

670) Click **OK** ✔ from the Mirror PropertyManager. Mirror1 is displayed in the FeatureManager.

Display a Trimetric view. Save the model.

671) Click **Trimetric view** 🔲 from the Heads-Up View toolbar.

672) Click **Save** 💾.

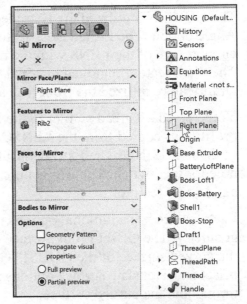

Close all parts.
673) Click **Window**, **Close All** from the Menu bar.

The parts for the FLASHLIGHT are complete. Review the HOUSING before moving on to the FLASHLIGHT assembly.

🔍 Additional information on Extrude Boss/Base, Extrude Cut, Swept, Loft, Helix/Spiral, Rib, Mirror and Reference Planes are found in SOLIDWORKS Help Topics.

Review of the HOUSING Part

The HOUSING utilized the Extruded Boss/Base feature with the Draft Angle option. The Lofted Boss feature was created to blend the circular face of the LENS with the rectangular face of the BATTERY. The Shell feature removed material with a constant wall thickness. The Draft feature utilized the front face as the Neutral plane.

You created a Thread similar to the LENSCAP Thread. The Thread profile was copied from the LENSCAP and inserted into the Top Plane of the HOUSING. The Extruded Cut feature was utilized to create a hole for the Switch. The Rib features were utilized in a Linear Pattern and Mirror feature.

Each feature has additional options that are applied to create different geometry. The Offset From Surface option creates an Extruded Cut on the curved surface of the HOUSING and LENSCAP. The Reverse offset and Translate surface options produce a cut depth constant throughout the curved surface.

Utilize Tools, Sketch Entities, Text to create the text profile on an Offset Plane.

The Instance to Vary option in the Linear Pattern PropertyManager allows you to vary the dimensions and locations of instances in a feature pattern *after it is created*. You can vary the dimensions of a series of instances, so that each instance is larger or smaller than the previous one. You can also change the dimensions of a single instance in a pattern and change the position of that instance relative to the seed feature of the pattern. For linear patterns, you can change the spacing between the columns and rows in the pattern.

FLASHLIGHT Assembly

Plan the Sub-assembly component layout diagram.

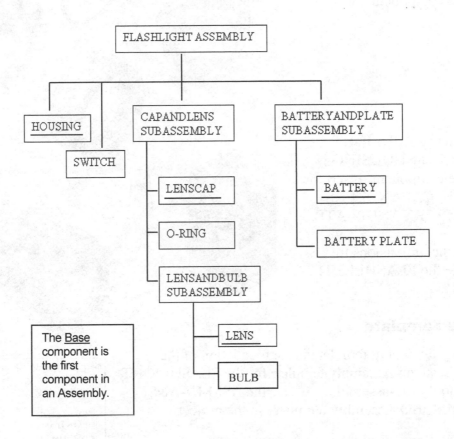

Assembly Layout Structure

The FLASHLIGHT assembly steps are as follows:

- Create the LENSANDBULB sub-assembly from the LENS and BULB components. The LENS is the Base component.

- Create the BATTERYANDPLATE sub-assembly from the BATTERY and BATTERYPLATE.

- Create the CAPANDLENS sub-assembly from the LENSCAP, O-RING and LENSANDBULB sub-assembly. The LENSCAP is the Base component.

- Create the FLASHLIGHT assembly. The HOUSING is the Base component. Insert the SWITCH, CAPANDLENS and BATTERYANDPLATE.

- Modify the dimensions to complete the FLASHLIGHT assembly.

Assembly Template

An Assembly document template is the foundation of the assembly. Create an Assembly template for the FLASHLIGHT assembly and its sub-assemblies. Create the ASM-IN-ANSI and ASM-MM-ISO Assembly templates in the project.

New SOLIDWORKS Document

Templates   Tutorial   MY-TEMPLATES

- PART-ANSI-IN
- PART-IN-ANSI
- PART-MM-ISO
- ASM-IN-ANSI
- ASM-IN-ISO
- ASM-MM-ANSI
- ASM-MM-ISO
- A-ANSI-MM
- B-ANSI-MM

Save the "Click OK" step. Double-click on the Template icon to open the Template in one step instead of two. Double-click on a part or assembly name in the Open Dialog box to open the document.

LENSANDBULB Sub-assembly

Create the LENSANDBULB sub-assembly. Utilize three Coincident mates to assemble the BULB component to the LENS component. When geometry is complex, select the planes in the Mate Selection box.

Suppress the Lens Shield feature to view all inside surfaces during the mate process.

Activity: Create two Assembly Templates

Create an Assembly Template.

674) Click **New** ⬜ from the Menu bar.

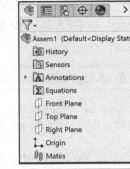

675) Double-click the **Assembly** icon from the default Templates tab.

676) Click **Cancel** ✖ from the Begin Assembly PropertyManager. The Assem1 FeatureManager is displayed.

Set Document Properties. Set drafting standard, units and precision.

677) Click **Options** ⚙, **Document Properties** tab from the Menu bar.

Select drafting standard, units, and precision.

678) Select **ANSI** for Overall drafting standard.

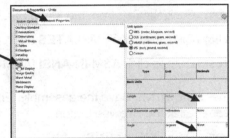

679) Click **Units**. Select **IPS** for Unit system.

680) Select **.123** for basic units length decimal places.

681) Select **None** for basic units angle decimal places.

682) Click **OK** from the dialog box.

Save the Assembly template. Enter name.

683) Click **Save As** from the Menu bar.

684) Click **Assembly Templates (*asmdot)** from the Save As type box. Select **ENGDESIGN-W-SOLIDWORKS\MY-TEMPLATES** for the Save in folder.

685) Enter **ASM-IN-ANSI** in the File name.

686) Click **Save**.

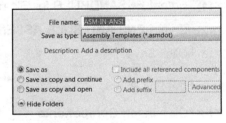

Create the ASM-MM-ISO Assembly template.

687) Click **New** ⬜ from the Menu bar.

688) Double-click **Assembly** from the default Templates tab.

689) Click **Cancel** ✖ from the Begin Assembly PropertyManager.

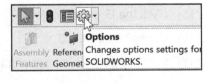

Select drafting standard, units and precision.

690) Click **Options** ⚙, **Document Properties** tab from the Menu bar.

691) Select **ISO** for Overall drafting standard.

692) Click **Units**.

693) Click **MMGS** for Unit system.

694) Select **.12** for basic units length decimal places.

695) Select **None** for basic units angle decimal places.

696) Click **OK**.

Save the Assembly template. Enter name.

697) Click **Save As** from the Menu bar menu.

698) Select **Assembly Templates (*asmdot)** from the Save As type drop-down menu.

699) Select **ENGDESIGN-W-SOLIDWORKS\MY-TEMPLATES** for the Save in folder.

700) Enter **ASM-MM-ISO** for File name.

701) Click **Save**.

Activity: LENSANDBULB Sub-assembly

Close all documents.

702) Click **Windows**, **Close All** from the Menu bar.

Create the LENSANDBULB sub-assembly.

703) Click **New** 🗋 from the Menu bar.

704) Click the **MY-TEMPLATES** tab.

705) Double-click **ASM-IN-ANSI** [ASM-MM-ISO].

Insert the LENS. Save the assembly. Enter name and description.

706) Click the **Browse** button.

707) Double-click **LENS**.

708) Click **OK** ✔ from the Begin Assembly PropertyManager.   The
LENS is fixed to the Origin.

709) Click **Save**.

710) Click **Save All**.

711) Enter **LENSANDBULB** for File name in the PROJECTS folder.

712) Enter **LENS AND BULB ASSEMBLY** for Description.

713) Click **Save**.

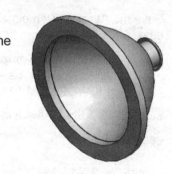

Insert the BULB.

714) Click **Insert Components** 🗇 from the Assembly toolbar.

715) Click the **Browse** button.

716) Double-click **BULB**.

717) Click a **position** in front of the LENS as illustrated.

Fit the model to the Graphics window.

718) Press the **f** key.

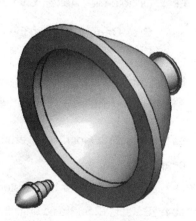

Move the BULB.

719) Click and drag the **BULB** in the Graphics window.

Save the LENSANDBULB.

720) Click **Save**. View the Assembly FeatureManager.

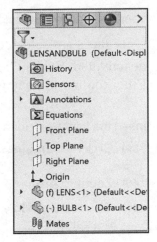

Suppress the LensShield feature.

721) **Expand** LENS in the FeatureManager.

722) Right-click **LensShield** in the FeatureManager.

723) Click Feature Properties.

724) Check the **Suppressed** box if needed.

725) Click **OK** from the Feature Properties dialog box.

Insert a Coincident mate.

726) Click the **Mate** Assembly tool. The Mate PropertyManager is displayed.

727) Pin the Mate PropertyManager. Click the **Keep Visible** icon.

728) Click **Right Plane** of the LENS from the fly-out FeatureManager.

729) **Expand** BULB in the fly-out FeatureManager.

730) Click **Right Plane** of the BULB. Coincident is selected by default.

731) Click **OK** from the Coincident PropertyManager. Coincident1 is created.

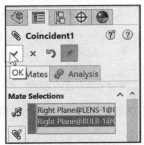

Insert the second Coincident mate.

732) Click **Top Plane** of the LENS from the fly-out FeatureManager.

733) Click **Top Plane** of the BULB from the fly-out FeatureManager. Coincident is selected by default.

734) Click **OK** from the Coincident PropertyManager. Coincident2 is created.

735) Click **OK** from the Mate PropertyManager.

Select face geometry efficiently. Position the mouse pointer in the middle of the face. Do not position the mouse pointer near the edge of the face. Zoom in on geometry. Utilize the Face Selection Filter for narrow faces.

Activate the Face Selection Filter.

736) Click **Right-click** in the Graphics window.

737) Click **Selection Filters**. The Selection Filter toolbar is displayed.

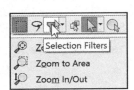

738) Click **Filter Faces**. The Filter icon is displayed.

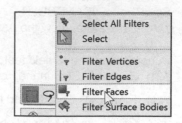

Only faces are selected until the Face Selection Filter is deactivated. Select Clear All Filters ⌦ from the Selection Filter toolbar to deactivate all filters.

Insert the third Coincident mate.

739) Click **Hidden Lines Visible** from the Heads-up View toolbar.

740) **Zoom to Area** and **Rotate** on the CBORE.

741) Click the **BulbHole face** of the LENS in the Graphics window as illustrated.

742) Hold the **Ctrl** key down.

743) Click the **bottom back flat face**, Revolve1 of the BULB.

744) Release the **Ctrl** key. The Mate pop-up menu is displayed.

745) Click **Coincident** from the Mate pop-up menu. Coincident3 is created.

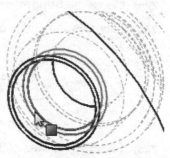

Clear the Face filter.

746) Click **Clear All Filters** from the Selection Filter toolbar.

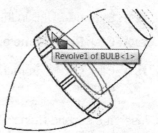

Display the Mate types.

747) **Expand** the Mates folder in the FeatureManager. View the inserted mates. The Mates under each sub-component in the FeatureManager are displayed.

Display a Right view in the Wireframe mode.

748) Click **Right view**.

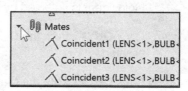

749) Click **Wireframe** from the Heads-up View toolbar.

Display an Isometric view with Shaded With Edges.

750) Click **Isometric view** from the Heads-up View toolbar.

751) Click **Shaded With Edges** from the Heads-up View toolbar.

Save the LENSANDBULB.

752) Click **Save**. View the results in the Graphics window.

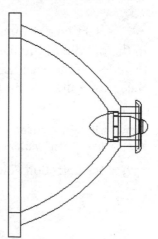

If the wrong face or edge is selected, click the face or edge again to remove it from the Mate Selections text box. Right-click Clear Selections to remove all geometry from the Mate Selections text box. To delete a mate from the FeatureManager, right-click on the mate, click Delete.

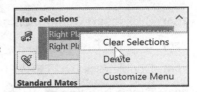

BATTERYANDPLATE Sub-assembly

Create the BATTERYANDPLATE sub-assembly. Utilize two Coincident mates and one Concentric mate to assemble the BATTERYPLATE component to the BATTERY component.

Use the Pack and Go option to save an assembly or drawing with references. The Pack and Go tool saves either to a folder or creates a zip file to e-mail. View SOLIDWORKS help for additional information.

Activity: BATTERYANDPLATE Sub-assembly

Create the BATTERYANDPLATE sub-assembly.

753) Click **New** from the Menu bar.

754) Click the **MY-TEMPLATES** tab.

755) Double-click **ASM-IN-ANSI**.

Insert the BATTERY part.

756) Click the **Browse** button.

757) Double-click **BATTERY** from the PROJECTS folder.

758) Click **OK** from the Begin Assembly PropertyManager. The BATTERY is fixed to the Origin.

Save the BATTERYANDPLATE sub-assembly. Enter name and description.

759) Click **Save**.

760) Select the **PROJECTS** folder.

761) Enter **BATTERYANDPLATE** for File name.

762) Enter **BATTERY AND PLATE FOR 6-VOLT FLASHLIGHT** for Description.

763) Click **Save**. The BATTERYANDPLATE FeatureManager is displayed.

Insert the BATTERYPLATE part.

764) Click **Insert Components**  from the Assembly toolbar.

765) Click the **Browse** button.

766) Double-click **BATTERYPLATE** from the PROJECTS folder.

767) Click a **position** above the BATTERY as illustrated.

Insert a Coincident mate.

768) Click the **outside bottom face** of the BATTERYPLATE as illustrated.

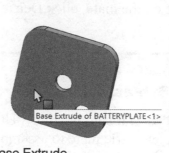

769) **Rotate** the model to view the top narrow flat face of the BATTERY Base Extrude feature as illustrated.

770) Hold the **Ctrl** key down.

771) Click the **top narrow flat face** of the BATTERY Base Extrude feature.

772) Release the **Ctrl** key. The Mate pop-up menu is displayed.

773) Click **Coincident** from the Mate pop-up menu. Coincident1 is created.

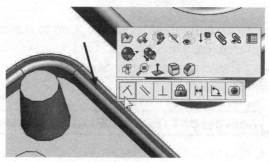

Insert a Coincident mate.

774) Click **Right Plane** of the BATTERY from the FeatureManager.

775) Hold the **Ctrl** key down.

776) Click **Right Plane** of the BATTERYPLATE from the FeatureManager.

777) Release the **Ctrl** key. The Mate pop-up menu is displayed.

778) Click **Coincident** from the Mate pop-up menu. Coincident2 is created.

Insert a Concentric mate.

779) Click the center Terminal feature **cylindrical face** of the BATTERY as illustrated.

780) Hold the **Ctrl** key down.

781) Click the Holder feature **cylindrical face** of the BATTERYPLATE.

782) Release the **Ctrl** key. The Mate pop-up menu is displayed.

783) Click **Concentric** from the Mate pop-up menu. Concentric1 is created.

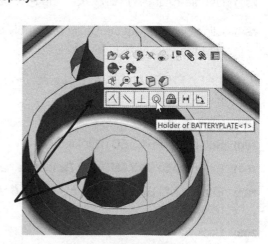

784) Expand the Mates folder. View the created mates.

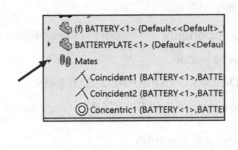

Display an Isometric view. Save the BATTERYANDPLATE.

785) Click **Isometric view** .

786) Click **Save** .

CAPANDLENS Sub-assembly

Create the CAPANDLENS sub-assembly. Utilize two Coincident mates and one Distance mate to assemble the O-RING to the LENSCAP. Utilize three Coincident mates to assemble the LENSANDBULB sub-assembly to the LENSCAP component.

Caution: Select the correct reference. Expand the LENSCAP and O-RING. Click the Right Plane within the LENSCAP. Click the Right plane within the O-RING.

Activity: CAPANDLENS Sub-assembly

Create the CAPANDLENS sub-assembly.

787) Click **New** from the Menu bar.

788) Click the **MY-TEMPLATES** tab.

789) Double-click **ASM-IN-ANSI**.

Insert the LENSCAP sub-assembly.

790) Click the **Browse** button.

791) Double-click **LENSCAP** from the PROJECTS folder.

792) Click **OK** from the Begin Assembly PropertyManager. The LENSCAP is fixed to the Origin.

Save the CAPANDLENS assembly. Enter name and description.

793) Click **Save**.

794) Select the **PROJECTS** folder.

795) Enter **CAPANDLENS** for File name.

796) Enter **LENSCAP AND LENS** for Description.

797) Click **Save**. The CAPANDLENS FeatureManager is displayed.

Insert the O-RING part.

798) Click **Insert Components**  from the Assembly toolbar.

799) Click the **Browse** button.

800) Double-click **O-RING** from the PROJECTS folder.

801) Click a **position** behind the LENSCAP as illustrated.

Insert the LENSANDBULB assembly.

802) Click **Insert Components** from the Assembly toolbar.

803) Click the **Browse** button.

804) Double-click **LENSANDBULB**.

805) Click a **position** behind the O-RING as illustrated.

806) Click **Isometric view** .

Move and hide components.

807) Click and drag the **O-RING** and **LENSANDBULB** as illustrated in the Graphics window.

808) Right-click **LENSANDBULB** in the FeatureManager.

809) Click **Hide Components** from the Context toolbar.

Insert three mates between the LENSCAP and O-RING.

810) Click **Right Plane** of the LENSCAP from the FeatureManager.

811) Hold the **Ctrl** key down.

812) Click **Right Plane** of the O-RING from the FeatureManager.

813) Release the **Ctrl** key. The Mate pop-up menu is displayed.

814) Click **Coincident** from the Mate pop-up menu. Coincident1 is created.

Insert a second Coincident mate.

815) Click **Top Plane** of the LENSCAP from the FeatureManager.

816) Hold the **Ctrl** key down.

817) Click **Top Plane** of the O-RING from the FeatureManager.

818) Release the **Ctrl** key. The Mate pop-up menu is displayed.

819) Click **Coincident** from the Mate pop-up menu. Coincident2 is created.

Insert a Distance mate.

820) Click the **Mate** ✎ Assembly tool. The Mate PropertyManager is displayed.

821) Click the Shell1 **back inside face** of the LENSCAP as illustrated.

822) Click **Front Plane** of the O-RING in the fly-out FeatureManager.

823) Click **Distance**.

824) Enter **.125/2**in, [**3.175/2mm**].

825) Click **OK** ✔ from the Distance Mate PropertyManager.

826) Click **OK** ✔ from the Mate PropertyManager.

Display an Isometric view. View the created mates.

827) Click **Isometric view** 🧊 from the Heads-up View toolbar.

828) **Expand** the Mates folder. View the created mates.

Save the model.

829) Click **Save** 💾.

Shell1 of LENSCAP<1>

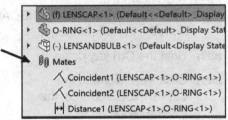

How is the Distance mate, .0625in, [1.588] calculated? Answer:

O-RING Radius (.1250in/2) = .0625in.

O-RING Radius [3.175mm/2] = [1.588mm].

💡 Utilize the Section view 📖 tool from the Heads-up View toolbar to locate internal geometry for mating and verify position of components.

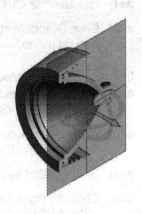

💡 Build flexibility into the mate. A Distance mate offers additional flexibility over a Coincident mate. You can modify the value of a Distance mate.

Show the LENSANDBULB.

830) Right-click **LENSANDBULB** in the FeatureManager.

831) Click **Show Components** from the Contexts toolbar.

Fit the model to the Graphics window. Create the first mate.

832) Press the **f** key.

Insert a Coincident mate.

833) Click **Right Plane** of the LENSCAP from the FeatureManager.

834) Hold the **Ctrl** key down.

835) Click **Right Plane** of the LENSANDBULB from the FeatureManager.

836) Release the **Ctrl** key. The Mate pop-up menu is displayed.

837) Click **Coincident** from the Mate pop-up menu.

Insert a Coincident Mate.

838) Click **Top Plane** of the LENSCAP from the FeatureManager.

839) Hold the **Ctrl** key down.

840) Click **Top Plane** of the LENSANDBULB from the FeatureManager.

841) Release the **Ctrl** key.

842) Click **Coincident** from the Mate pop-up menu.

Insert a Coincident Mate.

843) Click the flat inside **narrow back face** of the LENSCAP as illustrated.

844) **Rotate** the model to view the front flat face of the LENSANDBULB.

845) Hold the **Ctrl** key down.

846) Click the **front flat face** of the LENSANDBULB.

847) Release the **Ctrl** key.

848) Click **Coincident** from the Mate pop-up menu.

View the created mates.

849) **Expand** the Mates folder. View the created mates.

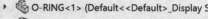

Confirm the location of the O-RING.

850) Click **Right Plane** of the CAPANDLENS from the FeatureManager.

851) Click **Section view** 🗐 from the Heads-up View toolbar.

852) Click **Isometric view** 🧊 from the Heads-up View toolbar.

853) **Expand** the Section 2 box.

854) Click **inside** the Reference Section Plane box.

855) Click **Top Plane** of the CAPANDLENS from the fly-out FeatureManager.

856) Click **OK** ✔ from the Section View PropertyManager.

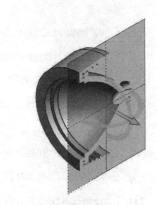

Return to a full view. Save the CAPANDLENS sub-assembly.

857) Click **Section view** 🗐 from the Heads-up View toolbar.

858) Click **Save** 💾 .

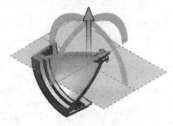

The LENSANDBULB, BATTERYANDPLATE and CAPANDLENS sub-assemblies are complete. The components in each assembly are fully defined. No minus (-) sign or red error flags exist in the FeatureManager. Insert the sub-assemblies into the final FLASHLIGHT assembly.

FLASHLIGHT Assembly

Create the FLASHLIGHT assembly. The HOUSING is the Base component. The FLASHLIGHT assembly mates the HOUSING to the SWITCH component. The FLASHLIGHT assembly mates the CAPANDLENS and BATTERYANDPLATE.

Activity: FLASHLIGHT Assembly

Create the FLASHLIGHT assembly.

859) Click **New** 🗋 from the Menu bar.

860) Click the **MY-TEMPLATES** tab.

861) Double-click **ASM-IN-ANSI**.

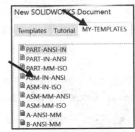

Insert the HOUSING and SWITCH.

862) Click the **Browse** button.

863) Double-click **HOUSING** from the PROJECTS folder.

864) Click **OK** ✔ from the Begin Assembly PropertyManager. The HOUSING is fixed to the Origin.

865) Click **Insert Components**  from the Assembly toolbar.

866) Click the **Browse** button.

867) Double-click **SWITCH** from the PROJECTS folder.

868) Click a **position** in front of the HOUSING as illustrated.

Save the FLASHLIGHT assembly. Enter name and description.
869) Click **Save**. Select the **PROJECTS** folder.

870) Enter **FLASHLIGHT** for File name.

871) Enter **FLASHLIGHT ASSEMBLY** for Description.

872) Click **Save**. The FLASHLIGHT FeatureManager is displayed.

Insert a Coincident mate.
873) Click the **Mate** Assembly tool.

874) Pin the Mate PropertyManager. Click the **Keep Visible** icon.

875) Click **Right Plane** of the HOUSING from the fly-out FeatureManager.

876) Click **Right Plane** of the SWITCH from the fly-out FeatureManager. Coincident is selected by default.

877) Click **OK** from the Coincident Mate PropertyManager.

Insert a Coincident mate.
878) Click **View**, **Hide/Show,** check **Temporary Axes** from the Menu bar.

879) Click the **Temporary axis** inside the Switch Hole of the HOUSING.

880) Click **Front Plane** of the SWITCH from the fly-out FeatureManager. Coincident is selected by default. Click **OK** from the Coincident Mate PropertyManager.

Insert a Distance mate.
881) Click the **top face** of the Handle.

882) Click the **Vertex** on the Loft top face of the SWITCH.

883) Click **Distance**. Enter **.100**in, **[2.54]**.

884) Check the **Flip Direction** box if needed.

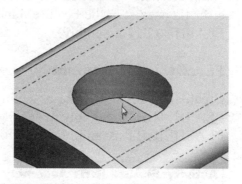

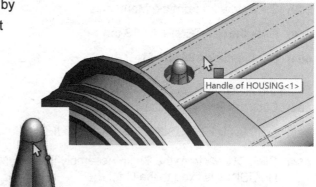

885) Click **OK** ✔ from the Distance Mate PropertyManager.

886) Un-pin the Mate PropertyManager. Click the **Keep Visible** 📌 icon.

887) Click **OK** ✔ from the Mate PropertyManager.

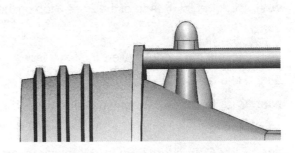

Insert the CAPANDLENS assembly.

888) Click **View**, **Hide/Show**, un-check **Temporary Axis** from the Menu bar.

889) Click **Insert Components** 📦 from the Assembly toolbar.

890) Click the **Browse** button.

891) Double-click **CAPANDLENS** from the PROJECTS folder.

Place the sub-assembly.

892) Click a **position** in front of the HOUSING as illustrated.

Insert Mates between the HOUSING component and the CAPANDLENS sub-assembly. Insert a Coincident mate.

893) Click **Right Plane** of the HOUSING from the FeatureManager.

894) Hold the **Ctrl** key down.

895) Click **Right Plane** of the CAPANDLENS from the FeatureManager.

896) Release the **Ctrl** key. The Mate pop-up menu is displayed.

897) Click **Coincident** from the Mate pop-up menu.

Insert a Coincident mate.

898) Click **Top Plane** of the HOUSING from the FeatureManager.

899) Hold the **Ctrl** key down.

900) Click **Top Plane** of the CAPANDLENS from the FeatureManager.

901) Release the **Ctrl** key.

902) Click **Coincident** from the Mate pop-up menu.

Insert a Coincident mate.
903) Click the **front face** of the Boss-Stop on the HOUSING.

Rotate the view.
904) **Rotate** the model to view the back face.

905) Hold the **Ctrl** key down.

906) Click the **back face** of the CAPANDLENS.

907) Release the **Ctrl** key.

908) Click **Coincident** from the Mate pop-up menu.

Display an Isometric view. Save the FLASHLIGHT assembly.
909) Click **Isometric view** 🧊.

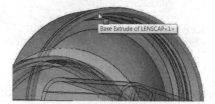

910) Click **Save** 💾. Click **Save All**.

Insert the BATTERYANDPLATE sub-assembly.

911) Click **Insert Components** 📂 from the Assembly toolbar.

912) Double-click **BATTERYANDPLATE** from the PROJECTS folder.

913) Click a **position** to the top of the HOUSING as illustrated.

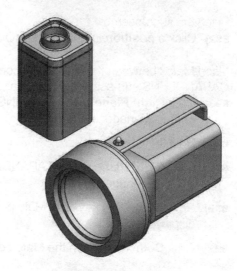

Rotate the part.
914) Click **BATTERYANDPLATE** in the FeatureManager.

915) Click **Rotate Component** 🔄 from the Assembly toolbar.

916) Rotate the **BATTERYANDPLATE** until it is approximately parallel with the HOUSING.

917) Click **OK** ✔ from the Rotate Component PropertyManager.

Insert a Coincident mate.
918) Click **Right Plane** of the HOUSING from the FeatureManager.

919) Hold the **Ctrl** key down.

920) Click **Front Plane** of the BATTERYANDPLATE from the FeatureManager.

921) Release the **Ctrl** key.

922) Click **Coincident** from the Mate pop-up menu.

923) Move the **BATTERYANDYPLATE** in front of the HOUSING.

Insert a Coincident mate.

924) Click **Top Plane** of the HOUSING from the FeatureManager.

925) Hold the **Ctrl** key down.

926) Click **Right Plane** of the BATTERYANDPLATE from the FeatureManager.

927) Release the **Ctrl** key. The Mate pop-up menu is displayed.

928) Click **Coincident** from the Mate pop-up menu.

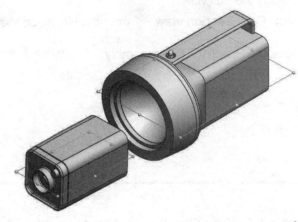

Display the Section view.

929) Click **Right Plane** in the FLASHLIGHT Assembly FeatureManager.

930) Click **Section view** 📰 from the Heads-up View toolbar.

931) Click **OK** ✔ from the Section View PropertyManager.

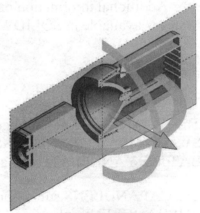

Move the BATTERYANDPLATE in front of the HOUSING.

932) Click and drag the **BATTERYANDPLATE** in front of the HOUSING.

Insert a Coincident mate.

933) Click the **back center Rib1 face** of the HOUSING as illustrated.

934) **Rotate** the model to view the bottom face of the BATTERYANDPLATE.

935) Hold the **Ctrl** key down.

936) Click the **bottom face** of the BATTERYANDPLATE.

937) Release the **Ctrl** key.

938) Click **Coincident** from the Mate pop-up menu.

939) Click **Isometric view** 🔷 from the Heads-up View toolbar.

Display the Full view.

940) Click **Section view** 🗔 from the Heads-up View toolbar.

Save the FLASHLIGHT assembly.

941) Click **Save** 🔲 .

🔍 Additional information on Assembly, Move Component, Rotate Component and Mates is available in SOLIDWORKS Help Topics.

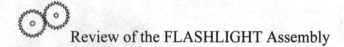 Review of the FLASHLIGHT Assembly

The FLASHLIGHT assembly consisted of the HOUSING part, SWITCH part, CAPANDLENS sub-assembly and the BATTERYANDPLATE sub-assembly.

The CAPANDLENS sub-assembly contained the BULBANDLENS sub-assembly, the O-RING and the LENSCAP part. The BATTERYANDPLATE sub-assembly contained the BATTERY and BATTERYPLATE part.

You inserted eight Coincident mates and a Distance mate. Through the Assembly Layout illustration you simplified the number of components into a series of smaller assemblies. You also enhanced your modeling techniques and skills.

You still have a few more areas to address. One of the biggest design issues in assembly modeling is interference. Let's investigate the FLASHLIGHT assembly.

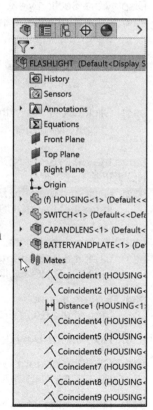

💡 Clearance Verification checks the minimum distance between components and reports any value that fails to meet your input value of the minimum clearance. View SOLIDWORKS Help for additional information.

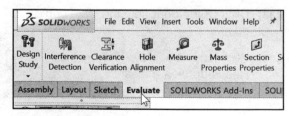

Addressing Interference Issues

There is an interference issue between the FLASHLIGHT components. Address the design issue. Adjust Rib2 on the HOUSING. Test with the Interference Check command. The FLASHLIGHT assembly is illustrated in inches.

Activity: Addressing Interference Issues

Check for interference.

942) Click the **Interference Detection** tool from the Evaluate tab in the CommandManager.

943) Delete **FLASHLIGHT.SLDASM** from the Selected Components box.

944) Click **BATTERYANDPLATE** from the fly-out FeatureManager.

945) Click **HOUSING** from the fly-out FeatureManager.

946) Click the **Calculate** button. The interference is displayed in red in the Graphics window.

947) Click each **Interference** in the Results box to view the interference in red with Rib2 of the HOUSING.

948) Click **OK** ✔ from the Interference Detection PropertyManager.

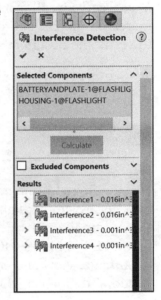

Modify the Rib2 dimension to address the interference issue.
949) **Open** the HOUSING in the FeatureManager.

950) Double-click on the **Rib2** feature.

951) Double click **1.300**in, [**33.02**].

952) Enter **1.350**in, [**34.29**].

953) **Rebuild** the model.

954) **Return** to the FLASHLIGHT assembly.

Recheck for Interference.

955) Click **Interference Detection** from the Evaluate tab. The Interference dialog box is displayed.

956) Delete **FLASHLIGHT.SLDASM** from the Selected Components box.

957) Click **BATTERYANDPLATE** from the FeatureManager.

958) Click **HOUSING** from the FeatureManager.

959) Click the **Calculate** button. No Interference is displayed in the Results box. The FLASHLIGHT design is complete.

960) Click **OK** ✔ from the Interference Detection PropertyManager.

Save the FLASHLIGHT assembly.

961) Click **Save** 💾 .

Export Files and eDrawings

You receive a call from the sales department. They inform you that the customer increased the initial order by 200,000 units. However, the customer requires a prototype to verify the design in six days. What do you do? Answer: Contact a Rapid Prototype supplier. You export three SOLIDWORKS files:

- HOUSING

- LENSCAP

- BATTERYPLATE

Use the Stereo Lithography (STL) format. Email the three files to a Rapid Prototype supplier. Example: Paperless Parts Inc. (www.paperlessparts.com). A Stereolithography (SLA) supplier provides physical models from 3D drawings. 2D drawings are not required. Export the HOUSING. SOLIDWORKS eDrawings provides a facility for you to animate, view and create compressed documents to send to colleagues, customers and vendors. Publish an eDrawing of the FLASHLIGHT assembly.

Activity: Export Files and eDrawings

Open and Export the HOUSING.

962) Right-click **HOUSING** from the FeatureManager.

963) Click **Open Part** from the Context toolbar.

964) Click **Save As** from Menu bar.

965) Click **Save as copy and open**.

966) Select **STL (*.stl)** from the Save as type drop-down menu.

967) Click the **Options** button.

968) Click the **Binary** box from the Output as format box.

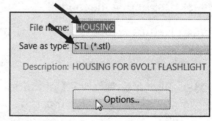

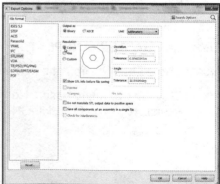

969) Click the **Coarse** box for Resolution.

Create the binary STL file.
970) Click **OK** from the Export Options dialog box.

971) Click **Save** from the Save dialog box. A status report is provided.

972) Click **Yes**.

Publish an eDrawing and email the document to a colleague.

Create the eDrawing and animation.
973) Click **File**, **Publish to eDrawings** from the Menu bar.

974) Click the **Run Animation** button. Click **Play**. View the results.

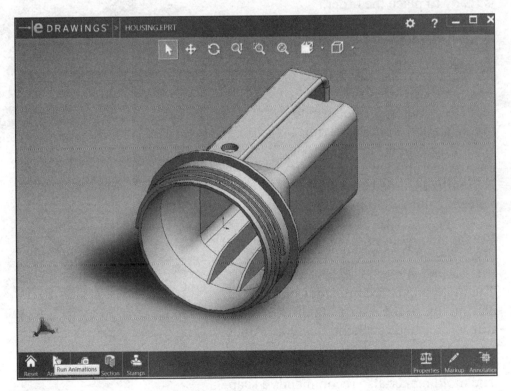

Stop the animation.
975) Click the **Stop** button. Click the **Reset** button to return to the original position.

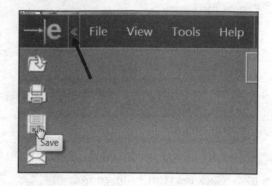

Save the eDrawing.
976) Click **Save** from the eDrawing Main menu.

977) Select the **PROJECTS** folder.

978) Enter **HOUSING** for File name.

979) Click **Save**.

980) **Close** ☒ the eDrawing dialog box.

981) **Close** all models in the session.

It is time to go home. The telephone rings. The customer is ready to place the order. Tomorrow you will receive the purchase order.

The customer also discusses a new purchase order that requires a major design change to the handle. You work with your industrial designer and discuss the options. One option is to utilize Guide Curves on a Sweep feature.

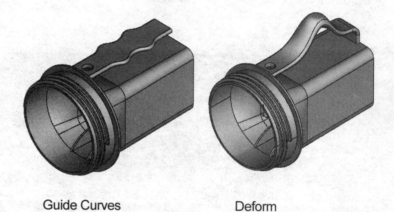

Guide Curves   Deform

You contact your mold maker and send an eDrawing of the LENSCAP.

The mold maker recommends placing the parting line at the edge of the Revolved Cut surface and reversing the Draft Angle direction.

The mold maker also recommends a snap fit versus a thread to reduce cost. The Core-Cavity mold tooling is explored in the project exercises.

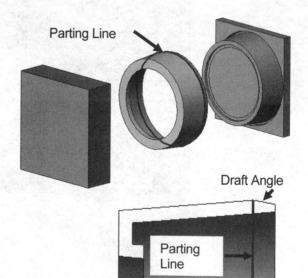

Additional information on Interference Detection, eDrawings, STL files (stereolithography), Guide Curves, Deform and Mold Tools are available in SOLIDWORKS Help Topics.

Project Summary

The FLASHLIGHT assembly contains over 100 features, reference planes, sketches and components. You organized the features in each part. You developed an assembly layout structure to organize your components.

The O-RING utilized a Swept Base feature. The SWITCH utilized a Loft Base feature. The simple Swept feature requires two sketches: a path and a profile. A complex Swept requires multiple sketches and guide curves. The Loft feature requires two or more sketches created on different planes.

The LENSCAP and HOUSING utilized a variety of features. You applied design intent to reuse geometry through Geometric relationships, symmetry and patterns.

The assembly required an Assembly Template. You utilized the ASM-IN-ANSI Template to create the LENSANDBULB, CAPANDLENS, BATTERYANDPLATE and FLASHLIGHT assemblies.

You created an STL file of the Housing and an eDrawing of the Housing to communicate with your vendor, mold maker and customer. Review the project exercises before moving on to the next Project.

Questions

1. Identify the function of the following features:

    *   Swept Boss/Base.

    *   Revolved Thin Cut.

    *   Lofted Boss/Base.

    *   Rib.

    *   Draft.

    *   Linear Pattern.

2. Describe a Suppressed feature. Why would you suppress a feature?

3. The Rib features require a sketch, thickness and a _____ direction.

4. What is a Pierce relation?

5. Describe how to create a thread using the Swept feature. Provide an example.

6. Explain how to create a Linear Pattern feature. Provide an example.

6a. Identify 5 proven assembly modeling techniques.

6b. How do you determine the Interference between components in an assembly?

7. Identify two advantages of utilizing Convert Entities in a sketch to obtain the profile.

8. How is symmetry built into a:

    *   Sketch. Provide a few examples.

    *   Feature. Provide a few examples.

9. Define a Guide Curve. Identify the features that utilize Guide Curves.

10. Identify the differences between a Draft feature and the Draft Angle option in the Extruded Boss/Base feature.

11. Describe the differences between a Circular Pattern feature and a Linear Pattern feature.

12. Identify the advantages of the Convert Entities tool.

13. True or False. A Lofted feature can only be inserted as the first feature in a part. Explain your answer.

14. A Swept feature adds and removes material. Identify the location on the Main Pull down menu that contains the Swept Cut feature. Hint: SOLIDWORKS Help Topics.

15. Describe the difference between a Distance Mate and a Coincident Mate. Provide an example.

Exercises

Exercise 6.1: Simple Wrap feature using the Sketch Entities text tool.

Create a simple Wrap feature using the Deboss option and the Sketch Text tool.

- Copy and open Wrap 6-1 from the Chapter 6 Homework folder. The Wrap 6-1 FeatureManager is displayed.

- Create a sketch on the Right plane with a horizontal construction line midpoint on Extrude1 to center the Wrap Sketch text.

Create text for the Wrap feature.
- Click the Tools, Sketch Entities, Text from the Main menu. The Sketch Text PropertyManager is displayed.

- Click the construction line from the Graphics window. Line1 is displayed in the Curves box. Enter Made in USA in the Text box.

- Click the Center Align button. Uncheck the Use document font box. Enter 150% in the Width Factor box. Enter 120% in the Spacing box.

- Click OK from the Sketch Text PropertyManager. Rebuild the model.

- Click Sketch2 from the FeatureManager. Click the Wrap Features tool. The Wrap PropertyManager is displayed. Sketch2 is displayed in the Source Sketch box.

- Select the Deboss option.

- Click the right cylindrical face of Extrude1. Face<1> is displayed in the Face for Wrap Sketch box. Enter 10mm for Depth.

- Click OK from the Wrap PropertyManager. View the results.

- Close the model.

This feature wraps a sketch onto a planar or non-planar face. You can create a planar face from cylindrical, conical, or extruded models. You can also select a planar profile to add multiple, closed spline sketches. The Wrap feature supports contour selection and sketch reuse.

Exercise 6.2: Wrap feature with Circular pattern.

Create a Wrap feature using the Emboss option and the Sketch Text tool.

- Copy and open Wrap 6-2 from the Chapter 6 Homework folder.

- Create a sketch on Plane1. This is your Sketch plane. Click Sketch.

Create text for the Wrap feature around the top section of the cone.

- Click the Tools, Sketch Entities, Text from the Main menu. The Sketch Text

  PropertyManager is displayed.

- Click inside the Text box. Enter Eat it all in the Text box.

- Click the top section (face) of the cone (shell3).

- Click OK from the Sketch Text PropertyManager. Rebuild the model.

Create the Wrap feature.

- Click the Wrap Features tool. Click Sketch6 from the fly-out FeatureManager. The Wrap PropertyManager is displayed. Sketch6 is displayed in the Source Sketch box.

- Select the Emboss option. Click the top section of the cone (shell3). Face<1> is displayed in the Face for Wrap Sketch box.

- Enter .03in for thickness.

Create a Circular pattern of the Wrap feature.

- Click Circular Pattern from the Features tab.

- Check the Equal spacing box. Enter 5 instances.

- Click inside the Features to Pattern box. Click Wrap1 from the fly-out FeatureManager.

- Click inside the Pattern Access box. Click the Temporary Axis. Click OK from the Circular Pattern PropertyManager. View the results.

- Close the model.

Exercise 6.3: HOOK Part.

Create the HOOK Part. Utilize the Swept Boss/Base, Dome and Thread feature. View the illustrated FeatureManagers and model.

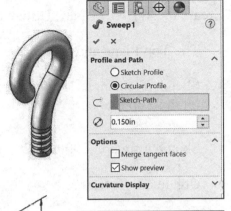

Note: Not all dimensions are provided. Your HOOK part will vary.

- Create a Swept Base feature. The Swept Base adds material by moving a profile along a path. The Swept Base feature in this exercise requires a path sketch and a circular profile diameter. The first sketch (Right Plane) is called the path.

- Utilize a Dome feature (050in, [1.27]) to create a spherical feature on a circular face.

- Create the Thread Feature.

- Add material (Plain Carbon Steel).

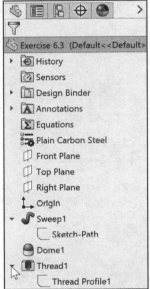

Exercise 6.4: WEIGHT Part.

Create the WEIGHT part. Utilize the Loft Base Feature.

- The Top Plane and Plane1 are 0.5in, [12.7mm] apart.

- Sketch a rectangle 1.000in, [25.4mm] x .750in, [19.05] on the Top Plane.

- Sketch a square .500in, [12.7mm] on Plane1.

- Create a Loft feature.

- Add a centered ⌀.150in, [3.81mm] Thru Hole.

- Add material.

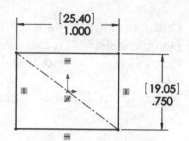

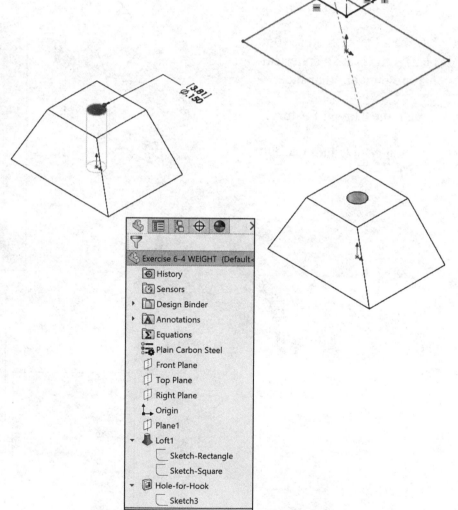

Exercise 6.5: Hole Wizard, Rib and Linear Pattern Feature.

Create the part from the illustrated A-ANSI Third Angle drawing: Front, Top, Right and Isometric views.

- Apply 6061 Alloy material.

- Calculate the volume of the part and locate the Center of mass.

- Think about the steps that you would take to build the model. **Note: ANSI standard states, "Dimensioning to hidden lines should be avoided wherever possible." However, sometimes it is necessary as below.**

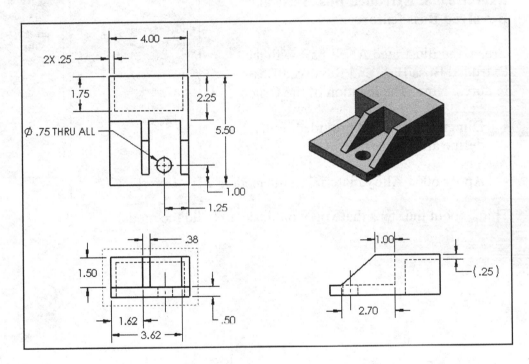

Exercise 6.6: Shell feature.

Create the illustrated part with the Extruded Boss/Base, Fillet and Shell features.

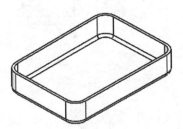

- Dimensions are not provided. Design your case to hold a bar of soap.

- Apply ABS material to the model. Think about the steps that you would take to build the model.

Exercise 6.7: Revolved Base, Hole Wizard and Circular Pattern features.

Create the illustrated ANSI part with the Revolved Base, Hole Wizard, and Circular Pattern features.

- Dimensions are not provided.

- Apply PBT General Purpose Plastic material to the model.

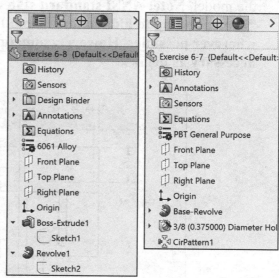

Think about the steps that you would take to build the model.

Exercise 6.8: Extruded Boss/Base and Revolved Boss features.

Create the illustrated ANSI part with the Extruded Boss/Base and Revolved Boss features. Note: The location of the Origin.

- Dimensions are not provided. Fully define all sketches.

- Apply 6061 Alloy material to the model.

Think about the steps that you would take to build the model.

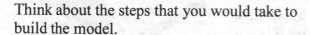

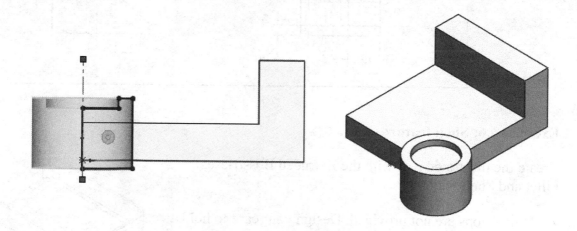

Exercise 6.9: - Exploded View Drawing.

Create an Exploded View of the FLASHLIGHT assembly. Create a FLASHLIGHT assembly drawing.

- Insert a Bill of Materials into the drawing.

- Insert Balloons.

- Utilize Custom Properties to add Part No. and Material for each FLASHLIGHT component.

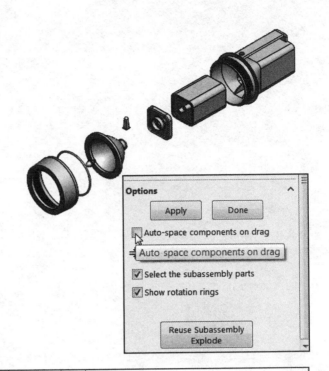

ITEM NO.	PART NUMBER	DESCRIPTION	MATERIAL	QTY.
1	A1-1-10	HOUSING FOR 6 VOLT FLASHLIGHT	ABS PC	1
2	A1-1-C10	BUTTON STYLE	PBT General Purpose	1
3	55-099	LENSCAP AND LENS		1
4	445-88	BATTERY AND PLATE FOR 6-VOLT FLASHLIGHT		1
5	55-55	ORING	RUBBER	1

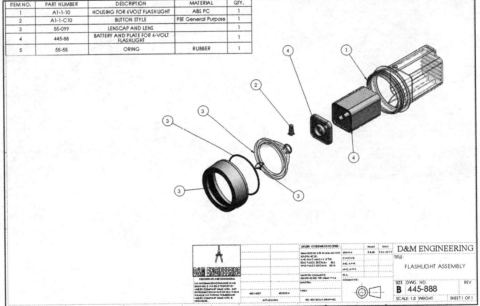

Notes:

Project 7

Top-Down Assembly Modeling and Sheet Metal Parts

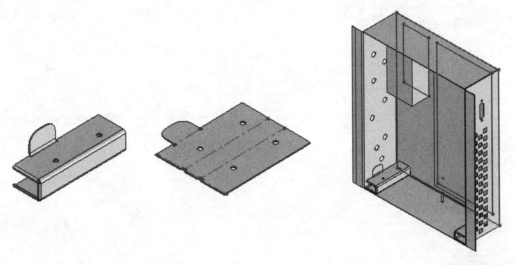

Below are the desired outcomes and usage competencies based on the completion of Project 7.

Project Desired Outcomes:	Usage Competencies:

• Three BOX configurations:
 	
   o Small
 	
   o Medium
 	
   o Large
	• Understand three key methods for the Top-down assembly modeling. Develop components In-Context with InPlace Mates.

	• Ability to import parts using the Top-down assembly method.
	• Knowledge to develop a Sheet Metal Design Table and equations.

• CABINET drawing with two configurations:
 	
   o Default - 3D formed state
   o Flat - 2D flatten state
	• Ability to convert a solid part into a Sheet Metal part and insert Sheet Metal features.

	• Knowledge of Sheet Metal configurations in the drawing: 3D Formed and 2D Flat Pattern states.
• BRACKET part.	• Ability to insert Sheet Metal features: Base, Edge, Miter Flange, Hem and Flat Pattern.

Notes:

Project 7 - Top-Down Assembly Modeling and Sheet Metal Parts

Project Objective

Create three BOX configurations utilizing the Top-down assembly approach. The Top-down approach is a conceptual approach used to develop products from within the assembly. Major design requirements are translated into sub-assemblies or individual components and key relationships.

Model Sheet metal components in their 3D formed state. Manufacture the Sheet metal components in a flatten state. Control the formed state and flatten state through configurations.

Create the parameters and equations utilized to control the configurations in a Design Table.

On the completion of this project, you will be able to:

- Set Document Properties and System Options.

- Create a new Top-down assembly and Layout sketch.

- Choose the proper Sketch plane.

- Insert Sheet Metal bends to transform a solid part into a sheet metal part.

- Use the Linear Pattern feature to create holes.

- Insert IGES format PEM® self-clinching fasteners.

- Use the Component Pattern feature to maintain a relationship between the holes and the fasteners.

- Create Global Variables.

- Create and Edit Equations.

- Insert a Sheet Metal Library feature and Forming tool feature.

- Address and understand general sheet metal manufacturing considerations.

- Insert Geometric relations.

- Insert Sheet Metal assembly features.

- Replace Components in an assembly.

- Insert a Design Table.

- Create a sheet metal part by starting with a solid part and inserting a Shell feature, Rip feature, and Sheet Metal Bends feature.

- Create a Sheet Metal part by starting with a Base Flange feature.

- Use the following SOLIDWORKS features: Extruded Boss/Base, Extruded Cut, Shell, Hole Wizard, Linear Pattern, Sketch Driven Pattern, Pattern Driven Component Pattern, Rip, Insert Bends, Flatten, Hem, Edge Flange, Jog, Flat Pattern, Break Corner/Corner Trim, Miter Flange, Base Flange/Tab, Unfold/Fold, Mirror Components and Replace Component.

Project Situation

You now work for a different company. Life is full with opportunities. You are part of a global project design team that is required to create a family of electrical boxes for general industrial use. You are the project manager.

You receive a customer request from the Sales department for three different size electrical boxes.

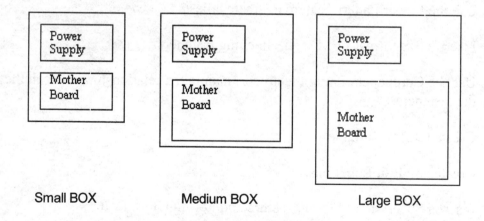

Small BOX   Medium BOX   Large BOX

Delivery time to the customer is a concern. You work in a concurrent engineering environment. Your company is expecting a sales order for 5,000 units in each requested BOX configuration.

The BOX contains the following key components:

- Power supply.

- Motherboard.

The size of the power supply is the same for all three boxes. The size of the power supply is the constant. The size of the BOX is a variable. The available space for the motherboard is dependent on the size of the BOX. The depth of the three boxes is 100mm.

You contact the customer to discuss and obtain design options and product specifications. Key customer requirements:

Three different BOX sizes:

- 300mm x 400mm x 100mm   Small.

- 400mm x 500mm x 100mm   Medium.

- 550mm x 600mm x100mm   Large.

- Adequate spacing between the power supply, motherboard and internal walls.

- Field serviceable.

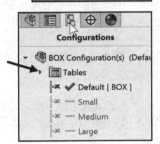

You are responsible to produce a sketched layout from the provided critical dimensions. You are also required to design the three boxes. The BOX is used in an outside environment.

Top-Down Assembly Modeling

Top-down assembly is a conceptual method used to develop products from within the assembly. In Top-down assembly modeling, one or more features of the part are defined by something in the assembly. Example: A Layout sketch or the geometry of another part. The design intent of the model, for example the size of the features, location of the components in the assembly, etc. take place from the top level of the assembly and translate downward from the assembly to the component.

A few advantages of the Top-down assembly modeling are that design details of all components are not required and much less rework is required when a design change is needed. The model requires individual relationships between components. The parts know how to update themselves based on the way you created them.

Designers usually use Top-down assembly modeling to lay their assemblies out and to capture key design aspects of custom parts specific in the assemblies. There are three key methods to use for the Top-down assembly modeling approach:

- *Individual features method*. The Individual features method provides the ability to reference the various components and sub-components in an existing assembly. Example: Creating a structural brace in a box by using the Extruded Boss/Base feature tool. You might use the Up to Surface option for the End Condition and select the bottom of the box, which is a different part. The Individual features method maintains the correct support brace length, even if you modify the box in the future. The length of the structural brace is defined in the assembly. The length is not defined by a static dimension in the part. The Individual features method is useful for parts that are typically static but have various features which interface with other assembly components in the model.

- *Entire assembly method*. The Entire assembly method provides the ability to create an assembly from a layout sketch. The layout sketch defines the component locations, key dimensions, etc. A major advantage of designing an assembly using a layout sketch is that if you modify the layout sketch, the assembly and its related parts are automatically updated. The entire assembly method is useful when you create changes quickly, and in a single location.

You can create an assembly and its components from a layout of sketch blocks.

- *Complete parts method*. The Complete parts method provides the ability to build your model by creating new components In-Context of the assembly. The component you build is actually mated to another existing component in the assembly. The geometry for the component you build is based upon the existing component.

This method is useful for parts like brackets and fixtures, which are mostly or completely dependent on other parts to define their shape and size.

Whenever you create a part or feature using the Top-Down approach, external references are created to the geometry you referenced.

The Top-down assembly approach is also referred to as "In-Context design."

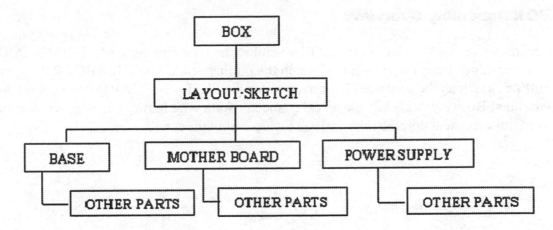

Conceptual Top-down design approach

Use a combination of the three assembly methods to create the BOX assembly. Consider the following in a preliminary design product specification:

- What are the major design components?

- The motherboard and power supply are the major design components.

- What are the key design constraints?

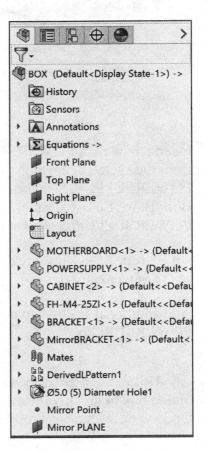

The customer specified three different BOX sizes. How does each part relate to the other? From experience and discussions with the electrical engineering department, a 25mm physical gap is required between the power supply and the motherboard. A 20mm physical gap is required between the internal components and the side of the BOX.

How will the customer use the product? The customer does not disclose the specific usage of the BOX. The customer is in a very competitive market.

What is the most cost-effective material for the product? Aluminum is cost-effective, strong, relatively easy to fabricate, corrosion resistant, and non-magnetic.

Incorporate the design specifications into the BOX assembly. Use a sketch, solid parts and sheet metal parts. Obtain additional parts from your vendors.

BOX Assembly Overview

Create the sketch "layout" in the BOX assembly. Insert a new part, MOTHERBOARD. Convert edges from the sketch to develop the outline of the MOTHERBOARD. Use the outline sketch as the Extruded Base feature for the MOTHERBOARD component. An Extruded Boss locates a key electrical connector for a wire harness. Design the location of major electrical connections early in the assembly process.

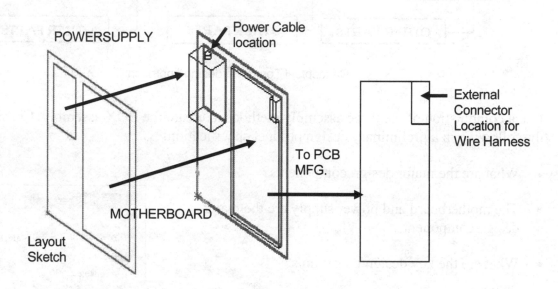

Insert a new part, POWERSUPPLY. Convert edges from the sketch to develop the outline of the POWERSUPPLY. Use the outline as the Extruded Boss/Base (Boss-Extude1) feature for the POWERSUPPLY component.

An Extruded Boss/Base feature represents the location of the power cable.

Add additional features to complete the POWERSUPPLY.

POWER SUPPLY:
Send the part to a colleague to add additional features.

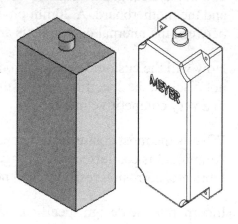

Utilize the Search feature in the Systems Options dialog box to quickly locate information for System Options or Document Properties.

Insert a new part,
CABINET. Convert edges
from the sketch.

Use the outline sketch as the
Extruded Base feature for
the CABINET component.

Shell the Extruded feature.
Add a Rip feature to the
four CABINET corners.

Insert sheet metal bends to
transform a solid part into a
sheet metal part.

Sheet metal components
utilize features to create
flanges, cuts and forms.

Save Model sheet metal
components in their 3D
formed state.

Manufacture sheet metal components in their
2D flatten state.

Work with your sheet metal manufacturer.
Discuss cost effective options.

Create a sketched pattern to add square
cuts instead of formed louvers.

Create a Linear Pattern feature of holes for
the CABINET.

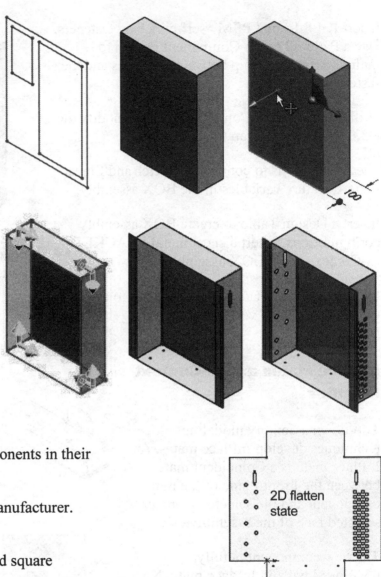

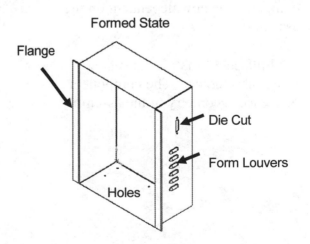

2D flatten state

Formed State

Flange

Die Cut

Form Louvers

Holes

Insert IGES format PEM® self-clinching fasteners. Use a Pattern Driven Component Pattern to maintain a relationship between the holes and the fasteners.

Utilize the Replace Component tool to modify the Ø5.0mm fastener to an Ø 4.0mm fastener.

Create equations to control the sketch and Linear Pattern feature variables in the BOX assembly.

Insert a Design Table to create BOX assembly configurations. Insert a sheet metal BRACKET In-Context of the BOX assembly.

Utilize the Mirror Components feature to create a right and left hand version of the BRACKET.

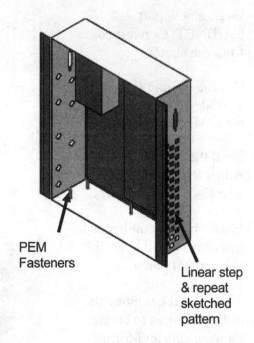

PEM Fasteners

Linear step & repeat sketched pattern

InPlace Mates and In-Context Features

Top-down assembly modeling techniques develop InPlace mates. An InPlace mate is a Coincident mate between the Front Plane of the new component and the selected plane or selected face of the assembly.

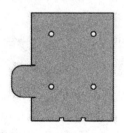

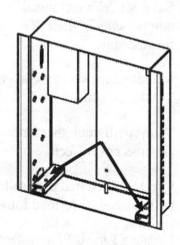

The new component is fully positioned with an InPlace mate. No additional mates are required. The component is now dependent on the assembly.

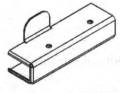

An InPlace ✎ mate creates an External reference. The component references geometry in the assembly.

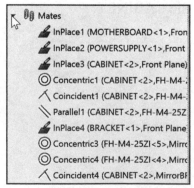

If the referenced document changes, the dependent document changes. The InPlace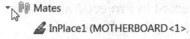
mates are listed under the Mates folder in the FeatureManager. Example:

Mates
    InPlace1 (MOTHERBOARD<1>

In a Top-down assembly, you create In-Context features. An In-Context feature is a
feature that references geometry from another component. The In-Context feature has an
External reference to that component. If you change geometry on the referenced
component, the associated In-Context feature also changes.

An In-Context feature is indicated in the FeatureManager with the ">" symbol. The
External references exist with the feature, Sketch plane or sketch geometry. The update
path to the referenced component is contained in the assembly. When the update path is
not valid or the referenced component is not found the "?" symbol is displayed in the
FeatureManager.

In the BOX assembly, develop InPlace Mates for the MOTHERBOARD,
POWERSUPPLY and CABINET components.

Create External references with the Convert Entities Sketch tool and extracting geometry
from the sketch of the BOX assembly.

Working with In-Context features requires practice and time. Planning and selecting the
correct reference and understanding how to incorporate changes are important. Explore
various techniques using InPlace mates and External references developed In-Context of
another component.

Assembly Modeling Techniques with InPlace Mates:
Plan the Top-down assembly method. Start from a sketch or with a component in the assembly.
Prepare the references. Utilize descriptive feature names for referenced features and sketches.
Utilize InPlace Mates sparingly. Load all related components into memory to propagate changes. Do not use InPlace Mates for purchased parts or hardware.
Group references. Select references from one component at a time.
Ask questions. Will the part be used again in a different assembly? If the answer is yes, do not use InPlace Mates. If the answer is no, use InPlace Mates.
Will the part be used in physical dynamics or multiple configurations? If the answer is yes, do not use InPlace mates.
Work in the Edit Component tool to obtain the required External references in the assembly. Create all non-referenced features in the part, not in the assembly.
Obtain knowledge of your company's policy on InPlace mates or develop one as part of an engineering standard.

Part Template and Assembly Template

The parts and assemblies in this project require Templates created in Project 2 and Project 3. The MY-TEMPLATES tab is located in the New dialog box. If your MY-TEMPLATES folder contains the ASM-MM-ANSI and PART-MM-ANSI Templates, skip the next few steps.

Activity: Create an Assembly and Part Template

Create the Assembly Template for the BOX.

1) Click **New** from the Menu bar.

2) Double-click the **Assembly** icon from the defaults Templates tab.

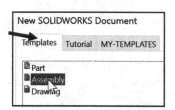

3) Click **Cancel** ✖ from the Begin Assembly PropertyManager. The Assembly FeatureManager is displayed.

Set the Assembly Document Template options.

4) Click **Options** ⚙, **Document Properties** tab from the Menu bar. Select **ANSI [ISO]** for the Overall drafting standard.

5) Click **Units**. Click **MMGS** for Units system.

6) Select **.12** (two decimal places) for Length Basic Units.

7) Select **None** for Angular units Decimal places.

8) Click **OK** from the Document Properties - Units dialog box.

Save the Assembly template. Enter name.

9) Click **Save As** from the Menu bar.

10) Click the **Assembly Templates (*.asmdot)** from the Save As type box.

11) Select **ENGDESIGN-W-SOLIDWORKS\MY-TEMPLATES** for Save in folder.

12) Enter **ASM-MM-ANSI, [ASM-MM-ISO]** for File name. Click **Save**. The ASM-MM-ANSI FeatureManager is displayed.

Create the Part Template for the BOX.

13) Click **New** from the Menu bar.

14) Double-click the **Part** icon from the default Templates tab. The Part FeatureManager is displayed.

15) Click **Options** ⚙, Document Properties tab from the Menu bar.

16) Select **ANSI, [ISO]** for the Overall drafting standard.

Set units and precision.
17) Click **Units**. Click **MMGS** for Unit system.

18) Select **.12** (two decimal places) for Length Basic Units. Select **None** for Angular units Decimal places.

19) Click **OK** from the Document Properties - Units dialog box.

Save the Part Template. Enter name.
20) Click **Save As** from the Menu bar.

21) Click **Part Templates (*.prtdot)** from the Save As type box.

22) Select **ENGDESIGN-W-SOLIDWORKS\MY-TEMPLATES** for Save in folder.

23) Enter **PART-MM-ANSI, [PART-MM-ISO]** in the File name text box. Click **Save**.

Close all active documents.
24) Click **Windows**, **Close All** from the Menu bar.

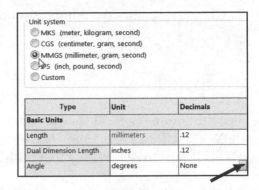

Type	Unit	Decimals
Basic Units		
Length	millimeters	.12
Dual Dimension Length	inches	.12
Angle	degrees	None

BOX Assembly and Sketch

The BOX assembly utilizes both the Top-down assembly design approach and the Bottom-up assembly design approach. Begin the BOX assembly with a sketch.

Create a New assembly named BOX. Insert a sketch to develop component space allocations and relations in the BOX assembly.

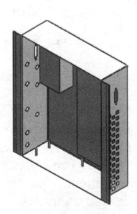

Add dimensions and relations to the sketch. Components and assemblies reference the sketch.

🔅 In Layout-based assembly design, you can switch back and forth between Top-down and Bottom-up design methods.

The BOX assembly contains the following key components:

- POWERSUPPLY.

- MOTHERBOARD.

The minimum physical spatial gap between the MOTHERBOARD and the POWERSUPPLY is 25mm. The minimum physical spatial gap between the MOTHERBOARD, POWERSUPPLY and the internal sheet metal BOX wall is 20mm.

After numerous discussions with the electrical engineer, you standardize on a POWERSUPPLY size: 150mm x 75mm x 50mm. You know the overall dimensions for the BOX, POWERSUPPLY and MOTHERBOARD. You also know the dimensional relationship between these components.

Now you must build these parameters into the design intent of the sketch. There is no symmetry between the major components. Locate the sketch with respect to the BOX assembly Origin.

Activity: Box Assembly and Sketch Layout

Create the BOX assembly.

25) Click **New** ⬜ from the Menu bar.

26) Click the **MY-TEMPLATES** tab.

27) Double-click **ASM-MM-ANSI**.

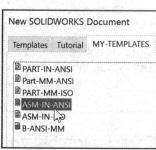

28) Click **Cancel** ✖ from the Begin Assembly PropertyManager. The Assembly FeatureManager is displayed.

Save the BOX assembly. Enter name.

29) Click **Save** 💾.

30) Select **ENGDESIGN-W-SOLIDWORKS\PROJECTS** for folder.

31) Enter **BOX** for File name.

32) Click **Save**. The BOX FeatureManager is displayed.

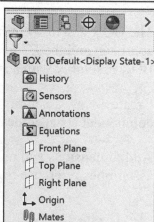

Each part in the BOX assembly requires a template. Utilize the PART-MM-ANSI [PART-MM-ISO] and ASM-MM-ANSI [ASM-MM-ISO] templates. Under the Option, System Options, Default Templates section prompts you to select a different document template. Otherwise, SOLIDWORKS utilizes the default templates.

Set System Options.

33) Click **Options** ⚙, **Default Templates** from the Menu bar.

34) Click the **Prompt user to select document template** button. Click **OK**.

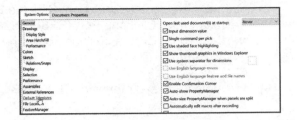

The 2D sketch contains the 2D relationships between all major components in the BOX assembly.

Select the Sketch plane for the sketch.

35) Right-click **Front Plane** from the FeatureManager.

Sketch the profile of the BOX.

36) Click **Sketch** 🖉 from the Context toolbar. The Sketch toolbar is displayed. Click **Corner Rectangle** ⬚ from the Sketch toolbar.

37) Click the **Origin**. Sketch a **rectangle** as illustrated. The first point is coincident with the Origin.

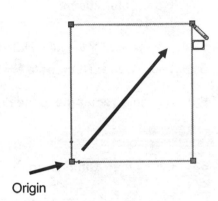

Origin

Add dimensions.

38) Click **Smart Dimension** ⟋ from the Sketch toolbar.

39) Click the **left vertical line**. Click a **position** to the left of the profile. Enter **400**mm.

40) Click the **bottom horizontal line**. Click a **position** below the profile. Enter **300**mm.

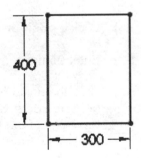

Fit the model to the Graphics window.

41) Press the **f** key.

Sketch the 2D profile for the POWER SUPPLY.

42) Click **Corner Rectangle** ⬚ from the Sketch toolbar.

43) Sketch a **small rectangle** inside the BOX as illustrated.

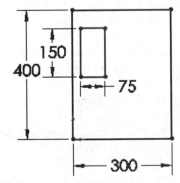

Add dimensions.

44) Click **Smart Dimension** ⟋ from the Sketch toolbar.

45) Create a vertical dimension. Enter **150**mm.

46) Create a horizontal dimension. Enter **75**mm.

Dimension the POWERSUPPLY 20mm from the left and top edges of the BOX.

47) Create a horizontal dimension. Click the **BOX left vertical** edge. Click the **POWERSUPPLY left vertical** edge. Click a position **above** the BOX. Enter **20mm**.

48) Create a vertical dimension. Click the **BOX top** horizontal edge.

49) Click the **POWERSUPPLY top horizontal** edge.

50) Click a **position above** and to the right of the BOX. Enter **20**mm.

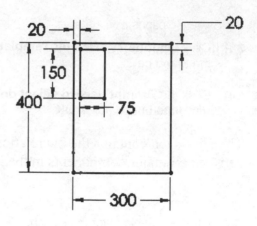

Sketch the profile for the MOTHERBOARD.

51) Click **Corner Rectangle** ▭ from the Sketch toolbar.

52) Sketch a **rectangle** to the right of the POWERSUPPLY as illustrated.

Add dimensions.

53) Click **Smart Dimension** ⟋ from the Sketch toolbar.

54) Click the **left vertical edge** of the MOTHERBOARD.

55) Click the **right vertical edge** of the POWERSUPPLY.

56) Click a **position** above the BOX.

57) Enter **25**mm.

58) Click the horizontal **top edge** of the MOTHERBOARD. Click the horizontal **top edge** of the BOX. Click a position off the profile. Enter **20**mm.

59) Click the vertical **right edge** of the MOTHERBOARD.

60) Click the vertical **right edge** of the BOX. Click a **position below** the Box.

61) Enter **20**mm.

62) Click the horizontal **bottom edge** of the MOTHERBOARD.

63) Click the horizontal **bottom edge** of the BOX.

64) Click a **position** to the right of the BOX. Enter **20**mm.

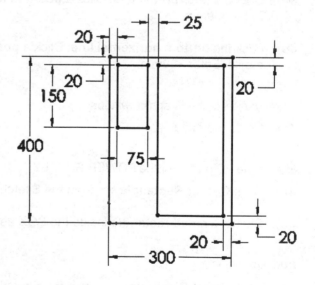

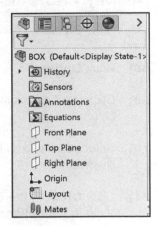

Sketch1 is fully defined and is displayed in black. Insert a Collinear relation between the top lines of the two inside rectangles if required.

Close and rename the sketch.

65) Right-click **Exit Sketch** .

66) Rename **Sketch1** to **Layout**.

Save the BOX assembly.

67) Click **Save** .

What happens when the size of the MOTHERBOARD changes? How do you ensure that the BOX maintains the required 20mm spatial gap between the internal components and the BOX boundary? How do you design for future revisions? Answer: Through Global Variables and equations.

Global Variables and Equations

In previous versions of SOLIDWORKS, linked or shared values were used to link two or more dimensions without using equations or relations. Changing any one of the linked values would change the other to which it was linked. Existing linked values are still supported, but you cannot create new linked values in SOLIDWORKS 2012 and later.

Instead, use Global Variables for the same purpose as linked values. Global Variables are much easier to find, change and manage than linked values.

To use a Global Variable to link dimensions: 1.) Create a Global Variable in the Equations dialog box or the Modify dialog box for dimensions, and 2.) Set two or more dimensions equal to the Global Variable.

The project goal is to create three boxes of different sizes. Ensure that the models remain valid when dimensions change for various internal components. This is key for proper design intent. Apply Global Variables and equations.

Activity: Create Global Variables and Equations

Display the Equations folder in the FeatureManager and display the dimensions in the Graphics window.

68)   Click **Options** ⚙, **System Options** tab from the Main menu.

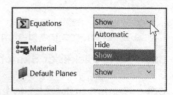

69)   Click the **FeatureManager** folder from the System Options column.

70)   Select **Show** from the Equations drop-down menu as illustrated.

71)   Click **OK** from the Systems Option - FeatureManager. View the Equations folder in the FeatureManager.

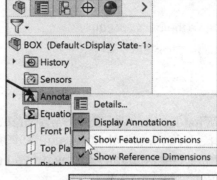

72)   Right-click the **Annotations** folder in the FeatureManager.

73)   Click **Show Feature Dimensions**.

74)   Double-click **Layout** from the FeatureManager.

Insert a Global Variable.

75)   Right-click the **Equations** folder from the FeatureManager.

76)   Click **Manager Equations**. The Equations, Global Variables, and Dimensions dialog box is displayed.

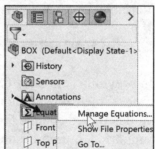

77)   Click the **Equation view** Σ⌀ button icon.

78)   Click a **position** in the first cell under Global Variables.

79)   Enter the Global Variable name **Gap.**

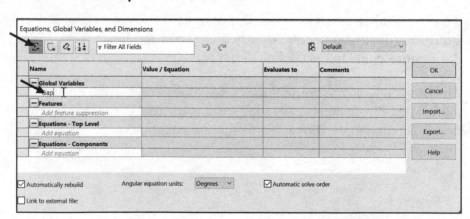

80)   Press the **Tab** key.

81) Click the **lower right horizontal dimension 20** in the Graphics window. The dimension name is displayed in the Value/Equation column. Note: Your dimension name may be different. A green check mark indicates the value can be calculated.

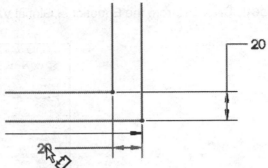

82) Press the **Tab** key. The Evaluates to column displays 20.

83) Enter **Gap from the Cabinet wall** for comment. The first Global Variable Gap is defined. Next define equations to equate the other variables to Gap.

84) Click **inside** the first cell below Equations-Top Level.

85) Click the **lower right vertical dimension 20** in the Graphics window.

86) Slide the **mouse pointer** to the right of Global Variables to display Gap (20mm).

87) Click **Gap (20mm).**

88) Press the **Enter** key to create a new row under Equations-Top Level.

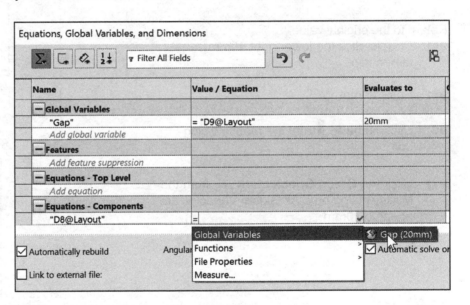

89) Repeat the **above procedure** to enter the rest of the 20mm dimensions on the Layout sketch.

90) Click **OK** from the Equations, Global Variables, and Dimensions dialog box.

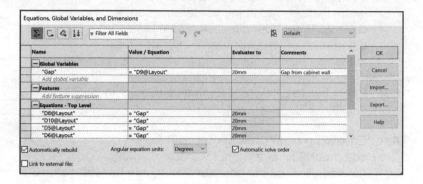

Test the Global Variables.

91) Double-click the lower right **20mm** dimension as illustrated.

92) Enter **10**mm.

93) Click **Rebuild** from the Modify dialog box. The four Global variables change and display the equation symbol Σ.

Return to the original value.

94) Double-click the lower right **10**mm dimension.

95) Enter **20**mm.

96) Click **Rebuild** from the Modify dialog box.

97) Click the **Green Check mark** ✔ from the Modify dialog box. All Global Variables are equal to 20mm.

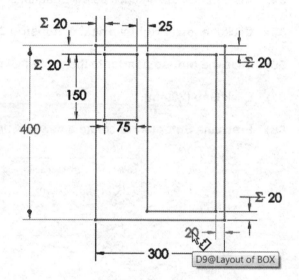

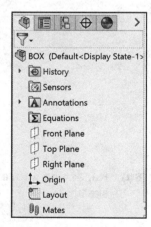

Additional dimensions are required for Equations. Each dimension has a unique variable name. The names are used as Equation variables. The default names are based on the Sketch, Feature or Part. Feature names do not have to be changed. Rename variables for clarity when creating numerous Equations.

Rename the BOX assembly width and height. The full variable name is "box-width@ Layout." The system automatically appends Layout. If features are created or deleted in a different order, your variable names will be different.

Rename for overall box width.
98) Click the horizontal dimension **300** in the Graphics window. The PropertyManager is displayed.

99) Delete **D2@Layout**.

100) Enter **box_width@**.

101) Click **OK** ✔ from the PropertyManager.

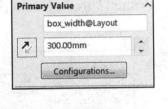

Rename the overall box height.
102) Click the vertical dimension **400** in the Graphics window. The PropertyManager is displayed.

103) Delete **D1@Layout**.

104) Enter **box height**.

105) Click **OK** ✔ from the PropertyManager.

Display an Isometric view.
106) Click **Isometric view** from the Heads-up View toolbar.

Hide all dimensions.
107) Right-click **the Annotations** folder.

108) Uncheck **Show Feature Dimensions**.

Save the BOX assembly.
109) Click **Save** 💾.

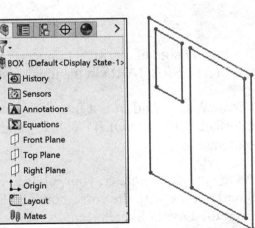

The sketch is complete. Develop the MOTHERBOARD part and the POWERSUPPLY part In-context of the BOX assembly Layout sketch.

Select the BOX assembly Front Plane to create InPlace Mates. An InPlace Mate is a Coincident Mate developed between the Front Plane of the component and the selected plane in the assembly. Utilize Convert Entities and extract geometry from the Layout Sketch to create external references for both the MOTHERBOARD part and POWERSUPPLY part. Utilize Ctrl-Tab to switch between part and assembly windows.

MOTHERBOARD-Insert Component

The MOTHERBOARD requires the greatest amount of lead-time to design and manufacture. The outline of the MOTHERBOARD is created. Create the MOTHERBOARD component from the Layout Sketch. An electrical engineer develops the Logic Diagram and Schematic required for the Printed Circuit Board (PCB). The MOTHERBOARD rectangular sketch represents the special constraints of a blank Printed Circuit Board (PCB).

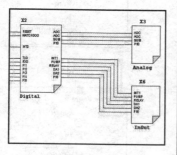
Logic Diagram (partial)

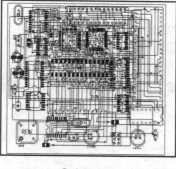
Schematic

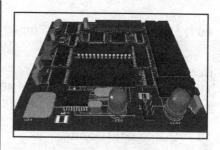
PCB

Courtesy of Electronic Workbench
(www.electronicworkbench.com)

A rough design of a critical connector is located on the MOTHERBOARD in the upper right corner.

CircuitWorks Add-in is a fully integrated data interface between SOLIDWORKS and PCB Design systems.

As the project manager, your job is to create the board outline with the corresponding dimensions from the Layout Sketch. Export the MOTHERBOARD data in an industry-standard Intermediate Data Format (IDF) from SOLIDWORKS.

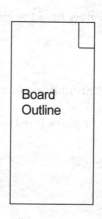
Board Outline

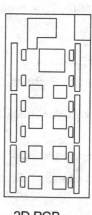
2D PCB

The IDF file is sent to the PCB designer to populate the board with the correct 2D electronic components.

CircuitWorks utilizes industry-standard IDF files or PADS files, and produces the 3D SOLIDWORKS assembly of the MOTHERBOARD. IDF and PADS are common file formats utilized in the PCB industry.

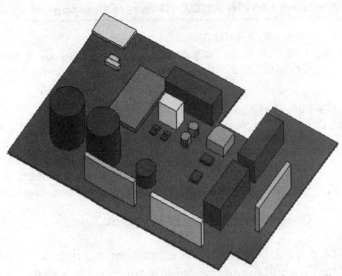

The MOTHERBOARD is fully populated with the components at the correct height. Your colleagues use the MOTHERBOARD assembly to develop other areas of the BOX assembly. An engineer develops the wire harness from the MOTHERBOARD to electrical components in the BOX.

SOLIDWORKS assembly developed with CircuitWorks
Courtesy of Computer Aided Products, Inc.

A second engineer uses the 3D geometry from each electrical component on the MOTHERBOARD to create a heat sink. As the project manager, you move components that interfere with other mechanical parts, cables and or wire harness. You distribute the updated information to your colleagues and manufacturing partners.

To save a virtual component to its own external file, right-click the component and select **Save Part (in External File)** or **Save Assembly (in External file)**.

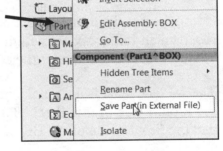

To rename the virtual component in the assembly, right-click the **virtual component** and click **Rename Part**.

Remember, the **Prompt user to select document template** option is selected in System Options, Default Templates.

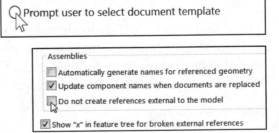

If you do not want to create any External references in your part, click **Options**, **External References** and check the **Do not create references external to the model** box.

Activity: MOTHERBOARD-Insert Component

Create the MOTHERBOARD component.

110) The BOX assembly is the open document. Click **New Part** from the Consolidated Insert Components toolbar as illustrated.

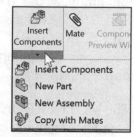

Select the Part Template.

111) Double-click **PART-MM-ANSI** from the MY-TEMPLATES tab in the New SOLIDWORKS Document dialog box. The new part is displayed in the FeatureManager design tree with a name in the form [Partn^assembly_name]. The square brackets indicate that the part is a virtual component. The new

Component pointer ✔ icon is displayed.

The default component is empty and requires a Sketch plane. The Front Plane of the default component is mated with the Front Plane of the BOX.

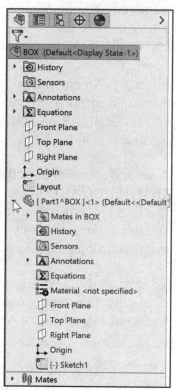

Select the Sketch plane for the default component.

112) Click **Front Plane** in the BOX assembly FeatureManager.

113) Expand the new inserted component [Part1^BOX] in the FeatureManager. The FeatureManager is displayed in light blue.

Components added In-Context of the assembly automatically received an InPlace Mate within the Mates entry in the FeatureManager and under the component.

The InPlace mate is a Coincident mate between the Front Plane of the [Part1^BOX] component and the Front Plane of the BOX.

The [Part1^BOX] entry is added to the FeatureManager. The system automatically selects the Edit Component 🗗 mode.

The [Part1^BOX] text is displayed in blue to indicate that the component is actively being edited. The current Sketch plane is the Front Plane. The system automatically selects the

Sketch 🖊 icon.

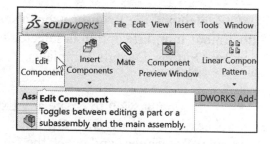

The current sketch name is 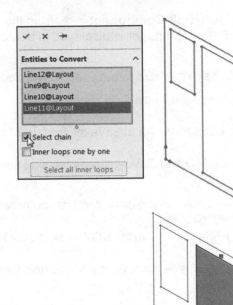 (-) Sketch1 .

Create the sketch.

114) Click **Convert Entities** ⬚ from the Sketch toolbar.

115) Click the **right vertical edge** of the right inside rectangle as illustrated.

116) Click the **three other sketch** lines of the inside rectangle.

117) Check the **Select chain** box.

118) Click **OK** ✔ from the Convert Entities PropertyManager. The outside perimeter of the Layout Sketch is the current sketch.

Insert an Extruded Boss/Base feature for the [Part1^BOX] component.

119) Click **Extruded Boss/Base** 🔲 from the Features toolbar. The Boss-Extrude PropertyManager is displayed.

120) Enter **10**mm for Depth in Direction 1. Accept the default settings.

121) Click **OK** ✔ from the Boss-Extrude PropertyManager. Boss-Extrude1 is displayed in the FeatureManager.

Rename Boss-Extrude1.
122) Rename **Boss-Extrude1** to **Base Extrude**.

The default component, "[Part1^BOX]<#> ->", Base Extrude -> and Sketch1 -> all contain the "->" symbol indicating External references to the BOX assembly.

The Edit Component feature acts as a switch between the assembly and the component edited In-Context.

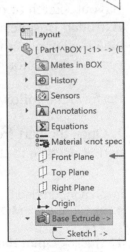

💡 If you do not want to create any External references in your model, click **Options**, **External References** and check the **Do not create references external to the model** box.

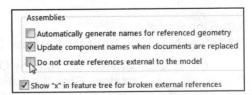

Return to the BOX assembly.

123) Click **Edit Component** 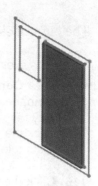 from the Assembly toolbar. The FeatureManager is displayed in black. Rebuild the model.

Fit and Save the BOX to the Graphics window.

124) Press the **f** key. Click **Save** .

125) Click **Save All** to save all modified documents.

126) Click **OK** to Save internally (inside the assembly).

Save and rename the default virtual component [Part1^BOX]<#> -> to MOTHERBOARD.

127) Right-click **[Part1^BOX]<#> ->**. Click **Open Part**.

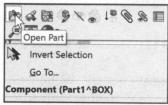

128) Click **Save As** from the Menu bar. Click **OK** from the dialog box.

129) Enter **MOTHERBOARD** in the File name box. Note: Save in the VENDOR-COMPONENTS folder. Click **Save as**. Click **Save**. **Close** the MOTHERBOARD part. Return to the **BOX** assembly. View the updated FeatureManager.

The Reference models are the MOTHERBOARD and the BOX assembly. Additional features are required that do not reference the Layout Sketch or other components in the BOX assembly. An Extruded Boss feature indicates the approximate position of an electrical connector. The actual measurement of the connector or type of connector has not been determined.

Perform the following steps to avoid unwanted assembly references:

1. Open the part from the assembly. **2.** Add features. **3.** Save the part. **4.** Return to the assembly. **5.** Save the assembly.

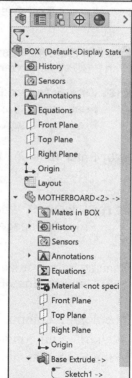

Insert an Extruded Boss/Base feature for the MOTHERBOARD.

130) Right-click **MOTHERBOARD** in the FeatureManager.

131) Click **Open Part**. The MOTHERBOARD FeatureManager is displayed. Press the **f** key.

Insert the Sketch.

132) Right-click the **front face** of Base Extrude. This is your Sketch plane. Click **Sketch** from the Context toolbar. The Sketch toolbar is displayed. Click the **outside right vertical edge**. Hold the **Ctrl** key down. Click the **top horizontal edge**. Release the **Ctrl** key.

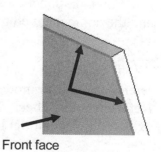

Front face

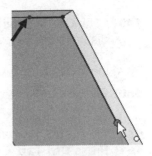

133) Click **Convert Entities** 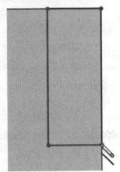 from the Sketch toolbar.

134) Drag the **bottom end point** of the converted line three quarters upward.

135) Drag the **left end point** of the converted line three quarters of the way to the right. Click **Front view** from the Heads-up View toolbar.

136) Click **Line** from the Sketch toolbar.

137) Sketch a **vertical line** as illustrated.

138) Sketch a **horizontal line** to complete the rectangle.

Add dimensions.

139) Click **Smart Dimension** from the Sketch toolbar.

140) Enter **30**mm for the horizontal dimension. Enter **50**mm for the vertical dimension. The sketch is fully defined.

Extrude the sketch.

141) Click **Extruded Boss/Base** from the Features toolbar. The Boss-Extrude PropertyManager is displayed.

142) Enter **10**mm for Depth in Direction 1.

143) Click **OK** from the Boss-Extrude PropertyManager.

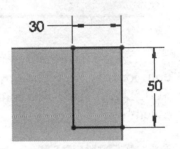

Rename Boss-Extrude2.
144) Rename **Boss-Extrude2** to **Connector1**.

Display an Isometric view and save the model.
145) Click **Isometric view**. Click **Save**.

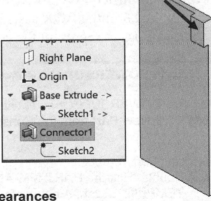

Use color to indicate electrical connectors. Color the face of the Extruded Boss feature.

Color the Front face of Connector1.
146) Right-click the **Front face** of Connector1. Click the **Appearances** drop-down arrow.

147) Click **Connector1** as illustrated.

148) Select **yellow** as the color.

149) Click **OK** ✔ from the Color PropertyManager. The selected face is displayed in yellow.

Save the MOTHERBOARD.
150) Click **Save** 💾.

Return to the BOX assembly.
151) Press **Ctrl Tab**. Click **Yes** to rebuild the assembly.

The MOTHERBOARD contains the new Extruded Boss feature. For now, the MOTHERBOARD is complete. The POWERSUPPLY is the second key component to be defined In-Context of the BOX assembly.

💡 A learning goal of this book is to expose the new user to various tools, techniques and procedures. It may not always use the most direct tool or process.

💡 Locate External references. Open the BOX assembly before opening individual components referenced by the assembly when you start a new session of SOLIDWORKS. The components load and locate their external references to the original assembly.

POWERSUPPLY-Insert Component

Create a simplified POWERSUPPLY part In-Context of the BOX assembly. Utilize the Layout Sketch in the BOX assembly. Open the POWERSUPPLY part and insert an Extruded Boss to represent the connection to a cable.

Activity: POWERSUPPLY-Insert Component

Insert the POWERSUPPLY component.
152) Click **New Part** from the Consolidated Insert Components toolbar as illustrated.

Select the Part Template.
153) Double-click the **PART-MM-ANSI** template from the MY-TEMPLATES tab. A new default virtual component is added to the FeatureManager. Note: Save the new component and name it POWERSUPPLY later in the procedure.

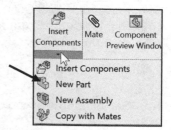

The Component Pointer ✔ is displayed. The new default component is empty and requires a Sketch plane.

Select the Sketch plane.

154) Click the **Front Plane** of the BOX in the FeatureManager. An InPlace mate is added to the BOX FeatureManager.

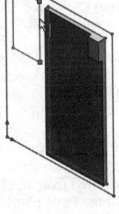

The Front Plane of the default component [Part#^BOX]<#> is mated to the Front Plane of the BOX. Sketch1 of the [Part#^BOX]<1> is the active sketch.

The system automatically selects the Edit Component tool. The default component [Part#^BOX]<#> entry in the FeatureManager is displayed in blue. The current Sketch plane is the Front Plane. The current sketch name is Sketch1. The name is indicated on the current document window title.

[Part#^BOX]<#> is the default name of the component created In-Context of the BOX assembly. Create the first Extruded Base feature In-Context of the Layout Sketch.

Create the sketch.

155) Click **Convert Entities** from the Sketch toolbar. Click the **Select chain** box.

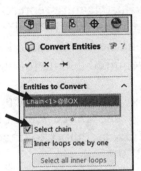

156) Click the **right vertical edge** of [Part#^BOX]<1> as illustrated. The **three other sides** of the inside small rectangle are selected.

157) Click **OK** from the Convert Entities PropertyManager.

Extrude the sketch.

158) Click **Extruded Boss/Base** from the Features toolbar. Blind is the default End Condition in Direction 1.

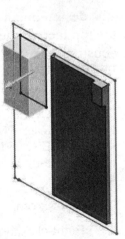

159) Enter **50**mm for Depth in Direction 1. Accept the default settings.

160) Click **OK** from the Boss-Extrude PropertyManager. Boss-Extrude1 is displayed in the FeatureManager under [Part#^BOX]<#>.

The [Part#^BOX]<#> default name is displayed in blue. The component is being edited In-Context of the BOX assembly.

Return to the BOX assembly.

161) Click **Edit Component** 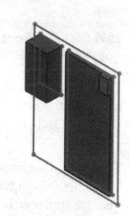 from the Assembly toolbar. The [Part#^BOX]<#> displayed in black.

Save the BOX assembly.

162) Click **Save** . Click **Save All**. Click **OK** to save internally.

Rename the default component [Part2^BOX]<1> to POWERSUPPLY.
163) Right-click **[Part#^BOX]<1> ->**. Click **Open Part**.
The [Part#^BOX]<1> -> FeatureManager is displayed.

164) Click **Save As** from the Menu bar. Enter **POWERSUPPLY** in the File name box. Save in the VENDOR-COMPONENTS folder. Click **Save as**.

165) Click **Save**. **Close** the part. **Return** to the BOX assembly. View the updated FeatureManager.

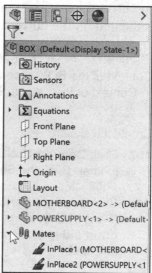

The Extruded Boss/Base feature represents the location of the cable that connects to the POWERSUPPLY. Think about where the cables and wire harness connect to key components. You do not have all of the required details for the cables.

In a concurrent engineering environment, create a simplified version early in the design process. No other information is required from the BOX assembly to create additional features for the POWERSUPPLY.

The design intent for the POWERSUPPLY is for the cable connection to be centered on the top face of the POWERSUPPLY. Utilize a centerline and Midpoint relation to construct the Extruded Boss/Base feature centered on the top face. Open the POWERSUPPLY from the BOX assembly.

Open the POWERSUPPLY. Fit the model to the Graphics window.
166) Right-click **POWERSUPPLY** in the Graphics window. Click **Open Part**. The POWERSUPPLY FeatureManager is displayed. Press the f key. Click **Top view** from the Heads-up View toolbar.

Create the sketch.

167) Right-click the **top face** of Boss-Extrude1. Click **Sketch** from the Context toolbar. Click **Centerline** from the Sketch toolbar. The Insert Line PropertyManager is displayed.

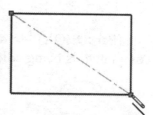

168) Sketch a **diagonal centerline** from the top left to the bottom right as illustrated. Align the endpoints with the corners of the POWERSUPPLY.

Sketch a circle.

169) Click **Circle** 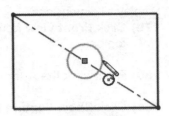 from the Sketch toolbar. Click the **Midpoint** of the centerline. Sketch a circle as illustrated. The centerline midpoint is the center of the circle.

Add a dimension.

170) Click **Smart Dimension** from the Sketch toolbar.

171) Click the **circumference** of the circle. Click a **position off** the profile. Enter **15**mm for diameter. The circle is displayed in black.

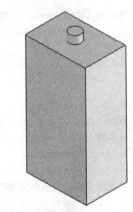

Extrude the sketch.

172) Click **Extruded Boss/Base** from the Features toolbar. The Boss-Extrude PropertyManager is displayed. Blind is the default End Condition in Direction 1.

173) Enter **10**mm for End Condition in Direction 1.

174) Click **OK** ✔ from the Boss-Extrude PropertyManager. The feature is displayed in the FeatureManager.

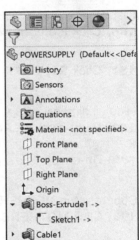

Rename the Boss-Extrude2 feature. Display an Isometric view.

175) Rename the **Boss-Extrude2** to **Cable1**.

176) Click **Isometric view**.

Save the POWERSUPPLY.

177) Click **Save**.

Return to the BOX assembly.

178) Press **Ctrl Tab**. Click **Yes** to update the assembly.

The POWERSUPPLY contains the new Extruded Boss feature, Cable1. Recall the initial design parameters. The requirement calls for three different size boxes. Test the Layout Sketch dimensions for the three configurations: *Small, Medium and Large.*

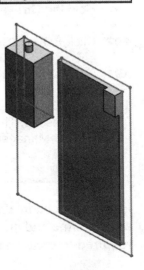

Do the Layout Sketch, MOTHERBOARD and POWERSUPPLY reflect the design intent of the BOX assembly? Review the next steps to confirm the original design intent.

Display all dimensions.
179) Click **Front view** from the Heads-up View toolbar.

180) Right-click the **Annotations** folder.

181) Click **Show Feature Dimensions**.

182) Double-click **300**. Enter **550** for the horizontal dimension.

183) Double-click **400**. Enter **600** for the vertical dimension.

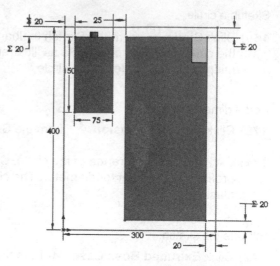

Return to the original dimensions.
184) Double-click **550**. Enter **300** for width dimension.

185) Double-click **600**. Enter **400** for height dimension.

Display an Isometric view. Save the assembly.
186) Click **Isometric view** from the Heads-up View toolbar.

187) Click **Save** 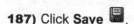.

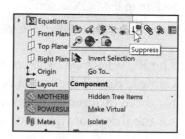

Hide all dimensions.
188) Right-click the **Annotations** folder.

189) Un-check **Show Feature Dimensions**.

Suppress the MOTHERBOARD and POWERSUPPLY.
190) Click **MOTHERBOARD** from the FeatureManager.

191) Hold the **Ctrl** key down.

192) Click **POWERSUPPLY** from the FeatureManager.

193) Release the **Ctrl** key.

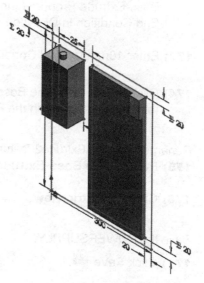

194) Right-click **Suppress**. Both components and their Mates are Suppressed. They are displayed in light gray in the FeatureManager.

Electro-mechanical assemblies contain hundreds of parts. Suppressing components saves rebuild time and simplifies the model. Only display the components and geometry required to create a new part or to mate a component.

Improve rebuild time and display performance for large assemblies. Review the System Options listed in Large Assembly Mode. Adjust the large assembly threshold to match your computer performance.

Check Remove detail during zoom/pan/rotate. Check Hide all planes, axes and sketches.

Review the System Option, Image Quality. Drag the resolution slider towards Low to improve computer performance.

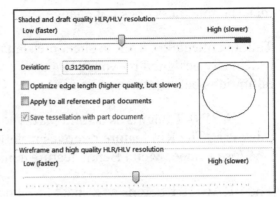

Review the BOX Assembly

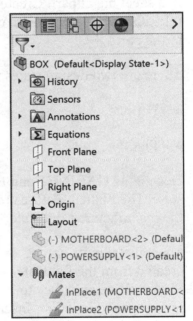

The BOX assembly utilized the Top-down assembly approach by incorporating a Layout Sketch. The Layout Sketch consisted of 2D profiles and dimensions of the BOX, MOTHERBOARD and POWERSUPPLY.

The MOTHERBOARD was developed as a new part In-Context of the BOX assembly. An InPlace mate was created between the MOTHERBOARD Front Plane and the BOX assembly Front Plane. In the Edit Component mode, you extracted the Layout Sketch geometry to create the first feature of the MOTHERBOARD part. The first feature, sketch and Sketch plane contained External references to the BOX assembly.

You opened the MOTHERBOARD from the BOX assembly. Additional features were added to the MOTHERBOARD part. The POWERSUPPLY was developed as a new part In-Context of the BOX assembly. The BOX assembly design intent and requirements were verified by modifying the dimensions to the different sizes of the BOX configurations.

The MOTHERBOARD and POWERSUPPLY are only two of numerous components in the BOX assembly. They are partially complete but represent the overall dimensions of the final component. Electro-Mechanical assemblies contain parts fabricated from sheet metal. Fabricate the CABINET and BRACKET for the BOX assembly from Sheet metal.

Sheet Metal Overview

Sheet metal manufacturers create sheet metal parts from a flat piece of raw material. To produce the final part, the material is cut, formed and folded. In SOLIDWORKS, the material thickness does not change.

Talk to colleagues. Talk to sheet metal manufacturers. Review other sheet metal parts previously designed by your company. You need to understand a few basic sheet metal definitions before starting the next parts, CABINET and BRACKET.

The CABINET begins as a solid part. The Shell feature determines the material thickness. The Rip feature removes material at the edges to prepare for sheet metal bends. The Insert Sheet Metal Bends feature transforms the solid part into a sheet metal part.

The BRACKET begins as a sheet metal part. The Base Flange feature is the first feature in a sheet metal part. Create the Base Flange feature from a sketch. The material thickness and bend radius of the Base Flange are applied to all other sheet metal features in the part.

There are two design states for sheet metal parts:

- Formed.

- Flat.

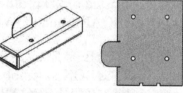

Design the CABINET and BRACKET in the formed state. The Flatten feature creates the Flat Pattern for the manufactured flat state.

Formed State   Flat State
BRACKET

Sheet metal parts can be created from the flat state. Sketch a line on a face to indicate bend location. Insert the Sketched Bend feature to create the formed state.

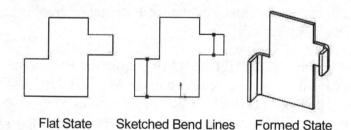

Flat State   Sketched Bend Lines   Formed State

Bends

Example: Use a flexible eraser, 50mm or longer. Bend the eraser in a U shape. The eraser displays tension and compression forces.

The area where there is no compression or tension is called the neutral axis or neutral bend line.

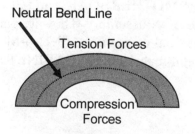

Neutral Bend Line

Tension Forces

Compression Forces

Eraser Example

Assume the material has no thickness. The length of the material formed into a 360° circle is the same as its circumference.

The length of a 90° bend would be ¼ of the circumference of a circle.

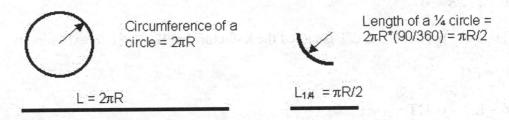

Circumference of a circle = 2πR

$L = 2πR$

Length of a ¼ circle = 2πR*(90/360) = πR/2

$L_{¼} = πR/2$

Length of 90° bend

In the real world, materials have thickness. Materials develop different lengths when formed in a bend depending on their thickness.

There are three major properties which determine the length of a bend:

- Bend radius.

- Material thickness.

- Bend angle.

The distance from the inside radius of the bend to the neutral bend line is labeled δ.

The symbol 'δ' is the Greek letter delta. The amount of flat material required to create a bend is greater than the inside radius and depends on the neutral bend line.

The true developed flat length, L, is measured from the endpoints of the neutral bend line.

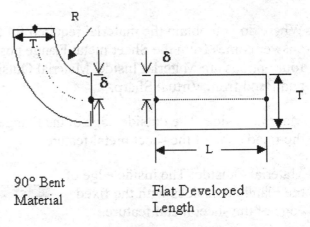

90° Bent Material

Flat Developed Length

True developed flat length

Example:

T is the Material thickness.

Create a 90° bend with an inside radius of R:

$L = \frac{1}{2} \pi R + \delta T$

The ratio between δ and T is called the K-factor. Let K = 0.41 for Aluminum:

$K = \delta/T$

$\delta = KT = 0.41T$

$L = \frac{1}{2} \pi R + 0.41T$

U.S. Sheet metal shops use their own numbers.

Example:

One shop may use K = 0.41 for Aluminum, versus another shop that uses K = 0.45.

Use tables, manufacturer's specifications or experience.

Save design time and manufacturing time. Work with your Sheet metal shop to know their K-factor. Material and their equipment produce different values. Build these values into the initial design.

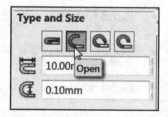

Where do you obtain the material required for the Bend? The answer comes from the Sheet metal Flange position option. The four options are Material Inside, Material Outside, Bend Outside and Bend from Virtual Sharp.

Material Inside: The outside edge of the Flange coincides with the fixed edge of the sheet metal feature.

Material Outside: The inside edge of the Flange coincides with the fixed edge of the sheet metal feature.

Bend Outside: The Flange is offset by the bend radius.

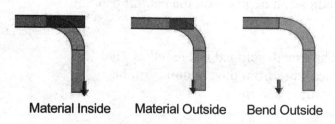

Bend from Virtual Sharp: The Flange maintains the dimension to the original edge and varies the bend material condition to automatically match with the flange's end condition.

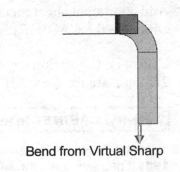

Bend from Virtual Sharp

Relief

Sheet metal corners are subject to stress. Excess stress will tear material. Remove material to relieve stress with Auto Relief. The three options are Rectangular, Tear and Obround.

- Rectangular: Removes material at the bend with a rectangular shaped cut.

- Tear: Creates a rip at the bend, a cut with no thickness.

- Obround: Removes material at the bend with a rounded shaped cut.

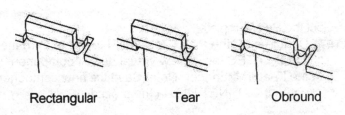

Rectangular   Tear   Obround

CABINET-Insert Component

Create the CABINET component inside the assembly and attach it to the Front Plane. The CABINET component references the Layout sketch.

Create the CABINET component as a solid box. Shell the box to create the constant sheet metal thickness. Utilize the Rip feature to cut the solid edges. Utilize the Insert Bends feature to create a sheet metal part from a solid part.

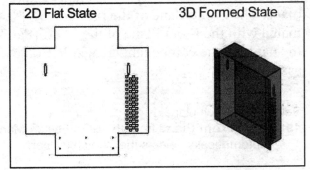

2D Flat State   3D Formed State

Add additional sheet metal Edge Flange and Hem features. Add dies, louvers, and cuts to complete the CABINET.

Utilize configurations to create the 3D formed state for the BOX assembly and the 2D flat state for the CABINET drawing.

Activity: CABINET-Insert Component

Insert the CABINET Component.
195) If needed, open the **BOX** assembly.

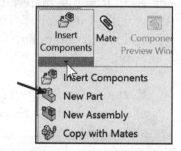

196) Click **New Part** from the Consolidated Insert Components toolbar as illustrated.

Select the Part Template.
197) Double-click the **PART-MM-ANSI** template from the MY-TEMPLATES tab. A new virtual default component is added to the FeatureManager. Note: Save the new component and name it CABINET later in the procedure.

The Component Pointer ✔ is displayed. The default component [Part3^BOX]<1> is empty and requires a Sketch plane. The Front Plane of the [Part3^BOX]<1> component is mated with the Front Plane of the BOX. [Part3^BOX]<1> is the name of the component created In-Context of the BOX assembly.

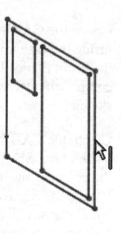

Select the Sketch plane.
198) Click **Front Plane** from the BOX FeatureManager. The system automatically selects the Sketch toolbar.

Create the sketch.
199) Click the **Convert Entities** Sketch tool.

200) Select the **Select chain** box.

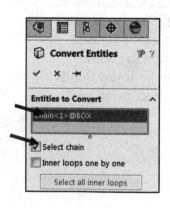

201) Click the **right vertical edge** of the outside BOX as illustrated. The **three other sides** of the outside BOX are selected.

202) Click **OK** ✔ from the Convert Entities PropertyManager. The outside perimeter is the current sketch.

Extrude the sketch.

203) Click **Extruded Boss/Base**  from the Features toolbar. The Boss-Extrude PropertyManager is displayed.

204) Enter **100**mm for Depth in Direction 1. Accept the default settings.

205) Click **OK** ✔ from the Boss-Extrude PropertyManager.

Return to the BOX assembly.

206) Click **Edit Component** from the Assembly toolbar.

Rename the default component [Part3^BOX]<1> to CABINET.
207) Right-click **[Part3^BOX]<1> ->**. Click **Open Part**. The [Part3^BOX]<1> -> FeatureManager is displayed.

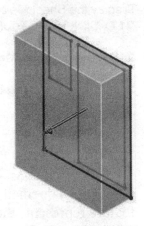

208) Click **Save As** from the Menu bar. Click **OK**. Enter **CABINET** in the File name box. Note: Save in the VENDOR-COMPONENTS folder.

209) Click **Save**. **Close** the part.

210) Return to the **BOX** assembly. **View** the updated FeatureManager

Caution: Do not create unwanted geometry references. Open the part when creating features that require no references from the assembly. The features of the CABINET require no additional references from the BOX assembly, MOTHERBOARD or POWERSUPPLY.

Open the CABINET part.
211) Right-click **CABINET** from the FeatureManager.

212) Click **Open Part**. The CABINET FeatureManager is displayed.

Fit the model to the Graphics window. Display an Isometric view.
213) Press the **f** key. Click the **Isometric view** from the Heads-up View toolbar.

Create the Shell.
214) Click the **front face** of the Boss-Extrude1 feature.

215) Click **Shell** from the Features toolbar. The Shell1 PropertyManager is displayed. Check the **Shell outward** box. Enter **1.00**mm for Thickness.

216) Click **OK** ✔ from the Shell1 PropertyManager. Shell1 is displayed in the FeatureManager.

Display the Sheet Metal toolbar.
217) Click **View**, **Toolbars** from the Main menu.

218) Check **Sheet Metal**. The Sheet Metal toolbar is displayed.

219) **Click and drag** to a location inside the Graphics window.

Maintain constant thickness. Sheet metal features only work with constant wall thickness. The Shell feature maintains constant wall thickness. The solid Extruded Base feature represents the overall size of the sheet metal CABINET. The CABINET part is solid. The Rip feature and Insert Sheet Metal Bends feature convert a shelled solid part into a sheet metal part.

CABINET-Rip Feature and Sheet Metal Bends

The Rip feature creates a cut of no thickness along the edges of the Extruded Base feature. Rip the Extruded Base feature along the four edges. The Bend feature creates sheet metal bends.

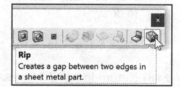

Specify bend parameters: bend radius, bend allowance and relief. Select the bottom face to remain fixed during bending and unbending. In the next example, you will create the Rip and the Insert Bend features in two steps. The Bends PropertyManager contains the Rip parameter. Utilize the Rip feature to create a simple Rip and Bend in a single step.

Activity: CABINET-Rip Feature and Sheet Metal Bends

Fit the model to the Graphics window.
220) Press the **f** key.

Insert a Rip feature. A Rip feature creates a gap between two edges in a sheet metal part.

221) Click **Rip** from the Sheet Metal toolbar. The Rip PropertyManager is displayed.

222) **Rotate** the part to view the inside edges. Click the **inside lower left** edge as illustrated.

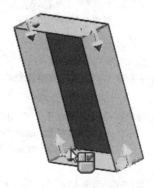

223) Click the **inside upper left** edge. Click the **inside upper right** edge.

224) Click the **inside lower right** edge. The selected entities are displayed in the Edges to Rip box.

225) Enter **.10mm** for Rip Gap.

226) Click **OK** ✔ from the Rip PropertyManager. Rip1 is displayed in the FeatureManager.

Insert the Sheet Metal Bends.
227) Click the **inside bottom face** to remain fixed.

228) Click **Insert Bends** ◇ from the Sheet Metal toolbar. The Bends PropertyManager is displayed. Face<1> is displayed.

229) Enter **2.00**mm for Bend radius.

230) Enter **.45** for K-Factor. Enter **.5** for Rectangular Relief.

231) Click **OK** ✔ from the Bends PropertyManager.

232) Click **OK** to the message, "Auto relief cuts were made for one or more bends."

233) **Zoom in** 🔍 on the Rectangular relief in the upper back corner.

Display an Isometric view. Save the assembly.
234) Click **Isometric view** from the Heads-up View toolbar.

235) Click **Save** 💾. View the created features in the FeatureManager.

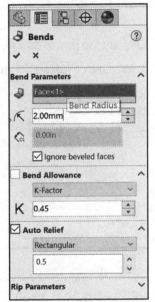

You just created your first sheet metal part. The .45 K-Factor is based on your machine shop's parameters for Aluminum.

🔅 Save manufacturing cost and reduce setup time. A sheet metal manufacturer maintains a turret of standard relief tools for Rectangular and Obround relief. Obtain the dimensions of these tools to utilize in your design.

The CABINET part is in its 3D formed state. Display its 2D flat manufactured state. Test every feature to determine if the part can be manufactured as a sheet metal part.

Alternate between 3D formed and 2D flat for every additional sheet metal feature you create. The Flatten feature alternates a sheet metal part between the flat state and formed state.

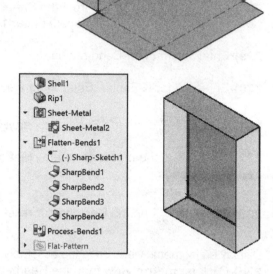

Display the Flat State.

236) Click **Flatten** from the Sheet Metal toolbar.

Fit the model to the Graphics window.
237) Press the **f** key.

Display the part in its fully formed state.

238) Click **Flatten** from the Sheet Metal toolbar.

Save the CABINET.
239) Click **Save** .

Note: To return to the solid part, utilize the No Bends feature to roll back the model before the first sheet metal bend.

CABINET-Edge Flange

Create the right flange wall and left flange wall with the Edge Flange feature. The Edge Flange feature adds a wall to a selected edge of a sheet metal part.

Select the inside edges when creating bends. Create the right hem and left hem with the Hem feature. The Hem feature folds back at an edge of a sheet metal part. A Hem can be closed, open, teardrop or rolled.

In the preliminary design stage, review Hem options with your sheet metal manufacturer.

Activity: CABINET-Edge Flange

Insert the front right Flange.

240) Click the **front vertical right** edge of the CABINET as illustrated.

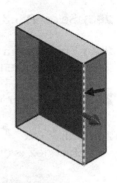

241) Click **Edge Flange** 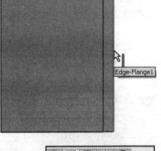 from the Sheet Metal toolbar. The Edge-Flange PropertyManager is displayed. Select **Blind** for Length Edge Condition.

242) Enter **30**mm for Length. Click the **Reverse Direction** button. The Direction arrow points towards the right. Click **Outer Virtual Sharp**. Accept all other defaults.

243) Click **OK** ✔ from the Edge-Flange PropertyManager. Edge-Flange1 is displayed in the FeatureManager.

Insert the right Hem.

244) Click **Front view** from the Heads-up View toolbar.

245) Click the **right edge** of the right flange.

246) Click **Hem**  from the Sheet Metal toolbar. The Hem PropertyManager is displayed.

247) Click **Isometric view** from the Heads-up View toolbar.

248) Click the **Reverse Direction** button.

249) Enter **10**mm for Length.

250) Enter **.10**mm for Gap Distance.

251) Click **OK** ✔ from the Hem PropertyManager. Hem1 is displayed in the FeatureManager.

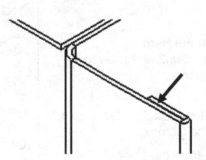

Insert the left Edge Flange wall.

252) Click **Front view** from the Heads-up View toolbar.

253) Click the **front vertical left** edge as illustrated.

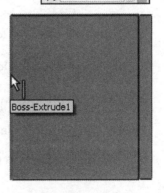

254) Click **Edge Flange** 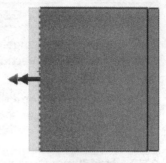 from the Sheet Metal toolbar. The Edge-Flange PropertyManager is displayed.

255) Select **Blind** for Length End Condition.

256) Enter **30**mm for Length. The feature direction arrow points towards the left. Reverse the arrow direction if required.

257) Click **Outer Virtual Sharp**. Accept the default settings.

258) Click **OK** ✔ from the Edge-Flange PropertyManager. Edge-Flange2 is displayed in the FeatureManager.

Insert the left edge Hem.
259) Select the **left edge** of the left flange as illustrated.

260) Click **Isometric view** from the Heads-up View toolbar.

261) Click **Hem**  from the Sheet Metal toolbar. The Hem PropertyManager is displayed.

262) Enter **10**mm for Length. Enter **.10**mm for Gap Distance. Click the **Reverse Direction** button.

263) Click **OK** ✔ from the Hem PropertyManager. Hem2 is displayed in the FeatureManager.

Display the part in its flat manufactured state and formed state. Rename the features.
264) Click **Flatten** from the Sheet Metal toolbar.

265) Click **Top view** from the Heads-up View toolbar.

266) Click **Flatten** to display the part in its fully formed state.

267) Click **Isometric view** from the Heads-up View toolbar.

268) Rename **Edge-Flange1** to **Edge-Flange1-Right**.

269) Rename **Edge-Flange2** to **Edge-Flange2-Left**.

270) Rename **Hem1** to **Hem1-Right**.

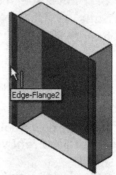
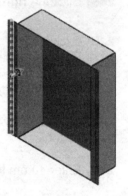

271) Rename **Hem2** to **Hem2-Left**.

Save the CABINET.

272) Click **Save** 💾.

Simplify the FeatureManager. Sheet metal parts contain numerous flanges and hems. Rename features with descriptive names.

Save design time. Utilize Mirror Feature to mirror sheet metal features about a plane. Example: In a second design iteration, the Edge Flange1-Right and Hem1-Right are mirrored about Plane1.

The Mirror feature was not utilized in the CABINET. A Mirror Plane and Equations are required to determine the flange and hem locations.

💡 Maintain the design intent. In the Top-down design process return to the Layout Sketch after inserting a new component to verify the assembly design intent.

Increase and decrease dimension values in the Layout Sketch. To avoid problems in a top level assembly, select the same face to remain fixed for the Flat Pattern and Bend/Unbend features. Save sheet metal parts in their 3D formed state.

The CABINET requires additional solid and sheet metal features. These features require no references in the BOX assembly. Work in the CABINET part.

CABINET - Hole Wizard and Linear Pattern features

Sheet metal holes are created through a punch or drill process. Each process has advantages and disadvantages: cost, time and accuracy.

Investigate a Linear Pattern feature of holes. Select a self-clinching threaded fastener that is inserted into the sheet metal during the manufacturing process. The fastener requires a thru hole in the sheet metal.

Holes should be of equal size and utilize common fasteners. Why? You need to ensure a cost effect design that is price competitive. Your company must be profitable with their designs to ensure financial stability and future growth.

Another important reason for fastener commonality and simplicity is the customer. The customer or service engineer does not want to supply a variety of tools for different fasteners.

A designer needs to be prepared for changes. You proposed two fasteners. Ask Purchasing to verify availability of each fastener. Select a Ø5.0mm hole and wait for a return phone call from the Purchasing department to confirm. Design flexibility is key!

Dimension the hole position. Do not dimension to sheet metal bends. If the sheet metal manufacturer modifies the bend radius, the hole position does not maintain the design intent. Reference the hole dimension by selecting the Origin in the FeatureManager.

Utilize the Hole Wizard feature to create a Simple Through All hole. The hole is positioned on the inside bottom face based on the selection point.

Activity: CABINET - Hole Wizard and Linear Pattern feature

Insert the first hole. Open the Hole Wizard.

273) Click **Hole Wizard** from the Features toolbar. The Hole Specification PropertyManager is displayed.

274) Click the **Hole** tab. Click **Reset Custom sizing values**.

275) Select **ANSI Metric** for Standard.

276) Select **Drill sizes** for Type.

277) Select **Ø5.0** for Size.

278) Select **Through All** for End Condition.

279) Click the **Positions** tab.

280) Click the **inside bottom face** as illustrated. The Point Sketch tool is active.

281) Click **again to place** the center point of the hole.

282) Right-click **Select** to deselect the Point Sketch tool.

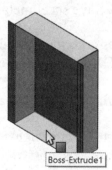

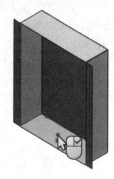

Dimension the hole. Create a horizontal and vertical dimension from the Origin.

283) Click **Smart Dimension** 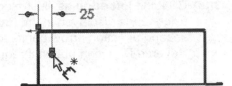.

284) Click **Top view** from the Heads-up View toolbar.

285) **Expand** CABINET from the fly-out FeatureManager.

286) Click the **Origin** from the FeatureManager.

287) Click the **center point of the hole** in the Graphics window.

288) Enter **25**mm for the horizontal dimension.

289) Click the **Origin** from the FeatureManager.

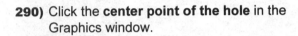
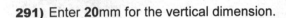

290) Click the **center point of the hole** in the Graphics window.

291) Enter **20**mm for the vertical dimension.

292) Click **OK** ✔ from the Dimension PropertyManager.

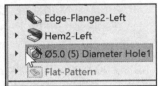

293) Click **OK** ✔ from the Hole Position PropertyManager. The Hole is displayed in the FeatureManager.

Display an Isometric view. Save the model.
294) Click **Isometric view** from the Heads-up View toolbar.

295) Click **Save** 💾.

Insert a Linear Pattern of holes.
296) Click **Linear Pattern** ⊞ from the Features toolbar. The Linear Pattern PropertyManager is displayed.

297) Click inside the **Features to Pattern** box.

298) Click **Hole1**.

299) Click inside the **Pattern Direction** box.

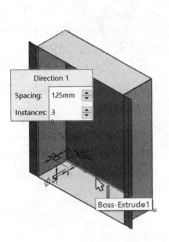

300) Click the **back inside edge** for Direction 1. Edge<1> is displayed in the Pattern Direction box. The direction arrow points to the right. Click the Reverse Direction button if required.

301) Enter **125**mm for Spacing.

302) Enter **3** for Number of Instances.

303) Click the **left edge** as illustrated for Direction 2. Edge<2> is displayed in the Pattern Direction box. The second direction arrow points to the front. Click the Reverse Direction button if required.

304) Enter **50**mm for Spacing.

305) Enter **2** for Number of Instances.

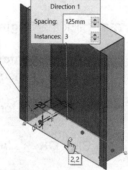
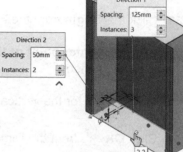

Remove an Instance.
306) Click **inside the Instances to Skip** box.

307) Click the **front middle hole, (2,2)** in the Graphics window as illustrated. (2,2) is displayed in the Instances to Skip box.

308) Check the **Geometry pattern** box. Accept the default settings.

309) Click **OK** ✔ from the Linear Pattern PropertyManager. LPattern1 is displayed in the FeatureManager.

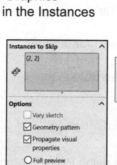
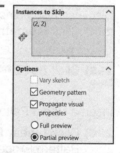

Display an Isometric view.
310) Click **Isometric view** from the Heads-up View toolbar.

💡 Use the Geometry pattern option to improve system performance for sheet metal parts. The Geometry pattern option copies faces and edges of the seed feature. Type options such as Up to Surface are not copied.

The Instances to Skip option removes a selected instance from the pattern. The value (2, 2) represents the instance position in the first direction - second instance, second direction - second instance.

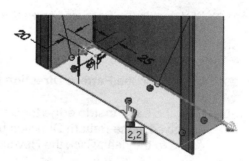

CABINET-Sheetmetal Design Library Feature

Sheet metal manufacturers utilize dies and forms to create specialty cuts and shapes. The Design Library contains information on dies and forms. The Design Library, features folder contains examples of predefined sheet metal shapes. Insert a die cut on the right wall of the CABINET. The die cut is for a data cable. The team will discuss sealing issues at a later date.

Activity: CABINET-Sheetmetal Design Library Feature

Insert a Sheet metal Library Feature.

311) Click **Design Library** 📚 from the Task Pane.

312) Expand the Design Library 📚 folder.

313) Expand the features folder.

314) Click the **Sheetmetal** folder.

315) Click and drag the **d-cutout** feature into the Graphics window.

316) Release the **mouse button on the right flange outside face** of the CABINET as illustrated. The d-cutout PropertyManager is displayed.

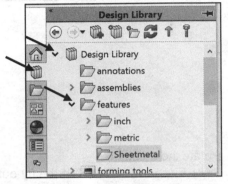

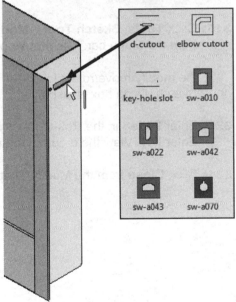

Edit the sketch.

317) Click the **Edit Sketch** button from the d-cutout PropertyManager. The Library Feature Profile dialog box is displayed. DO NOT SELECT THE FINISH BUTTON.

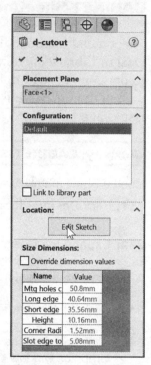

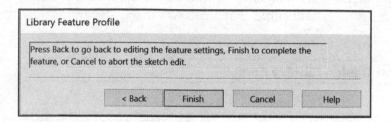

Library Feature Profile

Press Back to go back to editing the feature settings, Finish to complete the feature, or Cancel to abort the sketch edit.

      < Back   Finish   Cancel   Help

Rotate the d-cutout.

318) **Expand** CABINET in the fly-out FeatureManager.

319) Click **Tools**, **Sketch Tools**, **Modify** 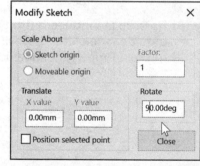 Modify... from the Menu bar. The mouse pointer displays the modify move/rotate icon . The Modify Sketch dialog box is displayed.

320) Enter **90deg** in the Rotate box. Note: Press the Enter key. View the d-cutout feature.

321) Click **Close** from the Modify Sketch dialog box.

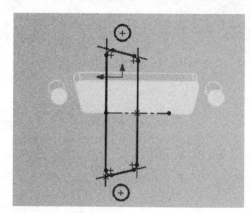

Display the Right view.
322) Click **Right view** from the Heads-up View toolbar.

323) Zoom to Area 🔍 on the d-cutout.

Dimension the d-cutout.
324) Click **Smart Dimension** ↙ from the Sketch toolbar.

Create a vertical dimension.
325) Click the **midpoint** of the d-cutout. The Point PropertyManager is displayed.

326) Expand CABINET from the fly-out FeatureManager.

327) Click the **Origin** of the CABINET.

328) Enter **320**mm.

Create a horizontal dimension.
329) Click the **left vertical line** of the d-cutout.

330) Expand the CABINET from the fly-out FeatureManager.

331) Click the **Origin** of the CABINET.

332) Enter **50**mm.

333) Click **Finish** from the Library Feature Profile dialog box. The d-cutout entry is displayed in the FeatureManager.

Display an Isometric view. Save the model.
334) Click **Isometric view** from the Heads-up View toolbar.

335) Click **Save** 💾.

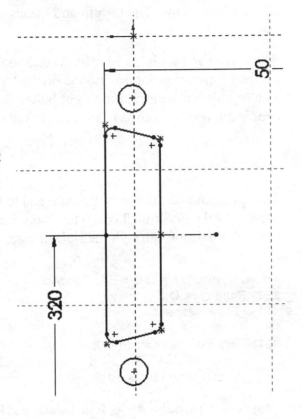

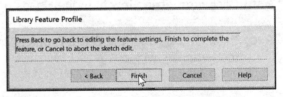

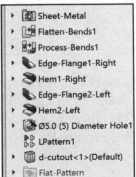

☀ With thin sheet metal parts, select the dimension references with care. Use the Zoom and Rotate commands to view the correct edge. Use reference planes or the Origin to create dimensions. The Origin and planes do not change during the flat and formed states.

☀ Save time with the Modify Sketch tool. Utilize the left and right mouse point positioned on the large black dots. The left button Pans and the right button Rotates. The center dot reorients the Origin of the sketch and Mirrors the

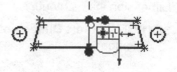

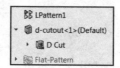

sketch about y, x or both.

The d-cutout goes through the right and left side. Through All is the current End Condition Type. This is not the design intent. Redefine the end condition through the right flange.

If needed, modify the D Cut End Condition.

336) Right-click **D Cut** in the FeatureManager.

337) Click **Edit Feature** from the Context toolbar. The d-cut PropertyManager is displayed.

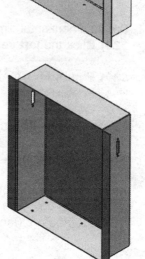

338) Select **Through All** for End Condition in Direction 1.

339) Click **Isometric view** from the Heads-up View toolbar.

340) Click **OK** ✔ from the D Cut PropertyManager. View the results in the Graphics window.

The D Cut feature is positioned before the Flat Pattern1 icon in the FeatureManager. The D Cut is incorporated into the Flat Pattern. The sheet metal manufacturing process is cost effective to perform cuts and holes in the flat state.

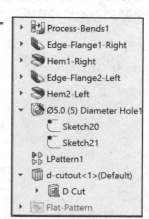

CABINET-Louver Forming Tool

Form features such as louvers are added in the formed state. Formed features are normally more expensive than cut features. Louvers are utilized to direct air flow. The louvers are used to dissipate the heat created by the internal electronic components.

The forming tools folder contains numerous sheet metal forming shapes. In SOLIDWORKS, the forming tools are inserted after the Bends are processed. Suppress forming tools in the Flat Pattern.

Activity: CABINET-Louver Forming Tool

Insert a Louver Forming tool.
341) Rotate the model to display the inside right flange.

342) Expand the forming tools folder in the Design Library.

343) Click the **louvers** folder. A louver is displayed.

344) Click and drag the **louver to the inside right flange** of the CABINET as illustrated. The Form Tool Feature PropertyManager is displayed.

345) Click **Right view**. **Rotate** the Louver as illustrated.

Dimension the location of the louver.
346) Click the **Position** tab as illustrated.

347) Click the **Smart Dimension** tool.

348) Click the **center point of the Louver**.

349) Click the **Origin** of the CABINET from the FeatureManager.

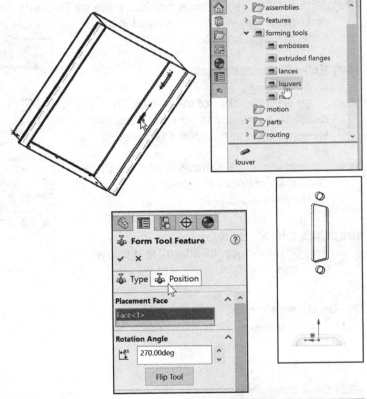

350) Enter **100**mm for the vertical dimension.

351) Click the **Origin** of the CABINET from the fly-out FeatureManager.

352) Click the **vertical centerline** of the Louver.

353) Enter **55**mm for the horizontal dimension.

Display the Louver.
354) Click **OK** from the FeatureManager. Press the **f** key.

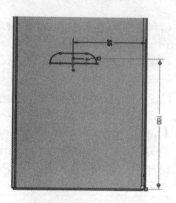

Add a Linear Pattern of louvers.
355) Click **Linear Pattern** from the Features toolbar.

356) Click the CABINET **back vertical edge** for Direction 1 as illustrated. Edge<1> is displayed in the Pattern Direction box.

357) Enter **25**mm for Spacing.

358) Enter **6** for Number of Instances. The direction arrow points upward. Click the Reverse Direction button if required.

359) Click inside the **Features and Faces** box. Click **louver1** from the fly-out FeatureManager.

360) Click **OK** from the Linear Pattern PropertyManager. LPattern2 is displayed in the FeatureManager.

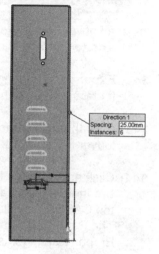

Display an Isometric view.
361) Click **Isometric view**.

Save the CABINET.
362) Click **Save**.

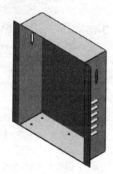

Manufacturing Considerations

How do you determine the size and shape of the louver form? Are additional die cuts or forms required in the project? Work with a sheet metal manufacturer. Ask questions. What are the standards? Identify the type of tooling in stock. Inquire about form advantages and disadvantages.

One company that has taken design and form information to the Internet is Lehi Sheetmetal, Westboro, MA (www.lehisheetmetal.com).

Standard dies, punches and manufacturing equipment such as brakes and turrets are listed in the Engineering helpers section.

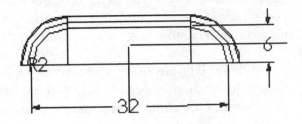

Dimensions are provided for their standard forms. The form tool you used for this project creates a 32mm x 6mm louver. The tool is commercially available.

Your manufacturer only stocks 3in. (75mm) and 4in. (100mm) louvers. Work with the sheet metal manufacturer to obtain a cost effective alternative. Create a pattern of standard square cuts to dissipate the heat.

If a custom form is required, most custom sheet metal manufacturers can accommodate your requirement. Example: Wilson Tool, Great Bear Lake, MN (www.wilsontool.com).

However, you will be charged for the tool. How do you select material? Consider the following:

- Strength.

- Fit.

- Bend Characteristics.

- Weight.

- Cost.

All of these factors influence material selection. Raw aluminum is a commodity. Large manufacturers such as Alcoa and Reynolds sell to a material supplier, such as *Pierce Aluminum.

**ALUMINUM SHEET
NON HEAT TREATABLE, 1100-0
QQ-A-250/1 ASTM B 209**

All thickness and widths available from coil for custom blanks

SIZE IN INCHES	WGT/SHEET
.032 x 36 x 96	11.05
.040 x 36 x 96	13.82
.040 x 48 x 144	27.65
.050 x 36 x 96	17.28
.050 x 48 x 144	34.56
.063 x 36 x 96	21.77
.063 x 48 x 144	43.55
.080 x 36 x 96	27.65
.080 x 48 x 144	55.30
.090 x 36 x 144	46.66
.090 x 48 x 144	62.26
.100 x 36 x 96	34.56
.125 x 48 x 144	86.40
.125 x 60 x 144	108.00
.190 x 36 x 144	131.33

*Courtesy of Piece Aluminum Co, Inc.
Canton, MA

Pierce Aluminum, in turn, sells material of different sizes and shapes to distributors and other manufacturers. Material is sold in sheets or cut to size in rolls. U.S. Sheet metal manufacturers work with standard 8ft - 12ft stock sheets. For larger quantities, the material is usually supplied in rolls. For a few cents more per pound, sheet metal manufacturers request the supplier to shear the material to a custom size.

Check Dual Dimensioning from Tools, Options Document Properties to display both Metric and English units. Do not waste raw material. Optimize flat pattern layout by knowing your sheet metal manufacturer's equipment.

Discuss options for large cabinets that require multiple panels and welds. In Project 1, you were required to be cognizant of the manufacturing process for machined parts.

In Project 4 and 5 you created and purchased parts and recognized the Draft feature required to manufacture plastic parts.

Whether a sheet metal part is produced in or out of house, knowledge of the materials, forms and layout provides the best cost effective design to the customer.

☼ Save manufacturing cost and time. Obtain a list of standard dies and forms from your sheet metal manufacturer. Utilize their standard dies and forms at the start of your design. Prepare for lead-time if custom dies and forms are required.

You require both a protective and cosmetic finish for the Aluminum BOX. Review options with the sheet metal vendor. Parts are anodized. Black and clear anodized finishes are the most common. In harsh environments, parts are covered with a protective coating such as Teflon® or Halon®.

The finish adds thickness to the material. A few thousandths could cause problems in the assembly. Think about the finish before the design begins.

Your sheet metal manufacturer suggests a pattern of square cuts replace the Louvers for substantial cost savings. Suppress the Louver. The Linear Pattern is a child of the Louver. The Linear Pattern is suppressed. Utilize the Sketch Linear Step and Repeat option.

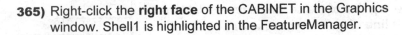

Activity: Manufacturing Considerations-Sketch Linear Pattern

Suppress the louver.
363) Right-click **louver1** from the FeatureManager. Click **Suppress**.

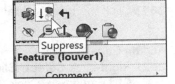

Sketch the profile of squares.
364) Click **Right view**.

365) Right-click the **right face** of the CABINET in the Graphics window. Shell1 is highlighted in the FeatureManager.

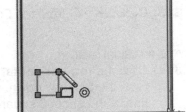

366) Click **Sketch** from the Context toolbar.

Create the first square.
367) Click **Corner Rectangle** from the Sketch toolbar. Sketch a square in the lower left corner as illustrated.

Add dimensions.
368) Click **Smart Dimension** from the Sketch toolbar.

369) Create a vertical and horizontal **10**mm dimension. Click the **Origin**.

370) Click the **bottom horizontal line** of the square.

371) Click a **position to the left** of the profile.

372) Enter **10**mm.

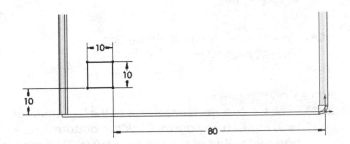

Create a horizontal dimension.
373) Click the **Origin**.

374) Click the **right vertical line** of the square.

375) Click a **position below** the profile.

376) Enter **80**mm.

Create a second square.
377) Click **Corner Rectangle** from the Sketch toolbar.

378) Sketch a **square diagonal to the right** of the first square as illustrated.

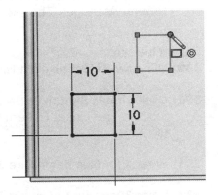

Create a horizontal dimension.

379) Click **Smart Dimension** from the Sketch toolbar.

380) Click the **right vertical line** of the first box.

381) Click the **left vertical line** of the second box. Enter **5mm**.

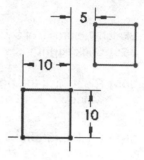

Add an Equal relation.
382) Right-click **Select**. Click the **top line** of the second square. Hold the **Ctrl** key down. Click the **left line** of the second square. The selected entities are displayed in the Selected Entities box. Release the **Ctrl** key. Click **Equal** from the Add Relations box.

383) Click **OK** from the Properties PropertyManager.

Add an Equal relation.
384) Click the **top horizontal line** of the first square. Hold the **Ctrl** key down. Click the **top horizontal line** of the second square. The selected entities are displayed in the Selected Entities box. Release the **Ctrl** key. Click **Equal** from the Add Relations box.

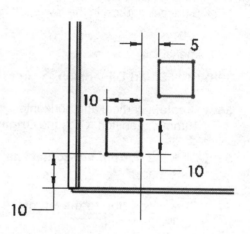

385) Click **OK** from the Properties PropertyManager.

Add a Collinear relation.
386) Click the **top horizontal line** of the first square. Hold the **Ctrl** key down. Click the **bottom horizontal line** of the second square. Release the **Ctrl** key.

387) Click **Collinear** from the Add Relations box.

388) Click **OK** from the Properties PropertyManager. The sketches are fully defined.

Repeat the sketch.
389) **Window-Select** all eight lines of the two squares.

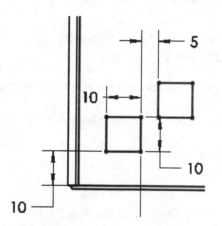

390) Click **Linear Sketch Pattern** from the Sketch toolbar. The Linear Pattern PropertyManager is displayed.

391) Enter **2** for Number in Direction 1.

392) Enter **30mm** for Spacing in Direction 1. The pattern direction arrow points to the right.

393) Enter **13** for Number in Direction 2. The pattern direction arrow points upward.

394) Enter **20**mm for Spacing in Direction 2.

395) Check the **Dimension angle dimension between axes** box.

396) Click **OK** ✔ from the Linear Pattern PropertyManager.

Extrude the sketch.

397) Click **Extruded Cut** from the Features toolbar. The Cut-Extrude PropertyManager is displayed.

398) Check the **Link to thickness** box. Accept the default settings.

399) Click **OK** ✔ from the Cut-Extrude PropertyManager.

Display an Isometric view. Rename the Cut-Extrude2 feature.
400) Click **Isometric view**.

401) Rename the **Cut-Extrude1** to **Vent**.

Save the CABINET in its formed state.
402) Click **Save** .

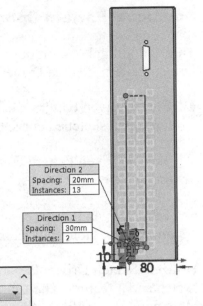

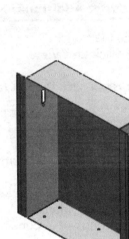

Before you utilize the Sketch Linear Pattern feature, determine the design intent of the pattern in the part and in a future assembly. If there are mating components, use Pattern features. Pattern features provide additional options compared to a simple sketch tool.

Additional Pattern Options

Three additional pattern options to discuss in this book are Curve Driven Pattern, Sketch Driven Pattern and Table Driven Pattern.

- Curve Driven Pattern creates instances of the seed feature through a sketched curve or edge of a sketch.

- Sketch Driven Pattern creates instances of the seed feature though sketched points.

- Table Driven Pattern creates instances of the seed feature through X-Y coordinates in an existing table file or text file.

Utilize a Sketch Driven Pattern for a random pattern. A Sketch Driven Pattern requires a sketch and a feature. The Library Feature Sheetmetal folder contained the d-cutout and other common profiles utilized in sheet metal parts. Create a sketch with random Points on the CABINET left face. Insert sw-b212 and create a Sketch Pattern. View SOLIDWORKS help for additional information on Pattern features.

Activity: Additional Pattern Options-Sketch Driven Pattern

Create the sketch.
403) Click **Left view**.

404) Right-click the **left face** of the CABINET for the Sketch plane.

405) Click **Sketch** from the Context toolbar.

406) Click **Point** □ from the Sketch toolbar.

407) Sketch a **random series of points on the left face** as illustrated. The Point PropertyManager is displayed. Remember the first point you select. This point is required to constrain the sw-b212 sheet metal feature.

Close the sketch. Save the model.
408) Right-click **Exit Sketch**.

409) Rename Sketch to **Sketch-Random**.

410) **Rotate** the CABINET to view the inside left face. Click **Save**.

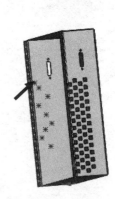

Insert the sw-b212 sheet metal tool.
411) **Expand** the Design Library folder in the Task Pane.

412) **Expand** the features folder. Click the **Sheetmetal** folder.

413) Drag the **sw-b212 feature** to the inside left face of the CABINET. The sw-b212 PropertyManager is displayed.

414) Click the **Edit Sketch** button. DO NOT SELECT THE FINISH BUTTON AT THIS TIME.

Add a Midpoint relation between the center point of the sw-b212 feature and the first random sketch point.
415) Click the **vertical centerline** of sw-b212.

416) Hold the **Ctrl** key down.

417) Click the **Point of the first random sketch point**. The selected entities are displayed in the Selected Entities box. Release the **Ctrl** key.

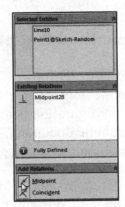

418) Click **Midpoint** from the Add Relations box.

419) Click the **Finish** button.

Edit the End Condition.
420) Right-click **Cut-Extrude1** in the FeatureManager.

421) Click **Edit Feature**.

422) Select **Blind** for End Condition in Direction 1.

423) Check the **Link to thickness** box.

424) Click **OK** from the Cut-Extrude1 PropertyManager.

Insert the Sketch Driven Pattern feature.
425) Click **Insert**, **Pattern/Mirror**, **Sketch Driven Pattern** from the Menu bar. The Sketch Driven Pattern PropertyManager is displayed.

426) Click inside the **Reference Sketch** box.

427) Expand the CABINET fly-out FeatureManager.

428) Click **Sketch-Random**. If required, click inside the Features to Pattern box. Click **Cut-Extrude1** from the fly-out FeatureManager.

429) Click **OK** from the Sketch Driven Pattern PropertyManager. Sketch-Pattern1 is displayed in the FeatureManager.

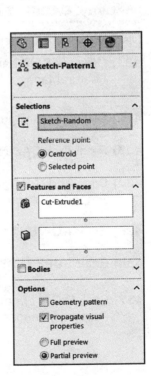

Display an Isometric view. Save the model.
430) Click **Isometric view**. Click **Save** .

The Sketch Driven Pattern is complete. Suppress the patterns when not required for future model development.

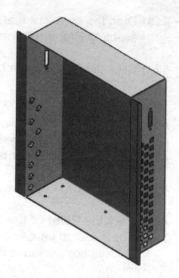

Utilize the Suppress option with configurations to save the formed and flat state of the CABINET.

CABINET-Formed and Flat States

How do you control the 3D formed state in the assembly and the 2D flat state in the drawing? Answer: Utilize configurations. A configuration is a variation of a part or assembly within a single document. Variations include different dimensions, features, and properties.

The BOX assembly requires the CABINET in the 3D formed state. Utilize the Default Configuration in the assembly.

The CABINET drawing requires the 2D flat state. The drawing requires a view with the Linear Pattern of square cuts and the random sketch pattern suppressed. Create a new configuration named NO-VENT. Suppress the patterns. Create a Derived configuration named NO-VENT-FLAT. Un-suppress the Flat Pattern feature to maintain the CABINET in its flat state. Create two configurations in the 3D formed and 2D flat state.

Activity: CABINET-Formed and Flat States

Display the ConfigurationManager and add a Configuration.

431) Drag the **Split bar** downward to divide the FeatureManager in half.

432) Click the **ConfigurationManager** tab.

433) Right-click **CABINET**.

434) Click **Add Configuration**. The Add Configuration PropertyManager is displayed.

435) Enter **NO-VENT** for Configuration name.

436) Enter **SUPPRESS VENT CUTS** for Description.

437) Click **OK** from the Add Configuration PropertyManager.

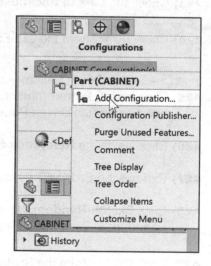

Suppress the Vent Extruded Cut feature.
438) Right-click **Vent** from the FeatureManager.

439) Click **Suppress** ↓ from the Context toolbar.

The NO-VENT Configuration contains the suppressed Vent Extruded Cut feature. A Flat Pattern exists for the Default Configuration. The Flat Pattern Configuration is called a Derived Configuration.

Create a Flat Pattern Derived Configuration for the NO-VENT configuration. Un-suppress the Flat Pattern feature.

Create a Derived Configuration.
440) Right-click **NO-VENT [CABINET]** from the ConfigurationManager.

441) Click **Add Derived Configuration**. The Add Configuration PropertyManager is displayed.

442) Enter **NO-VENT-FLAT-PATTERN** for Configuration name.

443) Click **OK** ✔ from the Add Configuration PropertyManager.

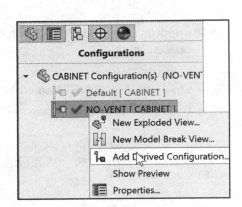

Un-suppress the Flat-Pattern1 feature.
444) Right-click **Flat-Pattern1** from the CABINET FeatureManager.

445) Click **Unsuppress** ↑ from the Context toolbar to display the Flat State.

446) Suppress the **Random Sketch points**.

447) Right-click **Sketch Random** from the CABINET FeatureManager.

448) Click **Suppress** ↓ from the Context toolbar.

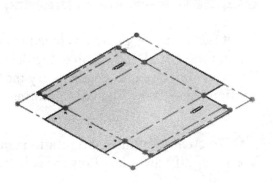

Display the four configurations.

449) Double-click **Default [CABINET]** from the ConfigurationManager.

450) UnSuppress ↑ **Flat-Pattern1** from the FeatureManager.

451) Double-click **NO-VENT [CABINET]** from the ConfigurationManager. View the results in the Graphics window.

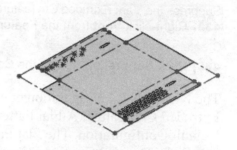

452) Double-click **NO-VENT-FLAT-PATTERN [CABINET]** from the ConfigurationManager.

453) Double-click **Default [CABINET]** from the ConfigurationManager to return to the Default Configuration.

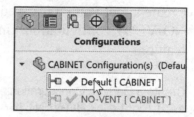

454) Suppress ↓ **Flat-Pattern1** from the CABINET FeatureManager.

Return to a single CABINET FeatureManager.

455) Click the **FeatureManager** 🔧 tab.

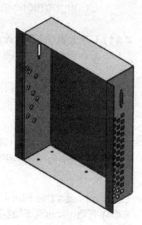

Save the CABINET.

456) Click **Save** 💾.

CABINET-Sheet Metal Drawing with Configurations

Sheet metal drawings are produced in the flat state. Create a new C-size [A2] drawing size for the CABINET. Utilize the default SOLIDWORKS Drawing Template. Insert the Top view into the drawing. Modify the Drawing View Properties from the Default configuration to the Flat configuration.

💡 To create a multi-configuration drawing, insert the Default configuration for Model View. Modify the View Properties in the drawing to change the configuration.

Activity: CABINET-Sheet Metal Drawing with Configurations

Create a new drawing.

457) Click **Make Drawing from Part/Assembly** from the Menu bar.

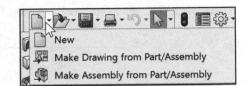

458) Double-click **Drawing** from the Templates tab.

459) Select **C (ANSI) Landscape [A2-Landscape]** for Sheet Format/Size.

460) Click **OK**. The View Palette is displayed on the right side of the Graphics window.

Insert an Isometric view using the View Palette.

461) Click and drag the **Isometric view** from the View Palette into the drawing sheet as illustrated.

462) Click **OK** from the Drawing View1 PropertyManager.

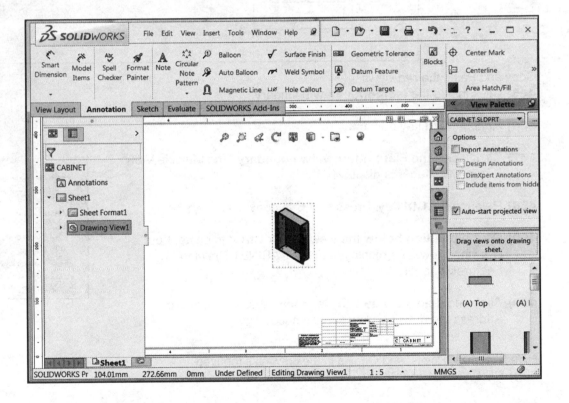

Insert the Default Flat pattern view.
463) Click and drag the **Flat pattern view** on the left side of the drawing.

464) Click **OK** ✔ from the Projected View PropertyManager.

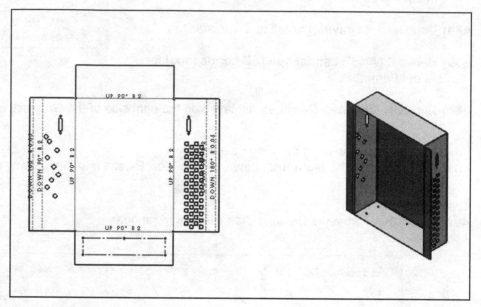

Copy and Paste the two views.
465) Click inside the **Isometric view** boundary.

466) Hold the **Ctrl** key down.

467) Click inside the **Flat pattern view boundary**. The Multiple View PropertyManager is displayed.

468) Release the **Ctrl** key. Press **Ctrl C** to copy.

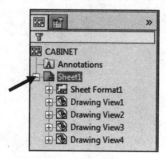

469) Click a **position below** the views. Click **Ctrl V** to paste. Four Drawing Views are displayed in the CABINET Drawing FeatureManager.

470) Click and drag the **views under each** other as illustrated. Address the drawing view scale if needed.

Hide Sketches, Points, and Origins in the drawing if required.
471) Click **View**, **Hide/Show**, un-check **Sketches** from the Menu bar.

472) Click **View**, **Hide/Show**, un-check **Points** from the Menu bar.

473) Click **View**, **Show/Hide**, un-check **Origins** from the Menu bar.

Modify the configurations.

474) Click **inside the lower left Flat pattern view** in Sheet1.

475) Right-click **Properties**.

476) Select **NO-VENT-FLAT-PATTERN** from the Configuration information drop-down menu.

477) Click **OK** from the Drawing View Properties dialog box.

478) Click **inside the lower right Isometric** view.

479) Right-click **Properties**.

480) Select **NO-VENT "SUPPRESS VENT CUTS"** from the Configuration information drop down menu.

481) Click **OK** from the Drawing View Properties dialog box.

Hide Origins, Planes, and Sketches in drawing views. Insert centerlines to represent sheet metal bend lines.

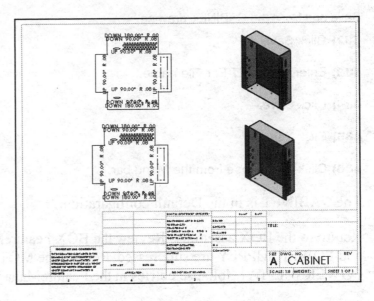

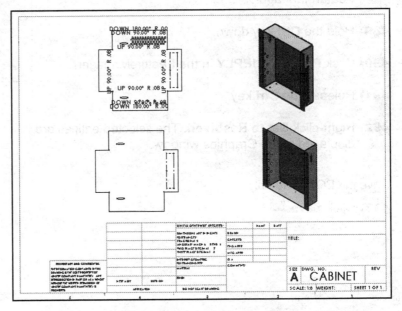

Save and close the CABINET drawing.

482) Click **Save** 🖫.

483) Enter **CABINET** For File name.

484) Click **Save**.

485) Click **Yes**.

486) Click **File**, **Close** from the Menu bar.

The CABINET is in the Default configuration.

Return to the BOX assembly. Review the BOX FeatureManager. The MOTHERBOARD and POWERSUPPLY are suppressed. Restore the MOTHERBOARD and POWERSUPPLY. Collapse entries in the FeatureManager.

Return to the BOX assembly.
487) Open the **BOX** assembly.

Display the components.
488) Click **MOTHERBOARD** in the BOX FeatureManager.

489) Hold the **Ctrl** key down.

490) Click **POWERSUPPLY** in the FeatureManager.

491) Release the **Ctrl** key.

492) Right-click **Set to Resolved**. The selected entities are displayed in the Graphics window.

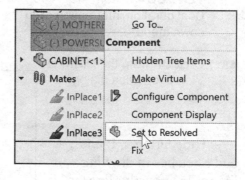

Save the BOX assembly.
493) Click **Save** 🖫.

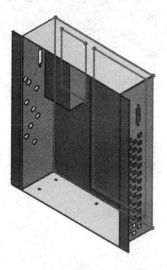

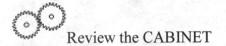 Work in SOLIDWORKS from the design through the manufacturing process to avoid data conversion issues and decreased shop time. The SOLIDWORKS Manufacturing Partner Network (www.SOLIDWORKS.com) lists Sheet Metal manufacturers that utilize SOLIDWORKS.

🔍 Additional details on Sheet Metal, Rip, Insert Bends, Flat Pattern, Forming Tools, Edge Flange, Hem, Library Feature, Linear Step and Repeat and Configuration Properties are available in SOLIDWORKS Help.

⚙️ Review the CABINET

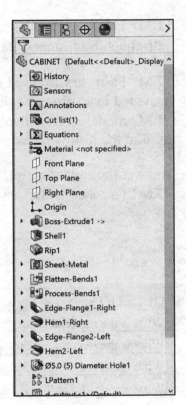

You created the CABINET component inside the BOX assembly. The InPlace Mate inserted a Coincident Mate between the Front Plane of the CABINET and the Front Plane of the BOX. The CABINET component referenced the outside profile of the Layout Sketch.

The first feature of the CABINET was the Extruded Boss/Base (Boss-Extrude1) feature that created a solid box. You utilized the Shell feature to create the constant wall thickness.

The Rip feature cut removed the solid edges. The Insert Bends feature created a sheet metal part from a solid part. The Flat Pattern feature displayed the sheet metal part in the manufactured 2D flat state.

The Edge Flange and Hem features were added to the right and left side of the CABINET. Sheet metal dies and form louvers were inserted from the Library Features.

The Sketch Linear Pattern was utilized to replace the form louvers with more cost effective cuts. You created a configuration to represent the 2D flat state for the CABINET drawing. The 3D formed state was utilized in the BOX assembly.

⚬ The Vent feature located in the Sheet Metal toolbar allows you to create a vent by selecting a closed profile as the outer vent boundary.

⚬ The Fill Pattern feature allows you to fill a closed boundary to create a grid for a sheet-metal perforation - style pattern.

PEM® Fasteners and IGES Components

The BOX assembly contains an additional support BRACKET that is fastened to the CABINET. To accommodate the new BRACKET, the Linear Pattern of five holes is added to the CABINET.

How do you fasten the BRACKET to the CABINET?

Answer: Utilize efficient and cost effective self-clinching fasteners. Example: PEM® Fasteners.

PEM® Fastening Systems, Danboro, PA, USA manufactures self-clinching fasteners called PEM® Fasteners.

PEM® Fasteners are used in numerous applications and are provided in a variety of shapes and sizes. IGES standard PEM® models are available at (www.pennfast.com) and are used in the following example.

Answer the following questions in order to select the correct PEM® Fastener for the assembly:

QUESTION:	ANSWER:
What is the material type?	Aluminum
What is the material thickness of the CABINET?	1mm
What is the material thickness of the BRACKET?	19mm
What standard size PEM® Fastener is available?	FHA-M5-30ZI
What is the hole diameter required for the CABINET?	5mm ± 0.08
What is the hole diameter required for the BRACKET?	5.6mm maximum
Total Thickness = CABINET Thickness + BRACKET Thickness + NUT Thickness Total Thickness = 1mm + 19mm + 5mm = 25mm (minimum)	25mm (minimum)

The selected PEM® Fastener for the project is the FHA-M5-30ZI.

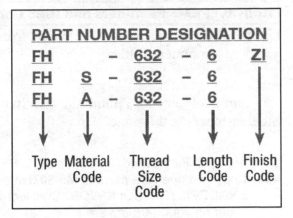

- **FHA** designates the stud type and material, Flush-head Aluminum.

- **M5** designates the thread size.

- **30** designates length code: 30mm.

- **ZI** designates finish code.

Part Number General Designation for PEM®

The overall dimensions of the FH PEM® Fasteners series are the same for various materials. Do not utilize PEM® Fasteners as a feature in the Flat State of a component. In the sheet metal manufacturing process, PEM® Fasteners are added after the material pattern has been fabricated.

Proper PEM® Fasteners material selection is critical. There are numerous grades of aluminum and stainless steel. Test the installation of the material fastener before final specification.

Note: When the fastener is pressed into ductile host metal, it displaces the host material around the mounting hole, causing metal to cold flow into a designed annular recess in the shank.

Ribs prevent the fastener from rotating in the host material once it has been properly inserted. The fasteners become a permanent part of the panel in which they are installed.

During the manufacturing process, a stud is installed by placing the fastener in a punched or drilled hole and squeezing the stud into place with a press. The squeezing action embeds the head of the stud flush into the sheet metal.

Obtain additional manufacturing information that directly affects the design.

Manufacturing Information:	FH-M4ZI:	FH-M5ZI:
Hole Size in sheet metal (CABINET).	4.0mm ± 0.8	5.0mm ± 0.8
Maximum hole in attached parts (BRACKET).	4.6mm	5.6mm
Centerline to edge minimum distance (from center point of hole to edge of CABINET).	7.2mm	7.2mm

---

Activity: PEM® Fasteners and IGES Components

Use the PEM Fasteners for this project from the
Vendor Component folder.

🔆 Part geometry for Aluminum and Stainless
Steel fasteners is the same.

Copy and open the needed files from the COMPONENTS folder.

494) Copy and open the files **FH-M5-30ZI** and **FH-M4-25ZI** to the
ENGDESIGN-W-SOLIDWORKS\Vendor Component
folder on your hard drive.

495) Open the **FH-M5-30ZI** part from the ENGDESIGN-W-
SOLIDWORKS\Vendor Component folder. Click **No**. The
model has imported geometry.

Insert the FH-M5-30ZI fastener into the BOX assembly.

496) Open the **BOX** assembly.

497) Click **Insert Components** 📦 from the Menu bar. The
Insert Component PropertyManager is displayed.

498) Double-click **FH-M5-30ZI** in the Open documents box.

499) Click inside the **Graphics window** in the inside bottom
left corner near the seed feature of the Linear Pattern of
holes.

500) Zoom to Area 🔍 on the hole and fastener.

Rotate the FH-M5-30ZI component.

501) Click **Rotate Component** 🔄 from the Assembly
toolbar. The Rotate Component PropertyManager is
displayed.

502) Rotate the **FH-M5-30ZI** component until the head faces
downward as illustrated.

503) Click **OK** ✔ from the Rotate Component
PropertyManager.

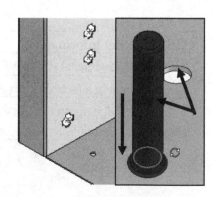

Insert a Concentric mate using the Quick mate procedure.

504) Hold the **Ctrl** key down.

505) Click the **cylindrical shaft face** of the FH-M5-30ZI fastener.

506) Click the **cylindrical face** of the back left hole.

507) Release the **Ctrl** key. The Mate Pop-up menu is displayed.

508) Click **Concentric** from the Pop-up menu. A Concentric mate is applied.

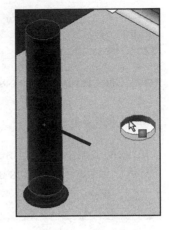

Insert a Coincident mate using the Quick mate procedure.

509) Rotate the **component** and click the flat **head face** of the FH-M5-30ZI fastener as illustrated.

510) Rotate **CABINET** to view the outside bottom face.

511) Hold the **Ctrl** key down.

512) Click the **outside bottom face** of the CABINET.

513) Release the **Ctrl** key. The Mate Pop-up menu is displayed.

514) Click **Coincident** ⌐ from the Pop-up menu. A Coincident mate is applied.

Insert a Parallel mate.

515) Click **Mate** ⊗ from the Assembly toolbar.

516) **Expand** BOX from the fly-out FeatureManager.

517) Click the **Front Plane** of the FH-M5-30ZI fastener.

518) Click the **Right Plane** of the CABINET. The selected planes are displayed in the Mate Selections box.

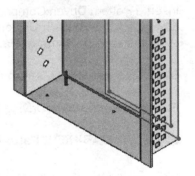

519) Click **Parallel**.

520) Click **OK** ✔ from the Mate PropertyManager.

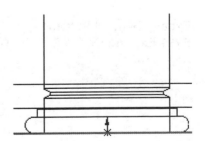

Fit to the Graphics window. Display an Isometric view. Save the
model.
521) Press the **f** key.

522) Click **Isometric view**.

523) Click **Save** 💾.

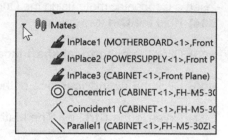

Three Mates fully define the FH-M5-30ZI fastener. The
PEM® fastener creates an interference fit with hole1.

Pattern Driven Component Pattern

A Pattern Driven Component Pattern is a pattern that utilizes an existing component in an
assembly and a driving feature from another component in the same assembly. Utilize the
Pattern Driven Component Pattern tool to create multiple copies of a component in an
assembly. The FH-M5-30ZI references the hole seed feature. The Pattern Driven
Component Pattern displays the five FH-M5-30ZI fasteners based on the CABINET
LPattern1 feature.

Activity: Pattern Driven Component Pattern

Insert a Pattern Driven Component Pattern.
524) Click the **Pattern Driven Component Pattern** tool. Click inside
the **Components to Pattern** box.

525) Click **FH-M5-30ZI** from the fly-out FeatureManager.

526) Click **inside** the Driving Feature or Component box.

527) Click **CABINET\LPattern1** from the BOX fly-out FeatureManager.

528) Click **OK** ✔ from the Pattern Driven PropertyManager.
DerivedLPattern1 is displayed in the FeatureManager.

Expand DerivedLPattern1.
529) **Expand** DerivedLPattern1 in the FeatureManager. View the
results.

Display an Isometric view. Save the BOX.
530) Click **Isometric view** from the Heads-up View toolbar.

531) Click **Save** 🖫.

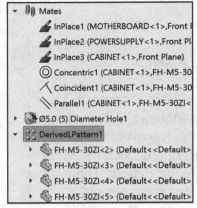

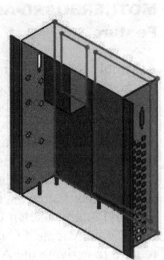

The additional FH-M5-30ZI instances are located under DerivedLPattern1 in the FeatureManager. The purchasing manager for your company determines that the FH-M5-30ZI fastener has a longer lead-time than the FH-M4-25ZI fastener.

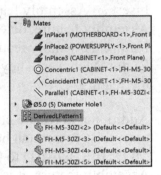

The FH-M4-25 fastener is in stock. Time is critical. You ask the engineer creating the corresponding BRACKET if the FH-M4-25ZI fastener is a reliable substitute.

You place a phone call. You get voice mail. You send email. You wait for the engineer's response and create the next assembly feature.

💡 The common PEM® fastener library is available in SOLIDWORKS Toolbox.

How do you locate a hole for a fastener that has to go through multiple components in an assembly? Answer: Utilize an Assembly Hole feature.

MOTHERBOARD-Assembly Hole Feature

Assembly features are holes and cuts created in the assembly. Utilize the Assembly Feature option from the Menu bar or from the Assembly tab after components have been assembled. The Assembly Feature option includes Hole Series, Hole Wizard, Simple Hole, Revolved Cut, Extruded Cut, Swept Cut, Fillet, Chamfer, Weld Bead, Linear Pattern, Circular Pattern, Table Driven Pattern, Sketch Driven Pattern and Belt/Chain. Create a Cut or Hole Assembly feature to activate the Assembly Feature Pattern tool.

Insert an Assembly feature into the BOX assembly. When creating an Assembly feature, determine the components to be affected by the feature. Assembly features are listed at the bottom of the FeatureManager. They are not displayed in the components affected by the feature.

Additional holes are created through multiple components in the BOX assembly. A through hole is required from the front of the MOTHERBOARD to the back of the CABINET. Since the hole is an Assembly feature, it is not displayed in the CABINET part.

Activity: MOTHERBOARD-Assembly Hole Feature

Insert an Assembly feature with the Hole Wizard.

532) Click **Hole Wizard** 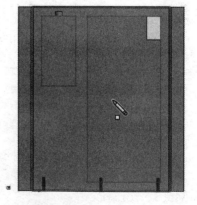 from the Assembly toolbar. The Hole Specification PropertyManager is displayed.

533) Click the **Hole** icon type.

534) Select **ANSI Metric** for Standard.

535) Select **Drill sizes** for Type. Select **Ø5.0** for Size.

536) Click **Through All** for End Condition.

537) Click the **Positions** tab. Click the **front face** of the MOTHERBOARD. The Point Sketch tool is displayed. Click **again** to position the center point of the hole.

538) Right-click **Select** to deselect the Sketch Point tool in the Graphics window.

539) Click **Front view**.

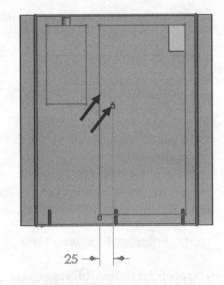

Dimension the hole on the MOTHERBOARD.
540) Click **Smart Dimension** from the Sketch toolbar.

Add a horizontal dimension.
541) Click the **left edge** of the MOTHERBOARD. Click the **centerpoint** of the hole. Click a **position** below the profile. Enter **25**mm.

Add a vertical dimension.
542) Click the **Origin** of the CABINET from the fly-out FeatureManager.

543) Click the **centerpoint** of the hole.

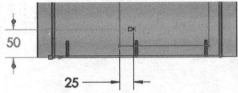

544) Click a **position** to the left of the profile. Enter **50**mm.

545) Click **OK** from the Dimension PropertyManager.

546) Click **OK** from the Hole Position PropertyManager. The hole is added to the FeatureManager.

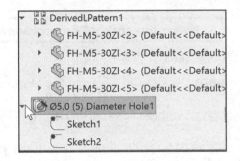

Display an Isometric view. Save the BOX.
547) Click **Isometric view**.

548) Click **Save** .

Close all parts and assemblies.
549) Click **Window**, **Close All** from the Menu bar.

Assembly FeatureManager and External References

The FeatureManager contains numerous entries in an assembly. Understanding the organization of the components and their Mates is critical in creating an assembly without errors. Errors occur within features such as a radius that is too large to create a Fillet feature. When an error occurs in the assembly, the feature or component is labeled in red.

Mate errors occur when you select conflicting geometry such as a Coincident Mate and a Distance Mate with the same faces.

In the initial mating, undo a mate that causes an error. Mate errors occur later on in the design process when you suppress required components and features. Plan ahead to avoid a problem.

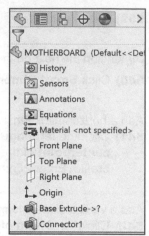

Use reference planes for Mates that will not be suppressed.

When a component displays a '->' symbol, an External reference exists between geometry from the assembly or from another component.

When the system does not locate referenced geometry, the part name displays a '->?' symbol to the right of the component name

in the FeatureManager.

Explore External references in the next activity. Open the CABINET without opening the BOX assembly. The '->?' symbol is displayed after the part name.

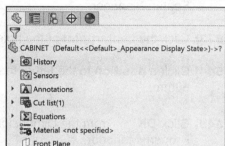

Activity: Assembly FeatureManager and External References

Open the CABINET part.
550) Click **Open** 📂 from the Menu bar.

551) Double-click **CABINET**. The CABINET FeatureManager is displayed.

The CABINET cannot locate the referenced geometry BOX. The "->?" is displayed to the right of the CABINET name and Extrude1 in the FeatureManager.

Open the BOX assembly.
552) Open the BOX assembly. The BOX FeatureManager is displayed.

Reload the CABINET.
553) Right-click **CABINET** in the FeatureManager.

554) Click the **More arrow** ⌄ at the bottom of the Pop-up menu.

555) Click **Reload**.

556) Click **OK** from the Reload dialog box.

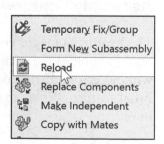

557) Return to the CABINET assembly.

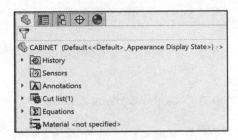

Save the CABINET.

558) Click **Save** 💾 .

The CABINET references the BOX assembly. The External references are resolved and the question marks are removed.

Replace Components

The Replace Components option replaces a component in the current assembly with another component. Replace a part with a new part or assembly or an assembly with a part. If multiple instances of a component exist, select the instances to replace.

The BRACKET engineer contacts you. You proceed to replace the FH-M5-30ZI fastener with the FH-M4-25ZI fastener. Use the Replace Components option to exchange the FH-M5-30ZI fastener with FH-M4-25ZI fastener in the BOX assembly.

Edit the Mates and select new Mate references. The Derived Component Pattern updates with the new fastener. Edit the CABINET and change the hole size of the seed feature from Ø5.0mm to Ø4.0mm. The CABINET Linear Pattern updates with the new hole.

Before you replace Component-A in an assembly with Component-B, close Component-B. The Replace Component option produces an error message with open components. Example: Close the FH-M4-25ZI part before replacing the FH-M5-30ZI part in the current assembly.

Activity: Replace Components

Replace the FH-M5-30ZI component with the FH-M4-25ZI component.

559) Open the **BOX** assembly.

560) Right-click **FH-M5-30ZI**

▸ 🔩 FH-M5-30ZI<1> from the FeatureManager.

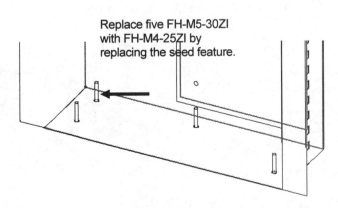

Replace five FH-M5-30ZI with FH-M4-25ZI by replacing the seed feature.

561) Click the **More arrow** ⹉ at the bottom of the Pop-up menu.

562) Click **Replace Components** 🐝. The Replace PropertyManager is displayed.

563) Click the **Browse** button.

564) Double-click **FH-M4-25ZI** from the folder on your hard drive. ENGDESIGN-W-SOLIDWORKS\Vendor Components folder.

565) Click **OK** ✔ from the Replace PropertyManager. The What's Wrong box is displayed.

566) Click **Close** from the What's Wrong dialog box.

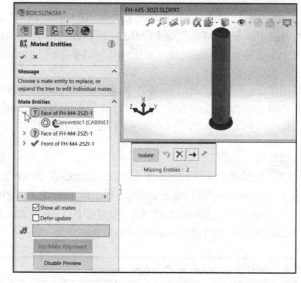

The five FH-M5-30ZI fasteners are replaced with the five FH-M4-25ZI fasteners. Mate Entity errors are displayed.

A red X is displayed on the Mated Entities PropertyManager. Redefine the Concentric Mate and Coincident Mate. Select faces from the FH-M4-25ZI component.

Edit the Concentric Mate.
567) **Expand** the first Mate entity.
⬚ ⍰ Face of FH-M4-25ZI-1

568) Double-click **Concentric1**.

569) Click the **FH-M4-25ZI cylindrical face** as illustrated in the Graphics window. View the results in the Mate Entities box.

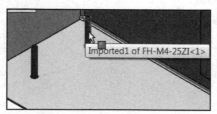

Edit the Coincident mate.
570) Expand the second Mate entity.

571) Click **Bottom view**.

572) Double-click **Coincident1**.

573) Click the **FH-M4-25 bottom face**. View the results in the Mate Entities box.

574) Click **OK** ✔ from the Mate Entities PropertyManager.

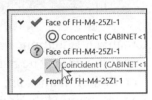

Edit the Ø5.0mm hole to a Ø4.0mm hole.
575) Click **CABINET** from the FeatureManager.

576) Click **Edit Component** from the Assembly toolbar. The CABINET entries are displayed in blue.

577) Expand CABINET in the FeatureManager.

578) Right-click **(5) Diameter Hole1** from the FeatureManager.

579) Click **Edit Feature**. The Hole Specification PropertyManager is displayed. Select **Ø4.0** for Size.

580) Click **OK** ✔ from the Hole Specification PropertyManager. The (4) Diameter Hole1 is displayed in the FeatureManager.

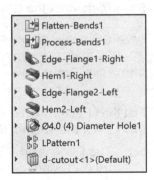

581) Click **OK**.

582) Click **Edit Component** from the Assembly toolbar to return to the BOX assembly. The CABINET entries are displayed in black.

Save the BOX.
583) Click **Save**. Click **Save All**.

The hole of the CABINET requires an Ø4mm ± 0.8. Add a limit dimension in the drawing and a revision note to document the change from an Ø5.0mm to an Ø4.0mm.

IGES imported geometry requires you to select new mate references during Replace Components. Utilize reference planes for Mate entities to save time. The Replace Components option attempts to replace all Mate entities. Reference planes provide the best security in replacing the appropriate Mate entity during the Replace Components option.

Equations

How do you ensure that the fasteners remain symmetrical with the bottom of the BOX when the box-width dimension varies?

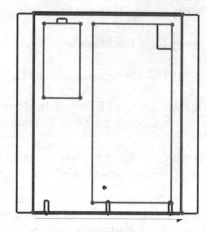

Answer: With an Equation. Equations use shared names to control dimensions. Use Equations to connect values from sketches, features, patterns and various parts in an assembly.

Create an Equation. Each dimension has a unique variable name. The names are used as Equation variables. The default names are based on the Sketch, Feature or Part. Rename default names for clarity.

Increase BOX width.

Fasteners are not symmetrical with the bottom of the BOX.

Resolve all Mate Errors before creating Equations. Mate Errors produce unpredictable results with parameters driven by equations.

Activity: Create an Equation

Display Feature dimensions.
584) Right-click the CABINET **Annotations** folder.

585) Check the **Show Feature Dimensions** box.

586) Click **Front view**.

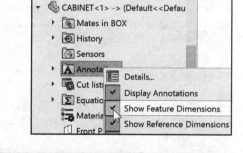

Edit the LPattern1 dimension name.
587) Click the **125**mm dimension in the Graphics window.

The Dimension PropertyManager is displayed.

Rename the D3@LPattern1 dimension.
588) Click inside the **D3@LPattern1** box.

589) Enter **lpattern-bottom** for Name.

590) Click **OK** ✔ from the Dimension PropertyManager.

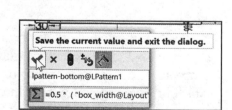

Edit the 4.0mm hole dimension name.
591) Rotate to display the back hole.

592) Click the **25**mm dimension. 25 is the Ø4.0mm hole horizontal dimension in the lower left corner. The Dimension PropertyManager is displayed.

Rename the Sketch dimension name.
593) Enter **hole-bottom** for Name.

594) Click **OK** ✔ from the Dimension PropertyManager.

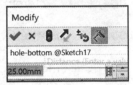

Display an Isometric view.
595) Click **Isometric view**.

Display the BOX feature dimensions.
596) Right-click the BOX **Annotations** folder.

597) Check the **Show Feature Dimensions** box.

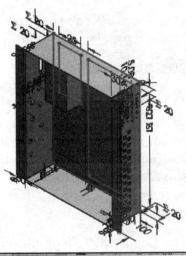

Create an Equation.
598) Click **Front view**.

599) Click **Tools**, **Equations** Σ Equations... from the Menu bar. The Equations, Global Variables, and Dimensions dialog box is displayed.

Create the first half of equation1.
600) Click a **position** in the fifth cell under Equations - Top Level.

601) Click the lpattern-bottom @Pattern1 horizontal dimension, **125**mm in the Graphics window. The variable "lpattern-bottom @LPattern1@ CABINET" is added.

Create the second half of the Equation.

602) Enter **0.5*(** from the keypad.

603) Click the box_width horizontal dimension, **300** from the Graphics window. The variable "box_width@layout" is added to the equation text box.

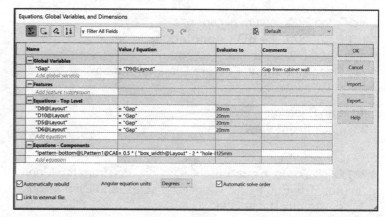

Equations - Top Level	
"D8@Layout"	= "Gap"
"D10@Layout"	= "Gap"
"D5@Layout"	= "Gap"
"D6@Layout"	= "Gap"
Add equation	
Equations - Components	=0.5
"lpattern-bottom@LPattern1@C	=0.5*(
Add equation	

604) Enter **–2*** from the keypad.

605) Click **Bottom view**.

606) Click the M4 hole-bottom dimension, **25**. The variable "hole-bottom@Sketch12@CABINET.Part" is added to the equation text box.

607) Enter **)** from the keypad.

608) Press the **Tab** key. View the results.

Click **OK** from the dialog box.

The Equations dialog box contains the complete equation. A green check mark indicates that the Equation is solved. The Equation evaluates to 125mm.

The Equation Σ icon placed in front of the 125 dimension indicates that an equation drives the value.

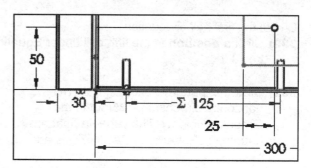

Remember the initial design parameters. There are three different size boxes. Remember the design intent. You build a Layout Sketch to control the different sizes.

Test the equation by modifying the Layout Sketch dimensions.

Verify the Equation. Modify the dimensions.
609) Click **Front view**.

610) Double-click the horizontal dimension box-width, **300**mm.

611) Enter **400**mm.

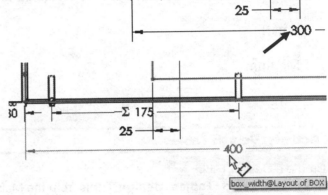

612) Click **Rebuild** 🛑. The Equation dimension displays 175.

613) Double-click the dimension box-height, **400**mm.

614) Enter **500**mm. Click **Rebuild** 🛑.

Return to the original dimensions.
615) Double-click on the horizontal dimension, **400**mm.

616) Enter **300**mm.

617) Double-click the vertical dimension, **500**mm.

618) Enter **400**mm.

619) Click **Rebuild** 🛑.

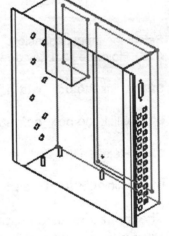

Hide the dimensions.
620) Right-click the CABINET **Annotations** folder, and un-check the **Show Feature Dimension** box.

621) Right-click the BOX **Annotations** folder, and uncheck the **Show Feature Dimension** box. View the Equations folder.

Display an Isometric view. Save the BOX.
622) Click **Isometric view**.

623) Click **Save** 💾.

The BOX assembly requires three configurations. Recall the CABINET drawing. You had two configurations that controlled the Flat Pattern feature state.

Now control dimensions, features and parts in an assembly. A Design Table is an efficient tool to control variations in assemblies and parts.

Design Tables

A Design Table is an Excel spreadsheet used to create multiple configurations in a part or assembly. Utilize the Design Table to create three configurations of the BOX assembly:

- Small.

- Medium.

- Large.

Activity: Design Tables

Create a Design Table.
624) Click **Insert**, **Tables**, **Design Table** from the Main menu. The Design Table PropertyManager is displayed. Accept the Auto-create default setting.

625) Click **OK** ✔ from the Design Table PropertyManager.

Select the input dimension.
626) Hold the **Ctrl** key down.

627) Click the **box_width@Layout** dimension.

628) Click the **box_height@Layout** dimension.

629) Release the **Ctrl** key.

630) Click **OK** from the Dimension dialog box.

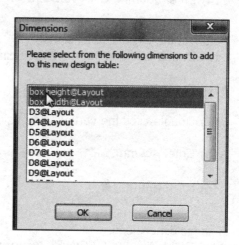

The input dimension names and default values are automatically entered into the Design Table.

The value Default is entered in Cell A3. The value 400 is entered in Cell B3 for box_height and the value 300 is entered in Cell C3 for box_width.

Enter the three configuration names.
631) Click Cell **A4**. Enter **Small**.

632) Click Cell **A5**. Enter **Medium**.

633) Click Cell **A6**. Enter **Large**.

Enter the dimension values for the Small configuration.
634) Click Cell **B4**. Enter **400** for box_height.

635) Click Cell **C4**. Enter **300** for box_width.

Enter the dimension values for the Medium configuration.
636) Click Cell **B5**. Enter **600**. Click Cell **C5**. Enter **400**.

Enter the dimension values for the Large configuration.
637) Click Cell **B6**. Enter **650**.

638) Click Cell **C6**. Enter **500**.

Build the three configurations.
639) Click a **position** outside the Excel Design Table in the Graphics window.

640) Click **OK** to generate the three configurations.

Display the configurations.
641) Click the **ConfigurationManager** 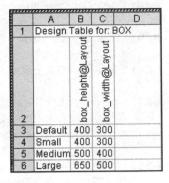 tab. The Design Table icon is displayed.

642) Double-click **Small**. Double-click **Medium**.

643) Double-click **Large**.

Return to the Default configuration.
644) Double-click **Default**.

Fit the model to the Graphics window.
645) Press the **f** key.

Edit to the Design Table.
646) Right-click **Design Table** from the BOX ConfigurationManager.

647) Click **Edit Table**.

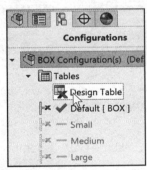

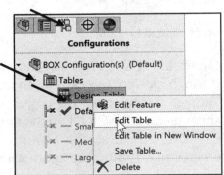

648) Click **OK** from the Add Rows and Columns dialog box.

649) Click Cell **D2**. Enter **$State@POWERSUPPLY<1>**.

650) Click Cell **D3**. Enter **R** for Resolved.

651) Copy Cell **D3**. Select Cell **D4** and Cell **D5**. Click **Paste**. Click Cell **D6**. Enter **S** for Suppressed. Note: If needed remove any hyperlinks.

652) Update the configurations. Click a **position** outside the EXCEL Design Table.

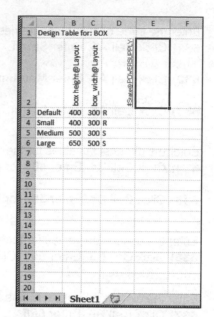

	A	B	C	D	E	F
1	Design Table for: BOX					
2		box height@Layout	box_width@Layout	$State@POWERSUPPLY		
3	Default	400	300	R		
4	Small	400	300	R		
5	Medium	500	300	S		
6	Large	650	500	S		

The Design Table component variables must match the FeatureManager names in order to create configurations. The Design Table variable $State@POWERSUPPLY<1> is exactly as displayed in the BOX FeatureManager. If the name in the BOX FeatureManager was P-SUPPLY<2> then the Design Table variable is $State@ P-SUPPLY<2>.

Display the Large Configuration.
653) Double-click **Large** from the ConfigurationManager. The POWERSUPPLY is suppressed in the Large configuration.

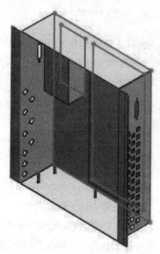

Return to the Default configuration and FeatureManager.
654) Double-click **Default**.

655) Click the **FeatureManager** tab.

Fit the model to the Graphics window.
656) Press the **f** key.

Save the BOX assembly.
657) Click **Save**.

658) Click **Save All**.

Dimensions, Color, Configurations, Custom Properties, and other variables are controlled through Design Tables. The engineer working on the sheet metal bracket had to visit another customer. As the project manager, it is up to you to get the job done. Create the sheet metal BRACKET in the next activity. The sheet metal BRACKET is created In-Context to illustrate the Top-down Assembly Modeling process. Sheet metal parts are created in a Bottom-up approach and then inserted and mated in the assembly.

BRACKET Part-Sheet Metal Features

The CABINET was converted from a solid part to a sheet metal part by using the Rip and Insert Bends feature. The sheet metal BRACKET part starts with the Base Flange feature. The Base Flange feature utilizes a U-shaped sketch profile. The material thickness and bend radius are set in the Base Flange.

Create the BRACKET In-Context of the BOX assembly. The BRACKET references the inside bottom edge and left corner of the CABINET.

⚡ Orient the Flat Pattern. The first sketched edge remains fixed with the Flatten tool.

Sketch the first line from left to right Coincident with the inside bottom edge. The BRACKET is not symmetrical and requires a right hand and left hand version. Utilize Mirror Component feature to create a right-hand copy of the BRACKET.

Activity: BRACKET Part-Sheet Metal Features

Insert the BRACKET component.
659) If required, open the **BOX** assembly.

660) Click **New Part** for the Consolidated Insert Components drop-down menu.

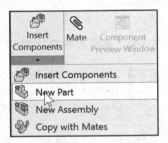

661) Click **OK** from the New SOLIDWORKS Document dialog box. A new virtual default component is added to the FeatureManager. Note: Save the new component and name it BRACKET later in the procedure.

The Component Pointer ✔ is displayed on the mouse pointer.

662) Click **Front Plane** from the BOX FeatureManager.

663) Expand the new inserted component [Part1^BOX]<1> in the FeatureManager. The FeatureManager is displayed in light blue.

Sketch the profile for the Base Flange.
664) Click **Front view**.

665) Check **View**, **Hide/Show**, **Temporary Axis** from the Menu bar.

666) Click **Hidden Lines Visible** from the Heads-up View toolbar.

667) **Zoom to Area** on the front lower inside edge of the BOX as illustrated. Click **Line** from the Sketch toolbar.

668) Sketch a **horizontal line** Collinear with the inside edge of the CABINET. Note: Sketch the horizontal line from left to right.

669) Sketch a **vertical line** and a **horizontal line** to complete the U-shaped profile.

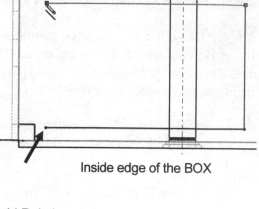

Add an Equal relation.

670) Right-click **Select**. Click the **top horizontal** line. Hold the **Ctrl** key down.

671) Click the **Bottom horizontal** line.

Inside edge of the BOX

672) Release the **Ctrl** key. Click **Equal** = from the Add Relations box.

673) Click **OK** ✔ from the PropertyManager.

The bottom horizontal line is displayed in black. The bottom line is Collinear with the inside bottom edge. Add a Collinear relation between the bottom line and the inside bottom edge if required.

Add dimensions.

674) Click **Smart Dimension** ✎ from the Sketch toolbar.

675) Click the **top horizontal line**. Click the **Top** of the fastener. Click a **position** to the right of the fastener. Enter **2mm**.

676) Click the **centerline** of the fastener. Click the **vertical line**. Click a **position** below the fastener.

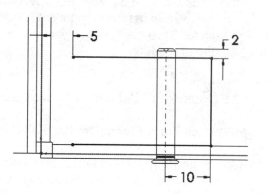

677) Enter **10**mm. Click the **left edge** of the CABINET.

678) Click the **left end point** of the top horizontal line.

679) Click a **position** above the profile.

680) Enter **5mm**.

Insert the Base Flange feature.

681) Click **Isometric view**.

682) Click **Base Flange/Tab** ⩗ from the Sheet Metal toolbar. The Base Flange PropertyManager is displayed.

683) Select **Up to Vertex** for End Condition in Direction 1.

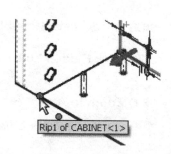

Rip1 of CABINET<1>

684) Click the CABINET **front left vertex** as illustrated.

685) Enter **1**mm for Thickness. If required, click the Reverse direction box. Enter **2**mm for Bend Radius.

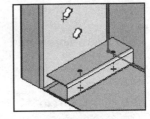

686) Enter **.45** for K factor. If needed, check the Reverse direction box. Accept the default settings.

687) Click **OK** ✔ from the Base Flange PropertyManager.

688) Click **Front view**. Note the location of the BRACKET.

The PEM® fasteners protrude through the BRACKET in the BOX assembly.

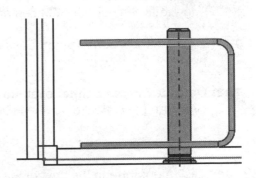

Save the BOX assembly.

689) Click **Edit Component** 🔧 from the Assembly toolbar.

690) Click **Save** 💾. Click **Save All**.

Rename the default component [Part1^BOX]<1> to BRACKET.

691) Right-click **[Part1^BOX]<1>** from the FeatureManager. Click **Open Part**. Click **Save As** from the Menu bar. Enter **BRACKET**. Click **Save**. The BRACKET FeatureManager is displayed. **Close** the BRACKET part.

692) Return to **BOX** assembly.

The BRACKET part contains no holes. The BRACKET requires two holes that reference the location of the fasteners. Utilize In-Context Extruded Cut feature to locate the holes. Display Wireframe mode to view the fastener position. Allow for a 0.8mm clearance.

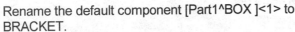
Activity: BRACKET Part-In-context Features

Insert the Extruded Cut feature In-Context of the BOX assembly.
693) Click **BRACKET** from the FeatureManager.

694) Right-click **Edit Part**. The Edit Component icon is activated and the BRACKET entry in the FeatureManager is displayed in blue.

695) Press the **f** key. **Rotate** to view the top face of the BRACKET.

696) Right-click the **top face** of the BRACKET. Click **Sketch** ✏️ from the Context toolbar.

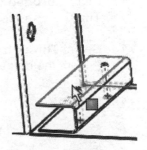

697) Click **Top view**.

698) Click **Wireframe** from the Heads-up View toolbar.

Sketch two circular profiles referencing the PEM center points.
699) Click **Circle** ⊙ from the Sketch toolbar.

700) Drag the **mouse pointer** over the centerpoint of the top PEM circular edge to "wake up" the PEM center point.

701) Click the **PEM centerpoint**. Click a **position** to the right of the center point.

702) Drag the **mouse pointer** over the centerpoint of the bottom PEM circular edge to "wake up" the PEM centerpoint.

703) Click the **PEM center point**.

704) Click a **position** to the right of the centerpoint.

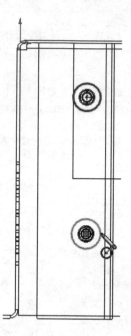

Deselect the Circle Sketch tool and add an Equal relation.
705) Right-click **Select**. Click the **circumference** of the top circle.

706) Hold the **Ctrl** key down. Click the **circumference** of the bottom circle.

707) Release the **Ctrl** key. Click **Equal** from the Add Relations box.

708) Click **OK** ✔ from the Properties PropertyManager.

Add dimensions.
709) Click **Smart Dimension** ⟋ from the Sketch toolbar.

710) Click the **circumference** of the top circle.

711) Click a **position** off the profile.

712) Enter **4.80**mm.

713) Click **Extruded Cut** ▣ from the Features toolbar. The Cut-Extrude PropertyManager is displayed.

714) Select **Through All** for End Condition in Direction1. Accept the default settings.

715) Click **OK** ✔ from the Cut-Extrude PropertyManager.

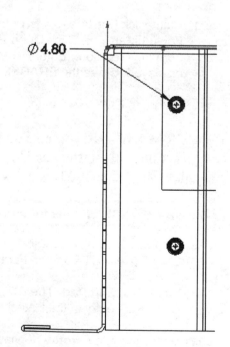

Ø 4.80

Return to the BOX assembly.

716) Click **Edit Component** from the Assembly toolbar.

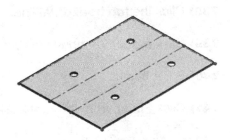

Display an Isometric view - Shaded With Edges.
717) Click **Isometric view**.

718) Click **Shaded With Edges** from the Heads-up View toolbar.

Save the Box assembly and its referenced components.
719) Click **Save** . Click **Save All**. Click **Yes**.

No additional references from the BOX components are required. The Edge Flange/Tab feature, Break Corner feature, and Miter Flange feature are developed in the BRACKET part.

The Edge Flange feature adds a flange on the selected edge. Edit the Flange Profile to create a 30mm tab, 20mm from the front face. The tab requires fillet corners. The Break Corner feature adds fillets or chamfers to sheet metal edges.

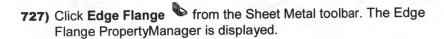

Activity: BRACKET Part-Edge Flange Tab, Break Corner, and Miter Flange Features

Open the BRACKET.
720) Right-click **BRACKET** in the Graphics window.

721) Click **Open Part** from the Context toolbar.

Display the 2D flat state.
722) Right-click **Flat-Pattern1** from the FeatureManager.

723) Click **Unsuppress** from the Context toolbar.

Display the 3D formed state.
724) Right-click **Flat-Pattern1** from the FeatureManager.

725) Click **Suppress** from the Context toolbar.

Insert a tab using the Edge Flange feature.
726) Click the **top left edge** of the BRACKET as illustrated.

727) Click **Edge Flange** from the Sheet Metal toolbar. The Edge Flange PropertyManager is displayed.

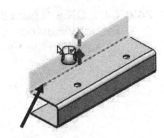

728) Click a **position** above the BRACKET for direction.

729) Enter **20**mm for Flange Length.

730) Click the **Edit Flange Profile** button. Do not click the Finish button at this time. Move the Profile Sketch dialog box off to the side of the Graphics window.

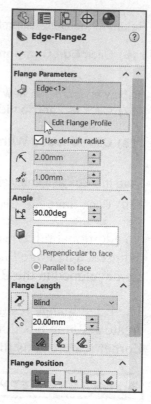

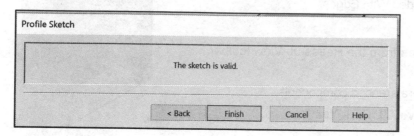

731) Drag the **front edge** of the tab towards the front hole as illustrated.

732) Drag the **back edge** of the tab towards the front hole as illustrated.

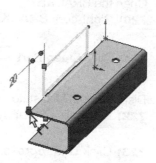

Add dimensions.

733) Click **Smart Dimension** from the Sketch toolbar.

734) Click the **front left vertex** of Base-Flange1.

735) Click the **front vertical** line as illustrated.

736) Click a **position** above the profile.

737) Enter **20**mm.

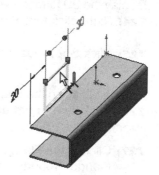

738) Click the **top horizontal** line.

739) Click a **position** above the profile.

740) Enter **30**mm.

741) Click **Finish** from the Profile Sketch box.

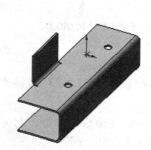

742) Click **OK** ✔ from the PropertyManager. Edge-Flange1 is displayed in the FeatureManager.

Insert the Break Corner/Corner Trim Feature.
743) Zoom in on the tab.

744) Click **Break Corner/Corner Trim** 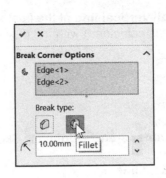 from the Sheet Metal toolbar. The Break Corner PropertyManager is displayed.

745) Click the small **front edge** of the flange as illustrated. Edge<1> is displayed in the Corner Edges box.

746) Click the small **back edge** of the flange. Edge<2> is displayed in the Corner Edges box. Select **Fillet** for Break type.

747) Enter **10**mm for Radius.

748) Click **OK** ✔ from the Break Corner PropertyManager. Break Corner1 is displayed in the FeatureManager.

Save the model.
749) Click **Save** .

The Miter Flange feature inserts a series of flanges to one or more edges of a sheet metal part. A Miter Flange requires a sketch. The sketch can contain lines or arcs and can contain more than one continuous line segment.

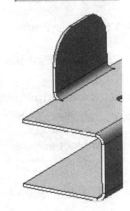

The Sketch plane is normal (perpendicular) to the first edge closest to the selection point. The geometry you create must be physically possible. New flanges cannot cross or deform. The current 2mm bend radius of the BRACKET is too small to insert a Miter Flange feature. Leave the area open for stress relief and utilize a Gap distance of 0.5mm.

Gap for stress relief

Insert the Miter Flange Feature.
750) Click **Miter Flange** from the Sheet Metal toolbar.

751) Click the BRACKET **front inside edge** as illustrated. The new sketch Origin is displayed in the left corner.

752) Zoom to Area on the left corner.

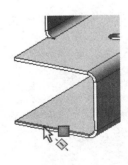

753) Click **Line** ✏ from the Sketch toolbar.

754) Sketch a **small vertical line** at the front inside left corner. The first point is coincident with the new sketch Origin as illustrated.

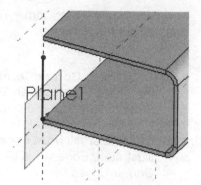

Add dimensions.

755) Click **Smart Dimension** ✎ from the Sketch toolbar.

756) Click the **vertical line**.

757) Click a **position** to the front of the BRACKET.

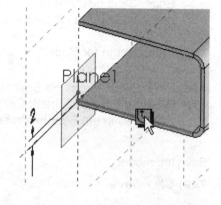

758) Enter **2**mm.

759) Click **Exit Sketch**.

Click the **Propagate** 🔲 icon to select inside tangent edges.

760) Uncheck the **Use default radius** box.

761) Enter **.5**mm for Bend Radius.

762) Check the **Trim side bends** box.

763) Enter **.5**mm for Rip Gap distance.

764) Click **OK** ✔ from the Miter Flange PropertyManager. Miter Flange1 is displayed in the FeatureManager.

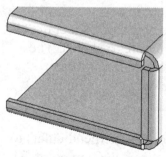

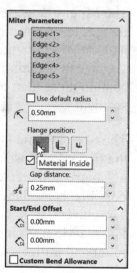

Display the flat and formed states.

765) Click **Isometric view**.

766) Click **Flatten**  from the Sheet Metal toolbar to display the 2D flat state.

767) Click **Flatten** from the Sheet Metal toolbar to display the 3D formed state.

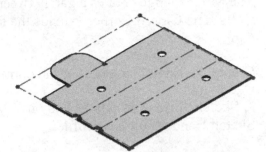

Save the BRACKET.

768) If required, click **View**, **Hide/Show**, un-check **Planes** from the Menu bar.

769) Click **Save** .

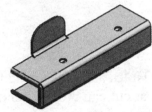

The Trim side bends option creates a Chamfer at the inside corners of the Miter Flange feature. Miter profiles produce various results. Combine lines and arcs for more complex miters.

Covers are very common sheet metal parts. Covers protect equipment and operators from harm.

Utilize the Base Flange, Miter Flange, Close Corner, and Edge Flange to create simple covers. Add holes for fasteners. The Base Flange is a single sketched line that represents the area to be covered.

Insert the Miter Flange feature and select the 4 edges on the top face. Utilize the Material Outside option for Flange position to add the walls to the outside of the Base Flange.

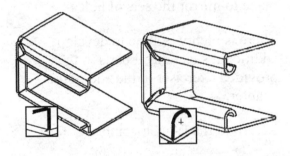

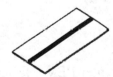

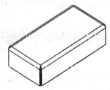

Base Flange
Sketch Line on Right Plane

Miter Flange
Select 4 edges around the top face.

The Miter Flange creates a gap between the perpendicular walls. The Closed Corner extends the faces to eliminate the gap.

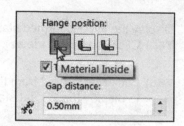

Create the Edge Flanges after the corners are closed.

A Sketched Bend inserts a bend on a flat face or plane. The Sketch Bend requires a profile.

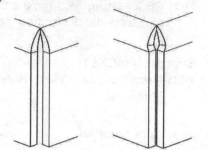

Holes and cuts are added to the cover. A hole or cut can cross a bend line.

The Base Flange utilized the MidPlane option. Utilize the same plane to mirror the sets of holes.

There is plenty of time to develop additional sheet metal parts in the provided exercises at the end of the chapter.

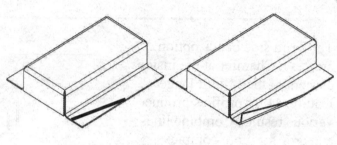

Return to the BOX assembly.

A second BRACKET is required for the right side of the CABINET. Design for the three configurations and reuse geometry.

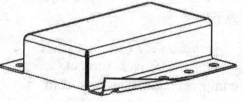

BRACKET-Mirror Component

Insert a reference plane through the center PEM® fastener. The right side BRACKET requires a right hand version of the left side BRACKET. Utilize the Mirror Component option to create the mirrored right side BRACKET.

Activity: BRACKET-Mirror Component

770) Open the **BOX** assembly.

Display Sketches and Temporary Axes.
771) Click **View**, **Hide/Show**, and check **Sketches** from the Menu bar.

772) Click **View**, **Hide/**Show, and check **Temporary Axes** from the Menu bar.

773) **Zoom to Area** on the top circular face of the middle PEM® fastener.

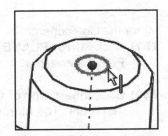

Insert a Reference Point.
774) Click **Insert**, **Reference Geometry**, **Point** from the Menu bar. The Point PropertyManager is displayed.

775) Click the **top circular edge** of the back middle fastener. The Reference Point is located at the center point of the top circle.

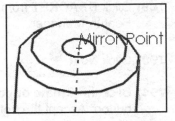

776) Click **OK** ✔ from the Point PropertyManager. Point1 is displayed.

Fit the model to the Graphics window. Rename Point1.
777) Press the **f** key.

778) Rename **Point1** to **Mirror Point**.

Insert a Reference Plane.
779) Click **Insert**, **Reference Geometry**, **Plane** from the Menu bar. The Plane PropertyManager is displayed.

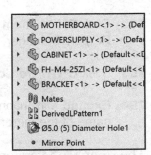

780) Click **Mirror Point** from the fly-out FeatureManager. Mirror Point is displayed in the First Reference box.

781) Click **Right Plane** in the BOX assembly from the fly-out FeatureManager. Right Plane is displayed in the Second Reference box.

782) Click **OK** ✔ from the Plane PropertyManager. PLANE1 is displayed in the FeatureManager.

Fit the model to the Graphics window.
783) Press the **f** key.

Rename PLANE1. Save the assembly.
784) Rename **PLANE1** to **Mirror PLANE**.

785) Click **Save** 💾.

786) Click **Save All**.

Mirror the Component.

787) Click **Mirror PLANE** from the
FeatureManager.

788) Click **Insert**, **Mirror Components** from the
Menu bar. The Mirror PLANE is selected.

789) Click **BRACKET** from the BOX fly-out
FeatureManager.

790) Click **Next** .

791) Click the **Create opposite hand version**
button. Accept the default conditions.

792) Click **OK** ✔ from the Mirror Components
PropertyManager.

793) Click **OK**. MirrorComponent1 is displayed in the FeatureManager.

794) **Expand** MirrorComponent1 in the FeatureManager. View the new
name of the mirror component.

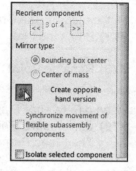

The right hand version of the BRACKET is named
MirrorBRACKET<#>. An interference exists between the
MOTHERBOARD and MirrorBRACKET<#>. Address the
interference in the next section.

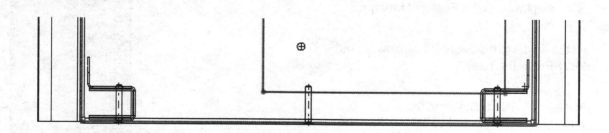

MirrorBRACKET Part-Bends, Unfold and Fold and Jog Features

Open the MirrorBRACKET part. The MirrorBRACKET is not a sheet metal part. The MirrorBRACKET references only the BRACKET's geometry. The MirrorBRACKET must be manufactured in a 2D Flat State. Insert Sheet Metal Bends to create the Flat State.

Control the unfold/fold process. Insert an Extruded Cut across a bend radius. Utilize the Unfold and Fold features to flatten and form two bends. Perform cuts across bends after the Unfold. Utilize unfold/fold steps for additional complex sheet metal components.

Sketch the cut profile across the Bend Radius. The Extruded Cut is linked to the Thickness. A Jog feature creates a small, short bend. Insert a Jog feature to offset the right tab and to complete the MirrorBRACKET.

> **Activity: MirrorBRACKET Part-Bends, Unfold and Fold and Jog Features**

Open MirrorBRACKET and Insert Bends.
795) Right-click **MirrorBRACKET** in the FeatureManager.

796) Click **Open Part** 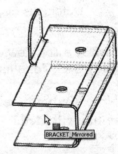 from the Context toolbar. The MirrorBRACKET FeatureManager is displayed.

Fit the Model to the Graphics window.
797) Press the **f** key.

798) Click the **inside bottom face** of MirrorBRACKET.

799) Click **Insert Bends** from the Sheet Metal toolbar. The Bends PropertyManager is displayed. Face<1> is displayed in the Fixed Face box. Accept the default conditions.

800) Click **OK** from the Bends PropertyManager. Process-Bends1 is displayed in the FeatureManager.

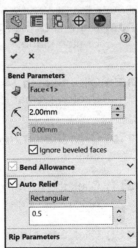

Unfold two Bends.

801) Click **Unfold**  from the Sheet Metal toolbar. The Unfold PropertyManager is displayed.

802) **Zoom to Area** on the right inside bottom face.

803) Click the right **inside bottom face** as illustrated.

804) Rotate the **model** to view the two long bends.

805) Click the two long **bends** to unfold as illustrated. RoundBend5 and RoundBend6 is displayed in the Bends to unfold box.

806) Click **OK** ✔ from the Unfold PropertyManager. Unfold1 is displayed in the PropertyManager.

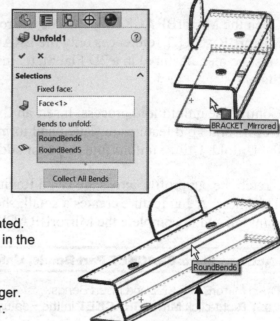

Display a Top view.

807) Click **Top view** from the Heads-up View toolbar.

Note: Only the selected two bends unfold. The tab remains folded.

Create the Sketch for an Extruded Cut.

808) Right-click the **right bottom face**. This is your Sketch plane.

809) Click **Sketch** from the Context toolbar. The Sketch toolbar is displayed.

810) Click **Corner Rectangle** from the Sketch toolbar.

811) Click the **lower right corner** of the Mirror BRACKET.

812) Click a **position** coincident with the left bend line as illustrated.

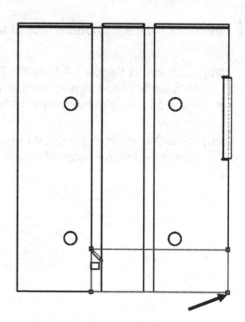

Add a dimension.

813) Click **Smart Dimension** from the Sketch toolbar.

814) Click the **top horizontal edge** of the MirrorBRACKET.

815) Click the **bottom horizontal line** of the rectangle.

816) Enter **15**mm.

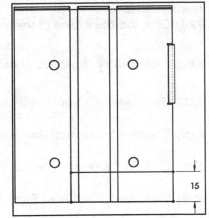

Extrude the sketch.

817) Click **Extruded Cut** from the Sheet Metal toolbar. The Cut-Extrude PropertyManager is displayed.

818) Check the **Link to Thickness** box.

819) Click **OK** from the Cut-Extrude PropertyManager. The Cut-Extrude feature is displayed in the FeatureManager.

820) Click **Wireframe** from the Heads-up View toolbar.

Fold the bends.

821) Click **Fold** from the Sheet Metal toolbar. The Fold PropertyManager is displayed.

822) Click the **Collect All Bends** button. All Unfolded bends are displayed in the Bends to fold box.

823) Click **OK** from the Fold PropertyManager. Fold1 is displayed in the FeatureManager. View the Cut-Extrude1 feature in the Graphics window.

Display Shaded With Edges - Isometric view.
824) Click **Shaded With Edges**.

825) Click **Isometric view**.

Insert a Jog feature.

826) Click the **back face** of the tab as illustrated.

827) Click **Jog** from the Sheet Metal toolbar.

828) Click **Line** from the Sketch toolbar.

829) Sketch a **horizontal line** across the midpoints of the tab.

830) Click **Exit Sketch**. The Jog PropertyManager is displayed.

831) Click inside the **Fixed Face** box.

832) Click the **back face** of the tab below the horizontal line. The direction arrow points to the right.

833) Enter **5**mm for Offset Distance.

834) Click the **Reverse Direction** button. The direction arrow points to the left.

835) Click **OK** from the Jog PropertyManager. Jog1 is displayed in the FeatureManager.

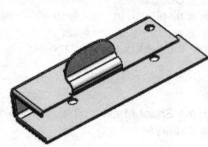

Display the Jog.

836) Use the **arrow keys** to rotate and view the Jog feature.

Display the flat/formed state.

837) Click **Flatten** from the Sheet Metal toolbar to display the flat state.

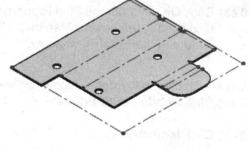

838) Click **Flatten** from the Sheet Metal toolbar to display the formed state.

Display an Isometric view. Save the model.

839) Click **Isometric view**.

840) Click **Save**.

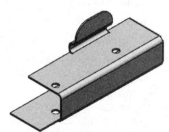

Return to the BOX assembly.
841) Open the **BOX** assembly.

Save the BOX assembly.
842) Click **Save** 💾.

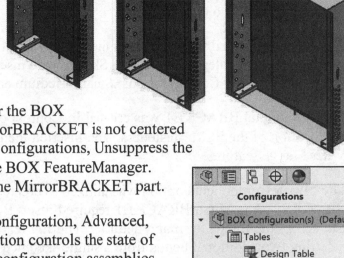

The BRACKET and
MirrorBRACKET remain
centered around the Mirror Plane for the BOX
assembly configurations. If the MirrorBRACKET is not centered
for the Small, Medium, and Large Configurations, Unsuppress the
Mirror Point and Mirror Plane in the BOX FeatureManager.
Un-suppress the Mates created for the MirrorBRACKET part.

The ConfigurationManager, Add Configuration, Advanced,
Suppress new features and mates option controls the state of
new features and mates in multiple configuration assemblies.

The BOX configurations are complete.

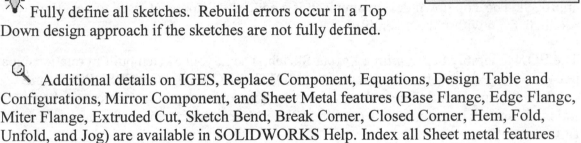

🔆 Fully define all sketches. Rebuild errors occur in a Top
Down design approach if the sketches are not fully defined.

🔍 Additional details on IGES, Replace Component, Equations, Design Table and
Configurations, Mirror Component, and Sheet Metal features (Base Flange, Edge Flange,
Miter Flange, Extruded Cut, Sketch Bend, Break Corner, Closed Corner, Hem, Fold,
Unfold, and Jog) are available in SOLIDWORKS Help. Index all Sheet metal features
under the keyword sheet metal.

⚙️ Review the PEM® fasteners, Configurations and Sheet Metal features

You obtained IGES format PEM® self-clinching fasteners either from the Internet or from
the enclosed CD in this book. IGES files are opened in SOLIDWORKS and saved as part
files. The PEM® fasteners were inserted into the BOX assembly seed hole. You
developed a Component Pattern for the fasteners that referenced the CABINET Linear
Pattern of Holes.

You added an Assembly Cut feature to the BOX assembly. The Cut feature created a hole
through the MOTHERBOARD and CABINET components.

Due to part availability, you utilized the Replace Components feature to exchange the Ø5.0mm fastener with the Ø4.0mm fastener. The new mates were redefined to update the Ø 4.0mm fastener. The Ø 4.0mm fastener required the CABINET holes to be modified from the Ø5.0mm to the Ø4.0mm.

Equations controlled the location of the CABINET holes based on the Layout Sketch dimensions. Variables in the Layout Sketch were inserted into a Design Table to create the BOX assembly Configurations: Small, Medium and Large.

The sheet metal BRACKET was created In-Context of the BOX assembly and referenced the location of the PEM® fasteners. You inserted the Base Flange, Edge Flange, and Miter Flange features.

The Mirror Components option was utilized to create a right hand version of the BRACKET. The MirrorBRACKET required Insert Bends, Fold/Unfold, and an Extruded Cut across the Bends for manufacturing. The Jog feature provided an offset for the tab by inserting two additional bends from a single sketched line.

Project Summary

You created three different BOX sizes utilizing the Top down assembly modeling approach. The Top down design approach is a conceptual approach used to develop products from within the assembly.

The BOX assembly began with a Layout Sketch. The Layout Sketch built in relationships between 2D geometry that represented the MOTHERBOARD, POWERSUPPLY, and CABINET components. These three components were developed In-Context of the Layout Sketch. The components contain InPlace Mates and External references to the BOX assembly.

Additional Extruded features were inserted in the MOTHERBOARD and POWERSUPPLY parts.

The CABINET part developed an Extruded Boss/Base feature In-Context of the BOX assembly. The Shell feature created a constant wall thickness for the solid part.

The Rip feature and Insert Bends feature were added to convert the solid part into a sheet metal part.

Sheet metal components utilized the Edge, Hem, Extruded Cut, Die Cuts, Forms, and Flat Pattern features. Model sheet metal components in their 3D formed state. Manufacture sheet metal components in their 2D flatten state.

The Component Pattern maintained the relationship between the Cabinet holes and the PEM® self-clinching fasteners. You utilized Replace Components to modify the Ø5.0mm fastener to the Ø 4.0mm fastener.

The new mates were redefined to update the Ø4.0mm fastener.

Equations controlled the location of the Cabinet holes. Variables in the Layout Sketch were inserted into a Design Table to create the BOX assembly Configurations: Small, Medium and Large.

The BRACKET was created utilizing sheet metal features: Base Flange, Edge Flange, and Miter Flange. You utilized Mirror Components to create a right hand version: MirrorBRACKET. The MirrorBRACKET required Insert Bends and an Extruded Cut feature through the sheet metal bends.

The Mirror Plane built in the design intent for the BRACKET and MirrorBRACKET to remain centered for each BOX assembly configuration.

Sheet metal parts can be created individually or In-Context of an assembly.

Questions

1. Explain the Top-down assembly approach. When do you create a Layout Sketch?

2. How do you create a new component In-Context of the assembly?

3. What is the difference between a Link Value and an Equation?

4. Name three characteristics unique to Sheet metal parts.

5. Identify the indicator for a part in an Edit Component state.

6. Where should you position the Layout Sketch?

7. For a solid extruded block, what features do you add to create a Sheet metal part?

8. Name the two primary states of a Sheet metal part.

9. Explain the procedure on how to insert a formed Sheet metal feature such as a dimple or louver.

10. Identify the type of information that a Sheet metal manufacturer provides.

11. What is an Assembly Feature?

12. What features are required before you create the Component Pattern in the assembly?

13. True or False. If you utilize Replace Component, you have to modify all the mate references.

14. Define a Design Table. Provide an example.

Exercises

Exercise 7.1: Insert an Extruded Cut feature In-Context of an assembly.

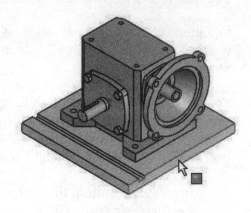

- Open the MOTOR assembly from the Chapter 7 Homework folder. Rotate the assembly and view the bottom plate. Insert mounting holes through the plate In-Context of the assembly.

- Edit the b5g-plate In-Context of the assembly. Right-click the right face of b5g-plate from the Graphics window.

- Click Edit Part.

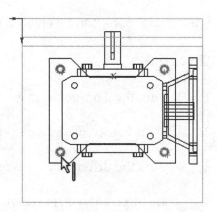

- Right-click the bottom face of b5g-plate for the sketch plane. Create a sketch.

- Display a Bottom, Wireframe view.

- Ctrl-select the 4 circular edges of the f718b-30-b5-g mounting holes.

- Apply the Convert Entities sketch tool.

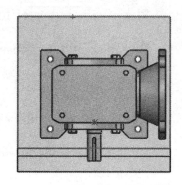

- Apply an Extruded Cut feature for the 4 selected circular holes thru the plate.

- Return to the assembly.

- Rebuild the model and view the results.

Exercise 7.2: Create a New Part In-Context of an assembly.

- Open the PLATE1 assembly from the Chapter 7 Homework folder located in the book.

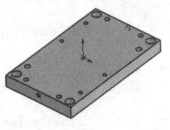

- Insert a new part into the assembly. Use the default part template. The mouse pointer displays the icon.

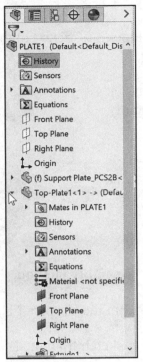

- The new virtual component name is displayed in the FeatureManager. Click the top face.

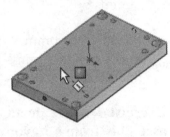

- Click the Convert Entities Sketch tool.

- Create a top plate from the bottom plate. Extrude upwards 50mm.

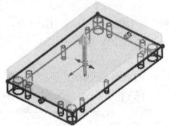

- Return to the assembly. Name the virtual component Top-Plate1.

Exercise 7.3: Layout Sketch.

Create components from a Layout sketch utilizing blocks.

- Open Block-Layout from the Chapter 7 - Homework folder. Three blocks are displayed in the Graphics window as illustrated.

Blocks can include the following items: Text (Notes), Dimensions, Sketch entities, Balloons, Imported entities and text, and Area hatch.

- Activate the Make Part from Block tool.

- Click the three blocks in the Graphics window.

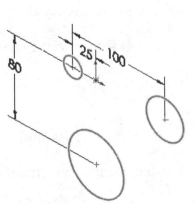

- Click the On Block option. Click OK. View the results.

- Edit the 20MM-6 block In-Context of the assembly.

- Extrude the 20MM-6 block 10mms.

- Return to the assembly.

- Extrude the 40MM-4 block In-Context of the assembly 10mms.

- Extrude the 60MM-4 block In-Context of the assembly 10mms.

- Modify the 80mm dimension to 70mms. View the results.

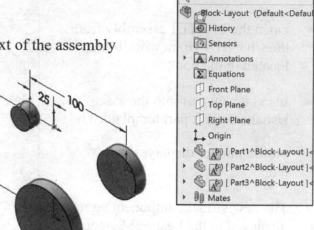

Exercise 7.4: Create an assembly from a Layout sketch. Utilize the Entire assembly method.

- Open Entire assembly from the Chapter 7 Homework folder.

- Double-click the Layout sketch from the FeatureManager to display the dimensions.

- Insert a New Part In-Context to the assembly. Select the PART-IN-ANSI template.

- Select the Top Plane and the bottom horizontal line as illustrated.

- Right-click Select chain.

- Select the Convert Entities Sketch tool.

- Extrude the Layout sketch in a downward direction 20mms.

- Return to the assembly.

- Rename the new virtual part, Bottom. View the results.

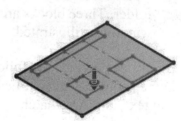

Exercise 7.5: L-BRACKET Part.

Create the L-BRACKET Sheet metal parts.

- Create a family of sheet metal L-BRACKETS. L-BRACKETS are used in the construction industry.

- Create an eight hole L-BRACKET.

- Use Equations to control the hole spacing.

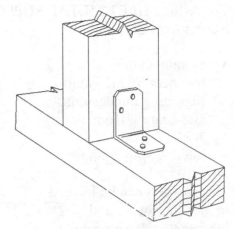

Strong-Tie Reinforcing Brackets
Courtesy of Simpson Strong Tie
Corporation of California

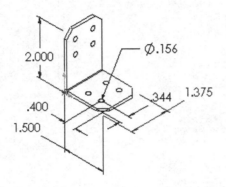

Review Design Tables in On-line Help and the SOLIDWORKS Tutorial.

	A	B	C	D	E
1	Design Table for: lbracket				
2		height@Sketch1	width@Sketch1	depth@Base-Extrude-Thin	hole1dia@Sketch2
3	First Instance	2	1.5	1.375	0.156
4	Small2x1.5x1	2	1.5	1	0.156
5	Medium3x2x2	3	2	2	0.156
6	Large4x3x4	4	3	4	0.25

L-BRACKET DESIGN TABLE

- Create a Design Table for the L-Bracket.

- Rename the dimensions of the L-Bracket.

- Create a drawing that contains the flat state and formed state of the L-BRACKET.

Exercise 7.6: SHEET METAL SUPPORTS.

Create the SHEET METAL SUPPORT parts. The following examples are courtesy of Strong Arm Corporation.

- Obtain additional engineering information regarding dimensions and bearing loads. Visit www.strongtie.com.

Similar parts can be located in local hardware and lumber stores. An actual physical model provides a great advantage in learning SOLIDWORKS Sheet metal functionality.

- Estimate the length, width and depth dimensions. Beams are 2" x 4" (50mmx100mm); posts are 4" x 4" (100mmx100mm).

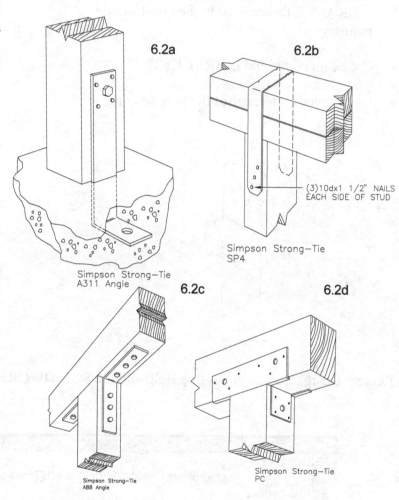

6.2a

Simpson Strong-Tie
A311 Angle

6.2b

(3)10dx1 1/2" NAILS
EACH SIDE OF STUD

Simpson Strong-Tie
SP4

6.2c

Simpson Strong-Tie
A88 Angle

6.2d

Simpson Strong-Tie
PC

Exercise 7.7: Create a lofted bend feature in a Sheet Metal part.

- Open the Lofted part from the Chapter 7 Homework folder.

- Apply a Lofted Bend feature on both sketches as illustrated.

- Thickness = .01in. The direction arrow is inward.

- View the results.

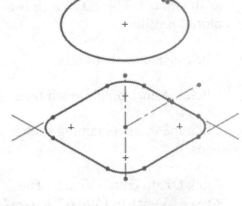

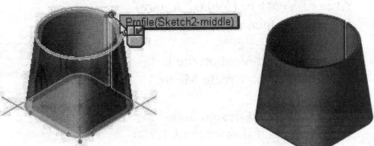

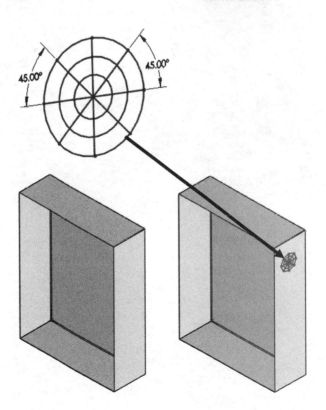

Exercise 7.8: Create a Vent in a Sheet Metal box.

- Open Vent from the Chapter 7 Homework folder. View the FeatureManager.

- Create the illustrated sketch for the Vent on the Right top side of the box.

- Click the Vent ⊞ tool.

- Click the large circle circumference as illustrated. The 2D Sketch is a closed profile.

- Click inside the Ribs box.

- Click the horizontal sketch line.

- Click the three remaining sketch lines.

- Click OK to create Vent1. The Area of Vent1 is 1590.67 square mms. The Open area is 51/13%.

- Insert a Mirror Vent on the left side of the Box. Create Mirror1.

- Modify the vent design and obtain an open flow area of 1,700 square mms.

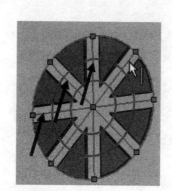

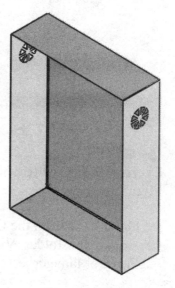

Flow Area
Area = 1590.67 square mm
Open area = 51.13 %

PROJECT 8: SOLIDWORKS SIMULATION

Project 8 Objective

Project 8 provides a general overview of SOLIDWORKS Simulation and the type of questions that are on the SOLIDWORKS Simulation Associate - Finite Element Analysis (CSWSA-FEA) exam. On the completion of this chapter, you will be able to:

Recognize the power of SOLIDWORKS Simulation.

Apply SOLIDWORKS Simulation to:

- Define a Static Analysis Study.

- Apply Material to a part model.

- Work with a Solid and Sheet Metal model.

- Define Solid, Shell and Beam elements.

- Define Standard and Advanced Fixtures and External loads.

- Define Local and Global coordinate systems.

- Understand the axial forces, shear forces, bending moments and factor of safety.

- Work with Multi-body parts as different solid bodies.

- Select different solvers as directed to optimize problems.

- Determine if the result is valid.

- Ability to use SOLIDWORKS Simulation Help and the on-line Tutorials.

- Understand the type of problems and questions that are on the CSWSA-FEA exam.

Basic FEA Concepts

SOLIDWORKS Simulation uses the Finite Element Method (FEM). FEM is a numerical technique for analyzing engineering designs. FEM is accepted as the standard analysis method due to its generality and suitability for computer implementation.

FEM divides the model into many small pieces of simple shapes called elements, effectively replacing a complex problem by many simple problems that need to be solved simultaneously.

Elements share common points called nodes. The process of dividing the model into small pieces is called meshing.

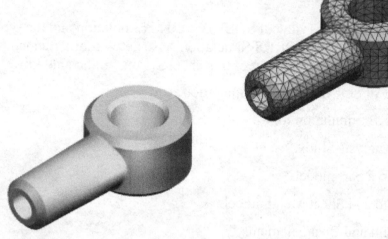

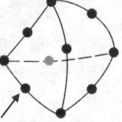

Node

Tetrahedral

The behavior of each element is well-known under all possible support and load scenarios. The finite element method uses elements with different shapes.

The response at any point in an element is interpolated from the response at the element nodes. Each node is fully described by a number of parameters depending on the analysis type and the element used.

For example, the temperature of a node fully describes its response in thermal analysis. For structural analyses, the response of a node is described, in general, by three translations and three rotations. These are called degrees of freedom (DOFs). Analysis using the FEM is called a Finite Element Analysis (FEA).

SOLIDWORKS Simulation formulates the equations governing the behavior of each element taking into consideration its connectivity to other elements. These equations relate the response to known material properties, restraints, and loads.

Next, SOLIDWORKS Simulation organizes the equations into a large set of simultaneous algebraic equations and solves for the unknowns.

In stress analysis, for example, the solver finds the displacements at each node and then the program calculates strains and finally stresses.

Static studies calculate displacements, reaction forces, strains, stresses, and factor of safety distribution. Material fails at locations where stresses exceed a certain level. Factor of safety calculations are based on one of the following failure criteria:

- **Maximum von Mises Stress**.

- **Maximum shear stress (Tresca)**.

- **Mohr-Coulomb stress**.

- **Maximum Normal stress**.

- **Automatic** (Automatically selects the most appropriate failure criterion across all element types).

Static studies can help avoid failure due to high stresses. A factor of safety less than unity indicates material failure. Large factors of safety in a contiguous region indicate low stresses and that you can probably remove some material from this region.

Simulation Advisor

Simulation Advisor is a set of tools that guide you through the analysis process. By answering a series of questions, these tools collect the necessary data to help you perform your analysis. Simulation Advisor includes:

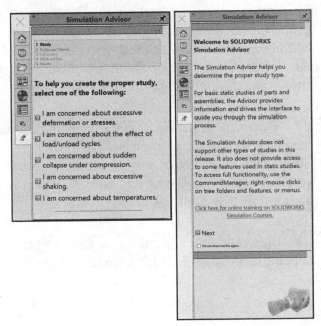

- **Study Advisor**. Recommends study types and outputs to expect. Helps you define sensors and creates studies automatically.

- **Bodies and Materials Advisor**. Specifies how to treat bodies within a part or an assembly and apply materials to components.

Apply material in SOLIDWORKS Simulation. Right-click on the part icon in the study. Click Apply/Edit Material.

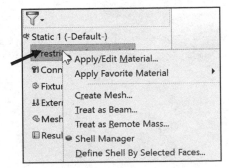

Interactions Advisor. Defines internal interactions between bodies in the model as well as external interactions between the model and the environment. Interactions can include loads, fixtures, connectors, and contacts.

- **Mesh and Run Advisor**. Helps you specify the mesh and run the study.

- **Results Advisor**. Provides tips for interpreting and viewing the output of the simulation. Also, helps determine if frequency or buckling might be areas of concern.

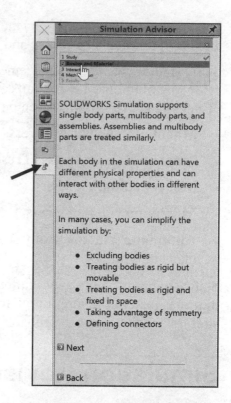

While taking the CSWSA-FEA exam, the Simulation Advisor and or SOLIDWORKS Simulation Help topics may provide required information to answer the exam questions.

Simulation Advisor works with the SOLIDWORKS Simulation interface by starting the appropriate PropertyManager and linking to online help topics for additional information. Simulation Advisor leads you through the analysis workflow from determining the study type through analyzing the simulation output.

The purpose of this chapter is not to educate a new or intermediate user on SOLIDWORKS Simulation but to cover and to inform you on the types of questions, layout and what to expect when taking the CSWSA-FEA exam.

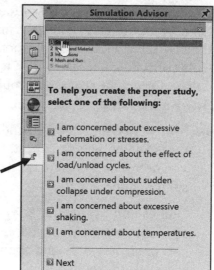

Illustrations and values may vary slightly depending on your SOLIDWORKS release.

SOLIDWORKS Simulation Help & Tutorials

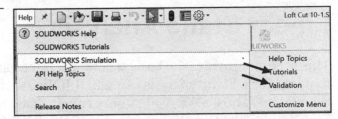

SOLIDWORKS Simulation is an Add-in. Use SOLIDWORKS Simulation during the CSWSA-FEA exam to discover information during the exam. Utilize the Contents and Search tabs to locate subject matter.

Review the SOLIDWORKS Simulation Tutorials on Static parts and assemblies. Understand your options and setup parameters. Questions in these areas will be on the exam.

Review the SOLIDWORKS Simulation Validation, Verification Problems, Static section and the SOLIDWORKS Simulation, Verification, NAFEMS Benchmarks, Linear Static section.

Access SOLIDWORKS Simulation directly from the SOLIDWORKS Add-Ins tab in the CommandManager.

A model needs to be open to access SOLIDWORKS Simulation Help or SOLIDWORKS Simulation Tutorials.

Linear Static Analysis

This section provides the basic theoretical information required for a Static analysis using SOLIDWORKS Simulation. The CSWSA-FEA only covers Static analysis. SOLIDWORKS Simulation and SOLIDWORKS Simulation Professional cover the following topics:

General Simulation.

- *Static Analysis*.
    - o Use 2D Simplification.
    - o Import Study Features.
- *Frequency Analysis*.

Design Insight.

- *Topology Study*.
- *Design Study*.

Advanced Simulation.

- *Thermal*.
- *Buckling*.
- *Fatigue*.
- *NonLinear*.
- *Linear Dynamic*.

Specialized Simulation.

- *Design Study*.
- *Drop Test*.
- *Pressure Vessel Design*.

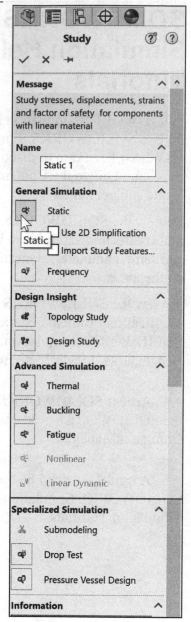

Linear Static Analysis

When loads are applied to a body, the body deforms and the effect of loads is transmitted throughout the body. The external loads induce internal forces and reactions to render the body into a state of equilibrium.

Linear Static analysis calculates displacements, strains, stresses, and reaction forces under the effect of applied loads.

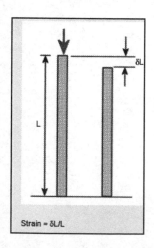

Strain ε is the ratio of change, δL, to the original length, L. Strain, is a dimensionless quantity. Stress σ is defined in terms of Force per unit Area.

Linear Static analysis makes the following assumptions:

- **Static Assumption**. All loads are applied *slowly* and gradually until they reach their full magnitudes. After reaching their full magnitudes, loads *remain constant* (time-invariant). This assumption neglects inertial and damping forces.

  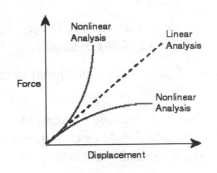

  Time-variant loads that induce considerable inertial and/or damping forces and require dynamic analysis.

  Dynamic loads change with time and in many cases induce considerable inertial and damping forces that cannot be neglected.

- **Linearity Assumption**. The relationship between loads and induced responses is linear. For example, if you double the loads, the response of the model (displacements, strains, and stresses) will also double. Apply the linearity assumption if:

  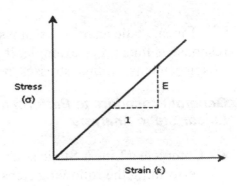

  - All materials in the model comply with Hooke's law; that is, stress is directly proportional to strain.
  - The induced displacements are small enough to ignore the change in stiffness caused by loading.
  - Boundary conditions do not vary during the application of loads. Loads must be constant in magnitude, direction, and distribution. They should not change while the model is deforming.

☀ In SOLIDWORKS Simulation for Static analysis, all displacements are small relative to the model geometry (unless the Large Displacements option is activated).

SOLIDWORKS Simulation assumes that the normals to contact areas do not change direction during loading. Hence, it applies the full load in one step. This approach may lead to inaccurate results or convergence difficulties in cases where these assumptions are not valid.

- Elastic Modulus, E is the stress required to cause one unit of strain. The material behaves linearly in the Elastic range. The slope of the Stress-Strain curve is in a linear portion.

- Elastic Modulus (Young's Modulus) is the slope defined as stress divided by strain. E = modulus of elasticity (Pa (N/m²), N/mm², psi).

- Stress σ is proportional to strain in a Linear Elastic Material. Units: (Pa (N/m²), N/mm², psi).

You must be able to work in SI and English units for the CSWSA-FEA exam within the same problem. For example, you apply a Force in Newton and then you determine displacement in inches.

Different materials have different stress property levels. Mathematical equations derived from Elasticity theory and Strength of Materials are utilized to solve for displacement and stress. These analytical equations solve for displacement and stress for simple cross sections.

Linear static analysis assumes that the relationship between loads and the induced response is linear. For example, if you double the magnitude of loads, the response (displacements, strains, stresses, reaction forces, etc.) will also double.

General Procedure to Perform a Linear Static Analysis

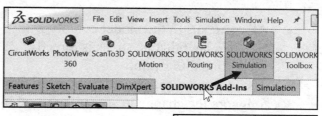

- Complete a Linear Static study by performing the following steps:

- Add-in SOLIDWORKS Simulation.

- Select the Simulation tab.

- Create a new Study. To create a new study, expand the Study Advisor and select New Study.

- Define Material. To define a material, right-click on the model icon in the Simulation study tree and select Apply/Edit Material.

- Define Restraints/Fixtures/Connections.

- Define External Loads. Right-click the External Loads icon in the Simulation study tree and select from the list.

- For assemblies and multi-body parts, use component contact and contact sets to simulate the behavior of the model.

- Mesh the model and Run the study. Select Mesh type and parameters. The Mesh PropertyManager lets you mesh models for solid, shell, and mixed mesh studies.

- View the results. In viewing the results after running a study, you can generate plots, lists, graphs, and reports depending on the study and result types.

- Double-click an icon in a results folder to display the associated plot.

- To define a new plot, right-click the Results folder, and select the desired option. You can plot displacements, stresses, strains, and deformation.

- To assess failure based on a yield criterion, right-click the Results folder, and select Define Factor of Safety Plot.

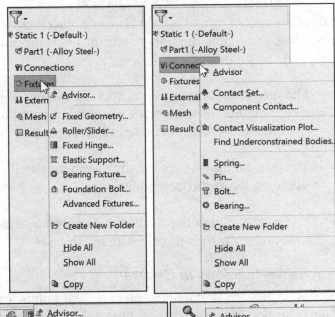

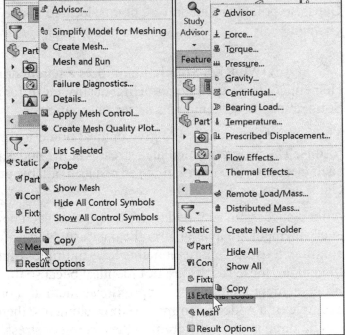

Sequence of Calculations in General

Given a meshed model with a set of displacement restraints and loads, the linear static analysis program proceeds as follows:

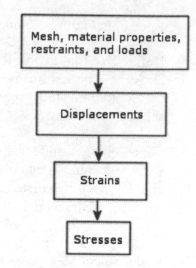

- The program constructs and solves a system of linear simultaneous finite element equilibrium equations to calculate displacement components at each node.

- The program then uses the displacement results to calculate the strain components.

- The program uses the strain results and the stress-strain relationships to calculate the stresses.

Stress Calculations in General

Stress results are first calculated at special points, called Gaussian points or Quadrature points, located inside each element.

☼ SOLIDWORKS Simulation utilizes a tetrahedral element containing 10 nodes (High quality mesh for a Solid). Each node contains a series of equations.

These points are selected to give optimal numerical results. The program calculates stresses at the nodes of each element by extrapolating the results available at the Gaussian points.

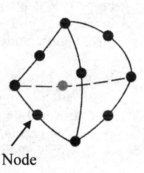

Node

Tetrahedral

After a successful run, nodal stress results at each node of every element are available in the database. Nodes common to two or more elements have multiple results. In general, these results are not identical because the finite element method is an approximate method. For example, if a node is common to three elements, there can be three slightly different values for every stress component at that node.

Overview of the Yield or Inflection Point in a Stress-Strain curve

When viewing stress results, you can ask for element stresses or nodal stresses. To calculate element stresses, the program averages the corresponding nodal stresses for each element.

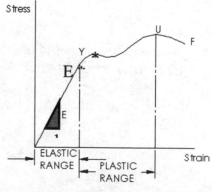

Stress versus Strain Plot
Linearly Elastic Material

To calculate nodal stresses, the program averages the corresponding results from all elements sharing that node.

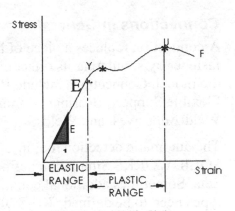

Stress versus Strain Plot
Linearly Elastic Material

- The material remains in the Elastic Range until it reaches the elastic limit.

- The point E is the elastic limit. The material begins Plastic deformation.

- Yield Stress is the stress level at which the material ceases to behave elastically.

- The point Y is called the Yield Point. The material begins to deform at a faster rate. In the Plastic Range the material behaves non-linearly.

- The point U is called the ultimate tensile strength. Point U is the maximum value of the non-linear curve. Point U represents the maximum tensile stress a material can handle before a fracture or failure.

- Point F represents where the material will fracture.

- Designers utilize maximum and minimum stress calculations to determine if a part is safe. Simulation reports a recommended Factor of Safety during the analysis.

- The Simulation Factor of Safety is a ratio between the material strength and the calculated stress.

Brittle materials do not have a specific yield point and hence it is not recommended to use the yield strength to define the limit stress for the criterion.

Material Properties in General

Before running a study, you must define all material properties required by the associated analysis type and the specified material model. A material model describes the behavior of the material and determines the required material properties. Linear isotropic and orthotropic material models are available for all structural and thermal studies. Other material models are available for nonlinear stress studies. The von Mises plasticity model is available for drop test studies. Material properties can be specified as function of temperature.

- For solid assemblies, each component can have a different material.

- For shell models, each shell can have a different material and thickness.

- For beam models, each beam can have a different material.

- For mixed mesh models, you must define the required material properties for solid and shell separately.

Connections in General

A connection replaces a piece of hardware or fastener by simulating its effect on the rest of the model. Connections include Bolts, Springs, Flexible Support, Bearings, Bonding - Weld/Adhesives, and Welds.

The automatic detection tool in SOLIDWORKS Simulation defines Contact Sets. Sometimes additional contact sets and types need to be defined. The SOLIDWORKS Simulation Study Advisor can help.

For example, the behavior of an adhesive depends on its strength and thickness. You can select the Type manually.

Fixtures: Adequate restraints to prevent the body from rigid body motion. If your model is not adequately constrained, check the Use soft springs to stabilize the model option in the Static dialog box.

When importing loads from SOLIDWORKS Motion, make sure that Use inertial relief option is checked. These options are available for the Direct Sparse solver and FFEPlus solver.

See SOLIDWORKS Simulation Help for additional information.

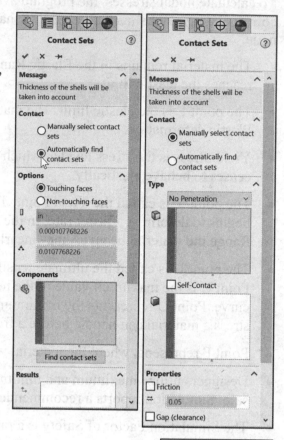

Restraint Types

The Fixture PropertyManager provides the ability to prescribe zero or non-zero displacements on vertices, edges, or faces for use with static, frequency, buckling, dynamic and nonlinear studies. This section will only address standard restraint types, namely Fixed Geometry and Immovable (No translation).

The Immovable option is displayed for Sheet Metal parts.

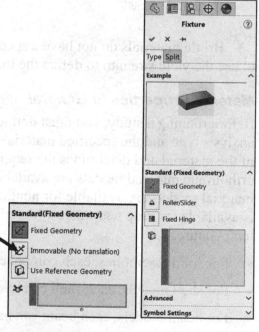

Fixed: For solids this restraint type sets all translational degrees of freedom to zero. For shells and beams, it sets the translational and the rotational degrees of freedom to zero. For truss joints, it sets the translational degrees of freedom to zero. When using this restraint type, no reference geometry is needed.

View the illustrated table for the attributes and input needed for this restraint.

Attribute	Value
DOFs restrained for solid meshes	3 translations
DOFs restrained for shells and beams	3 translations and 3 rotations
DOFs restrained for truss joints	3 translations
3D symbol (the arrows are for translations and the discs are for rotations)	
Selectable entities	Vertices, edges, faces and beam joints
Selectable reference entity	N/A
Translations	N/A
Rotations	N/A

Immovable (No translation): This restraint type sets all translational degrees of freedom to zero. It is the same for shells, beams and trusses. No reference geometry is used.

To access the immovable restraint, right-click on Fixtures in the Simulation study tree and select Fixed Geometry. Under Standard, select Immovable (No translation). View the illustrated table for the attributes and input needed for this restraint.

Attribute	Value
DOFs restrained for shell meshes	3 translations
DOFs restrained for beam and truss meshes	3 translations
3D symbol	
Selectable entities	Vertices, edges, faces and beam joints
Selectable reference entity	N/A
Translations	N/A
Rotations	N/A

There are differences for Shells and Beams between Immovable (No translation) and Fixed restraint types.

🔆 The Immovable option is not available for Solids.

The Fixture PropertyManager allows you to prescribe zero or non-zero displacements on vertices, edges, or faces for use with static, frequency, buckling, dynamic and nonlinear studies.

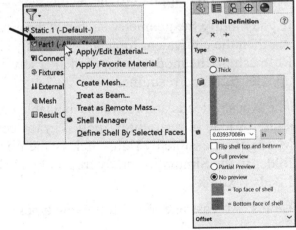

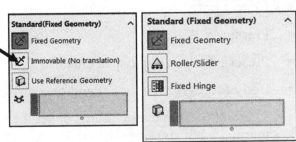

The Fixed Geometry Fixture allows for additional Advanced options: Symmetry, Circular Symmetry, User Reference Geometry, On Flat Faces, On Cylindrical Faces, and On Spherical Faces.

Attributes of each option are available in SOLIDWORKS Simulation Help.

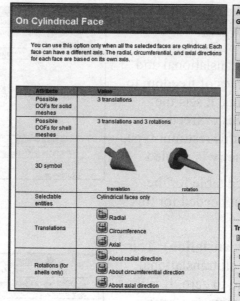

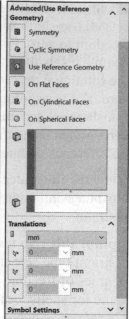

Loads and Restraints in General

Loads and restraints are necessary to define the service environment of the model. The results of analysis directly depend on the specified loads and restraints.

Loads and restraints are applied to geometric entities as features that are fully associative to geometry and automatically adjust to geometric changes.

For example, if you apply a pressure P to a face of area A_1, the equivalent force applied to the face is PA_1.

If you modify the geometry such that the area of the face changes to A_2, then the equivalent force automatically changes to PA_2. Re-meshing the model is required after any change in geometry to update loads and restraints.

The types of loads and restraints available depend on the type of the study. A load or restraint is applied by the corresponding Property Manager accessible by right-clicking the Fixtures or External Loads folder in the Simulation study tree, or by clicking Simulation, Loads/Fixture.

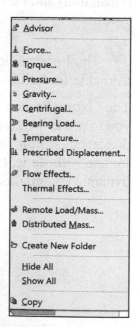

Loads: At least one of the following types of loading is required:

- Concentrated force.

- Pressure.

- Prescribed nonzero displacements.

- Body forces (gravitational and/or centrifugal).

- Thermal (define temperatures or get the temperature profile from thermal analysis).

- Imported loads from SOLIDWORKS Motion.

- Imported temperature and pressure from Flow Simulation.

In a linear static thermal stress analysis for an assembly it is possible to input different temperature boundary conditions for different parts.

Under the External Loads folder you can define Remote Load/Mass and Distributed Mass. In the Remote Load/Mass you define Load, Load/Mass or Displacement. Input values are required for the Remote Location and the Force.

By default, the Location is set to x=0, y=0, z=0. The Force is set to $F_x=0$, $F_y=0$, $F_z=0$. The Force requires you to first select the direction and then enter the value.

Meshing in General

Meshing splits continuous mathematical models into finite elements. The types of elements created by this process depend on the type of geometry meshed. SOLIDWORKS Simulation offers three types of elements:

- Solid elements - solid geometry.

- Shell elements - surface geometry.

- Beam elements - wire frame geometry.

In CAD terminology, "Solid" denotes the type of geometry. In FEA terminology, "Solid" denotes the type of element used to mesh the solid CAD Geometry.

Meshing Types

Meshing splits continuous mathematical models into finite elements. Finite element analysis looks at the model as a network of interconnected elements.

Meshing is a crucial step in design analysis. SOLIDWORKS Simulation automatically creates a mixed mesh of:

- **Solid**: The Solid mesh is appropriate for bulky or complex 3D models. In meshing a part or an assembly with solid elements, Simulation generates one of the following types of elements based on the active mesh options for the study:

  - Draft quality mesh. The automatic mesher generates linear tetrahedral solid elements **(4 nodes)**.

  - High quality mesh. The automatic mesher generates parabolic tetrahedral solid elements **(10 nodes)**.

Linear elements are also called first-order, or lower-order elements. Parabolic elements are also called second-order, or higher-order elements.

A linear tetrahedral element is defined by four corner nodes connected by six straight edges.

Linear solid element

A parabolic tetrahedral element assigns 10 nodes to each solid element: four corner nodes and one node at the middle of each edge (a total of six mid-side nodes).

In general, for the same mesh density (number of elements), parabolic elements yield better results than linear elements because:

- They represent curved boundaries more accurately.

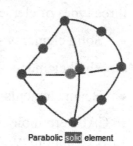

Parabolic solid element

- They produce better mathematical approximations. However, parabolic elements require greater computational resources than linear elements.

For structural problems, each node in a solid element has three degrees of freedom that represent the translations in three orthogonal directions.

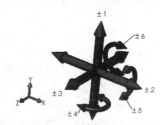

SOLIDWORKS Simulation uses the X, Y, and Z directions of the global Cartesian coordinate system in formulating the problem.

- **Shell**: Shell elements are suitable for thin parts (sheet metal models). Shell elements are 2D elements capable of resisting membrane and bending loads. When using shell elements, Simulation generates one of the following types of elements depending on the active meshing options for the study:

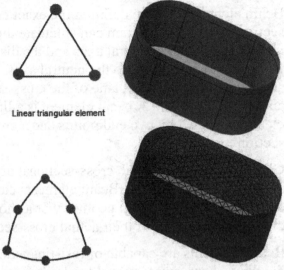

Linear triangular element

  - Draft quality mesh. The automatic mesher generates linear triangular shell elements **(3 nodes)**.

  - High quality mesh. The automatic mesher generates parabolic triangular shell elements **(6 nodes)**.

Parabolic triangular element

A linear triangular shell element is defined by three corner nodes connected by three straight edges.

A parabolic triangular element is defined by three corner nodes, three mid-side nodes, and three parabolic edges.

The Shell Definition PropertyManager is used to define the thickness of thin and thick shell elements. The program automatically extracts and assigns the thickness of the sheet metal to the shell. You cannot modify the thickness. You can select between the thin shell and thick shell formulations. You can also define a shell as a composite for static, frequency, and buckling studies. In general use thin shells when the thickness-to-span ratio is less than 0.05.

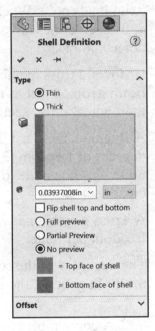

Surface models can only be meshed with shell elements.

- **Beam or Truss**: Beam or Truss elements are suitable for extruded or revolved objects and structural members with constant cross-sections. Beam elements can resist bending, shear, and torsional loads. The typical frame shown is modeled with beam elements to transfer the load to the supports. Modeling such frames with truss elements fails since there is no mechanism to transfer the applied horizontal load to the supports.

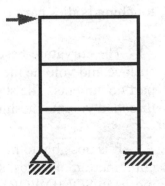

💡 A truss is a special beam element that can resist axial deformation only.

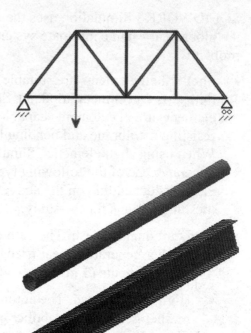

Beam elements require defining the exact cross section so that the program can calculate the moments of inertia, neutral axes and the distances from the extreme fibers to the neutral axes. The stresses vary within the plane of the cross-section and along the beam. A beam element is a line element defined by two end points and a cross-section.

Consider a 3D beam with cross-sectional area (A) and the associated mesh. Beam elements can be displayed on actual beam geometry or as hollow cylinders regardless of their actual cross-section.

Beam elements are capable of resisting axial, bending, shear, and torsional loads. Trusses resist axial loads only. When used with weldments, the software defines cross-sectional properties and detects joints.

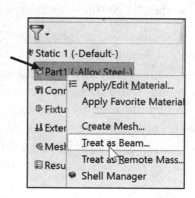

Mesh on cyclinders and beam geometry

Beam and truss members can be displayed on actual beam geometry or as hollow cylinders regardless of their actual cross-sectional shape.

A Beam element has 3 nodes (one at each end) with 6 degrees of freedom (3 translational and 3 rotational) per node plus one node to define the orientation of the beam cross section.

A Truss element has 2 nodes with 3 translational degrees of freedom per node.

The accuracy of the solution depends on the quality of the mesh. In general, the finer the mesh the better the accuracy.

💡 A compatible mesh is a mesh where elements on touching bodies have overlaying nodes.

💡 The curvature-based mesher supports multi-threaded surface and volume meshing for assembly and multi-body part documents. The standard mesher supports only multi-threaded volume meshing.

💡 It is possible to mesh a part or assembly with a combination of solids, shells and beam elements (mixed mesh) in SOLIDWORKS Simulation.

SOLIDWORKS Simulation Meshing Tips

SOLIDWORKS Simulation Help lists the following Meshing tips that you should know for the CSWSA-FEA exam.

When you mesh a study, the SOLIDWORKS Simulation meshes all unsuppressed solids, shells, and beams:

- Use Solid mesh for bulky objects.

- Use Shell elements for thin objects like sheet metals.

- Use Beam or Truss elements for extruded or revolved objects with constant cross-sections.

- Simplify structural beams to optimize performance in Simulation to be modeled with beam elements. The size of the problem and the resources required are dramatically reduced in this case. For the beam formulation to produce acceptable results, the length of the beam should be 10 times larger than the largest dimension of its cross section.

- Compatible meshing (a mesh where elements on touching bodies have overlaying nodes) is more accurate than incompatible meshing in the interface region. Requesting compatible meshing can cause mesh failure in some cases. Requesting incompatible meshing can result in successful results. You can request compatible meshing and select Re-mesh failed parts with incompatible mesh so that the software uses incompatible meshing only for bodies that fail to mesh.

- Check for interferences between bodies when using a compatible mesh with the curvature-based mesher. If you specify a bonded contact condition between bodies, they should be touching. If interferences are detected, meshing stops, and you can access the Interference Detection PropertyManager to view the interfering parts. Make sure to resolve all interferences before you mesh again.

- If meshing fails, use the Failure Diagnostics tool to locate the cause of mesh failure. Try the proposed options to solve the problem. You can also try different element size, define mesh control, or activate Enable automatic looping for solids.

- The SOLIDWORKS Simplify utility lets you suppress features that meet a specified simplification factor. In the Simulation study tree, right-click Mesh and select Simplify Model for Meshing. This displays the Simplify utility.

- It is good practice to check mesh options before meshing. For example, the Automatic transition can result in generating an unnecessarily large number of elements for models with many small features. The high quality mesh is recommended for most cases. The Automatic looping can help solve meshing problems automatically, but you can adjust its settings for a particular model. The Curvature-based mesher automatically uses smaller element sizes in regions with high curvature.

- To improve results in important areas, use mesh control to set a smaller element size. When meshing an assembly with a wide range of component sizes, default meshing results in a relatively coarse mesh for small components. Component mesh control offers an easy way to give more importance to the selected small components. Use this option to identify important small components.

- For assemblies, check component interference. To detect interference in an assembly, click Tools, Interference Detection. Interference is allowed only when using shrink fit. The Treat coincidence as interference and Include multi-body part interferences options allow you to detect touching areas. These are the only areas affected by the global and component contact settings.

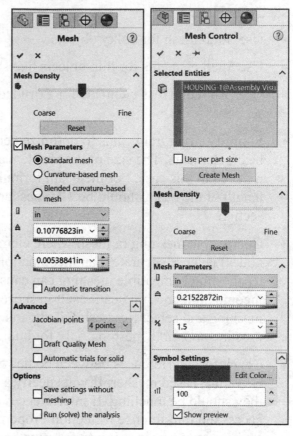

Use the mesh and displacement plots to calculate the distance between two nodes using SOLIDWORKS Simulation.

The Global element size parameter provides the ability to set the global average element size. SOLIDWORKS suggests a default value based on the model volume and surface area. This option is only available for a standard mesh.

The Ratio value in Mesh Control provides the geometric growth ratio from one layer of elements to the next. To access Mesh Control, right-click the Mesh folder in the Simulation study tree and click Apply Mesh Control.

Running the Study

When you run a study, Simulation calculates the results based on the specified input for materials, restraints, loads and mesh.

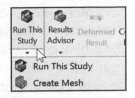

Set the default plots that you want to see in your Simulation Study tree under Simulation, Options from the Main menu.

When you run one or multiple studies, they run as background processes.

In viewing the results after running a study, you can generate plots, lists, graphs, and reports depending on the study and result types.

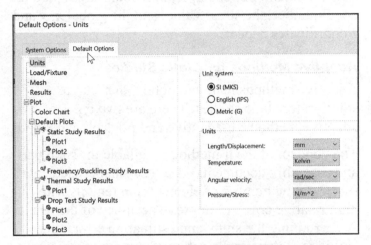

Run multiple studies (batches) either by using the SOLIDWORKS Task Scheduler or the Run all Studies command.

If you modify the study (force, material, etc.) you only need to re-run the study to update the results. You do not need to re-mesh unless you modified contact conditions.

Displacement Plot - Output of Linear Static Analysis

The Displacement Plot PropertyManager allows you to plot displacement and reaction force results for static, nonlinear, dynamic, drop test studies, or mode shapes for bucking and frequency studies. By default, directions X, Y, and Z refer to the global coordinate system.

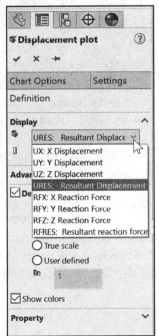

If you choose a reference geometry, these directions refer to the selected reference entity. Displacement components are:

UX = Displacement in the X-direction.

UY = Displacement in the Y-direction.

UZ = Displacement in the Z-direction.

URES = Resultant displacement.

RFX = Reaction force in the X-direction.

RFY = Reaction force in the Y-direction.

RFZ = Reaction force in the Z-axis.

RFRES = Resultant reaction force.

The Probe function allows you to query a plot and view the values of plotted quantities at defined nodes or centers of elements. When you probe a mesh plot, Simulation displays the node or element number and the global coordinates of the node. When you probe a result plot, SOLIDWORKS Simulation displays the node or element number, the value of the plotted result, and the global coordinates of the node or center of the element. For example, in a nodal stress plot, the node number, the stress value, and the global x, y and z coordinates appear.

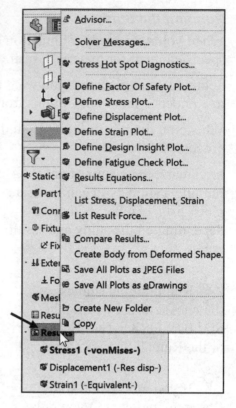

Adaptive Methods for Static Studies

Adaptive methods help you obtain an accurate solution for static studies. There are two types of adaptive methods: h-adaptive and p-adaptive method.

The concept of the h-method (available for solid part and assembly documents) is to use smaller elements (increase the number of elements) in regions with high relative errors to improve accuracy of the results. After running the study and estimating errors, the software automatically refines the mesh where needed to improve results.

The p-adaptive method (available for solid part and assembly documents) increases the polynomial order of elements with high relative errors. The p-method does not change the mesh. It changes the order of the polynomials used to approximate the displacement field using a unified polynomial order for all elements. See SOLIDWORKS Help for additional information.

A complete understanding of p-adaptive and h-adaptive methods requires advanced study; for the CSWSA- FEA exam, you require general knowledge of these two different methods.

Sample Exam Questions

These questions are examples of what to expect on the certification exam. The multiple choice questions should serve as a check for your knowledge of the exam materials.

1. What is the Modulus of Elasticity?

- The slope of the Deflection-Stress curve

- The slope of the Stress-Strain curve in its linear section

- The slope of the Force-Deflection curve in its linear section

- The first inflection point of a Strain curve

2. What is Stress?

- A measure of power

- A measure of strain

- A measure of material strength

- A measure of the average amount of force exerted per unit area

3. Which of the following assumptions are true for a static analysis in SOLIDWORKS Simulation with small displacements?

- Inertia effects are negligible and loads are applied slowly

- The model is not fully elastic. If loads are removed, the model will not return to its original position

- Results are proportional to loads

- All the displacements are small relative to the model geometry

4. What is Yield Stress?

- The stress level beyond which the material becomes plastic

- The stress level beyond which the material breaks

- The strain level above the stress level which the material breaks

- The stress level beyond the melting point of the material

5. A high quality Shell element has _____ nodes.

- 4
- 5
- 6
- 8

6. Stress σ is proportional to _____ in a Linear Elastic Material.

- Strain
- Stress
- Force
- Pressure

7. The Elastic Modulus (Young's Modulus) is the slope defined as _____ divided by
_____.

- Strain, Stress
- Stress, Strain
- Stress, Force
- Force, Area

8. Linear static analysis assumes that the relationship between loads and the induced
response is _____.

- Flat
- Linear
- Doubles per area
- Translational

9. In SOLIDWORKS Simulation, the Factor of Safety (FOS) calculations are based on
one of the following failure criterion.

- Maximum von Mises Stress
- Maximum shear stress (Tresca)
- Mohr-Coulomb stress
- Maximum Normal stress

10. The Yield point is the point where the material begins to deform at a faster rate than at the elastic limit. The material behaves _____ in the Plastic Range.

- Flatly

- Linearly

- Non-Linearly

- Like a liquid

11. What are the Degrees of Freedom (DOFs) restrained for a Solid?

- None

- 3 Translations

- 3 Translations and 3 Rotations

- 3 Rotations

12. What are the Degrees of Freedom (DOFs) restrained for Truss joints?

- None

- 3 Translations

- 3 Translations and 3 Rotations

- 3 Rotations

13. What are the Degrees of Freedom (DOFs) restrained for Shells and Beams?

- None

- 3 Translations

- 3 Translations and 3 Rotations

- 3 Rotations

14. Which statements are true for Material Properties using SOLIDWORKS Simulation?

- For solid assemblies, each component can have a different material.

- For shell models, each shell cannot have a different material and thickness.

- For shell models, the material of the part is used for all shells.

- For beam models, each beam cannot have a different material.

15. A Beam element has _____nodes (one at each end) with _____degrees of freedom per node plus_____ node to define the orientation of the beam cross section.

- 6, 3, 1

- 3, 3, 1

- 3, 6, 1

- None of the above

16. A Truss element has _____ nodes with _____ translational degrees of freedom per node.

- 2, 3

- 3, 3

- 6, 6

- 2, 2

17. In general, the finer the mesh the better the accuracy of the results.

- True

- False

18. How does SOLIDWORKS Simulation automatically treat a Sheet metal part with uniform thickness?

- Shell

- Solid

- Beam

- Mixed Mesh

19. Use the mesh and displacement plots to calculate the distance between two _____ using SOLIDWORKS Simulation.

- Nodes

- Elements

- Bodies

- Surfaces

20. Surface models can only be meshed with _____ elements.

- Shell
- Beam
- Mixed Mesh
- Solid

21. The shell mesh is generated on the surface (located at the mid-surface of the shell).

- True
- False

22. In general, use Thin shells when the thickness-to-span ratio is less than _____.

- 0.05
- 0.5
- 1
- 2

23. The model (a rectangular plate) has a length to thickness ratio of less than 5. You extracted its mid-surface to use it in SOLIDWORKS Simulation. You should use a _____.

- Thin Shell element formulation
- Thick Shell element formulation
- Thick or Thin Shell element formulation, it does not matter
- Beam Shell element formulation

24. The model, a rectangular sheet metal part, uses SOLIDWORKS Simulation. You should use a:

- Thin Shell element formulation
- Thick Shell element formulation
- Thick or Thin Shell element formulation, it does not matter
- Beam Shell element formulation

25. The Global element size parameter provides the ability to set the global average element size. SOLIDWORKS Simulation suggests a default value based on the model volume and _____ area. This option is only available for a standard mesh.

- Force

- Pressure

- Surface

- None of the above

26. A remote load applied on a face with a Force component and no Moment can result in: Note: Remember (DOFs restrain).

- A Force and Moment of the face

- A Force on the face only

- A Moment on the face only

- A Pressure and Force on the face

27. There are _____ DOFs restrain for a Solid element.

- 3

- 1

- 6

- None

28. There are _____ DOFs restrain for a Beam element.

- 3

- 1

- 6

- None

29. What best describes the difference(s) between a Fixed and Immovable (No translation) boundary condition in SOLIDWORKS Simulation?

- There are no differences

- There are no difference(s) for Shells but it is different for Solids

- There is no difference(s) for Solids, but it is different for Shells and Beams

- There are only differences(s) for a Static Study.

30. Can a non-uniform pressure of force be applied on a face using SOLIDWORKS Simulation?

- No.

- Yes, but the variation must be along a single direction only.

- Yes. The non-uniform pressure distribution is defined by a reference coordinate system and the associated coefficients of a second order polynomial.

- Yes, but the variation must be linear.

31. You are performing an analysis on your model. You select five faces, 3 edges and 2 vertices and apply a force of 20lbf. What is the total force applied to the model using SOLIDWORKS Simulation?

- 100lbf

- 1600lbf

- 180lbf

- 200lbf

32. Yield strength is typically determined at _____ strain.
- 0.1%

- 0.2%

- 0.02%

- 0.002%

33. There are four key assumptions made in Linear Static Analysis: 1: Effects of inertia and damping are neglected, 2. The response of the system is directly proportional to the applied loads, 3: Loads are applied slowly and gradually, and_____.

- Displacements are very small. The highest stress is in the linear range of the stress-strain curve.

- There are no loads

- Material is not elastic

- Loads are applied quickly

34. How many degrees of freedom does a physical structure have?

- Zero

- Three - Rotations only

- Three - Translations only

- Six - Three translations and three rotational

35. Brittle materials have little tendency to deform (or strain) before fracture and do not have a specific yield point. It is not recommended to apply the yield strength analysis as a failure criterion on brittle material. Which of the following failure theories is appropriate for brittle materials?

- Mohr-Columb stress criterion

- Maximum shear stress criterion

- Maximum von Mises stress criterion

- Minimum shear stress criterion

36. You are performing an analysis on your model. You select three faces and apply a force of 40lbf. What is the total force applied to the model using SOLIDWORKS Simulation?

- 40lbf

- 20lbf

- 120lbf

- Additional information is required

37. A material is orthotropic if its mechanical or thermal properties are not unique and independent in three mutually perpendicular directions.

- True

- False

38. An increase in the number of elements in a mesh for a part will:

- Decrease calculation accuracy and time

- Increase calculation accuracy and time

- Have no effect on the calculation

- Change the FOS below 1

39. SOLIDWORKS Simulation uses the von Mises Yield Criterion to calculate the Factor of Safety of many ductile materials. According to the criterion:

- Material yields when the von Mises stress in the model equals the yield strength of the material.

- Material yields when the von Mises stress in the model is 5 times greater than the minimum tensile strength of the material.

- Material yields when the von Mises stress in the model is 3 times greater than the FOS of the material.

- None of the above.

40. SOLIDWORKS Simulation calculates structural failure on:

- Buckling

- Fatigue

- Creep

- Material yield

41. Apply a uniform total force of 200lbf on two faces of a model. The two faces have different areas. How do you apply the load using SOLIDWORKS Simulation for a Linear Static Study?

- Select the two faces and input a normal to direction force of 200lbf on each face.

- Select the two faces and a reference plane. Apply 100lbf on each face.

- Apply equal force to the two faces. The force on each face is the total force divided by the total area of the two faces.

- None of the above.

42. Maximum and Minimum value indicators are displayed on Stress and Displacement plots in SOLIDWORKS Simulation for a Linear Static Study.

- True

- False

43. What SOLIDWORKS Simulation tool should you use to determine the result values at specific locations (nodes) in a model using SOLIDWORKS Simulation?

- Section tool

- Probe tool

- Clipping tool

- Surface tool

44. What criteria are best suited to check the failure of ductile materials in SOLIDWORKS Simulation?

- Maximum von Mises Strain and Maximum Shear Strain criterion

- Maximum von Misses Stress and Maximum Shear Stress criterion

- Maximum Mohr-Coulomb Stress and Maximum Mohr-Coulomb Shear Strain criterion

- Mohr-Coulomb Stress and Maximum Normal Stress criterion

45. Set the scale factor for plots_____ to avoid any misinterpretation of the results, after performing a Static analysis with gap/contact elements.

- Equal to 0

- Equal to 1

- Less than 1

- To the Maximum displacement value for the model

46. It is possible to mesh _____ with a combination of Solids, Shells and Beam elements in SOLIDWORKS Simulation.

- Parts and Assemblies

- Only Parts

- Only Assemblies

- None of the above

47. SOLIDWORKS Simulation supports multi-body parts. Which of the following is a true statement?

- You can employ different mesh controls to each Solid body

- You can classify Contact conditions between multiple Solid bodies

- You can classify a different material for each Solid body

- All of the above are correct

48. Which statement best describes a Compatible mesh?

- A mesh where only one type of element is used

- A mesh where elements on touching bodies have overlaying nodes

- A mesh where only a Shell or Solid element is used

- A mesh where only a single Solid element is used

49. The Ratio value in Mesh Control provides the geometric growth ratio from one layer of elements to the next.

- True

- False

50. The structures displayed in the following illustration are best analyzed using:

- Shell elements

- Solid elements

- Beam elements

- A mixture of Beam and Shell elements

51. The structure displayed in the following illustration is best analyzed using:

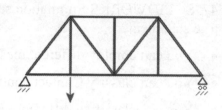

- Shell elements

- Solid elements

- Beam elements

- A mixture of Beam and Shell elements

52. The structure displayed in the following illustration is best analyzed using:

- Shell elements

- Solid elements

- Beam elements

- A mixture of Beam and Shell elements

Sheet metal model

53. The structure displayed in the following illustration is best analyzed using:

- Shell elements

- Solid elements

- Beam elements

- A mixture of Beam and Shell elements

54. Surface models can only be meshed with _____ elements.

- Shell elements

- Solid elements

- Beam elements

- A mixture of Beam and Shell elements

55. Use the _____ and _____ plots to calculate the distance between two nodes using SOLIDWORKS Simulation.

- Mesh and Displacement

- Displacement and FOS

- Resultant Displacement and FOS

- None of the above

56. You can simplify a large assembly in a Static Study by using the _____ or _____ options in your study.

- Make Rigid, Fix

- Shell element, Solid element

- Shell element, Compound element

- Make Rigid, Load element

57. A force "F" applied in a static analysis produces a resultant displacement URES. If the force is now 2x F and the mesh is not changed, then URES will:

- Double if there is no contact specified and there are large displacements in the structure

- Be divided by 2 if contacts are specified

- The analysis must be run again to find out

- Double if there is no source of nonlinearity in the study (like contacts or large displacement options)

58. To compute thermal stresses on a model with a uniform temperature distribution, what type/types of study/studies are required?

- Static only

- Thermal only

- Both Static and Thermal

- None of these answers is correct

59. In an h-adaptive method, use smaller elements in mesh regions with high errors to improve the accuracy of results.

- True

- False

60. In a p-adaptive method, use elements with a higher order polynomial in mesh regions with high errors to improve the accuracy of results.

- True

- False

61. Where will the maximum stress be in the illustration?

- A

- B

- C

- D

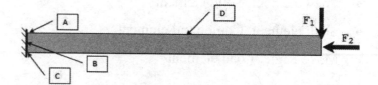

 The purpose of this chapter is not to educate a new or intermediate user on SOLIDWORKS Simulation, but to cover and to inform you on the types of questions, layout and what to expect when taking the CSWSA-FEA exam.

 The CSWSA-FEA only covers Linear Static analysis.

FEA Modeling Section

Tutorial FEA Model 8-1

An exam question in this category could read:

In the figure displayed, what is the vertical displacement in the Global Y direction in (inches) at the location of the dot? Calculate the answer to 3 decimal places.

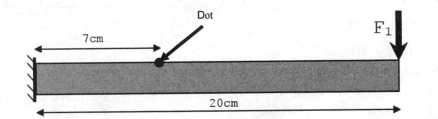
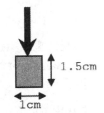

Given Information:

Material: Alloy Steel (SS) from the SOLIDWORKS Simulation Library.

ENGDESIGN-W-SOLIDWORKS ➤ CSWSA-FEA Model Folder 2018

Elastic modulus = 2.1e11 N/m²

Poisson's ratio = 0.28

F₁ = 200lbf

Use the default high quality element size to mesh.

Let's start.

1. **Open** Model 8-1 from the CSWSA-FEA Model folder.

Think about the problem. Think about the model.

The bar which you opened was created on the Front Plane.

The upper left corner of the rectangle is located at the origin. This simplifies the actual deformation of the part. The height dimension references zero in the Global Y direction.

Split Lines were created to provide the ability to locate the needed Joints in this problem. To add the force at the center of the beam, a split line in the shape of a small circle on the top face of the right end of the beam is used.

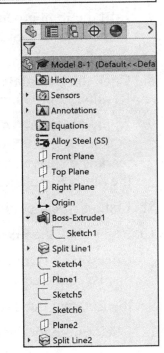

The proper model setup is very important to obtain the correct mesh and to obtain the correct final results.

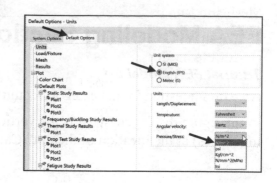

Set Simulation Options and start a Static SOLIDWORKS Simulation Study.

2. **Add-In** SOLIDWORKS Simulation.

3. Click **Simulation, Options** from the Main menu. Click the **Default Options** tab.

Set the Unit system and Mesh quality.

4. Select **English (IPS)** and **Pressure/Stress (N/m²)** as illustrated.

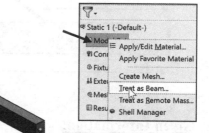

5. Click the **Mesh folder**. Select **High** for Mesh quality.

6. **Create** a new Static Study. Accept the default name (Static #). Treat the model as a **Beam**.

7. **Edit** the Joint groups folder from the Study Simulation tree. Split Line 2 is selected by default.

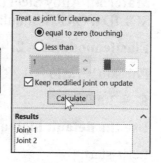

8. Click the **Calculate** button and accept the Results: Joint 1, Joint 2.

9. Apply **Material** - Alloy Steel (SS) from the SOLIDWORKS Simulation Library. A check mark is displayed next to the model name in the Study Simulation tree.

Set Fixture type. Set Fixed Geometry.

10. Right-click the **Fixtures** folder.

11. Click **Fixed Geometry**. The Fixture PropertyManager is displayed. Select the **joint on the left side** of the beam.

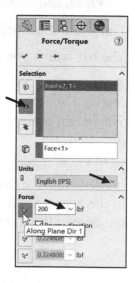

Set the External Load (Force) at the end of the beam.

12. Click **Force** for the External load.

13. Click the **Joints** option. Click the **joint on the right side** of the beam.

14. Click the **end face** of the beam as the plane for direction. Enter **200**lbf Along Plane Direction1. The arrow is displayed downwards; reverse direction if needed.

Mesh and Run the model.

15. **Mesh and Run** the model. Use the standard default setting for the mesh.

16. Double-click the **Stress1** folder. View the results.

17. Double-click the **Displacement1** folder. View the results. If needed, Right-click the Displacement1 folder, click Edit Definition. Set Chart Options to floating.

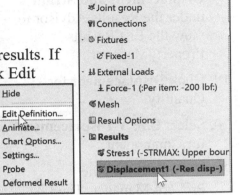

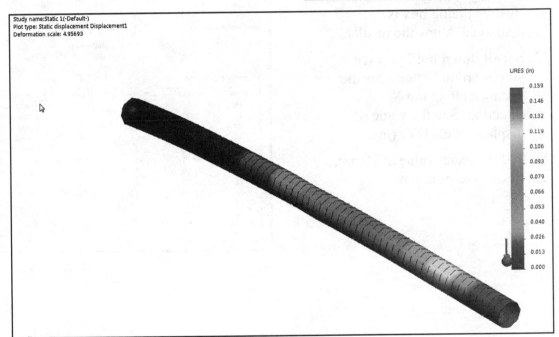

💡 Illustrations may vary slightly depending on your SOLIDWORKS version.

Locate the displacement at 7cm.

18. Click **List Stress, Displacement and Strain** under the Results Advisor to view the List Results PropertyManager.

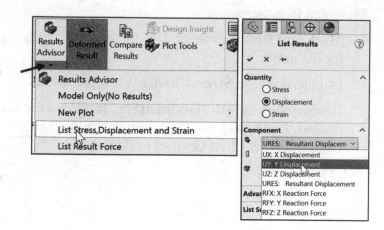

19. Click **Displacement** for Quantity.

20. Select **UY: Y Displacement** for Component.

21. Click **OK** from the PropertyManager. The List Results dialog box is displayed. View the results.

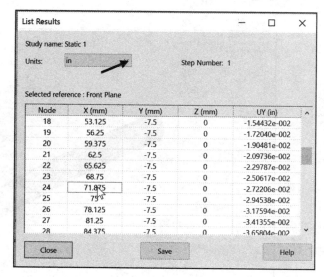

22. **Scroll down** until you see values around 70mm for the distance along the X direction. See the value of displacement (UY) (in).

To find the exact value at 70mm, use linear interpolation.

Node	X (mm)	Y (mm)	Z (mm)	UY (in)
18	53.125	-7.5	0	-1.54432e-002
19	56.25	-7.5	0	-1.72040e-002
20	59.375	-7.5	0	-1.90481e-002
21	62.5	-7.5	0	-2.09736e-002
22	65.625	-7.5	0	-2.29787e-002
23	68.75	-7.5	0	-2.50617e-002
24	71.875	-7.5	0	-2.72206e-002
25	75	-7.5	0	-2.94538e-002
26	78.125	-7.5	0	-3.17594e-002
27	81.25	-7.5	0	-3.41355e-002
28	84.375	-7.5	0	-3.65804e-002

This method is shown below and uses the values greater and less than the optimal one to find the actual displacement.

$$\frac{X_U - X_L}{UY_U - UY_L} = \frac{X_U - X_O}{UY_U - UY_O}$$

$$\frac{71.875 - 68.75}{-0.0272206 - -0.0250617} = \frac{71.875 - 70}{-0.02718 - UY_O}$$

$$UY_o = 0.0258$$

At the distance of 7cm (70mm) the displacement is found to be approximately 0.026in.

The correct answer is **B**.

A = 0.034in

B = 0.026in

C = 0.043in

D = 0.021in

Tutorial FEA Model 8-2

Below is a second way to address the first problem (Tutorial FEA Model 8-1) with a different model, without using Split lines, using the Study Advisor and the Probe tool.

In the figure displayed, what is the vertical displacement in the Global Y direction in (inches) at the location of the red dot? Calculate the answer to 3 decimal places.

Given Information:

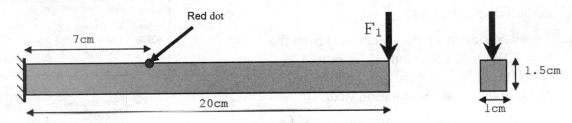

Material: Alloy Steel (SS) from the SOLIDWORKS Simulation Library.

Elastic modulus = 2.1e11 N/m²

Poisson's ratio = 0.28

F₁ = 200lbf

Use the default high quality element size to mesh.

Let's start.

1. **Open** Model 8-2 from the CSWSA-FEA Model folder.

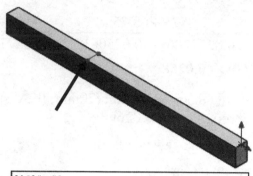

Think about the problem. Think about the model. The bar that you opened is created on the Front Plane.

The bar was created so that the origin is the point at which the force is applied. A construction line is created across the part at a distance of 7cm from the end that is to be fixed.

Set Simulation Options and start a Static SOLIDWORKS Simulation Study.

2. **Add-In** SOLIDWORKS Simulation.

3. Click **Simulation**, **Options** from the Main menu.

4. Click the **Default Options** tab.

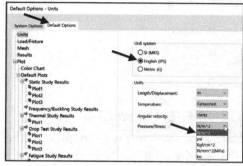

Set the Unit system and Mesh quality.

5. Select **English (IPS)** and **Pressure/Stress (N/m²)** as illustrated.

6. Click the **Mesh folder**. Select **High** for Mesh quality.

7. **Create** a new Static Study. Accept the default name (Static #).

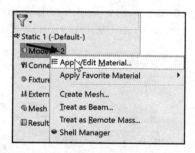

8. Apply **Material** - Alloy Steel (SS) from the SOLIDWORKS Simulation Library. A check mark is displayed next to the model name in the Study Simulation tree.

Set Fixture type. Use the Study Advisor. Set as Fixed Geometry.

9. Right-click the **Fixtures** folder from the Simulation study tree.

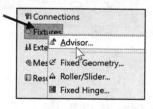

10. Click **Advisor**.

11. Click **Add a fixture**.

12. Select the **proper face (left end)** as illustrated.

13. Set as **Fixed Geometry**.

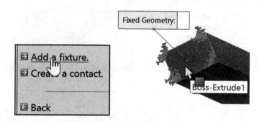

Set the Force.

14. Right-click the **External Loads** folder from the Simulation study tree. Select **Remote Load/Mass**. The Remote Loads/Mass PropertyManager is displayed.

15. Click **Load (Direct transfer)** for Type.

16. Select the **top face of the part** - for Faces for Remote Load.

17. Set the Reference Coordinate system to **Global**.

18. Leave all of the location boxes at **zero** value.

19. Select the **Force** check box.

20. Click the **Y-Direction** box.

21. Enter **200lbf**. Use the X-, Y-, and Z-direction boxes to direct the load; reverse direction as necessary.

Mesh and Run the model.

22. **Mesh and Run** the model. Use the standard default setting for the mesh.

23. Double-click the **Stress1** folder.

24. **View** the results.

25. Double-click the **Displacement1** folder.

26. **View** the results. If needed, Right-click the Displacment1 folder, click Edit Definition. Set Chart Options to floating.

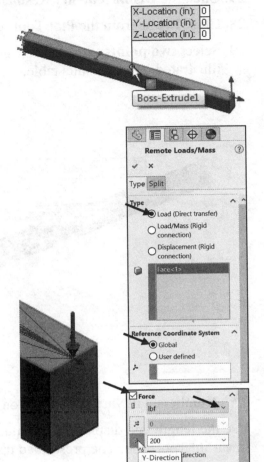

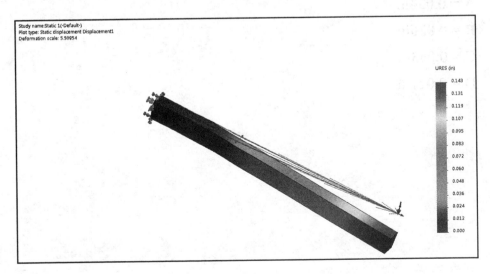

Locate the displacement at 7cm using the Probe tool.

27. Click the **Displacement1** Results folder if needed.

28. Click **Probe** from the Plot Tools drop-down menu.

29. Select **two points**: one on each side of the construction line as illustrated in the Results table.

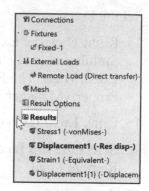

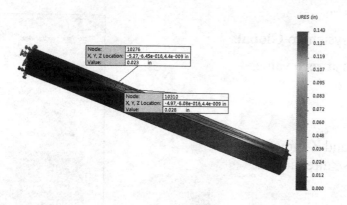

Numbers will vary depending on the selection location.

Use the length and displacement values of the selected points to find the answer through linear interpolation. The prescribed distance of 7cm is equal to 2.755in.

At the distance of 7cm (70mm) the displacement is found to be approximately 0.026in.

The correct answer is **B**.

A = 0.034in

B = 0.026in

C = 0.043in

D = 0.021in

Tutorial FEA Model 8-3

Proper model setup is very important to obtain the correct mesh and to obtain the correct final results. In the last two examples, you needed to manually apply linear interpolation to locate your final answer.

How can you eliminate the need to manually apply linear interpolation for your final answer? Create a Sensor or use Split Lines at the 7cm point.

Below is the Tutorial FEA Model 8-2 problem. Let's use the same model, but apply a Split line at the 7cm point before we begin the Linear Static study.

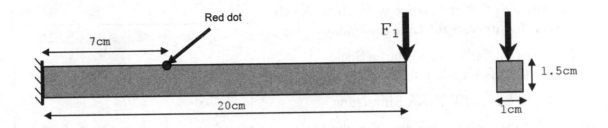

In the figure displayed, what is the vertical displacement in the Global Y direction in (inches) at the location of the red dot? Calculate the answer to 3 decimal places.

Given Information:

Material: Alloy Steel (SS) from the SOLIDWORKS Simulation Library.

ENGDESIGN-W-SOLIDWORKS › CSWSA-FEA Model Folder 2018

Elastic modulus = 2.1e11 N/m^2

Poisson's ratio = 0.28

F_1 = 200lbf

Use the default high quality element size to mesh.

Let's start.

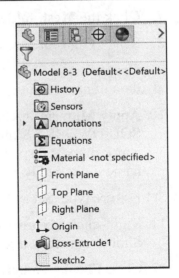

1. **Open** Model 8-3 from the CSWSA-FEA Model Folder.

Think about the problem. This is the same model that you opened in the second example.

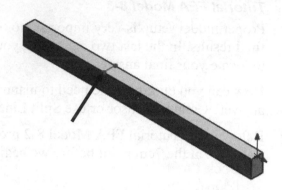

How can you eliminate the need to manually apply linear interpolation for your final answer?

Address this in the initial setup of the provided model.

Create a Split line at 70mm.

Apply the Probe tool and select the Split Line for the exact point.

2. Create a **Split line feature** with a new sketch **(Insert, Curve, Split Line)** at 70mm.

Set Simulation Options and start a Static SOLIDWORKS Simulation Study.

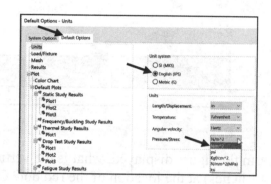

3. **Add-In** SOLIDWORKS Simulation.

4. Click **Simulation, Options** from the Main menu.

5. Click the **Default Options** tab.

Set the Unit system and Mesh quality.

6. Select **English (IPS)** and **Pressure/Stress (N/m²)**.

7. Click the **Mesh folder**.

8. Select **High** for Mesh quality.

9. **Create** a new Static Study. Accept the default name (Static #).

10. Apply **Material** - Alloy Steel (SS) from the SOLIDWORKS Simulation Library. A check mark is displayed next to the model name in the Study Simulation tree.

Set Fixture type. Set as Fixed Geometry.

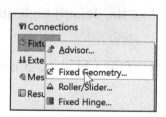

11. Right-click **Fixtures** folder from the Simulation study tree.

12. Click **Fixed Geometry**.

13. Select the **proper face (left end)** as illustrated.

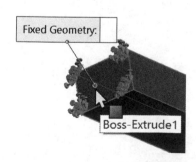

Set the External Load.

14. Right-click the **External Loads** folder from the Simulation study tree.

15. Click **Remote Load/Mass**. The Remote Loads/Mass PropertyManager is displayed.

16. Click **Load (Direct transfer)** for Type.

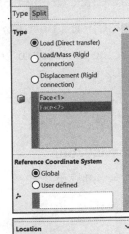

17. Select the **top face of the part** - for Faces for Remote Load. Note: Click **on both sides** of the Split line.

18. Set the Reference Coordinate system to **Global**.

19. Leave all of the location boxes at **zero** value.

20. Select the **Force** check box. Click the **Y-Direction** box.

21. Enter **200lb**. Use the X-, Y- and Z-direction boxes to direct the load; reverse direction as necessary.

Mesh and Run the model.

22. **Mesh and Run** the model. Use the standard default setting for the mesh.

23. Double-click the **Stress1** folder. View the results.

24. Double-click the **Displacement1** folder. View the results.

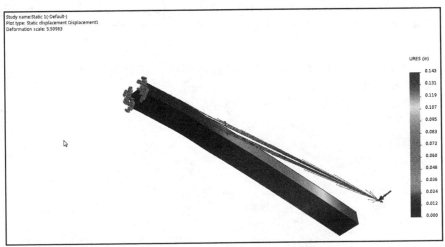

Study name:Static 1 (-Default-)
Plot type: Static displacement Displacement1
Deformation scale: 5.50983

Locate the displacement at 7cm using the Probe tool.

25. Click the **Displacement1** Results folder if needed.

26. Click **Probe** from the Plot Tools drop-down menu.

27. Click a **position** on the Split Line as illustrated. View the results.

At the distance of 7cm (70mm) the displacement is found to be 0.026in.

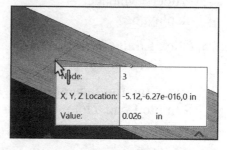

The correct answer is **B**.

A = 0.034in

B = 0.026in

C = 0.043in

D = 0.021in

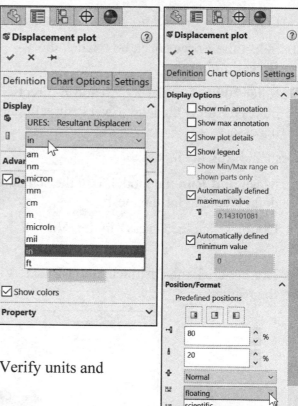

The purpose of this chapter is not to educate a new or intermediate user on SOLIDWORKS Simulation, but to cover and to inform you on the types of questions, layout and what to expect when taking the CSWSA-FEA exam.

The CSWSA-FEA exam requires that you work quickly. You can modify the units in the Plot menu. Right-click on the required Plot, and select Edit Definition.

Each hands-on problem in the CSWSA-FEA exam requires a single answer. To save time, set the Chart value to floating and the number of decimal places required. Right-click on the required Plot, and select Chart Options.

Decimal places require change often. Verify units and decimal places.

Tutorial FEA Model 8-4

An exam question in the Solid category could read:

In the figure displayed, what is the maximum resultant displacement in millimeters on the annular face of the model? The three holes are fixed.

Calculate the answer to 3 decimal places.

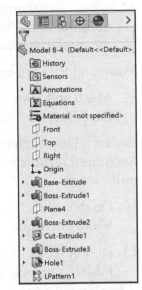

Given Information:

Material: Alloy Steel (SS).

A normal force, F_1 is applied to the annular face. F_1 = 3000lbf.

The three holes are **fixed**.

Use the default high quality element size to mesh.

Let's start.

1. **Open** Model 8-4 from the CSWSA-FEA Model folder. Think about the problem. Think about the model.

Set Simulation Options and start a Static SOLIDWORKS Simulation Study.

2. **Add-In** SOLIDWORKS Simulation.

3. Click **Simulation**, **Options** from the Main menu.

4. Click the **Default Options** tab.

Set the Unit system and Mesh quality.

5. Select **English (IPS)** and **Pressure/Stress (psi)**.

6. Click the **Mesh folder**.

7. Select **High** for Mesh quality.

8. **Create** a new Static Study. Accept the default name (Static #).

9. Apply **Material** - Alloy Steel (SS) from the SOLIDWORKS Simulation Library. A check mark is displayed next to the model name in the Study Simulation tree.

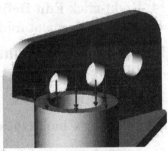

10. Apply Fixed Geometry **Fixtures**. Select the three cylindrical faces of the hole pattern as illustrated.

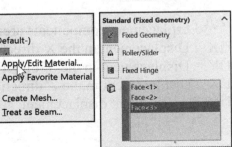

11. Apply an External Load (Force) of **3000lbf** on the annular face of the model as illustrated.

Mesh and Run the model.

12. **Mesh** and **Run** the model. Use the default setting for the mesh.

Create a Displacement Plot for the maximum resultant displacement in millimeters.

10. Double-click the **Stress1** folder. View the results.

13. Double-click the **Displacement1** folder. View the results.

14. Right-click **Edit Definition** from the Displacement folder.

15. Select **URES, Resultant Displacement**.

16. Select **mm** for units.

17. Click the **Chart Options** tab.

18. Select **Show max annotation**. The maximum displacement is displayed: 1.655mm.

The correct answer is **±1%** of this value.

The correct answer is **C**.

A = 1.112mm

B = 1.014mm

C = 1.655mm

D = 1.734mm

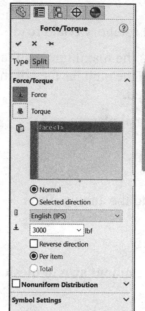

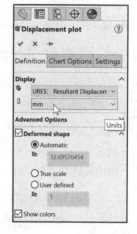

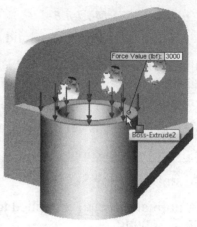

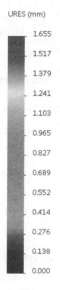

💡 Displacement components are:
UX = Displacement in the X-direction,
UY = Displacement in the Y-direction,
UZ = Displacement in the Z-direction,
URES = Resultant displacement.

After you calculate displacement or other parameters in a Simulation Study, the CSWSA-FEA exam will also deliver a series of successive questions to test the understanding of the results.

In the first question you calculate Displacement; in the second question you calculate Resultant Force. Do not re-mesh and re-run. Create the required parameter in the Results folder.

In the third question, you are asked to determine if the results are valid or invalid. If the materials yield strength was passed, then the results are invalid.

Use the Define Factor of Safety Plot to determine if your results are valid or invalid. The CSWSA-FEA exam requires you to apply Finite Element Method theory and review displacement values, factory of safety, mesh refinement, and material properties such as yield strength.

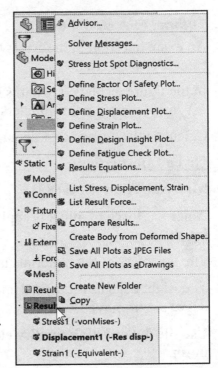

The following are some statements you will encounter in the CSWSA-FEA exam:

- The results are invalid because the material's yield strength was passed.

- The results area is invalid because the displacement is more than ½ the plate's thickness.

- The results are valid as they are, even if mesh refinement was better.

- The results are invalid because for sure a dynamic study is required.

In general use Thin Shells when thickness to span ratio < 0.05.

Displacement components are UX = Displacement in the X-direction, UY = Displacement in the Y-direction, UZ = Displacement in the Z-direction, URES = Resultant displacement.

Tutorial FEA Model 8-5

An exam question in the Sheet Metal category could read:

In the figure displayed, what is the maximum *UX* displacement in millimeters?

Calculate the answer to 4 decimal places.

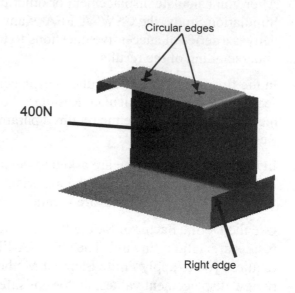

Circular edges

400N

Right edge

Given Information:

Material: Alloy Steel (SS).

A normal **force of 400N** is applied to the **inside** face.

The thickness of the materials is **0.15in**.

The Right edge as illustrated and circular edges are **immovable**.

Use the default high quality element size to mesh.

Let's start.

You need to define the thickness. The Fixture option Immovable is added for shell models. Thin models created with no Sheet Metal feature require you to define the use of Shell elements in SOLIDWORKS Simulation.

Use the models from the CSWSA-FEA Model folder for this section. Models created with the Sheet Metal feature automatically create Shell elements in SOLIDWORKS Simulation.

1. **Open** Model 8-5 from the CSWSA-FEA Model folder.

Set Simulation Options and start a Static SOLIDWORKS Simulation Study.

2. **Add-In** SOLIDWORKS Simulation.

3. Click **Simulation, Options** from the Main menu. Click the Default Options tab.

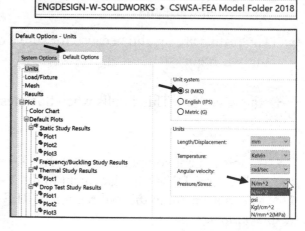

Set the Unit system and Mesh quality.

4. Select **SI (MKS)** and **Pressure/Stress (N/m²)**.

5. Click the **Mesh folder**.

6. Select **High** for Mesh quality.

7. **Create** a new Static Study. Accept the default name (Static #).

8. Apply **Material -** Alloy Steel (SS) from the SOLIDWORKS
    Simulation Library. A check mark is displayed next to the
    model name in the Study Simulation tree.

Define the Shell thickness.

9. Right-click **Model 8-5** in the Study tree.

10. Click **Edit Definition** from the drop-down menu.

11. Enter **0.15in** for **Thin** Type.

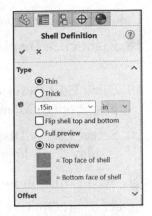

💡 In general use Thin Shells
when thickness to span ratio
< 0.05.

Apply Fixed Geometry Fixtures.

12. Click the **Immovable (No
    translation)** option.

13. Select the **edge** and the **two
    circular edges** of the model
    as illustrated.

Apply an External Load (Force)
of 400N.

14. Apply an **External Load** of
    400N normal to the inside
    face as illustrated.

Mesh and Run the model.

15. **Mesh** and **Run** the model.
    Use the standard default
    setting for the mesh. Review
    the Results.

View the Results.

16. Double-click the **Stress1**
    folder. View the results.

17. Double-click the
    Displacement1 folder. View
    the results.

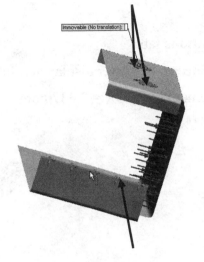

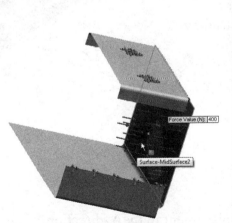

Calculate Displacement in X.

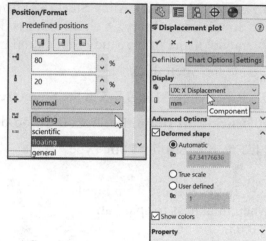

18. Right-click **Displacement1** from the study tree.

19. Click **Edit Definition**. The Displacement plot PropertyManager is displayed.

20. Select **UX: X Displacement** from the display drop-down menu.

21. Select Units in **mm**.

Select Chart Options.

22. Click the **Chart Options** tab.

23. Select **4 decimal places**, millimeter display and floating.

The maximum displacement in X, UX = 0.4370mm.

The correct answer is ±**1%** of this value.

The correct answer is **D**.

A = 0.4000mm

B = 0.4120mm

C = 1.655mm

D = 0.4370mm

When you enter a value in the CSWSA-FEA exam, include the required number of decimal places and leading and trailing zeroes.

🔆 In general use Thin Shells when thickness to span ratio < 0.05.

🔆 Displacement components are UX = Displacement in the X-direction, UY = Displacement in the Y-direction, UZ = Displacement in the Z-direction, URES = Resultant displacement.

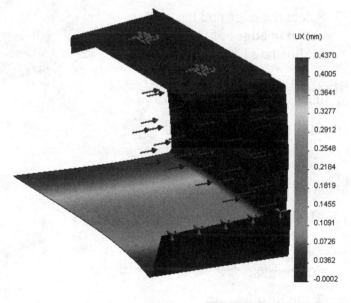

Screen shot from the exam

Definitions:

The following are a few key definitions for the exam:

Axisymmetry: Having symmetry around an axis.

Brittle: A material is brittle if, when subjected to stress, it breaks without significant deformation (strain). Brittle materials, such as concrete and carbon fiber, are characterized by failure at small strains. They often fail while still behaving in a linear elastic manner, and thus do not have a defined yield point. Because strains are low, there is negligible difference between the engineering stress and the true stress. Testing of several identical specimens will result in different failure stresses; this is due to the Weibull modulus of the brittle material.

Compatible meshing: A mesh where elements on touching bodies have overlaying nodes.

Cyclic Symmetry: To define the number of sectors and the axis of symmetry in a cyclic symmetric structure for use in a cyclic symmetry calculation.

Deflection: is a term to describe the magnitude to which a structural element bends under a load.

Deformation: is the change in geometry created when stress is applied (in the form of force loading, gravitational field, acceleration, thermal expansion, etc.). Deformation is expressed by the displacement field of the material.

Distributed Mass Load: Distributes a specified mass value on the selected faces for use with static, frequency, buckling, and linear dynamic studies. Use this functionality to simulate the effect of components that are suppressed or not included in the modeling when their mass can be assumed to be uniformly distributed on the specified faces. The distributed mass is assumed to lie directly on the selected faces, so rotational effects are not considered.

Ductile Material: In materials science, ductility is a solid material's ability to deform under tensile stress; this is often characterized by the material's ability to be stretched into a wire. Stress vs. Strain curve typical of aluminum.

Maximum Normal Stress criterion: The maximum normal stress criterion also known as Coulomb's criterion is based on the Maximum normal stress theory. According to this theory, failure occurs when the maximum principal stress reaches the ultimate strength of the material for simple tension.

This criterion is used for brittle materials. It assumes that the ultimate strength of the material in tension and compression is the same. This assumption is not valid in all cases. For example, cracks decrease the strength of the material in tension considerably while their effect is far smaller in compression because the cracks tend to close.

Brittle materials do not have a specific yield point, and hence it is not recommended to use the yield strength to define the limit stress for this criterion.

This theory predicts failure to occur when:

$$\sigma_1 \geq \sigma_{limit}$$

where σ_1 is the maximum principal stress.

Hence:

Factor of safety = $\sigma_{limit} / \sigma_1$

Maximum Shear Stress criterion: The maximum shear stress criterion, also known as Tresca yield criterion, is based on the Maximum Shear stress theory.

This theory predicts failure of a material to occur when the absolute maximum shear stress (τ_{max}) reaches the stress that causes the material to yield in a simple tension test. The Maximum shear stress criterion is used for ductile materials.

$$\tau_{max} \geq \sigma_{limit}/2$$

τ_{max} is the greatest of τ_{12}, τ_{23} and τ_{13}

Where:

$$\tau_{12} = (\sigma_1 - \sigma_2)/2; \; \tau_{23} = (\sigma_2 - \sigma_3)/2; \; \tau_{13} = (\sigma_1 - \sigma_3)/2$$

Hence:

Factor of safety (FOS) = $\sigma_{limit}/(2 * \tau_{max})$

Maximum von Mises Stress criterion: The maximum von Mises stress criterion is based on the von Mises-Hencky theory, also known as the Shear-energy theory or the Maximum distortion energy theory.

In terms of the principal stresses s1, s2, and s3, the von Mises stress is expressed as:

$$\sigma_{vonMises} = \{[(s1 - s2)2 + (s2 - s3)2 + (s1 - s3)2]/2\}(1/2)$$

The theory states that a ductile material starts to yield at a location when the von Mises stress becomes equal to the stress limit. In most cases, the yield strength is used as the stress limit. However, the software allows you to use the ultimate tensile or set your own stress limit.

$$\sigma_{vonMises} \geq \sigma_{limit}$$

Yield strength is a temperature-dependent property. This specified value of the yield strength should consider the temperature of the component. The factor of safety at a location is calculated from:

Factor of Safety (FOS) = $\sigma_{limit} / \sigma_{vonMises}$

Modulus of Elasticity or Young's Modulus: The Elastic Modulus (Young's Modulus) is the slope defined as stress divided by strain. E = modulus of elasticity (Pa (N/m^2), N/mm^2, psi). The Modulus of Elasticity can be used to determine the stress-strain relationship in the linear-elastic portion of the stress-strain curve. The linear-elastic region is either below the yield point, or if a yield point is not easily identified on the stress-strain plot it is defined to be between 0 and 0.2% strain, and is defined as the region of strain in which no yielding (permanent deformation) occurs.

Force is the action of one body on another. A force tends to move a body in the direction of its action.

Mohr-Coulomb: The Mohr-Coulomb stress criterion is based on the Mohr-Coulomb theory, also known as the Internal Friction theory. This criterion is used for brittle materials with different tensile and compressive properties. Brittle materials do not have a specific yield point, and hence it is not recommended to use the yield strength to define the limit stress for this criterion.

Mohr-Coulomb Stress criterion: The Mohr-Coulomb stress criterion is based on the Mohr-Coulomb theory, also known as the Internal Friction theory. This criterion is used for brittle materials with different tensile and compressive properties. Brittle materials do not have a specific yield point and hence it is not recommended to use the yield strength to define the limit stress for this criterion.

This theory predicts failure to occur when:

$$\sigma_1 \geq \sigma_{TensileLimit} \quad \text{if } \sigma_1 > 0 \text{ and } \sigma_3 > 0$$

$$\sigma_3 \geq -\sigma_{CompressiveLimit} \quad \text{if } \sigma_1 < 0 \text{ and } \sigma_3 < 0$$

$$\sigma_1 / \sigma_{TensileLimit} + \sigma_3 / -\sigma_{CompressiveLimit} \geq 1 \text{ if } \sigma_1 \geq 0 \text{ and } \sigma_3 \leq 0$$

The factor of safety is given by:

$$\text{Factor of Safety (FOS)} = \{\sigma_1 / \sigma_{TensileLimit} + \sigma_3 / -\sigma_{CompressiveLimit}\}^{(-1)}$$

Stress: Stress is defined in terms of Force per unit Area:

$$Stress = \frac{f}{A}.$$

Stress vs. Strain diagram: Many materials display linear elastic behavior, defined by a linear stress-strain relationship, as shown in the figure up to point 2, in which deformations are completely recoverable upon removal of the load; that is, a specimen loaded elastically in tension will elongate but will return to its original shape and size when unloaded. Beyond this linear region, for ductile materials, such as steel, deformations are plastic. A plastically deformed specimen will not return to its original size and shape

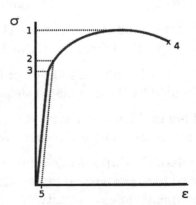

Stress vs. Strain curve typical of aluminum

when unloaded. Note that there will be elastic recovery of a portion of the deformation. For many applications, plastic deformation is unacceptable and is used as the design limitation.

1 - Ultimate Strength

2 - Yield Strength

3 - Proportional Limit Stress

4 - Rupture

5 - Offset Strain (usually 0.002)

Tensile strength: Ultimate tensile strength (UTS), often shortened to tensile strength (TS) or ultimate strength, is the maximum stress that a material can withstand while being stretched or pulled before necking, which is when the specimen's cross-section starts to significantly contract. Tensile strength is the opposite of compressive strength and the values can be quite different.

Yield Stress: The stress level beyond which the material becomes plastic.

Yield Strength: is the lowest stress that produces a permanent deformation in a material. In some materials, like aluminum alloys, the point of yielding is difficult to identify, thus it is usually defined as the stress required to cause 0.2% plastic strain. This is called a 0.2% proof stress.

Young's Modulus, or the "Modulus of Elasticity: The Elastic Modulus (Young's Modulus) is the slope defined as stress divided by strain. E = modulus of elasticity (Pa (N/m2), N/mm2, psi). The Modulus of Elasticity can be used to determine the stress-strain relationship in the linear-elastic portion of the stress-strain curve. The linear-elastic region is either below the yield point, or if a yield point is not easily identified on the stress-strain plot it is defined to be between 0 and 0.2% strain, and is defined as the region of strain in which no yielding (permanent deformation) occurs.

Thermal Expansion: The change in length per unit length per one degree change in temperature (change in normal strain per unit temperature) (K).

Specific Heat: Quantity of heat needed to raise the temperature of a unit mass of the material by one degree of temperature (J/kg K).

Density: Mass per unit volume (kg/m^3).

Elastic Modulus (Young's Modulus): Ratio between the stress and the associated strain in a specified direction (N/m^2).

Shear Modulus (Modulus of Rigidity): Ratio between the shearing stress in a plane divided by the associated shearing strain (N/m^2).

Thermal Conductivity: Rate of heat transfer through a unit thickness of the material per unit temperature difference (W/m K).

Poisson's Ratio: Ratio between the contraction (traverse strain), normal to the applied load to the extension (axial strain), in the direction of the applied load. Poisson's ratio is a dimensionless quantity.

Tensile Strength: The maximum amount of tensile stress that a material can be subjected to before failure (N/m^2).

Yield Strength: Stress at which the material becomes permanently deformed (N/m^2).

Notes:

Project 9

Intelligent Modeling Techniques

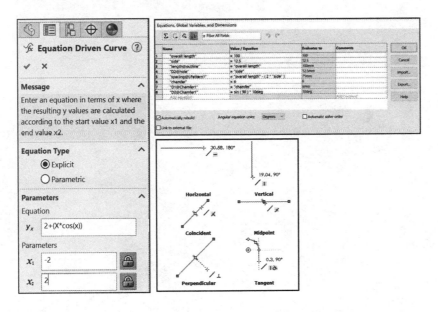

Below are the desired outcomes and usage competencies based on the completion of Project 9.

Project Desired Outcomes:	Usage Competencies:
• Utilize and understand various Design Intent tools and Intelligent Modeling Techniques.	• Apply Design Intent and Intelligent Modeling Techniques in a Sketch, Feature, Part, Plane, Assembly and Drawing. • Product Assembly Visualization tables and plots.
• Awareness of the Equation Driven Sketch Curve tool.	• Create Explicit Dimension Driven and Parametric Equation Driven Curves. • Generate curves using the Curve Through XYZ Points tool for a NACA aerofoil and CAM data text file.
• Knowledge of a few SOLIDWORKS Xpert tools.	• Apply SketchXpert, DimXpert and FeatureXpert.

Notes:

Project 9 - Intelligent Modeling Techniques

Project Objective

Understand some of the available tools in SOLIDWORKS to perform intelligent modeling. Intelligent modeling is incorporating design intent into the definition of the sketch, feature, part, and assembly or drawing document. Intelligent modeling is most commonly addressed through design intent using a:

- Sketch:
  - Geometric relations
  - Fully defined Sketch tool
  - SketchXpert
  - Equations:
    - Explicit Dimension Driven
    - Parametric Driven Curve
  - Curves:
    - Curve Through XYZ Points
    - Projected Composite
- Feature:
  - End Conditions:
    - Blind, Through All, Up to Next, Up to Vertex, Up to Surface, Offset from Surface, Up to Body and Mid Plane
  - Along a Vector
  - FeatureXpert (Constant Radius)
  - Symmetry:
    - Mirror
- Plane
- Assembly:
  - Symmetry (Mirror/Pattern)
  - Assembly Visualization
  - SOLIDWORKS Sustainability
  - MateXpert
- Drawing:
  - DimXpert (Slots, Pockets, Machined features, etc.)

This project uses short step-by-step tutorials to practice and reinforce the subject matter and objectives. Learn by doing, not just by reading. All models are located in the Chapter 9 Intelligent Modeling folder.

Design Intent

What is design intent? All designs are created for a purpose. Design intent is the intellectual arrangement of features and dimensions of a design. Design intent governs the relationship between sketches in a feature, features in a part and parts in an assembly or drawing document.

The SOLIDWORKS definition of design intent is the process in which the model is developed to accept future modifications. Models behave differently when design changes occur.

Design for change. Utilize geometry for symmetry, reuse common features, and reuse common parts. Build change into the following areas that you create: sketch, feature, part, assembly and drawing.

Design Intent is how your part reacts as parameters are modified. Example: If you have a hole in a part that must always be .125≤ from an edge, you would dimension to the edge rather than to another point on the sketch. As the part size is modified, the hole location remains .125≤ from the edge.

Sketch

In SOLIDWORKS, relations between sketch entities and model geometry, in either 2D or 3D sketches, are an important means of building in design intent. In this chapter - we will only address 2D sketches.

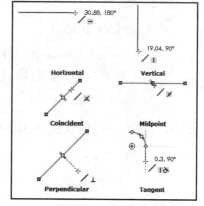

Apply design intent in a sketch as the profile is created. A profile is determined from the Sketch Entities. Example: Rectangle, Circle, Arc, Point, Slot etc.

Develop design intent as you sketch with Geometric relations. Sketch relations are geometric constraints between sketch entities or between a sketch entity and a plane, axis, edge, or vertex. Relations can be added automatically or manually.

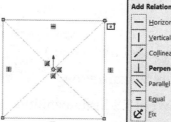

As you sketch, allow SOLIDWORKS to automatically add relations. Automatic relations rely on:

- Inferencing.
- Pointer display.
- Sketch snaps and Quick Snaps.

After you sketch, manually add relations using the Add Relations tool, or edit existing relations using the Display/Delete Relations tool.

Fully Defined Sketch

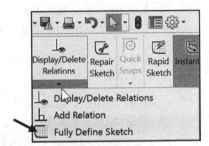

Sketches are generally in one of the following states:

- Under defined.
- Fully defined.
- Over defined.

Although you can create features using sketches that are not fully defined, it is a good idea to always fully define sketches for production models. Sketches are parametric, and if they are fully defined, changes are predictable. However, sketches in drawings, although they follow the same conventions as sketches in parts, do not need to be fully defined since they are not the basis of features.

SOLIDWORKS provides a tool to help the user fully define a sketch. The Fully Defined Sketch tool provides the ability to calculate which dimensions and relations are required to fully define under defined sketches or selected sketch entities. You can access the Fully Define Sketch tool at any point and with any combination of dimensions and relations already added. See Chapter 5 for additional information.

Your sketch should include some dimensions and relations before you use the Fully Define Sketch tool.

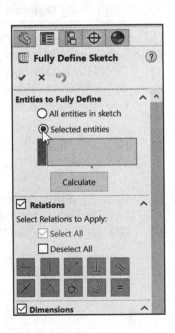

The Fully Define Sketch tool uses the Fully Define Sketch PropertyManager. The Fully Define Sketch PropertyManager provides the following selections:

- **Entities to Fully Define**. The Entities to Fully Define box provides the following options:

  - **All entities in sketch**. Fully defines the sketch by applying combinations of relations and dimensions.

  - **Selected entities**. Provides the ability to select sketch entities.

    - **Entities to Fully Define**. Only available when the Selected entities box is checked. Applies relations and dimensions to the specified sketch entities.

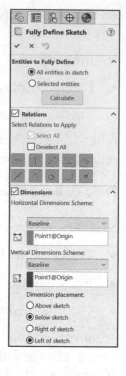

- **Calculate**. Analyzes the sketch and generates the appropriate relations and dimensions.

- *Relations*. The Relations box provides the following selections:

  - **Select All**. Includes all relations in the results.

  - **Deselect All**. Omits all relations in the results.

  - **Individual relations**. Include or exclude needed relations. The available relations are **Horizontal**, **Vertical**, **Collinear**, **Perpendicular**, **Parallel**, **Midpoint**, **Coincident**, **Tangent**, **Concentric** and **Equal radius/Length**.

- *Dimensions*. The Dimensions box provides the following selections:

  - **Horizontal Dimensions**. Displays the selected Horizontal Dimensions Scheme and the entity used as the Datum - Vertical Model Edge, Model Vertex, Vertical Line or Point for the dimensions. The available options are **Baseline**, **Chain** and **Ordinate**.

  - **Vertical Dimensions.** Displays the selected Vertical Dimensions Scheme and the entity used as the Datum - Horizontal Model Edge, Model Vertex, Horizontal Line or Point for the dimensions. The available options are **Baseline**, **Chain** and **Ordinate**.

  - **Dimension**. Below sketch and Left of sketch is selected by default. Locates the dimension. There are four selections: **Above sketch**, **Below the sketch**, **Right of sketch** and **Left of sketch**.

Activity: Fully Defined Sketch

Close all parts, assemblies and drawings. Apply the Fully Defined Sketch tool. Modify the dimension reference location in the sketch profile with control points.

1) Open **Fully Defined Sketch 9-1** from the ENGDESIGN-W-SOLIDWORKS\Chapter 9 Intelligent Modeling folder. View the FeatureManager. Sketch1 is under defined.

2) **Edit** Sketch1. The two circles are equal and symmetrical about the y axis. The rectangle is centered at the origin.

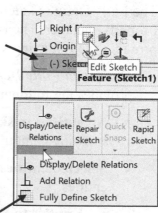

3) Click the **Fully Define Sketch** tool from the Consolidated Display/Delete Relations drop-down menu. The Fully Defined Sketch PropertyManager is displayed.

4) The All entities in the sketch are selected by default. Click **Calculate**. View the results. Sketch1 is fully defined to the origin.

5) Click **OK** ✔ from the PropertyManager. Drag all dimensions off the profile.

6) Modify the **vertical dimension to 50mm** and the **diameter dimension to 25mm** as illustrated. SOLIDWORKS suggests a dimension scheme to create a fully defined sketch.

7) Click **View**, **Hide/Show**, uncheck **Sketch Relations** from the Main menu.

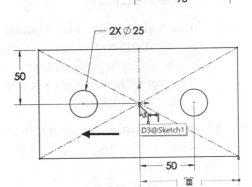

Modify the dimension reference location in the sketch profile with control points.

8) Click the **90**mm dimension in the Graphic window.

9) Click and drag the **left control point** as illustrated.

10) Release the mouse pointer on the **left vertical line** of the profile. View the new dimension reference location of the profile.

11) Repeat the same procedure for the horizontal **50mm** dimension. Select the new reference location to the left hole diameter as illustrated.

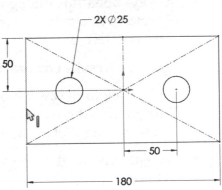

Close the model.

12) **Close** the model. View the results.

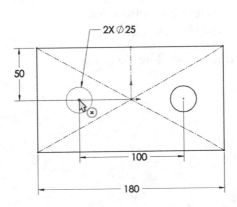

SketchXpert

SketchXpert resolves conflicts in over defined sketches and proposes possible solution sets. Color codes are displayed in the SOLIDWORKS Graphics window to represent the sketch states. The SketchXpert tool uses the SketchXpert PropertyManager. The SketchXpert PropertyManager provides the following selections:

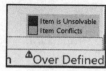

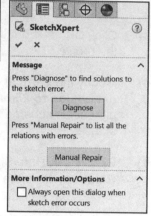

- *Message*. The Message box provides access to the following selections:

  - **Diagnose**. The Diagnose button generates a list of solutions for the sketch. The generated solutions are displayed in the Results section of the SketchXpert PropertyManager.

  - **Manual Repair**. The Manual Repair button generates a list of all relations and dimensions in the sketch. The Manual Repair information is displayed in the Conflicting Relations/Dimensions section of the SketchXpert PropertyManager.

- *More Information/Options*. Provides information on the relations or dimensions that would be deleted to solve the sketch.

  - **Always open this dialog when sketch error occurs**. Selected by default. Opens the dialog box when a sketch error is detected.

- *Results*. The Results box provides the following selections:

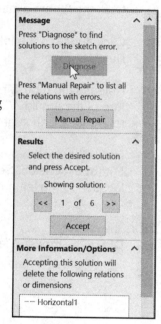

  - **Left or Right arrows**. Provides the ability to cycle through the solutions. As you select a solution, the solution is highlighted in the Graphics window.

  - **Accept**. Applies the selected solution. The sketch is fully defined.

- *More Information/Options*. The More Information/Options box provides the following selections:

  - **Diagnose**. The Diagnose box displays a list of the valid generated solutions.

  - **Always open this dialog when sketch error occurs**. Selected by default. Opens the dialog box when a sketch error is detected.

- *Conflicting Relations/Dimensions*. The Conflicting Relations/Dimensions box provides the ability to select a displayed conflicting relation or dimension. The selected item is highlighted in the Graphics window. The options include:

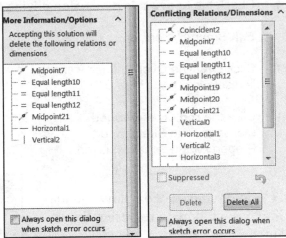

  - **Suppressed**. Suppresses the relation or dimension.

  - **Delete**. Removes the selected relation or dimension.

  - **Delete All**. Removes all relations and dimensions.

  - **Always open this dialog when sketch error occurs**. Selected by default. Opens the dialog box when a sketch error is detected.

Activity: SketchXpert

Close all parts, assemblies and drawings. Create an over defined sketch. Apply SketchXpert to select a solution.

1) Open **SketchXpert 9-1** from the ENGDESIGN-W-SOLIDWORKS\Chapter 9 Intelligent Modeling folder. View the FeatureManager.

2) **Edit** Sketch1. Sketch1 is fully defined. The rectangle has a midpoint relation to the origin, and an equal relation with all four sides. The top horizontal line is dimensioned. Insert a dimension to create an over defined sketch.

3) Click **Smart Dimension**.

4) Add a dimension to the **left vertical line**. This makes the sketch over-defined. The Make Dimension Driven dialog box is displayed.

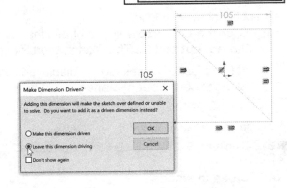

5) Check the **Leave this dimension driving** box option.

6) **Exit** Smart Dimension.

7) Click **OK**. The Over Defined warning is displayed.

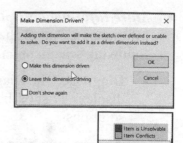

8) Click the **red Over Defined** message. The SketchXpert PropertyManager is displayed.

💡 Color codes are displayed in the Graphics window to represent the sketch states.

9) Click the **Diagnose** button. The Diagnose button generates a list of solutions for your sketch. You can either accept the first solution or click the Right arrow key in the Results box to view the section solution. The first solution is to delete the vertical dimension of 105mm.

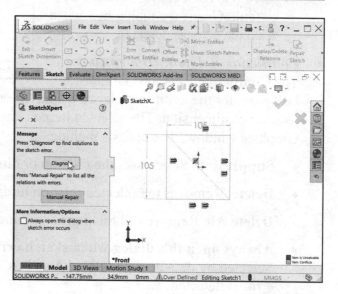

View the second solution.
10) Click the **Right arrow** key in the Results box. The second solution is displayed. The second solution is to delete the horizontal dimension of 105mm.

View the third solution.
11) Click the **Right arrow** key in the Results box. The third solution is displayed. The third solution is to delete the Equal relation between the vertical and horizontal lines.

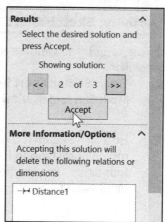

Accept the second solution.
12) Click the **Left arrow** key to obtain the second solution. Click the **Accept** button. The SketchXpert tool resolves the over-defined issue. A message is displayed. Click **OK** ✔ from the SketchXpert PropertyManager.

Rebuild and close the model.
13) **Rebuild** 🔋 the model. View the results.

14) **Close** the model.

Equations

Dimension driven by equations

You want to design a hinge that you can modify easily to make similar sizes. You need an efficient way to create multi sizes. Equations create a mathematical relation between dimensions. You can use equations to locate entities instead of setting explicit dimensions.

In the example below, you set one screw hole location to be one half the height of the hinge, and the other screw hole to be one third the length of the hinge. If you change the height or length hinge dimension, the screw holes maintain this mathematical relation. Below I used the Mirror feature with the equation to quickly copy the existing hole feature across the Front plane.

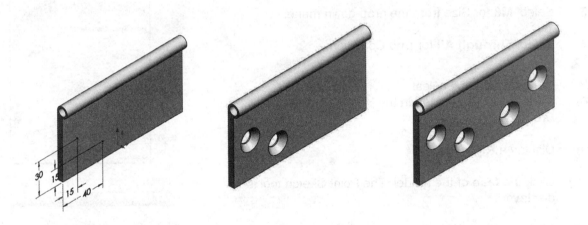

💡 You can add comments to equations to document your design intent. Place a single quote (') at the end of the equation, then enter the comment. Anything after the single quote is ignored when the equation is evaluated. Example: "D2@Sketch1" = "D1@Sketch1" / 2 'height is 1/2 width. You can also use comment syntax to prevent an equation from being evaluated. Place a single quote (') at the beginning of the equation. The equation is then interpreted as a comment, and it is ignored. See SOLIDWORKS Help for additional information.

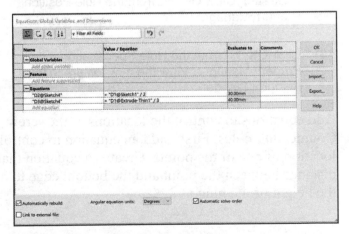

Activity: Equations

Close all parts, assemblies and drawings. Create two driven equations. Insert two Countersink holes. To position each hole on the hinge, one dimension is fixed, and the other is driven by an equation.

1) Open the **Hinge 9-1** part from the ENGDESIGN-W-SOLIDWORKS\Chapter 9 Intelligent Modeling folder.

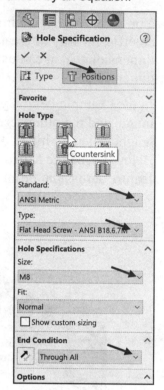

Add two Countersink holes using the Hole Wizard.

2) Click **Hole Wizard** from the Features toolbar. The Hole Specification PropertyManager is displayed.

3) Click **Countersink** for Hole Type as illustrated.

4) Select **ANSI Metric** for Standard. Accept the default Type.

5) Select **Flat Head Screw - ANSI B18.6.7M**.

6) Select **M8** for Size from the drop-down menu.

7) Select **Through All** for End Condition.

Place the location of the holes.

8) Click the **Positions** tab in the Hole Specification PropertyManager.

9) Display a **Right** view.

10) Click the **face** of the model. The Point Sketch tool is displayed.

11) Click to **place the holes** approximately as shown. **De-select** the Point Sketch tool.

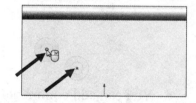

Dimension the holes.

12) Click **Smart Dimension**. **Dimension** the holes as illustrated. Click **OK** from the Hole Position PropertyManager.

13) Click **OK** from the Hole Specification PropertyManager. View the new feature in the FeatureManager.

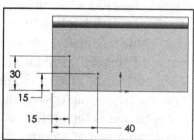

Add equations to control the locations of the screw Countersink holes. First - add an equation to control the location of one of the points. Create an equation that sets the distance between the point and the bottom edge to one-half the height of the hinge.

14) Right-click **Sketch4**. Click **Edit Sketch**.

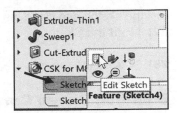

15) Double-click **Extruded-Thin1** from the FeatureManager. Dimensions are displayed in the Graphics window.

16) Click **Tools**, **Equations** from the Menu bar. The Equations, Global Variables, and Dimensions dialog box is displayed.

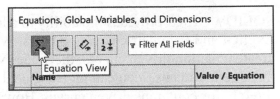

17) Click the **Equations View** Σ icon as illustrated. Click inside the **first empty** cell under Equations.

18) Click the **30mm** dimension in the Graphics window. An = sign is displayed in the Value/Equation cell. Click the **60mm** dimension in the Graphics window.

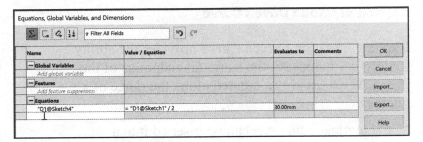

19) Enter **/2** in the dialog box to complete the dimension.

20) Click inside the **Value / Equation** box. This equation sets the distance between the point and the bottom edge to one-half the height of the hinge.

Create the second equation. Add an equation to control the location of the other point.
21) Click inside the **first empty** cell under Equations.

22) Click the **40mm** dimension from the Graphics window. An = sign is displayed in the Value / Equation cell.

23) Click the **120mm** dimension for the base.

24) Enter **/3** in the dialog box to complete the dimension.

25) Click inside the **Evaluate to** box. View the active equations. This sets the distance between the point and the side edge to one-third the length of the hinge.

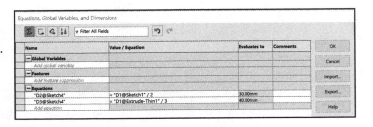

26) Click **OK** from the dialog box.

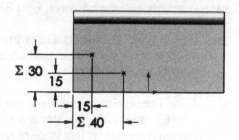

Exit the sketch and close the model.
27) **Exit** the Sketch.

28) **Close** the model.

Equation Driven Curve

SOLIDWORKS provides the ability to address
Explicit and Parametric equation types. When you
create equation driven curves, the values you use
must be in radians. You cannot use global variables
directly for equations driven curves. However, you
can create a global variable and associate it with a
dimension, then use the dimension in the equation
for the curve.

Explicit Equation Driven Curve

Explicit Equation Driven Curve provides the ability
to define x values for the start and endpoints of the
range. Y values are calculated along the range of x
values.

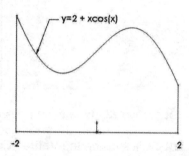

Mathematical equations can be inserted into a
sketch. Utilize parentheses to manage the order of
operations. For example, calculate the volume of a
solid bounded by a curve as illustrated. The region
bounded by the equation $y = (2+x*\cos(x))$, and the
x-axis, over the interval $x = -2$ to $x = 2$, is revolved
about the x-axis to generate a solid.

Activity: Explicit Equation Driven Curve

Create an Explicit Equation Driven Curve on the Front plane. Revolve the curve. Add material. Calculate the volume of the solid.

1) Create a **New** 📄 part. Use the default ANSI, IPS Part template.

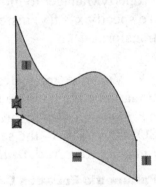

Create a 2D Sketch on the Front Plane.

2) Right-click **Sketch** from the Front Plane in the FeatureManager. The Sketch toolbar is displayed. Front Plane is your Sketch plane.

3) Click the **Equation Driven Curve Sketch** 𝒇𝓍 tool from the Consolidated drop-down menu. The Equation Driven Curve PropertyManager is displayed. Explicit is selected by default.

4) Enter the **Equation y$_x$** as illustrated.

5) Enter the **parameters x$_1$, x$_2$** that defines the lower and upper bound of the equation as illustrated.

6) Click **OK** ✔ from the Equation Driven Curve Sketch PropertyManager. View the curve in the Graphics window. Size the curve in the Graphics window. The Sketch is under defined.

7) Insert **three lines to close** the profile as illustrated. Insert a **Coincident** relation between the origin and the front left vertex as illustrated. Insert **dimensions** to fully define the sketch.

Create a Revolved feature.
8) Click the **horizontal line**.

9) Click the **Revolved Boss/Base** 🍥 tool from the Feature toolbar. 360 degrees is the default.

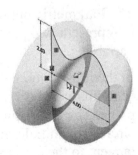

10) Click **OK** ✔ from the PropertyManager. View the results in the Graphics window. Revolve1 is displayed. Utilize the Section tool, parallel with the Right plane to view how each cross section is a circle.

Apply material to the model.
11) Apply **Brass** for material.

12) Calculate the **volume** of the part using the Mass Properties tool. View the results. Also note the surface area and the Center of mass.

Save the model.
13) **Save** the model.

Name the model. Close the model.
14) Name the model **Explicit Equation Driven Curve 9-1**.

15) **Close** the model.

💡 You can create parametric (in addition to explicit) equation-driven curves in both 2D and 3D sketches.

💡 Use regular mathematical notation and order of operations to write an equation. x_1 and x_2 are for the beginning and end of the curve. Use the transform options at the bottom of the PropertyManager to move the entire curve in x-, y- or rotation. To specify $x = f(y)$ instead of $y = f(x)$, use a 90 degree transform.

💡 View the .mp4 files to better understand the potential of the Equation Driven Curve tool. The first one: *Calculating Area of a region bounded by two curves (secx)^2 and sin x* in SOLIDWORKS - the second one: *Determine the Volume of a Function Revolved Around the x Axis* in SOLIDWORKS.

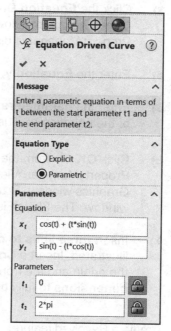

Parametric Equation Driven Curve

The Parametric option of Equation Driven Curve can be utilized to represent two parameters, x- and y-, in terms of a third variable, t.

In the illustration below, a string is wound about a fixed circle of radius 1 and is then unwound while being held taut in the x-y-plane. The end point of the string P traces an involute of the circle. The initial point (1,0) is on the x- axis and the string is tangent to the circle at Q. The angle t is measured in radians from the positive x- axis.

Use the Equation Driven Curve, Parametric option to illustrate how the parametric equations $x = \cos t + t \sin t$ and $y = \sin t - t \cos t$ represent the involute of the circle from $t = 0$ to $t = 2\pi$.

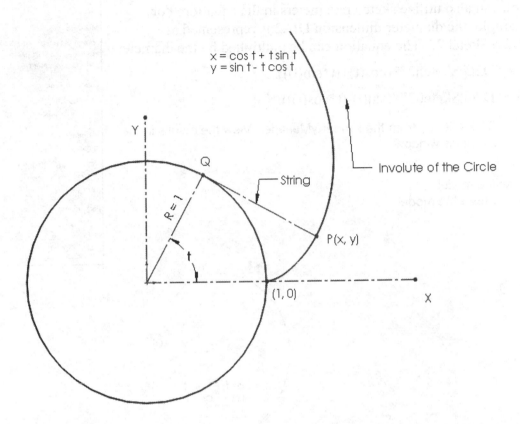

💡 Rename a feature or sketch for clarity. Slowly click the feature or sketch name twice and enter the new name when the old one is highlighted.

Activity: Parametric Equation Driven Curve

Close all parts, assemblies and drawings. Create a Parametric Equation Driven Curve.

1)   Open the **Parametric Equation 9.1** part from the ENGDESIGN-W-SOLIDWORKS\Chapter 9 Intelligent Modeling folder. The circle of radius 1 is displayed.

2)   **Edit** Sketch1 from the FeatureManager.

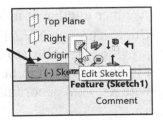

Activate the Equation Driven Curve tool.

3)   Click **Tools**, **Sketch Entities**, **Equation Driven Curve** from the Menu bar. The Equation Driven Curve PropertyManager is displayed.

4)   Click **Parametric** in the Equation Type dialog box.

5) Enter the parametric equations for x_t and y_t as illustrated.

6) Enter t_1 and t_2 as illustrated.

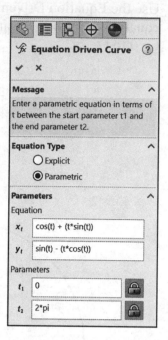

You can also utilize sketch parameters in the equation. For example, the diameter dimension DIA2 is represented as "D2@Sketch2." The equation can be multiplied by the diameter.

X = "D2@Sketch2"*(cos(t)+(t*sin(t))).

Y = "D2@Sketch2"*(sin(t)-(t*cos(t))).

7) Click **OK** ✔ from the PropertyManager. View the results in the Graphics window.

Close the model.
8) Close the model.

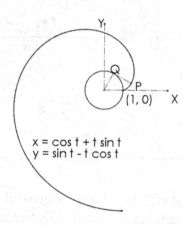

$$x = \cos t + t \sin t$$
$$y = \sin t - t \cos t$$

Curves

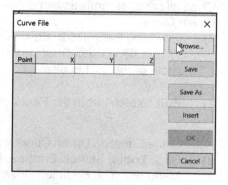

Another intelligent modeling technique is to apply the Curve Through XYZ Points ♌ feature. This feature provides the ability to either type in (using the Curve File dialog box) or click Browse and import a text file with x-, y-, z-, coordinates for points on a curve.

Generate the text file by any program which creates columns of numbers. The Curve 𝒰 feature reacts like a default spline that is fully defined.

☀ It is highly recommended that you insert a few extra points on either end of the curve to set end conditions or tangency in the Curve File dialog box.

Imported files can have an extension of either *.sldcrv or *.text. The imported data x-, y-, z-, must be separated by a space, a comma, or a tab.

The National Advisory Committee for Aeronautics (NACA) developed airfoil shapes for aircraft wings. The shape of the airfoil is defined by parameters in the numerical code that can be entered into equations to generate accurate cross-sections.

Activity: Curve Through XYZ Points

Create a curve using the Curve Through XYZ Points tool and the Composite curve tool. Import the x-, y-, z- data for an NACA airfoil for various cross sections.

1) Create a **New** 🗋 part. Use the default ANSI, MMGS Part template.

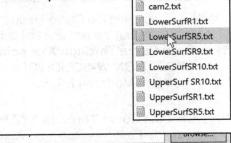

Browse to select the imported x-, y-, z- data.

2) Click **Insert**, **Curve**, **Curve Through XYZ Points** from the Menu bar. The Curve File dialog box is displayed.

3) **Browse** to the ENGDESIGN-W-SOLIDWORKS\Chapter 9 Intelligent Modeling folder.

4) Select **Text Files (*.txt)** for file type.

5) Double-click **LowerSurfSR5.txt**. View the results in the Curve File dialog box.

Point	X	Y	Z		
1	0in	0in	187.81in		Save
2	2.91in	-3.32in	187.81in		
3	5.81in	-4.53in	187.81in		
4	11.62in	-5.79in	187.81in		Save As
5	17.43in	-6.37in	187.81in		
6	23.24in	-6.65in	187.81in		Insert
7	34.86in	-6.69in	187.81in		
8	46.48in	-6.37in	187.81in		OK
9	58.1in	-5.81in	187.81in		
10	69.72in	-5.25in	187.81in		Cancel
11	92.96in	-4.18in	187.81in		

6) Click **OK** from the Curve File dialog box. Curve1 is displayed in the FeatureManager.

7) **Repeat** the above procedure to create Curve2 from the UpperSurfSR5.txt file. Both the Lower and Upper curves are on the same Z plane. They are separate entities.

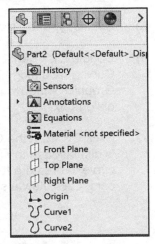

Use the Composite curve tool to join the Lower and Upper curves.

8)   Click **Insert**, **Curve**, **Composite** from the Menu bar.

9)   Select **Curve1** and **Curve2** from the fly-out FeatureManager as illustrated. Both curves are displayed in the Entities to Join box.

10)   Click **OK** ✔ from the PropertyManager. View the results in the FeatureManager.

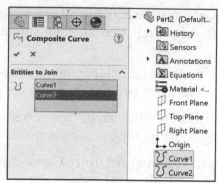

Close the model.

11)   **Close** the model.

Activity: Curve Through XYZ Points - Second Tutorial

Create a curve using the Curve Through XYZ Points tool. Import the x-, y-, z- data from a CAM program. Verify that the first and last points in the curve file are the same for a closed profile.

1)   Open **Curve Through XYZ points 9-2** from the ENGDESIGN-W-SOLIDWORKS\ Chapter 9 Intelligent Modeling folder.

2)   Click the **Curve Through XYZ Points** ℧ tool from the Features CommandManager. The Curve File dialog box is displayed.

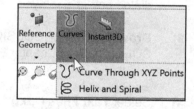

Import curve data. Additional zero points were added on either end of the curve to set end conditions.

3)   Click **Browse** from the Curve File dialog box.

4)   Browse to the **ENGDESIGN-W-SOLIDWORKS\Chapter 9 - Intelligent Modeling** folder.

5)   Set file type to **Text Files**.

6)   Double-click **cam2.text**. View the data in the Curve File dialog box. View the sketch in the Graphics window. Review the data points in the dialog box.

7)   Click **OK** from the Curve File dialog box. Curve1 is displayed in the FeatureManager. Use the curve to create a fully defined base sketch of a cam. Let's view a sample model using the curve.

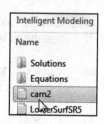

Close the existing model.

8)   Close the model.

9)   Open Curve Through XYZ points 9-3
from the ENGDESIGN-W-
SOLIDWORKS\Chapter 9 - Intelligent
Modeling folder to view the final results.
Curve1 is used to fully define Sketch1.

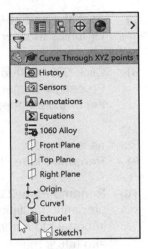

Close the model.

10)   Close the model.

Activity: Projected Composite Curves

Create a plane and sketch for two composite curves. Note: Before you create a Loft feature, you must insert a plane for each curve and add additional construction geometry to the sketch.

1)   Open the Projected Composite Curves 9**-1** part from the ENGDESIGN-W-SOLIDWORKS \Chapter 9 Intelligent Modeling folder. Two NACA airfoil profiles are displayed created from x-, y-, z- data.

Create a reference plane through each point of the composite curve.

2)   Click Insert, Reference Geometry, Plane from the Menu bar. The Plane PropertyManager is displayed.

3)   Click Front Plane for First
Reference from the fly-out
FeatureManager.

4)   Click the end point of CompCurve1
in the Graphics window as illustrated
for Second Reference. The Plane is
fully defined.

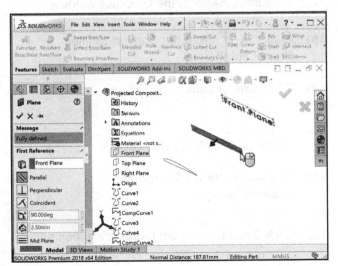

5)   Click OK ✔ from the Plane
PropertyManager. Plane1 is
displayed in the FeatureManager.

Create a Reference Plane through
CompCurve 2, parallel to Plane1.

6)   Click Insert, Reference Geometry,
Plane from the Menu bar. The Plane
PropertyManager is displayed.
Plane1 is the First Reference.

7) Click the **end point** of CompCurve2 in the Graphics window as illustrated.

8) Click **OK** ✔ from the Plane PropertyManager. Plane2 is displayed in the FeatureManager.

Rename features.

9) **Rename** Plane1 to PlaneSR5.

10) **Rename** Plane2 to PlaneSR9. When preparing multiple cross sections it is important to name the planes for clarity and future modification in the design.

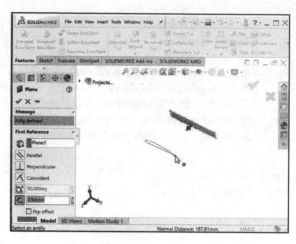

Convert CompCurve1 to Plane SR5 and CompCurve2 to Plane SR9.

11) Right-click **PlaneSR5** from the FeatureManager.

12) Click **Sketch** ⊞ from the Context toolbar.

13) Click the **Convert Entities** ⬒ Sketch tool.

14) Click **CompCurve1** from the fly-out FeatureManager in the Graphics window.

15) Click **OK** ✔ from the Convert Entities PropertyManager.

16) Click **Exit Sketch**.

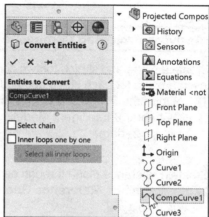

Rename Sketch1.

17) **Rename** Sketch1 to SketchSR5.

18) Right-click **PlaneSR9** from the FeatureManager.

19) Click **Sketch** ⊞ from the Context toolbar.

20) Click the **Convert Entities** ⬒ Sketch tool.

21) Click **CompCurve2** from the fly-out FeatureManager in the Graphics window.

22) Click **OK** ✔ from the Convert Entities PropertyManager.

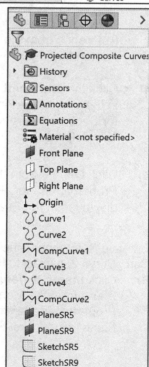

23) Click **Exit Sketch**.

24) **Rename** Sketch2 to SketchSR9.

25) Display an **Isometric** view.

26) **Show** Front Plane. View the results in the Graphics window.

27) **Close** the model.

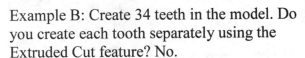

To save time in developing a series of curves, sketches can be copied and rotated.

Feature - End Conditions

Build design intent into a feature by addressing End Conditions (Blind, Through All, Up to Next, Up to Vertex, Up to Surface, Offset from Surface, Up to Body and Mid Plane) symmetry, feature selection, and the order of feature creation.

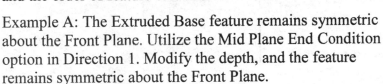

Example A: The Extruded Base feature remains symmetric about the Front Plane. Utilize the Mid Plane End Condition option in Direction 1. Modify the depth, and the feature remains symmetric about the Front Plane.

Example B: Create 34 teeth in the model. Do you create each tooth separately using the Extruded Cut feature? No.

Create a single tooth and then apply the Circular Pattern feature. Modify the Circular Pattern from 32 to 24 teeth. Think about Design Intent when you apply an End Condition in a feature during modeling. The basic End Conditions are:

- *Blind* - Extends the feature from the selected sketch plane for a specified distance - default End Condition.

- *Through All* - Extends the feature from the selected sketch plane through all existing geometry.

- *Up to Next* - Extends the feature from the selected sketch plane to the next surface that intercepts the entire profile. The intercepting surface must be on the same part.

- *Up To Vertex* - Extends the feature from the selected sketch plane to a plane that is parallel to the sketch plane and passing through the specified vertex.

- *Up To Surface* - Extends the feature from the selected sketch plane to the selected surface.

- *Offset from Surface* - Extends the feature from the selected sketch plane to a specified distance from the selected surface.

- *Up To Body* - Extends the feature up to the selected body. Use this option with assemblies, mold parts, or multi-body parts.

- *Mid Plane* - Extends the feature from the selected sketch plane equally in both directions.

Activity: Various End Condition options

Create Extruded Cut features using various End Condition options. Think about Design Intent of the model.

1) Open **End Condition 9-1** from the ENGDESIGN-W-SOLIDWORKS\Chapter 9 Intelligent Modeling folder. The FeatureManager displays two Extrude features, a Shell feature and a Linear Pattern feature.

2) Click the **circumference of the front most circle**. Sketch3 is highlighted in the FeatureManager.

Create an Extruded Cut feature using the Selected Contours and Through All option.

3) Click **Extruded Cut**  from the Features toolbar. The Cut-Extrude PropertyManager is displayed.

4) **Expand** the Selected Contours box if needed. Sketch3-Contour<1> is displayed in the Selected Contours box. The direction arrow points downward; if not, click the Reverse Direction button.

5) Select **Through All** for End Condition in Direction 1. Only the first circle of your sketch is extruded.

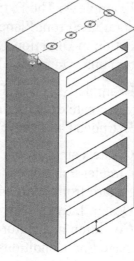

6) Click **OK** from the Cut-Extruded PropertyManager. Cut-Extrude1 is displayed in the FeatureManager.

Create an Extruded Cut feature using the Selected Contours and the Up To Next option. Think about Design Intent.

7) Click **Sketch3** from the FeatureManager. Note the icon type for a contour sketch and its relationship to Boss-Extrude1.

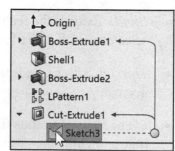

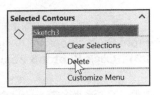

8) Click **Extruded Cut** from the Features toolbar. The Cut-Extrude PropertyManager is displayed.

9) **Expand** the Selected Contours box.

10) Delete **Sketch3** from the Selected Contours box.

11) Click the **circumference of the second circle** as illustrated. Sketch3-Contour<1> is displayed in the Selected Contours box.

12) Select **Up To Next** for End Condition in Direction 1. Only the second circle of your sketch is extruded.

13) Click **OK** from the Cut-Extrude PropertyManager. Cut-Extrude2 is displayed in the FeatureManager.

14) **Rotate** your model and view the created hole.

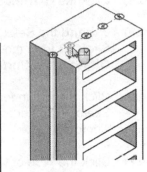

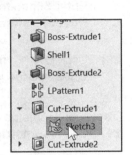

Create an Extruded Cut feature using the Selected Contours and the Up To Vertex option.

15) Click **Sketch3** from the FeatureManager. Note the icon (sharing a selected contour sketch).

16) Click the **Extruded Cut** Features tool. The Cut-Extrude PropertyManager is displayed.

17) **Expand** the Selected Contours box.

18) Delete **Sketch3** from the Selected Contours box.

19) Click the circumference of the **third circle**. Sketch3-Contour<1> is displayed in the Selected Contours box.

20) Select **Up To Vertex** for End Condition in Direction 1.

21) Click the **Vertex point** as illustrated. Only the third circle of your sketch is extruded.

22) Click **OK** from the PropertyManager. Cut-Extrude3 is displayed in the FeatureManager. The third circle has an Extruded Cut feature through the top two shelves.

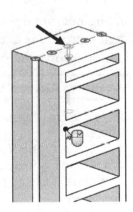

Create an Extruded Cut feature using the Selected Contours and the Offset
From Surface option.

23) Click **Sketch3** from the FeatureManager.

24) Click the **Extruded Cut** 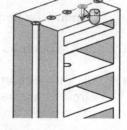 Features tool.

25) **Expand** the Selected Contours box. Delete
Sketch3 from the Selected Contours box. Select
Offset From Surface for End Condition.

26) Click the circumference of the **fourth circle**.
Sketch3-Contour<1> is displayed in the Selected
Contours box.

27) Click the **face** of the third shelf. Face<1> is
displayed in the Face/Plane box in Direction1.
Enter **60**mm for Offset Distance. Click the
Reverse offset box.

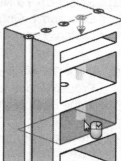

28) Click **OK** ✔ from the PropertyManager.
Cut-Extrude4 is displayed in the FeatureManager.

Display an Isometric view. Save the model.
29) Click **Isometric** view. View the created features. **Close** the model.

Along a Vector

In engineering design, vectors are utilized in modeling techniques. In an extrusion, the
sketch profile is extruded perpendicular to the sketch plane. The Direction of Extrusion
Option allows the profile to be extruded normal to the vector.

Activity: Along a Vector

Utilize a vector, created in a separate sketch, and the Direction of Extrusion to modify the extruded
feature.

1) Open **Along Vector 9-1** from the ENGDESIGN-W-SOLIDWORKS\Chapter 9 Intelligent
Modeling folder. View the model in the Graphics window. The
current sketch profile is extruded normal to the Top Sketch
Plane.

2) Edit **Boss-Extrude1** from the FeatureManager. The Boss-
Extrude1 PropertyManager is displayed. Select **Up to Vertex**
for End Condition in Direction 1.

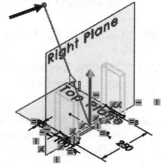

3) Click the **top endpoint** of Sketch Vector. Click inside the
Direction of Extrusion box. Click the **Sketch Vector** from the
Graphics window as illustrated. The feature is extruded along
the Sketch Vector normal to the Sketch Profile.

4)   Click **OK** ✔ from the Boss-Extrude1
PropertyManager. View the results in the
Graphics window.

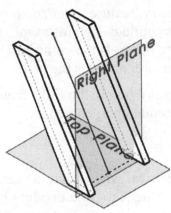

Close the model.
5)   **Close** the model.

FeatureXpert (Constant Radius)

FeatureXpert manages the interaction
between fillet and draft features when
features fail. The FeatureXpert, manages fillet and draft features for
you so you can concentrate on your design. See Xperts Overview in
SOLIDWORKS Help for additional information.

When you add or make changes to constant radius fillets and neutral
plane drafts that cause rebuild errors, the **What's Wrong** dialog
appears with a description of the error. Click **FeatureXpert** in the
dialog to run the FeatureXpert to attempt to fix the error.

The FeatureXpert can change the feature order in the FeatureManager
design tree or adjust the tangent properties so a part successfully
rebuilds. The FeatureXpert can also, to a lesser extent, repair reference
planes that have lost references.

Supported features:

- *Constant radius fillets.*

- *Neutral plane drafts.*

- *Reference planes.*

Unsupported items:

- *Other types of fillets or draft features.*

- *Mirror or pattern features. When mirror or pattern features contain
  a fillet or draft feature, the FeatureXpert cannot manipulate those
  features in the mirrored or patterned copies.*

- *Library features. Fillet or draft features in a library feature are ignored by the
  FeatureXpert and the entire Library feature is treated as one rigid feature.*

- *Configurations and Design Tables. The FeatureXpert is not available for parts that
  contain these items.*

💡 Utilize symmetry, feature order and reusing common features to build design intent into a part. Example A: Feature order. Is the entire part symmetric? Feature order affects the part.

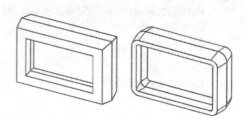

Apply the Shell feature before the Fillet feature and the inside corners remain perpendicular.

Symmetry

An object is symmetrical when it has the same exact shape on opposite sides of a dividing line (or plane) or about a center or axis. The simplest type of Symmetry is a "Mirror" as we discussed above in this chapter.

Symmetry can be important when creating a 2D sketch, a 3D feature or an assembly. Symmetry is important because:

- Mirrored shapes have symmetry where points on opposite sides of the dividing line (or mirror line) are the same distance away from the mirror line.

- For a 2D mirrored shape, the axis of symmetry is the mirror line.

- For a 3D mirrored shape, the symmetry is about a plane.

- Molded symmetrical parts are often made using a mold with two halves, one on each side of the axis of symmetry.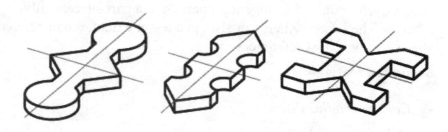

- The axis or line where two mold parts join is called a parting line.

- When items are removed from a mold, sometimes a small ridge of material is left on the object. Have you ever noticed a parting line on a molded object such as your toothbrush or a screwdriver handle?

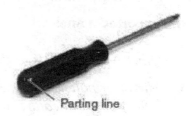
Parting line

Bodies to Mirror

When a model contains single point entities, the pattern and mirror features may create disjointed geometry and the feature will fail. For example, a cone contains a single point at its origin. To mirror the cone about a plane would create disjointed geometry. To resolve this issue, utilize Bodies to Pattern option.

Activity: Bodies to Mirror

Create a Mirror feature with the Top plane and utilize the Body to Mirror option.

1)   Open **Bodies to Mirror 9-1** from the ENGDESIGN-W-SOLIDWORKS\Chapter 9 Intelligent Modeling folder. View the model in the Graphics window.

2)   Select **Mirror**  from the Features toolbar. The Mirror PropertyManager is displayed.

3)   **Expand** the Bodies to Mirror dialog box.

4)   Click the **face of the Cone** in the Graphics window. Note the icon feedback symbol.

5)   Click inside the **Mirror Face/Plane** dialog box.

6)   Click **Top Plane** from the fly-out FeatureManager.

7)   Uncheck the **Merge solids** box. (You cannot merge these two cones at a single point).

8)   Click **OK** ✔ from the Mirror PropertyManager. View the results in the Graphics window. Mirror1 is displayed in the FeatureManager.

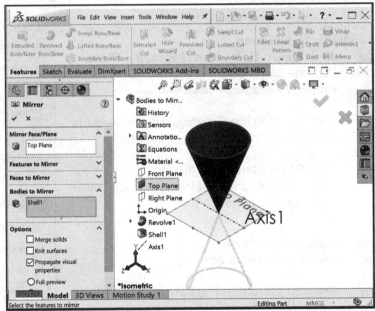

Planes

Certain types of models are better suited to design automation than others. When setting up models for automation, consider how they fit into an assembly and how the parts might change when automated.

Create planes so that sketches and features can be referenced to them (coincident, up to surface, etc.) This provides the ability to dimension for the plane to be changed and the extrusion extended with it. When placed in an assembly, other components can be mated to a plane so that they move with consideration to the parts altered.

Incorporating planes into the design process prepares the model for future changes. As geometry becomes more complex, additional sketch geometry (construction lines, circles) or reference geometry (axis, planes) may be required to construct planes.

For example, a sketched line coincident with the silhouette edge of the cone provides a reference to create a plane through to the outer cone's face. An axis, a plane and an angle creates a new plane through the axis at a specified angle.

Activity: Angle Plane

Create a plane at a specified angle through an axis.

1) Open **Angle Planes 9-1** from the ENGDESIGN-W-SOLIDWORKS\Chapter 9 - Intelligent Modeling folder. View the model in the Graphics window.

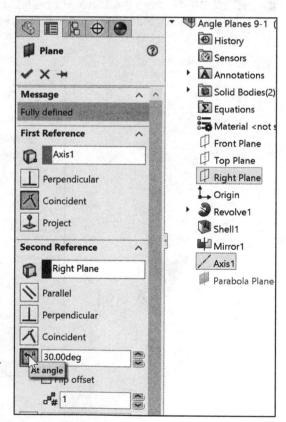

Create an Angle Plane.

2) Click **Insert, Reference Geometry, Plane** from the Menu bar. The Plane PropertyManager is displayed.

3) Click **Axis1** in the Graphics window. Axis1 is displayed in the First Reference box.

4) Click **Right Plane** from the fly-out FeatureManager. Right Plane is displayed in the Second Reference box.

Set the angle of the Plane.

5) Click the **At angle** box.

6) Enter **30** for Angle. The Plane is fully defined.

7) Click **OK** ✔ from the Plane PropertyManager. Plane1 is displayed in the FeatureManager.

8) **Close** the Model.

As geometry becomes more complex in mechanical design, planes and conical geometry can be combined to create circular, elliptical, parabolic and hyperbolic sketches with the Intersection Curve Sketch Curve.

Conic Sections and Planes

Conic sections are paths traveled by planets, satellites, electrons and other bodies whose motions are driven by inverse-square forces. Once the path of a moving body is known, information about its velocity and forces can be derived. Using planes to section cones creates circular, elliptical, parabolic, hyperbolic and other cross sections that are utilized in engineering design.

The Intersection Curve tool provides the ability to open a sketch and create a sketched curve at the following kinds of intersections: a plane and a surface or a model face, two surfaces, a surface and a model face, a plane and the entire part, and a surface and the entire part.

Activity: Conic Section

Apply the Intersection Curve tool. Create a sketched curve.

1) Open **Conic Section 9-1** from the ENGDESIGN-W-SOLIDWORKS\Chapter 9 Intelligent Modeling folder. View the model in the Graphics window. Create a sketch.

2) Right-click **Plane1 Circle** from the FeatureManager.

3) Click **Sketch** 📝 from the Context toolbar. The Sketch toolbar is displayed in the CommandManager.

4) Click **Tools, Sketch Tools, Intersection Curve** from the Menu bar. The Intersection Curve PropertyManager is displayed.

5) Click the **inside Shell face** as illustrated. Face<1> is displayed in the Selected Entities dialog box.

6) Click **OK** ✔ from the Intersection Curve PropertyManager.

7) Click **OK** ✔ from the Intersection Curve PropertyManager.

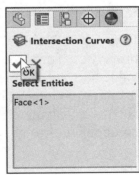

8) Click **Exit Sketch**.

9) **Hide** Plane1 Circle. The intersection of the cone with the plane creates a circle.

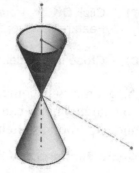

10) **Close** the model. You can use the resulting sketched intersection curve in the same way that you use any sketched curve.

Assembly

Utilizing symmetry, reusing common parts and using the Mate relation between parts builds the design intent into an assembly. For example: Reuse geometry in an assembly. The assembly contains a linear pattern of holes. Insert one screw into the first hole. Utilize the Component Pattern feature to copy the machine screw to the other holes.

Assembly Visualization

In an assembly, the designer selects material based on cost, performance, physical properties, manufacturing processes, sustainability, etc. The SOLIDWORKS Assembly Visualization tool includes a set of predefined columns to help troubleshoot assembly performance. You can view the open and rebuild times for the components, and the total number of graphics triangles for all instances of components. The Assembly Visualization tool provides the ability to rank components based on the default values (**weight, mass, density, volume, etc.**) or their custom properties (**cost, sustainability, density, surface area, volume, etc.**) or an equation and activate a spectrum of colors that reflects the relative values of the properties for each component.

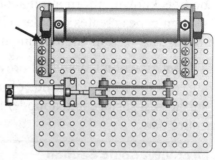

Hide/Show Value Bar ▮ **icon**. Available for numeric properties. Turns the value bars off and on. When the value bars are on, the component with the highest value displays the longest bar. You can set the length of the bars to be calculated relative to the highest-value component or relative to the entire assembly.

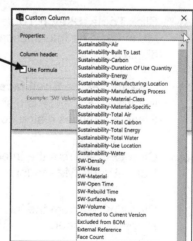

Flat Nested View icon. Nested view, where subassemblies are indented. Flat view, where subassembly structures are ignored (similar to a parts-only BOM).

Grouped/Ungrouped View icon. Groups multiple instances of a component into a single line item in the list. Grouped View is useful when listing values for properties that are identical for every instance of the component. Ungrouped views lists each instance of a component individually. Ungrouped View is useful when listing values for instance-specific properties, such as Fully mated, which might be different for different instances of the component.

erformance Analysis icon. Provides additional information on the open, display, and rebuild performance of models in an assembly.

Filter icon. Filters the list by text and by component show/hide state.

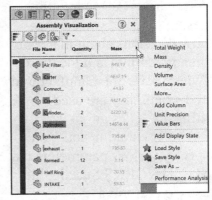

The Assembly Visualization tab in the FeatureManager design tree panel contains a list of all components in the assembly, sorted initially by file name. There are three default columns:

- File name.

- Quantity.

- Mass.

 Mass is the default column. View your options as illustrated other than Mass.

Activity: Assembly Visualization

Apply the Assembly Visualization tool to an assembly. View your options.

1) Open **Assembly Visualization 9-1** from the ENGDESIGN-W-SOLIDWORKS\Chapter 9 Intelligent Modeling folder. View the assembly in the Graphics window. Material was assigned to each component in the assembly. Click the **Assembly**

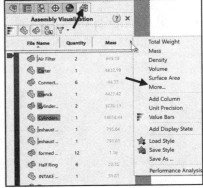

Visualization tool from the Evaluate Tab in the CommandManager. The Assembly Visualization PropertyManager is displayed. Click the **expand arrow** to the right of Mass as illustrated to display the default visualization properties. Mass is selected by default. Click the **More** tab. View your options.

2) **Explore** the available tabs and SOLIDWORKS Help for additional information.

3) **Close** the model.

SOLIDWORKS Sustainability - Assembly

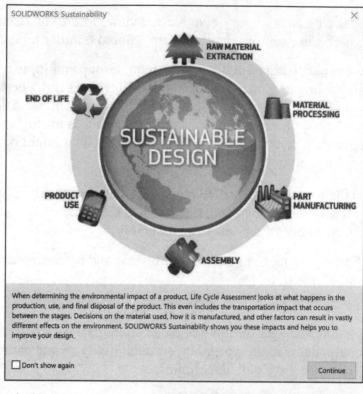

With SOLIDWORKS Sustainability you can determine Life Cycle Assessment (LCA) properties for a part or assembly. By integrating Life Cycle Assessment (LCA) into the design process, you can see how decisions about material, manufacturing, and location (where parts are manufactured and where they are used) influence a design's environmental impact. You specify various parameters that SOLIDWORKS Sustainability uses to perform a comprehensive evaluation of all the steps in a design's life. LCA includes:

- Ore extraction from the earth.

- Material processing.

- Part manufacturing.

- Assembly.

- Product usage by the end consumer.

- End of Life (EOL) - Landfill, recycling, and incineration.

- All the transportation that occurs between and within each of these steps.

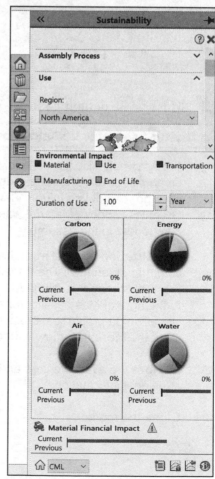

By combining SOLIDWORKS Sustainability and Assembly Visualization, you can determine the components in the assembly that contain the greatest carbon footprint and review materials with similar properties to reduce CO_2 emissions.

💡 You need SOLIDWORKS Professional or SOLIDWORKS Premium to access SOLIDWORKS Sustainability for an assembly.

MateXpert

The MateXpert is a tool that provides the ability to identify mate problems in an assembly. You can examine the details of mates that are not satisfied, and identify groups of mates which over define the assembly. If the introduction of a component leads to multiple mate errors, it may be easier to delete the component, review the design intent, reinsert the component and then apply new mates. See SOLIDWORKS Help for additional information.

Drawing

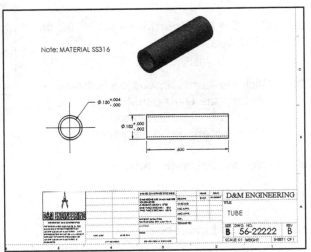

Utilize dimensions, tolerance and notes in parts and assemblies to build the design intent into a drawing.

Example A: Tolerance and material in the drawing. Insert an outside diameter tolerance +.000/-.002 into the Pipe part. The tolerance propagates to the drawing.

Define the Custom Property Material in the Part. The Material Custom Property propagates to your drawing.

DimXpert

DimXpert for parts is a set of tools you use to apply dimensions and tolerances to parts according to the requirements of the ASME Y14.41-2003 standard.

DimXpert dimensions show up in a different color to help identify them from model dims and reference dims. DimXpert dims are the dimensions that are used when calculating tolerance stack-up using TolAnalyst.

DimXpert applies dimensions in drawings so that manufacturing features (patterns, slots, pockets, etc.) are fully-defined.

DimXpert for parts and drawings automatically recognize manufacturing features. What are manufacturing features? Manufacturing features are *not SOLIDWORKS features*. Manufacturing features are defined in 1.1.12 of the ASME Y14.5M-1994 Dimensioning and Tolerancing standard as "The general term applied to a physical portion of a part, such as a surface, hole or slot."

The DimXpertManager provides the following selections:
Auto Dimension Scheme ⊕, **Basic Location Dimension**
⊢ᵒ⊣, **Basic Size Dimension** ⊏ **Show Tolerance Status**
±, **Copy Scheme** ⊕ and **TolAnalyst Study** ⊞.

💡 Care is required to apply DimXpert correctly on
complex surfaces or with some existing models. See SOLIDWORKS help for detailed
information on DimXpert with complex surfaces.

Activity: DimXpert 9-1

Apply the DimXpert tool. Apply Prismatic and Geometric options.

1) Open **DimXpert 9-1** from the ENGDESIGN-W-SOLIDWORKS\Chapter 9 Intelligent Modeling folder.

2) Click the **DimXpertManager** ⊕ tab as illustrated.

3) Click the **Auto Dimension Scheme** ⊕ tab from the DimXpertManager. The Auto Dimension Scheme PropertyManager is displayed.

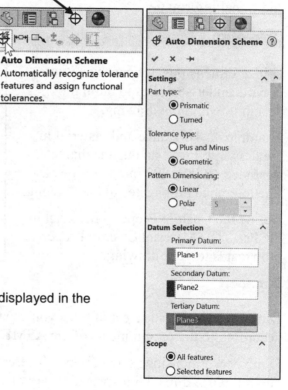

4) Click the **Prismatic** box.

5) Click the **Geometric** box.

6) Click the **Linear** box.

Select the Primary Datum.
7) Click the **front face** of the model. Plane1 is displayed in the Primary Datum box.

Select the Secondary Datum.
8) Click **inside** the Secondary Datum box.

9) Click the **left face** of the model. Plane2 is displayed in the Secondary Datum box.

Select the Tertiary Datum.
10) Click **inside** the Tertiary Datum box.

11) Click the **bottom face** of the model. Plane3 is displayed in the Tertiary Datum box. Accept the default options.

12) Click **OK** ✔ from the Auto Dimension Scheme PropertyManager. View the results in the Graphics window and in the DimXpertManager.

13) **Close** all models. Do not save the updates.

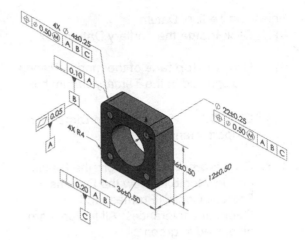

💡 Right-click Delete to delete the Auto Dimension Scheme in the DimXpert PropertyManager.

Activity: DimXpert 9-2

Apply DimXpert: Geometric option. Edit a Feature Control Frame.

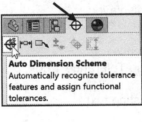

1) Open **DimXpert 9-2** from the ENGDESIGN-W-SOLIDWORKS\Chapter 9 Intelligent Modeling folder.

2) Click the **DimXpertManager** ⊕ tab.

3) Click the **Auto Dimension Scheme** ✪ tool from the DimXpertManager. The Auto Dimension PropertyManager is displayed.

4) Click the **Prismatic** box. Click the **Geometric** box. Click the **Linear** box.

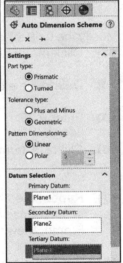

💡 DimXpert: Geometric option provides the ability to locate axial features with position and circular runout tolerances. Pockets and surfaces are located with surface profiles.

Create the Primary Datum.

5) Click the **back face** of the model. Plane1 is displayed in the Primary Datum box.

Select the Secondary Datum.

6) Click **inside** the Secondary Datum box.

7) Click the **left face** of the model. Plane2 is displayed in the Secondary Datum box.

Select the Tertiary Datum.
8) Click **inside** the Tertiary Datum box.

9) Click the **top face** of the model. Plane3 is displayed in the Tertiary Datum box.

10) Click **OK** ✔ from the Auto Dimension PropertyManager.

11) Click **Isometric view** from the Heads-up View toolbar. View the Datums, Feature Control Frames, and Geometric tolerances. All features are displayed in green.

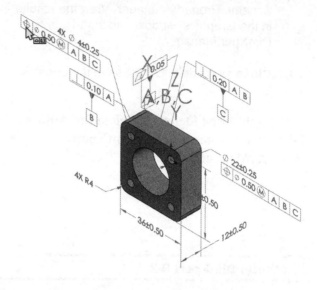

Edit a Feature Control Frame.
12) **Double-click** the illustrated Position Feature Control Frame. The Properties dialog box is displayed.

Modify the 0.50 tolerance.
13) Click **inside** the Tolerance 1 box.

14) **Delete** the existing text.

15) Enter **0.25**.

16) Click **OK** from the Properties dialog box.

17) **Repeat** the above procedure for the second Position Feature Control Frame. View the results.

18) **Close** the model. Do not save the model.

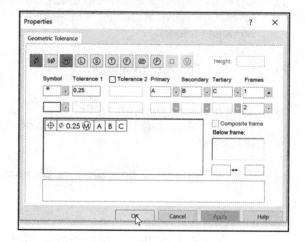

Project Summary

In this project, you performed short step by step tutorials to understand some of the available tools in SOLIDWORKS to perform intelligent modeling. Intelligent modeling is incorporating design intent into the definition of the Sketch, Feature, Part and Assembly or Drawing document.

All designs are created for a purpose. Design intent is the intellectual arrangement of features and dimensions of a design. Design intent governs the relationship between sketches in a feature, features in a part and parts in an assembly or drawing document.

The SOLIDWORKS definition of design intent is the process in which the model is developed to accept future modifications. Models behave differently when design changes occur.

You created Explicit Dimension Driven and Parametric Equation Driven Curves and generated curves using the Curve Through XYZ Points tool for a NACA aerofoil and CAM data text file.

You also applied SketchXpert, DimXpert and FeatureXpert.

View the .mp4 files located in the Chapter 9 - Intelligent Modeling folder to better understand the potential of the Equation Driven Curve tool.

Notes:

Project 10

Additive Manufacturing - 3D Printing

Below are the desired outcomes and usage competencies based on the completion of Project 10.

Desired Outcomes:	Usage Competencies:
• Knowledge of Additive Manufacturing. • Identify key features of a low cost $500 - $3,000 3D printer. • Create an STL (STereoLithography) file. • 3D Print directly from SOLIDWORKS. • Select the proper Slicer engine and print parameters. • Create a successful print.	• Discuss the advantages and disadvantages of Additive Manufacturing. • Describe the differences between a Cartesian and Delta printer. • Explain 3D printer technology: STereoLithography (STL), Fused Filament Fabrication (FFF), Fused Deposition Model (FDM), and Digital Light Process (DLP). • Recognize common 3D printer terminology: Raft, Skirt, Brim, Support, Touching Buildplate, Heated vs. Non-Heated build area. • Choose the proper Slicer and print parameters. • Select the proper filament. • Understand part orientation during the print cycle. • Define general 3D printing tips. • Address fit tolerance for interlocking parts.

Notes:

Project 10 - Additive Manufacturing - 3D Printing

Project Objective

Provide a basic understanding between the differences of Additive vs. Subtractive Manufacturing. Comprehend 3D printer terminology along with a working knowledge of preparing, saving, and printing a 3D CAD model on a low cost ($500 - $3,000) printer.

On the completion of this project, you will be able to:

- Discuss Additive vs Subtractive Manufacturing.

- Determine the differences between a Cartesian and Delta printer.

- Create a STereoLithography (STL) file in SOLIDWORKS.

- 3D print directly from SOLIDWORKS using an Add-In.

  o Save an STL file to G-code.

- Discuss printer hardware.

- Select the correct filament type:

  o PLA (Polylactic acid), ABS (Acrylonitrile butadiene styrene) or Nylon.

- Prepare the G-code.

  o Address model setup, print orientation, extruder temperature, bed temperature, support type, layer height, infill, and number of shells.

- Comprehend the following 3D printer terminology:

  o (STereoLithography) file - STL.

  o Fused Filament Fabrication - FFF.

  o Fused Deposition Model - FDM.

  o Digital Light Process - DLP.

  o Dissolvable Support System - DDS.

  o Fast Layer Deposition - FLD.

  o Raft, Skirt and Brim.

  o Support and Touching Buildplate.

  o Slicer.

  o G-code.

- Address fit tolerance for interlocking parts.

- Define general 3D Printing tips.

Additive vs. Subtractive Manufacturing

In April, 2012, *The Economist* published an article on 3D printing. In the article they stated that this was the "beginning of a third industrial revolution, offering the potential to revolutionize how the world makes just about everything."

Avi Reichental, President and CEO, 3D Systems, stated, "With 3D printing, complexity is free. The printer doesn't care if it makes the most rudimentary shape or the most complex shape, and that is completely turning design and manufacturing on its head as we know it."

Over the past five years, companies are now using 3D printing to evaluate more concepts in less time to improve decisions early in product development. As the design process moves forward, technical decisions are iteratively tested at every step to guide decisions big and small, to achieve improved performance, lower manufacturing costs, delivering higher quality and more successful product introductions. In pre-production, 3D printing is enabling faster first article production to support marketing and sales functions, and early adopter customers. And in final production processes, 3D printing is enabling higher productivity, increased flexibility, reduced warehouse and other logistics costs, economical customization, improved quality, reduced product weight, and greater efficiency in a growing number of industries.

Technology for 3D printing continues to advance in three key areas: **printers** and **printing methods**, **design software**, and **materials** used in printing.

Already, 3D printing is being used in the medical industry to help save lives and in some space exploration efforts. But how will 3D printing affect the average, middle-class person in the future? Low cost 3D printers are addressing this consumer market.

Additive manufacturing is the process of joining materials to create an object from a 3D model, usually adding layer over layer.

Subtractive manufacturing relies upon the removal of material to create something. The blacksmith hammered away at heated metal to create a product. Today, a Computer Numerical Control CNC machine cuts and drills and otherwise removes material from a larger initial block to create a product.

☀ Additive manufacturing, sometimes known as *rapid prototyping*, can be slower than Subtractive manufacturing. Both take skill in creating the G-code and understanding the machine limitations.

A few advantages of Additive manufacturing:

- Lower cost (different entry levels) into the manufacturing environment.

- Lowers the barriers (space, power, safety, and training) to traditional subtractive manufacturing.

- Reduce part count in an assembly from traditional subtractive manufacturing (complex parts vs. assemblies).

- Build complex features, shapes, and objects.

- Reduce prototyping time.

- Faster development cycle.

- Quicker customer feedback.

- Faster product to market.

- Quicker product customization and configuration.

- Parallel verticals: develop and prototype at the same time.

- Open source slicing engines (Slic3r, Skeinforge, Netfabb, KISSkice, Cura, etc.).

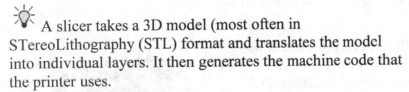 A slicer takes a 3D model (most often in STereoLithography (STL) format and translates the model into individual layers. It then generates the machine code that the printer uses.

A slicer program allows the user to calibrate printer settings: filament type, part orientation, extruder speed, extruder temperature, bed temperature, cooling fan rate, raft, support, percent infill, infill pattern type, etc.

3D printers are controlled either through a small on-board control screen, an external memory device (USB, Sims card, etc.) or through a computer interface. User interface/control software allows a user to generate the needed machine code (G-code, .gcode) file from the computer to the 3D printer.

A few disadvantages of Additive manufacturing:

- Slow build rates. Many printers lay down material at a speed of one to five cubic inches per hour. Depending on the part needed, other manufacturing processes may be significantly faster.

- Requires post-processing. The surface finish and dimensional accuracy may be lower quality than other manufacturing methods.

- Poor mechanical properties. Layering and multiple interfaces can cause defects in the product.

- Frequent calibration is required. Without frequent calibration, prints may not be the correct dimensions, they may not stick to the build surface, and a variety of other not-so-wanted effects can occur.

- Limited component size/small build volume. In most cases, polymer products are about 1 cubic yard in size, metal parts may only be one cubic foot. While larger machines are available, they come at a cost.

Cartesian Printer vs. Delta Printer

Cartesian and Delta style 3D printers are the most common styles of desktop 3D printers currently available. What are the main differences? To understand their unique abilities one first needs to understand the basic technology of a Fused Deposition Model printer.

A Fused Filament Fabrication (FFF) printer takes a filament of a material (usually plastic) and extrudes it through a print-head-like nozzle. This material is then laid down in thin layers to form a 3D object on a platform, built from the bottom up.

Cartesian 3D printers are named after the dimensional coordinate system - the X, Y, and Z-axis - which is used to determine where and how to move in three dimensions.

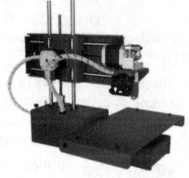

Cartesian 3D printers typically have a print bed which moves only in the Z-axis. The extruder sits on the X-axis and Y-axis, where it can move in four directions on a gantry. This principle can be seen in action on popular models from Ultimaker, Sindoh, and MakerBot.

Controlling a linear Cartesian system is relatively simple, which is why most low cost 3D printers on the market today use this type of design. The Cartesian coordinate system has long been used for tools like plotters, CNC milling machines, and 2D printers.

Delta 3D printers also work within the Cartesian plane, but use a very different approach to getting the print-nozzle where it needs to be.

Identifying characteristics start with the circular print bed. The extruder is suspended above the print bed by three arms in a triangular configuration, thus the name "Delta."

These arms move up and down independently to the print-nozzle keeping it precisely located throughout the print. Delta printers have the advantage in the ability to make taller objects due to the height of the printer and the tall arms.

Rather than using simple Cartesian geometry to calculate where the print-nozzle should go, Delta printers estimate the head position using trigonometric functions.

Create an STL file in SOLIDWORKS

STL (STereoLithography) is a file format native to the Stereolithography CAD software created by 3D Systems. STL has several after-the-fact backronyms such as "Standard Triangle Language" and "Standard Tessellation Language."

An STL file describes only the surface geometry of a three dimensional object without any representation of color, texture, or other common CAD model attributes. The STL format specifies both ASCII and Binary representations.

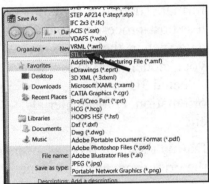

Binary files are more common, since they are more compact. An STL file describes a raw unstructured triangulated (point cloud) surface by the unit normal and vertices (ordered by the right-hand rule) of the triangles using a three-dimensional Cartesian coordinate system.

STL Save options allow you to control the number and size of the triangles by setting the various parameters in the CAD software.

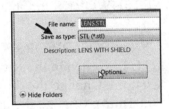

To save a SOLIDWORKS model as a STL file, Click **File**, **Save As** from the Main menu or **Save As** from the Main menu toolbar. The Save As dialog is displayed. Select **STL(*.stl)** as the Save as type.

Click the **Options** button. The Export Options dialog box is displayed. View your options. For most parts, utilize the default setting.

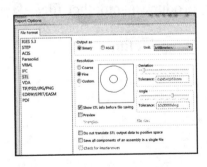

Close the Export Options dialog box. Click **OK**. View the generate point cloud of the part. Click **Yes**. The STL file is now ready to be imported into your 3D printer software.

3D Print Directly from SOLIDWORKS

Export part and assembly files to STL (.stl), 3D Manufacturing Format (.3mf), or Additive Manufacturing File Format (.amf) format.

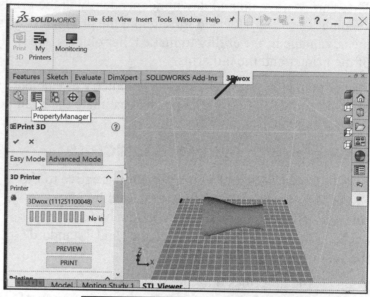

The 3MF (3D Manufacturing Format), and AMF (Additive Manufacturing File) formats provide additional model information over the .stl file format. Therefore, it requires less post-processing to define data such as the position of your model relative to the selected 3D printer, orientation, color, materials, etc.

During the print process, users can monitor the print time, needed filament, and estimated time to completion in the SOLIDWORKS graphics window.

In SOLIDWORKS for a smoother STL file, change the Resolution to Custom. Change the deviation to 0.0005in (0.01mm). Change the angle to 5. Smaller deviations and angles produce a smoother file but increase the file size and print time.

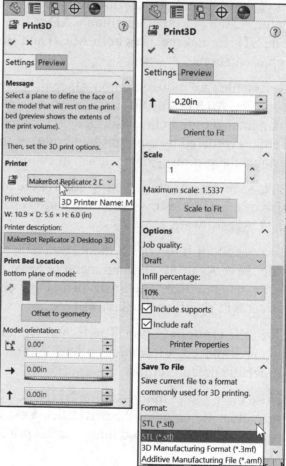

Print Material

As printed parts cool (PLA, ABS and Nylon), various areas of the object cool at different rates. Depending on the model being printed and the filament material, this effect can lead to warping, curling and or layer delamination.

There are many materials that are being explored for 3D printing; however, you will find that the two dominant plastics are ABS (Acrylonitrile butadiene styrene) and PLA (Polylactic acid). Both ABS and PLA are known as thermoplastics; that is, they become soft and moldable when heated and return to a solid when cooled.

This process can be repeated again and again. Their ability to melt and be processed again is what has made them so prevalent in society and is why most of the plastics you interact with on a daily basis are thermoplastics.

For a material to prove viable for 3D printing, it should pass four general criteria:

- Low cost and easily obtainable.

- Controllable extrusion.

- Second extrusion (two or more heads) and trace-binding during the 3D printing process.

- End use application.

☀ HIPS (High Impact Polystyrene) filament is a soluble support material used with dual head extruders.

ABS - Storage

ABS, PLA, and Nylon do best if, before use or when stored long term, they are sealed off from the atmosphere to prevent the absorption of moisture from the air. This does not mean your material will be ruined by a week of sitting on a bench in the shop, but long term exposure to a humid environment can have detrimental effects, both to the printing process and to the quality of finished parts.

Moisture laden ABS will tend to bubble and spurt from the tip of the extruder nozzle when printing, reducing the visual quality of the part, part accuracy, strength and introducing the risk of a stripping or clogging in the nozzle. ABS can be easily dried using a source of hot (preferably dry) air such as a food dehydrator.

ABS - Part Accuracy

For most, the single greatest hurdle for accurate parts in ABS is a curling upwards off the surface in direct contact with the 3D printer's print bed. A combination of heating the print surface and ensuring it is smooth, flat and clean helps to eliminate this issue. Additionally, some find various solutions can be useful when applied beforehand to the print surface. Example: A mixture of ABS dissolved in Acetone or a shot of hairspray on the build plate. Keep these solutions away from heat.

For fine features on parts involving sharp corners, such as gears, there may be a slight rounding of the corner. A fan, to provide a small amount of active cooling around the nozzle, can improve corners but one does also run the risk of introducing too much cooling and reducing adhesion between layers, eventually leading to cracks in the finished part.

PLA - Storage

PLA responds somewhat differently to moisture. In addition to bubbles or spurting at the nozzle, you may see discoloration and a reduction in 3D printed part properties. PLA (Polylactic acid) is a bioplastic derived from corn and is biodegradable and can react with water at high temperatures and undergo de-polymerization.

PLA can be dried using something as simple as a food dehydrator. It is important to note that this can alter the crystallinity ratio in the PLA and will lead to changes in extrusion temperature and other extrusion characteristics. For many 3D printers, moisture is not a major concern.

PLA - Part Accuracy

Compared to ABS and Nylon, PLA demonstrates much less part warping. For this reason it is possible to successfully print without a heated bed and use more commonly available "Blue" painter's tape as a print surface.

PLA undergoes more of a phase-change when heated and becomes much more liquid. If actively cooled, sharper details can be seen on printed corners without the risk of cracking or warping. The increased flow can also lead to stronger binding between layers, improving the strength of the printed part.

Nylon - Storage

Nylon is very hygroscopic, more so than PLA or ABS. Nylon can absorb more than 10% of its weight in water in less than 24 hours. Successful 3D printing with nylon requires dry filament. When you print with nylon that isn't dry, the water in the filament explodes causing air bubbles during printing that prevents good layer adhesion and greatly weakens the part. It also ruins the surface finish. To dry nylon, place it in an oven at 50-60C for 6-8 hours. After drying, store in an airtight container, preferably with desiccant.

Type 6, 6 Nylon requires a higher extruder temperature than ABS. Type 6, 6 Nylon melts at 255-265C (490-510F), type 6 nylon melts at 210-220C (410-428F) and polypropylene melts at 160-175C (320-347F).

Nylon - Part Accuracy

Compared to ABS and PLA, Nylon and ABS warp approximately the same. PLA demonstrates much less part warping. A heated plate (70 - 80C) is required for Nylon printing. It's best to apply a PVA (Polyvinyl acetate) Elmer's or Scotch based glue stick to the bed for adhesion in a cross-hatch pattern. Remember, less is more when applying the glue stick to the plate.

☼ Nylon is a flexible, stronger and more durable alternative to PLA and ABS. With Nylon, do not use layer cooling fans and avoid drafty or cool rooms for best results.

☼ **Flex PLA** - Common flexible filaments are polyester-based (non-toxic) with a low melting point. Print temp: 120C. It is highly recommended to drastically lower your printing speed to around 20% from standard PLA. To take best advantage of the filament's properties, print it with 10% infill or less. Most flexible filament adheres well to acrylic; it does not adhere well to painter's tape and glass.

Build Plates

PLA in general has a lower shrinkage factor than ABS and Nylon. They are common materials which can warp, curl, delaminate and potentially destroy the print. The following equipment and procedures deal with these potential areas.

Non-Heated

People have experimented with different build surfaces such as steel, glass and various kinds of plastic. However, from experience, both ABS and PLA material seem to stick fairly well to any of the above build surfaces when used with a thin even layer of Polymide Kapton tape or a good quality blue painter's tape. It is not recommended to print ABS or Nylon with a non-heated build plate. Using a raft is always a good idea until you fully understand your system limitations.

☼ Some users apply a small amount of acetone to the build plate while rubbing an old print on the plate. This applies a base layer of material to improve adhesion.

Polymide Kapton tape used in 3D printing.

Heated

A heated build plate or platform helps keep the lowest levels of a print warm as the higher layers are printed. This allows the overall print to cool more evenly. A heated build plate helps tremendously with most ABS prints and large PLA prints. You need a heated build plate for Nylon. Heated build plate temperatures should range between 90 - 95C for ABS, 60 - 65C for PLA and 70 - 80 for Nylon. Polymide Kapton tape works the best due to its temperature range capabilities for ABS and PLA. For Nylon, it is recommended to use a PVA (Elmer's or Scotch) based glue stick on the bed for adhesion in a cross-hatch pattern. Remember, less is more when applying the glue stick on the plate.

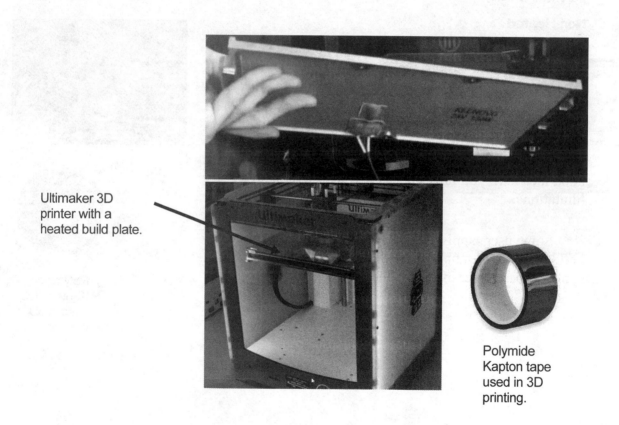

Ultimaker 3D printer with a heated build plate.

Polymide Kapton tape used in 3D printing.

Some users apply a small amount of acetone to the build plate while rubbing an old print to the plate. This applies a base layer of material to improve adhesion.

💡 Most PSUs (power supply units) supplied with non-heated build plate printers need to be upgraded to handle a heated plate upgrade.

Clean

Print materials stick better to a clean build plate. Clean your build plate after every build. Remove any build up material (hairspray, glue, old print residue, etc.). Replace needed tape (blue, Kapton, etc.).

Level

Level the build plate after every build. Do not print with a badly calibrated printer.

A level printing surface means every thin trace of support will be laid down at the intended height. A slope can cause some traces to be laid down too high from the print bed and prevent them from adhering.

Rule of thumb: For a low cost 3D printer, place the print surface (Z axis) approximately 0.1mm away from the nozzle at six points around the print bed.

Control Temperature

Eliminate all drafts (air conditioner, heater, windows) and control air flow that may cause a temperature gradient within the build area. Changes in temperature during a build cycle can cause curling, cracking and layer delamination, especially on long thin parts. Below are five fully enclosed 3D printers: A modified MakerBot Replicator 2, an Ultimaker 2 Go, a bq witbox, a Sindoh DP200 3DWox and an XYZ da Vinci.

PLA in general is more forgiving to temperature fluctuations and moisture than ABS and Nylon during a build cycle.

Filament Storage

Below is a filament roll. The filament roll is packed with a package of desiccant. A desiccant is a hygroscopic substance that induces or sustains a state of dryness (desiccation) in its vicinity. Do not open filament packages before you need them. Moisture is your enemy. The filament will swell with moisture and can be an issue (size and temperature) in the extruder head during the print cycle.

Darker colors in general (PLA and ABS) often require higher extruder temperatures. Extruding at higher temperatures may help prevent future delamination if you do not have a heated build plate or control build area.

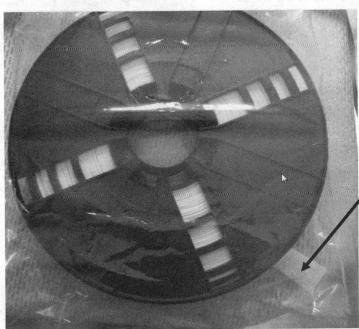

Desiccant package

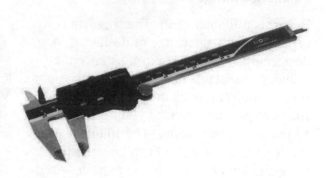

There are set defaults that come in many low cost printers. The default setting for the diameter of the filament in the Replicator G software is 1.85mm. If you are having build/extruder problems, use a pair of calipers to measure the diameter of the filament. Most calipers will have measurements down to .01mm. To measure your filament, take an average reading of at least three readings and re-set the default filament setting if needed.

Never focus too much on one single issue. These machines are complex, and trouble often arises from multiple reasons. A slipping filament may not only be caused by a bad gear drive, but also by an obstructed nozzle, a wrong feed value, a too low (or too high) temperature or a combination of all these.

Prepare the Model

There are many available open source software packages for preparing your model in an STL file format. To turn the .STL file into code for your printer, you'll need a slicing program. The MakerWare software (slicing program) for the MakerBot Replicator 2 Desktop and the Sindoh DP200 3DWox is displayed in this section.

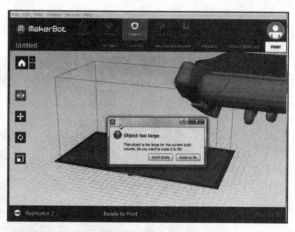

The first action is to add/insert or load your STL file using the slicing program in the build area.

Depending on your printer "slicing program," you may or may not receive a message indicating your object is too large for the current build volume.

If the object is too large, then you will need to scale it down or to redesign it into separate parts.

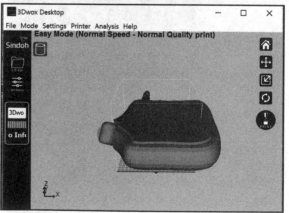

Use caution when scaling if you require fasteners or a minimum wall thickness.

You should center the part and have it lay flat on the build plate.

If you are printing more than one part, space them evenly on the plate or position them for a single build.

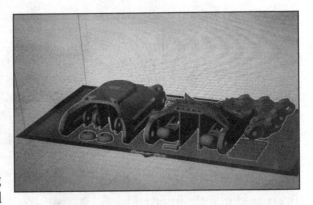

☀ In SOLIDWORKS, lay the parts out in an assembly. Save the assembly as a part. Save the part file as an STL file. Open the STL in the printer using your slicer.

Consideration should be used when printing an assembly. If the print takes 20 hours, and a failure happens after 19 hours, you just wasted a lot of time versus printing each part individually.

Example 1: Part Orientation

Part orientation is very important on build strength and the amount of raft and support material required for the build. Part orientations can also be related to warping, curling, and delamination.

If maximizing strength is an issue, select the part orientation on the build plate so that the "grain" of the print is oriented to maximize the strength of the part.

Example 1: First Orientation - Vertical

In the first orientation (vertical), due to the number of holes and slots, additional support material is required (with minimum raft material) to print the model.

Removing this material in these geometrics can be very time consuming.

Example 1: Second Orientation - Horizontal

In the second orientation (horizontal), additional raft material is used, and the support material is reduced.

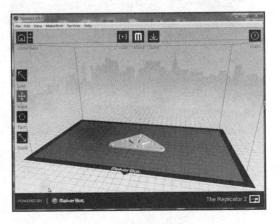

Second orientation - horizontal

The raft material can be easily removed with a pair of needle-nose pliers and no support material clean-up is required for the holes and slots. Note: In some cases, raft material is not needed.

Example 2: Part Orientation

The lens part is orientated in a vertical position with the large face flat on the build plate. This reduces the required support material and ensures proper contact with the build plate.

The needed support material is created mainly internal to the part to print the CBORE feature. Note: There is some outside support material on the top section of the part.

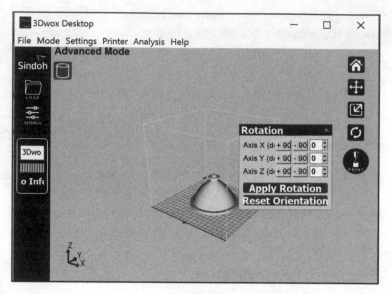

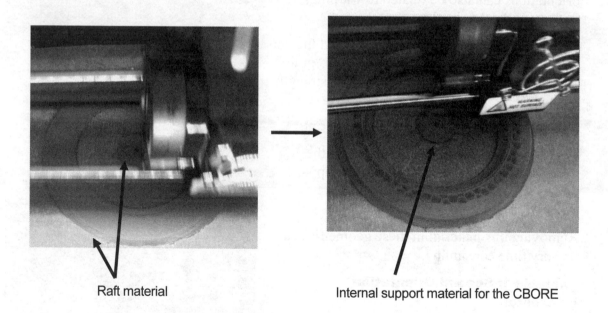

Raft material   Internal support material for the CBORE

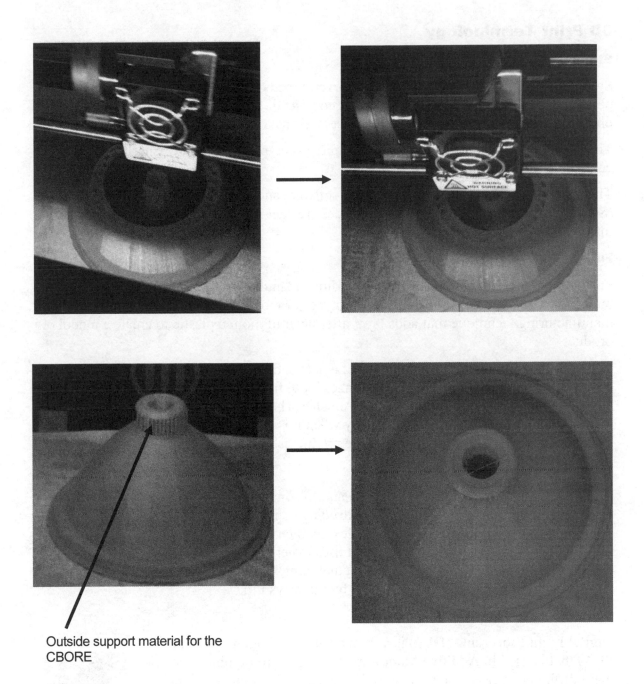

Outside support material for the
CBORE

3D Print Terminology

Stereolithography (SL or SLA)

Stereolithography (SL or SLA) is one of several methods used to create 3D printed objects. This is a liquid-based process that consists of the curing or solidification of a photosensitive polymer when an ultraviolet laser makes contact with the resin.

The process was patented as a means of rapid prototyping in 1986 by Charles Hull, co-founder of 3D Systems, Inc.

The process starts with a model in a CAD software and then it is translated to a STL (STereoLithography) file in which the pieces are "cut in slices" containing the information for each layer.

Fused Filament Fabrication (FFF)

Fused Filament Fabrication (FFF) is an additive manufacturing technology used for building three-dimensional products, prototypes or models. It is a rapid prototyping and manufacturing technique that adds layer after layer of molten plastic to create a model or product.

Typically, the FFF mechanism consists of a nozzle that emits material and deposits it onto a moving table. The FFF machine takes input from G-code and starts moving the nozzle and build plate to the needed coordinates. The material is heated in the nozzle to form liquid, which solidifies immediately when it is deposited onto the layer surface. The nozzle works layer-by-layer until the product is finished.

Fused Deposition Fabrication (FDM)

Fused Deposition Modeling (FDM) is an additive manufacturing technology that builds parts up layer-by-layer by heating and extruding thermoplastic filament. Ideal for building durable components with complex geometries in nearly any shape and size, FDM is the only 3D printing process that uses materials like ABS, PC-ISO polycarbonate, and ULTEM 9085. The actual term "Fused Deposition Modeling" and its abbreviation "FDM" are trademarked by Stratasys Ltd.

Digital Light Process (DLP)

Digital Light Processing (DLP) is a form of Stereolithography (SL, or SLA) that is used in Additive Manufacture or rapid prototyping. The main difference between DLP and SLA is the use of a projector light rather than a laser to cure photo-sensitive polymer resin. A DLP 3D printer projects the image of the object's cross section onto the surface of the resin. The exposed resin hardens while the machine's build platform descends, setting the stage for a new layer of fresh resin to be coated to the object and cured by light. Once a complete object is formed, additional post processing may be required such as removal of support material, chemical bath, and UV curing.

An example of a DLP 3D printer is the Formlabs Form1 printer as shown.

Raft, Skirt and Brim

A Raft is a horizontal latticework of filament located underneath the part. A Raft is used to help the part stick to the build plate (heated or non-heated).

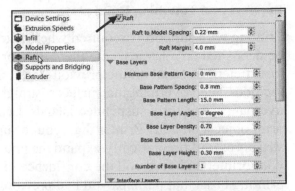

Rafts are also used to help stabilize thin tall parts with small build plate footprints.

When the print is complete, remove the part from the build plate. Peel the raft away from the part. If needed, use a scraper or spatula.

The Raft Settings dialog box illustrates numerous settings and options.

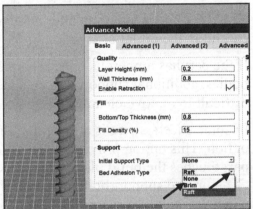

A Skirt is a layer of filament that surrounds the part with a 3-4mm offset. The layer does not connect the part directly to the build plate. The Skirt primes the extruder and establishes a smooth flow of filament. In some slicers, the skirt is added automatically when you select None for Bed Adhesion Type.

A Brim is basically like a Skirt for the part. A Brim has a zero offset from the part. It is a layer of filament that is laid down around the base of the part to increase its surface area. A Brim, however, does not extend underneath the part, which is the key difference between a brim and a raft.

Proper part orientation for thin parts will make the removal of the raft easier.

Part with a Skirt on the Build plate

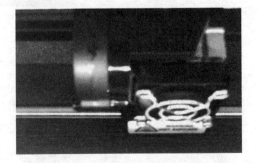

Part with a Raft on the Build plate

Support and Touching Build plate

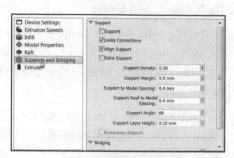

Printing with support is required when material must be deposited on a layer where there is no or insufficient material on the previous layer. This includes steep overhanging surfaces, straight overhangs, and fully suspended islands. Learning to print objects on a 3D Printer that require support structures will dramatically expand the potential of your printer and give you the confidence to undertake printing tasks that perhaps you had previously avoided.

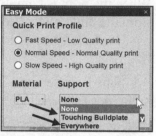

You can remove supports after building the object, but be careful in the initial orientation of the part.

If the part has numerous holes, sharp edges, steep angles or thin bodies, additional support will be added and can make it difficult to remove cleanly.

In some slicers, there is a Touching Buildplate option. This option provides supports only where the part touches the build plate. This reduces build time, clean up and support build material. Use this option when you have overhangs and tricky angles toward the bottom of a design, but do not wish to plug up holes, hollow spaces, or arches in the rest of the design.

See the below illustration of a support material example.

Support material

💡 My favorite tools to remove support and raft material are a good pair of needle nose pliers, angle tweezers and a small flat head screwdriver.

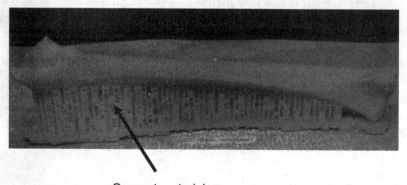

Slicer Engine

All 3D printers use a slicer engine. Slicing generates the G-code necessary to feed into your printer. Slicing is the process of turning your 3D model into a toolpath for your 3D printer. Most people call it slicing because the first thing the slicing engine does is cut your 3D model into thin horizontal layers. Open source slicing engines include Slic3r, Skeinforge, Netfabb, KISSkice, Cura, etc. to name a few.

G-code

There are different ways to prepare G-code for a printer. One is to use a slicer engine. These programs take a CAD model, slice it into layers, and output the G-code required for each layer. Slicers are an easy way to go from a 3D model to a printed part, but the user sacrifices some flexibility when using them. Another option for G-code generation is to use me-code, a lower level open source library. Me-code libraries give precise control over the toolpath providing a solution for a complex print that is not suitable for slicing. If you need to run a few test lines while calibrating your 3D printer, then the final option is to write your own G-code.

Infill

Infill is the internal structure of your object, which can be as sparse or as substantial as you would like it to be. A higher percentage will result in a more solid object, so 100% infill will make your object completely solid, while 0% infill will give you something completely hollow. The higher the infill percentage, the more material and longer the print time required. In general, use a 10% - 15% infill. 100% infill is very rarely used.

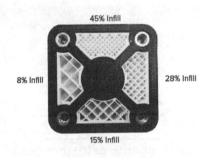

Infill Pattern/Shape

When using any infill percentage, a pattern is used to create a strong and durable structure inside the print. A few standard patterns are Rectilinear, Honeycomb, Circular and Triangular.

Shells/Parameters

Shells are the outer layers of a print which make the walls of an object, prior to the various infill levels being printed within.

The Number of Shells Display shows examples of cubes printed with 1, 2, 3, 4, 5, 10, 15, 20, 25, and 30 shells. The number of shells affects stability and translucency of the model.

Picture from 3D Printing.com
Duncan Smith

More shells result in a stronger object but longer printing time and more material.

Shells are also referred to as perimeters in some software and documents.

🔅 Do not use additional shells on fine featured models, such as small text. It will obscure the detail.

View the different infill and number of shells between the two models.

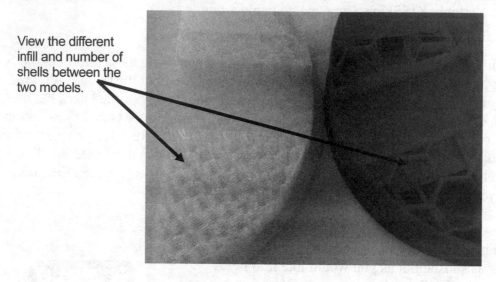

Layer Height

Traditionally, extrusion-based 3D printers have emphasized the layer height (such as 0.1mm or 0.3mm) as the main indicator of accuracy or quality. One reason for this may be that 3D prints made with this technology typically have visible ridges between different layers.

Layer height has a significant effect on the resolution of organic (curvy) parts. In general use 0.3mm for fast draft quality, 0.25mm for medium draft quality, 0.2mm for standard quality, and 0.1mm for high quality.

🔅 (1micron = 1μm = 0.001mm).

Influence of Percent Infill

How much infill is needed for your print? It depends. Infill refers to the structure that is printed inside the model. It is extruded in a designated percent and pattern. The designated percentage and pattern is set in the slicer.

The percent infill and pattern type (Rectilinear, Honeycomb, Circular), to name a few, influence material usage, strength, print weight, print time and sometimes flatness and decorative properties. As a general rule, the larger the percentage infill, the stronger the print, but the longer it takes to print. In most cases, 10% - 15% infill is sufficient. 100% infill is very rarely used.

💡 Strength corresponds to the maximum stress the print can take before breaking.

Remove the Model from the Build Plate

If you have a non-heated build plate with a sheet of good quality blue painter's tape, it is not too difficult to remove the part. Utilize a flat edge tool (a thin steel spatula), gently work under the part, and lift the raft or the part directly on the tape. You may rip the tape when taking the part off. Replace the tape. Re-level your build plate after every build.

If you have a heated build plate your temperatures can range between 60 - 90C (Kapton tape) depending on the material, so be careful. Again, as above, utilize a flat edge tool (a thin steel spatula), gently work under the part, and lift the part directly from the plate. Re-level your build plate after every build.

Know the Printer's Limitations

There are features that are too small to be printed in plastic on a desktop 3D printer. An important, but often overlooked, variable in what your printer can achieve is thread width.

Thread width is determined by the diameter of your printer's nozzle. Most printers have a 0.4mm or 0.5mm nozzle. Practically, this means that a circle drawn by a 3D printer is always two thread widths deep: 0.8mm thick with a 0.4mm nozzle to 1mm thick for a 0.5mm nozzle. A good rule of thumb is "The smallest feature you can create is double the thread width."

Tolerance for Interlocking Parts

For objects with multiple interlocking parts, design in your fit tolerance. Getting tolerances correct can be difficult. Typical minimum wall thickness for a model is roughly 1mm, and printed part accuracy is ±0.2mm. It is recommended to use a ±0.2mm offset for tight fit (press fit parts, connectors) and use a ±0.4mm offset for loose fit (hinges, box lids).

Keep in mind the underlying problems with low cost 3D printers. The same 3D model will produce prints with varying part tolerances when printed on different types and brands of 3D printers. Users need to check the specifications from the 3D printer and filament manufacturer and then adjust the model geometries accordingly. Also, with any 3D printer, make sure the printer is calibrated to ensure the resolution will be the same across multiple prints.

Test the fit yourself with the particular model to determine the right tolerance for the items you are creating and material you are using.

General Printing Tips

Reduce Infill

When printing a model, you can choose to print it hollow or completely solid, or some percentage between 0 and 100. Infill is a settable variable in most Slicer engines. The material inside the part exerts a force on the entire printed part as it cools. More material increases cost of the part and build time.

Parts with a lower percentage of Infill should have a lower internal force between layers and can reduce the chance of curling, cracking, and layer delamination along with a low build cost and time.

Control build area temperature

For a consistent quality build, control your build area and environment temperature. Eliminate all drafts and control air flow that may cause a temperature gradient within the build area.

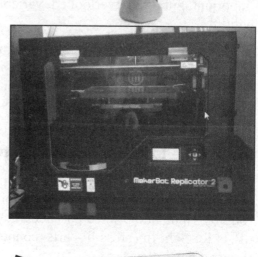

Changes in temperature during a build cycle can cause curling, cracking and layer delamination, especially on long thin parts. From 1000s of hours in 3D printing experience, having a top cover, sides and a heated build plate along with a consistent room temperature provides the best and repeatable builds.

When troubleshooting issues with your printer, it is always best to know if the nozzle and heated bed are achieving the desired temperatures. A thermocouple and a thermometer come in handy. I prefer a Type-K Thermocouple connected to a multi-meter. But a non-contact IR or Laser-based thermometer also works well.

One important thing to remember is that IR and Laser units are not 100% accurate when it comes to shiny reflective surfaces. A Type-K thermocouple can be taped to the nozzle or heated bed using Kapton tape, for a very accurate temperature measurement.

Cover

Add Pads

Sometimes, when you are printing a large flat object, such as a simple box container or a very long thin feature, you may view warping at the corners or extremities. One way to address this is to create small pads to your part during the modeling process. Create the model for the print. Think before you print and know your printer limitations. The pads can be any size and shape, but generally, diameter 10mm cylinders that are 1-2 layers thick work well. After the part is printed, remove them.

Makerbot Image

Unique Shape or a Large Part

If you need to make parts larger than your build area or create parts that have intricate projections, here are a few suggestions:

- Fuse smaller sections together using acetone (if using ABS). Glue if using PLA.

- Design smaller parts to be attached together (without hardware).

- Design smaller parts to be screwed together (with hardware).

Safe Zone Rule

Parts may have a safe zone. The safe zone is called "self-supporting" and no support material is required to build the part.

The safe zone can range between 30° to 150°. If the part's features are below 30° or greater than 150°, it should have support material during the build cycle. This is only a rule. Are there other factors to consider? Yes. They are layer thickness, extrusion speed, material type, length of the overhang along with the general model design of features.

Design your part for your printer. Use various modeling techniques (ribs, fillets, pads, etc.) during the design process to eliminate or to minimize the need for supports and clean up.

Wall Thickness

In 3D printing, wall thickness refers to the distance between one surface of your model and the opposite sheer surface. A model made in stereolithography with a minimal wall thickness of 1mm provides you with a strong solid surface.

If you scale your model with the 3D printer software, the wall thickness also scales proportionally. Make modifications in the CAD model to ensure the minimal wall thickness of 1mm.

Extruder Temperature

When working with a new roll of filament for the first time, I generally start out printing at about 200 - 210C (heated plate) and 225 - 230C (non-heated plate) for PLA and then adjust the temperature up or down by a few degrees until I obtain the quality of the print and the strength of the part to be in good balance with each other.

If the temperature is too high, you will see more strings between the separate parts of your print and you may notice that the extruder leaks out material while moving between separate areas of the print.

If the temperature is too cold, you will either see that the filament is not sticking to the (non-heated build plate) or the previous layer and you are getting a rough surface. You will get a part that is not strong and can be pulled apart easily.

Too cold – rough surface

The extruder temperature for PLA ranges between 200C - 215C (heated bed 60C - 65C), 225C - 230C (non-heated bed). For ABS they are 225C - 240 (heated bed 90C - 95C).

Various factors can influence the optimal extruding temperature:

- Moisture content.

- Temperature of the printing environment.

- Color.

- Glow in the dark ability.

- Elevation (from sea level) of the printing environment.

First Layer Not Sticking

One of the toughest aspects of 3D printing is to get your prints to stick to the build surface or bed platform. Investigate the following:

- Make sure the bed is level.

- Make sure the bed is hot enough. Do not use blue painter's tape with a heated plate.

- Make sure that the ambient temperature of the print environment isn't too hot or cold (or else adjust accordingly).

- Make sure you put the adhesive on the bed.

- Make sure the print head is close enough to make a nice squished first layer.

- Make sure you run the extruder enough before your print starts so there is filament going onto the bed during the entire first layer.

- Kapton/PET tape is a great way to print ABS. It makes a great shiny bottom layer and the heated bed ensures that your parts stay nice and flat.

Level Build Platform

An unleveled build platform will cause many headaches during a print. You can quickly check the platform by performing the business card test: use a single business card to judge the height of your extruder nozzle over the build platform. Achieve a consistent slight resistance when you position the business card between the tip of the extruder and the platform for all leveling positions.

Minimize Internal Support

Design your part for your printer. Use various modeling techniques (ribs, fillets, pads, etc.) during the design process to eliminate or to minimize the need for support and final clean up.

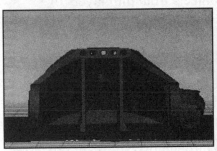 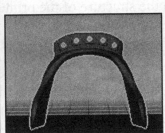 

Water Tight Mesh

A water-tight mesh is achieved by having closed edges creating a solid volume. If you were to fill your geometry with water, would you see a leak? You may have to clean up any internal geometry that could have been left behind accidentally from Booleans.

Clearance

If you are creating separate or interlocking parts, make sure there is a large enough distance between tight areas. 3D Printing production, such as Selective Laser Sintering (SLS), makes moving parts without assembly a possibility that was not there before. Take advantage of this strength by creating enough clearance that the model's pieces do not fuse together or trap support material inside.

In General

- Keep your software and firmware up to date.

- Think before you print. Design your model for your printer.

- Understand the printer's limitations. Adjust one thing at a time between prints and keep notes about the settings and effect on the print. Label test prints with a Sharpie and take photographs.

- Always level and clean the build plate before a build.

- Select the correct material (filament) for the build and application. Materials are still an area of active exploration. Use ABS, PLA and Nylon, each of which has specific printing requirements.

- Select the correct part orientation on the build plate. Many of these devices require some attention when printing. 3D printers fabricate objects from thin layers of material; there is a grain to the structure of the printed parts, much like there is grain in wood. Try to print in an orientation so that the "grain" of the print is oriented to maximize the strength of the part.

- Make sure that the bottom surface of each part of the print is firmly secured to the build platform, both for the success of the print and for the surface finish on the bottom face of the object.

- Select the correct Slicing engine settings. Models go through two software processes on their way to becoming finished prints: slicing and sending. Slicing divides a model into printable layers and plots the toolpaths to fill them in. The printer client then sends these movements to the hardware and provides a control interface for its other functions.

- Control the environment for the filament and build area (temperature, humidity, etc.).

- If in doubt, create your first build with a raft and support.

- 3D printers cannot print perfectly, so there will always be a bit of variation between the model and the output. If you make a round hole that's 10mm in diameter and a 10mm diameter drive shaft to fill that hole, it won't have enough clearance to fit into the hole. The best current commercially available printers have a resolution of 0.1mm. This means there could be up to 0.2mm difference between the actual surface of your print and the 3D model. Models also shrink a little after being printed. The best thing to do after a print has been completed is to cool down your machine entirely before switching it off. But, if you are in a rush, make sure to at least get it under 100C before switching it off.

Summary

With 1000s of hours using multiple low cost 3D printers, learning about additive print technology is a great experience. Students face a multitude of obstacles with their first 3D prints. Understanding what went wrong and knowing the capabilities of the 3D printer produces positive results.

Never focus too much on one single issue. These machines are complex, and trouble often arises from multiple reasons. A slipping filament may not only be caused by a bad gear drive, but also by an obstructed nozzle, a wrong feed value, a too low (or too high) temperature or a combination of all these.

As printed parts cool (PLA, ABS and Nylon), various areas of the object cool at different rates. Depending on the model being printed and the filament material, this effect can lead to warping, curling and or layer delamination.

Design your part for your printer. Use various modeling techniques (ribs, fillets, pads, etc.) during the design process to eliminate or to minimize the need for supports and clean up.

Avi Reichental, President and CEO, 3D Systems, stated, "With 3D printing, complexity is free. The printer doesn't care if it makes the most rudimentary shape or the most complex shape, and that is completely turning design and manufacturing on its head as we know it."

Additive manufacturing, sometimes known as **rapid prototyping**, can be slower than Subtractive manufacturing. Both take skill in creating the G-code and understanding the machine limitations.

Notes:

Project 11

Introduction to the Certified Associate - Mechanical Design (CSWA) Exam

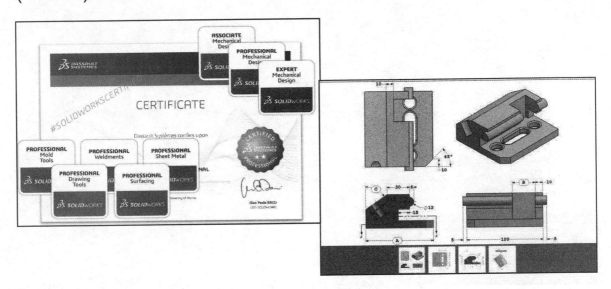

Below are the desired outcomes and usage competencies based on the completion of Project 11.

Desired Outcomes:	Usage Competencies:
<ul><li>Procedure and process knowledge</li><li>Exam categories:<ul><li>Drafting Competencies, Basic Part Creation and Modification, Intermediate Part Creation and Modification, Advanced Part Creation and Modification, and Assembly Creation and Modification.</li></ul></li></ul>	<ul><li>Familiarity of the CSWA exam.</li><li>Comprehension of the skill sets to pass the CSWA exam.</li><li>Awareness of the question types.</li><li>Capability to locate additional CSWA exam information.</li></ul>

Notes:

Project 11 - Certified Associate - Mechanical Design (CSWA) Exam

Project Objective

Provide a basic introduction into the curriculum and categories of the Certified Associate - Mechanical Design (CSWA) exam. Awareness of the exam procedure, process, and required model knowledge needed to take the CSWA exam. The five exam categories are:

- Drafting Competencies.

- Basic Part Creation and Modification.

- Intermediate Part Creation and Modification.

- Advanced Part Creation and Modification.

- Assembly Creation and Modification.

Introduction

DS SOLIDWORKS Corp. offers various types of certification. Each stage represents increasing levels of expertise in 3D CAD: *Certified SOLIDWORKS Associate CSWA, Certified SOLIDWORKS Professional CSWP* and *Certified SOLIDWORKS Expert CSWE* along with specialty fields.

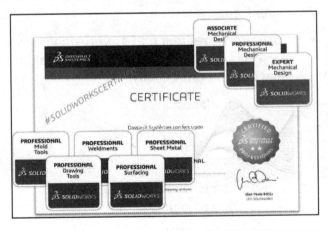

The CSWA certification indicates a foundation in and apprentice knowledge of 3D CAD design and engineering practices and principles. The main requirement for obtaining the CSWA certification is to take and pass the two part on-line proctored exams.

This first exam (Part 1) is 90 minutes; minimum passing score is 80, with 6 questions.

The second exam (Part 2) is 90 minutes; minimum passing score is 80 with 8 questions.

Part 1:

Basic Part Creation and Modification, Intermediate Part Creation and Modification

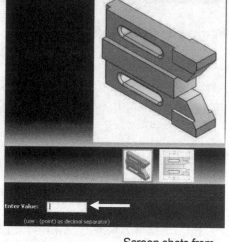

There are **two questions** on the CSWA exam (Part 1) in the *Basic Part Creation and Modification* category and **two questions** in the *Intermediate Part Creation and Modification* category.

The first question is in a multiple choice single answer format. You should be within 1% of the multiple choice answer before you move on to the modification single answer section (fill in the blank format).

Each question is worth fifteen (15) points for a total of thirty (30) points. You are required to build a model with six or more features and to answer a question either on the overall mass, volume, or the location of the Center of mass for the created model relative to the default part Origin location. You are then requested to modify the part and answer a fill in the blank format question.

Screen shots from an exam

* *Basic Part Creation and Modification*: (Two questions - one multiple choice/one single answer - 15 points each).

  * Sketch Planes:

    * Front, Top, Right

  * 2D Sketching:

    * Geometric Relations and Dimensioning

  * Extruded Boss/Base Feature

  * Extruded Cut feature

  * Modification of Basic part

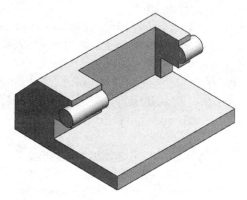

☀ In the *Basic Part Creation and Modification* category there is a dimension modification question based on the first (multiple choice) question. You should be within 1% of the multiple choice answer before you go on to the modification single answer section.

- *Intermediate Part Creation and Modification*: (Two questions - one multiple choice/one single answer - 15 points each).

  - Sketch Planes:

    - Front, Top, Right

  - 2D Sketching:

    Geometric Relations and Dimensioning

  - Extruded Boss/Base Feature

  - Extruded Cut Feature

  - Revolved Boss/Base Feature

  - Mirror and Fillet Feature

  - Circular and Linear Pattern Feature

  - Plane Feature

  - Modification of Intermediate part:

    - Sketch, Feature, Pattern, etc.

  - Modification of Intermediate part

In the *Intermediate Part Creation and Modification* category, there are two dimension modification questions based on the first (multiple choice) question. You should be within 1% of the multiple choice answer before you go on to the modification single answer section.

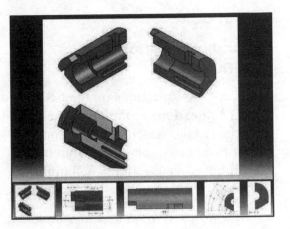

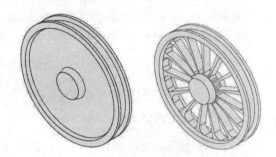

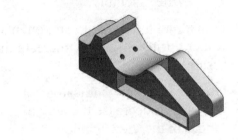

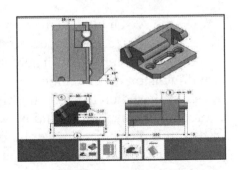

Hint: If you don't find an option within 1% of your answer please re-check your model(s).

Part 1: Cont.

Assembly Creation and Modification

There are four questions on the CSWA exam (**2 questions** in Part 1, 2 questions in Part 2) in the Assembly Creation and Modification category: (2) different assemblies - (4) questions - (2) multiple choice/(2) single answer - 30 points each.

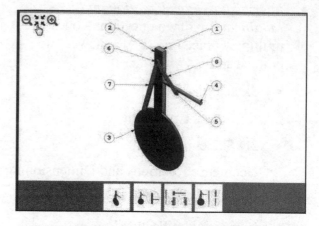

- *Assembly Creation and Modification*: (Two different assemblies - four questions - two multiple choice/two single answers - 30 points each).

  - Insert the first (fixed) component.

  - Insert all needed components.

  - Standard Mates.

  - Modification of key parameters in the assembly.

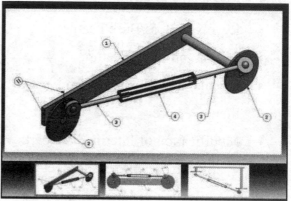

Download the needed components in a zip folder during the exam to create the assembly.

Use the new view indicator to increase or decrease the active model in the view window.

View indicator

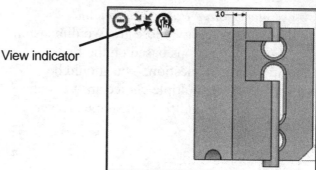

🔅 In the Assembly Creation and Modification category, expect to see five to seven components. There are two dimension modification questions based on the first (multiple choice) question. You should be within 1% of the multiple choice answer before you go on to the modification single answer section.

Part 2:

Introduction and Drafting Competencies

There are **three questions** on the CSWA exam in the *Drafting Competencies* category. Each question is worth five (5) points. Drafting Competency questions are addressed in Part 2 of the CSWA exam.

The three questions are in a multiple choice single answer format. You are allowed to answer the questions in any order you prefer.

In the *Drafting Competencies* category of the exam, you are **not required** to create or perform an analysis on a part, assembly, or drawing but you are required to have general drafting/drawing knowledge and understanding of various drawing view methods.

The questions are on general drawing views: Projected, Section, Break, Crop, Detail, Alternate Position, etc.

Advanced Part Creation and Modification

There are three questions on the CSWA exam (Part 2) in this category.

The first question is in a multiple choice single answer format.

The other two questions (Modification of the model) are in the fill in the blank format.

The main difference between the Advanced Part Creation and Modification and the Basic Part Creation and Modification or the Intermediate Part Creation and Modification category is the complexity of the sketches and the number of dimensions and geometric relations along with an increased number of features.

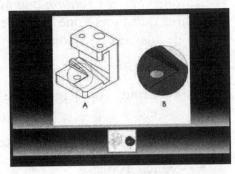

Screen shot from

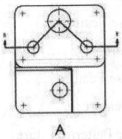

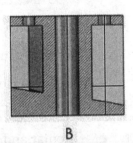

Drafting Competencies - To create drawing view 'B' it is necessary to select drawing view 'A' and insert which SolidWorks view type?

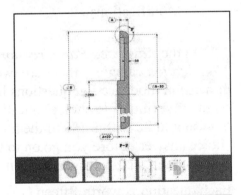

Advanced Part (Bracket) - Step 1
Build this part in SolidWorks.
(Save part after each question in a different file in case it must be reviewed)

Unit system: MMGS (millimeter, gram, second)
Decimal places: 2
Part origin: Arbitrary
All holes through all unless shown otherwise.
Material: AISI 1020 Steel
Density = 0.0079 g/mm^3

A = 64.00
B = 20.00
C = 26.50

What is the overall mass of the part (grams)?

- *Advanced Part Creation and Modification:*
  (Three questions - one multiple choice/two
  single answers - 15 points each).

  - Sketch Planes:

    - Front, Top, Right, Face, Created Plane,
      etc.

    - 2D Sketching or 3D Sketching

  - Sketch Tools:

    - Offset Entities, Convert Entitles, etc.

  - Extruded Boss/Base Feature

  - Extruded Cut Feature

  - Revolved Boss/Base Feature

  - Mirror and Fillet Feature

  - Circular and Linear Pattern Feature

  - Shell Feature

  - Plane Feature

  - More Difficult Geometry
    Modifications

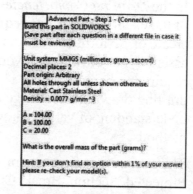

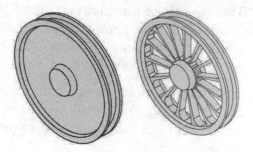

Advanced Part - Step 1 - (Connector)
Build this part in SOLIDWORKS.
(Save part after each question in a different file in case it
must be reviewed)

Unit system: MMGS (millimeter, gram, second)
Decimal places: 2
Part origin: Arbitrary
All holes through all unless shown otherwise.
Material: Cast Stainless Steel
Density = 0.0077 g/mm^3

A = 104.00
B = 100.00
C = 20.00

What is the overall mass of the part (grams)?

Hint: If you don't find an option within 1% of your answer
please re-check your model(s).

In the *Advanced Part Creation and
Modification* category, there are two
dimension modification questions based
on the first (multiple choice) question.
You should be within 1% of the multiple
choice answer before you go on to the
modification single answer section.

Each question is worth fifteen (15)
points for a total of forty five (45)
points.

You are required to build a model, with
six or more features and to answer a
question either on the overall mass, volume, or the location
the Center of mass for the created model relative to the
default part Origin location. You are then requested to modify
the model and answer fill in the blank format questions.

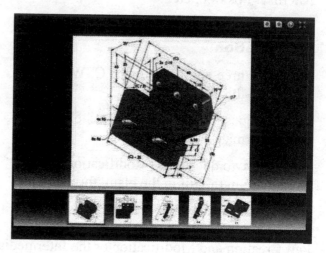

Screen shots from the exam

Assembly Creation and Modification

There are four questions on the CSWA exam (2 questions in part 1, **2 questions** in part 2) in the Assembly Creation and Modification category: (2) different assemblies - (4) questions - (2) multiple choice/(2) single answer - 30 points each.

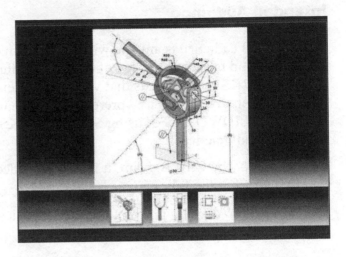

The first question is in a multiple choice single answer format. You should be within 1% of the multiple choice answer before you move on to the modification single answer section (fill in the blank format).

You are required to download the needed components from a provided zip file and insert them correctly to create the assembly.

In the Assembly Creation and Modification category, expect to see five to seven components. There are two dimension modification questions based on the first (multiple choice) question. You should be within 1% of the multiple choice answer before you go on to the modification single answer section.

No Surfacing questions are on the CSWA exam at this time.

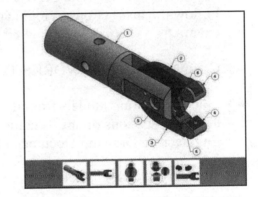

Screen shots from the exam

Intended Audience

The intended audience is anyone with a minimum of 6 - 9 months of SOLIDWORKS experience and basic knowledge of engineering fundamentals and practices. SOLIDWORKS recommends that you review their SOLIDWORKS Tutorials on Parts, Assemblies and Drawings as a prerequisite and have at least 45 hours of classroom time learning SOLIDWORKS or using SOLIDWORKS with basic engineering design principles and practices.

To prepare for the CSWA exam, it is recommended that you first perform the following:

- Take a CSWA exam preparation class or review a text book written for the CSWA exam.

- Download and open the CSWA Sample Exam folder. Follow the instructions to login and take a sample exam.

- Complete the SOLIDWORKS Tutorials.

- Practice creating models from the isometric working drawings sections of any Technical Drawing or Engineering Drawing Documentation text books.

Additional references to help you prepare are as follows:

- **Official Guide to Certified SOLIDWORKS Associate Exams: CSWA, CSDA, CSWSA-FEA, Version 3; 2015 - 2017,** Version 2; 2015 - 2012, Version 1; 2012, 2013.

- **Engineering Drawing and Design**, Jensen & Helsel, Glencoe, 1990.

During the Exam

During the exam, SOLIDWORKS provides the ability to click on a detail view below (as illustrated) to obtain additional details and dimensions during the exam.

💡 No Simulation questions are on the CSWA exam at this time.

💡 No Sheetmetal questions are on the CSWA exam at this time.

FeatureManager names were changed through various revisions of SOLIDWORKS. Example: Extrude1 vs. Boss-Extrude1. These changes do not affect the models or answers in this book.

During the exam, use the control keys at the bottom of the screen to:

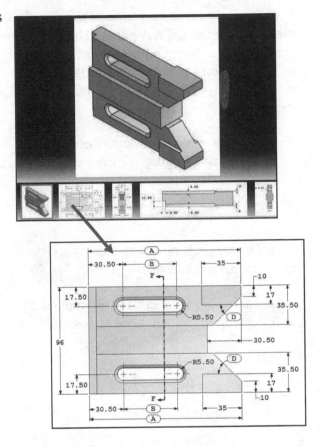

- *Show the Previous Question.*
- *Reset the Question.*
- *Show the Summary Screen.*
- *Move to the Next Question.*

When you are finished, press the End Examination button. The tester will ask you if you want to end the text. Click Yes.

If there are any unanswered questions, the tester will provide a warning message as illustrated.

💡 If you do not pass the certification exam, you will need to wait 30 days until you can retake each part of the exam.

Use the clock in the tester to view the amount of time that you used and the amount of time that is left in the exam.

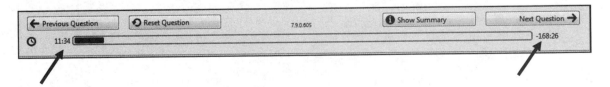

Examples: Drafting Competencies

Drafting Competencies is one of the five categories on the CSWA exam. There are three questions - multiple choice format - 5 points each that require general knowledge and understanding of drawing view methods and basic 3D modeling techniques.

Spend no more than 10 minutes on each question in this category for the exam. Manage your time.

Sample Questions in the category

In the *Drafting Competencies* category, an exam question could read:

Question 1: Identify the view procedure. To create the following view, you need to insert a:

- A: Open Spline

- B: Closed Spline

- C: 3 Point Arc

- D: None of the above

The correct answer is B.

Question 2: Identify the illustrated view type.

- A: Crop view

- B: Section view

- C: Projected view

- D: None of the above

The correct answer is A.

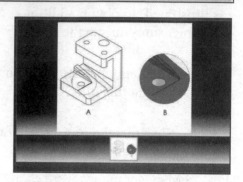

Drafting Competencies - To create drawing view 'B' it is necessary to select drawing view 'A' and insert which SolidWorks view type?

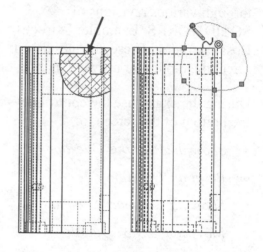

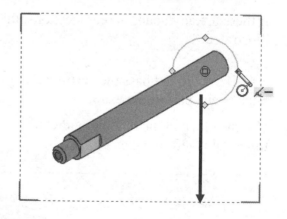

Question 3: Identify the illustrated Drawing view.

- A: Projected View

- B: Alternative Position View

- C: Extended View

- D: Aligned Section View

The correct answer is B.

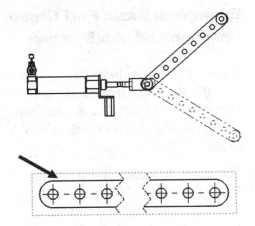

Question 4: Identify the illustrated Drawing view.

- A: Crop View

- B: Break View

- C: Broken-out Section View

- D: Aligned Section View

The correct answer is B.

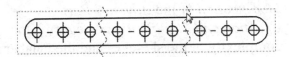

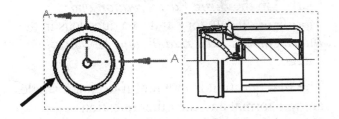

Question 5: Identify the illustrated Drawing view.

- A: Section View

- B: Crop View

- C: Broken-out Section View

- D: Aligned Section View

The correct answer is D.

Question 6: Identify the view procedure. To create the following view, you need to insert a:

- A: Rectangle Sketch tool

- B: Closed Profile: Spline

- C: Open Profile: Circle

- D: None of the above

The correct answer is B.

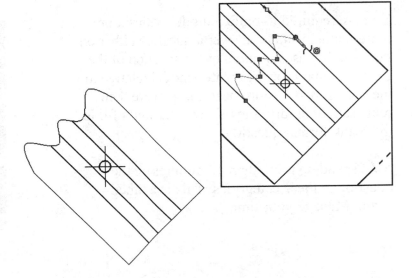

Examples: Basic Part Creation and Modification and Intermediate Part Creation and Modification

Basic Part Creation and Modification and *Intermediate Part Creation and Modification* are two of the five categories on the CSWA exam.

The main difference between the *Basic Part Creation and Modification* category and the *Intermediate Part Creation and Modification* or the *Advance Part Creation and Modification* category is the complexity of the sketches and the number of dimensions and geometric relations along with an increase in the number of features.

There are two questions on the CSWA exam (part 1) in the *Basic Part Creation and Modification* category and two questions in the *Intermediate Part Creation and Modification* category.

The first question is in a multiple choice single answer format and the other question (Modification of the model) is in the fill in the blank format.

Each question is worth fifteen (15) points for a total of thirty (30) points.

You are required to build a model with six or more features and to answer a question either on the overall mass, volume, or the location of the Center of mass for the created model relative to the default part Origin location. You are then requested to modify the part and answer a fill in the blank format question.

💡 Spend no more than 40 minutes on the question in these categories. This is a timed exam. Manage your time.

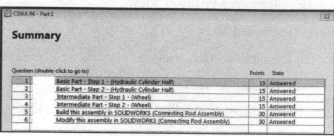

Question (double-click to go to)	Points	State
1 Basic Part - Step 1 - (Hydraulic Cylinder Half)	15	Answered
2 Basic Part - Step 2 - (Hydraulic Cylinder Half)	15	Answered
3 Intermediate Part - Step 1 - (Wheel)	15	Answered
4 Intermediate Part - Step 2 - (Wheel)	15	Answered
5 Build this assembly in SOLIDWORKS (Connecting Rod Assembly)	30	Answered
6 Modify this assembly in SOLIDWORKS (Connecting Rod Assembly)	30	Answered

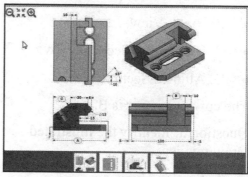

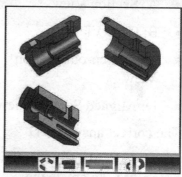

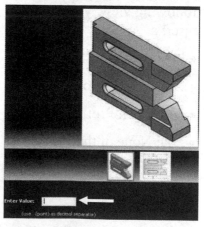

Screen shots from an exam

Sample Questions in this category

Question 1: Build the illustrated model from the provided information. Locate the Center of mass relative to the default coordinate system, Origin.

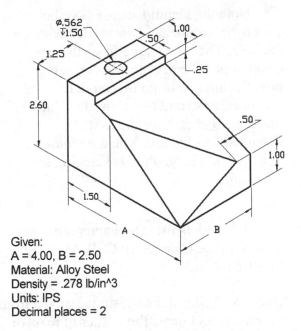

Given:
A = 4.00, B = 2.50
Material: Alloy Steel
Density = .278 lb/in^3
Units: IPS
Decimal places = 2

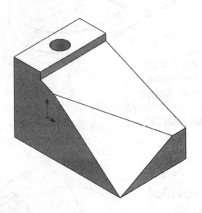

- A: X = -1.63 inches, Y = 1.48 inches, Z = -1.09 inches

- B: X = 1.63 inches, Y = 1.01 inches, Z = -0.04 inches

- C: X = 43.49 inches, Y = -0.86 inches, Z = -0.02 inches

- D: X = 1.63 inches, Y = 1.01 inches, Z = -0.04 inches

The correct answer is B.

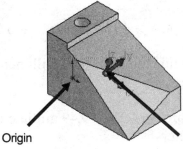

Origin   Center of mass relative to the part Origin

In the *Basic Part Creation and Modification* and *Intermediate Part Creation and Modification* category of the exam, you are required to read and understand an engineering document, set document properties, identify the correct Sketch planes, apply the correct Sketch and Feature tools and apply material to build a part.

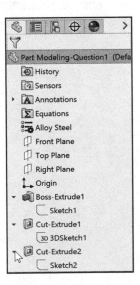

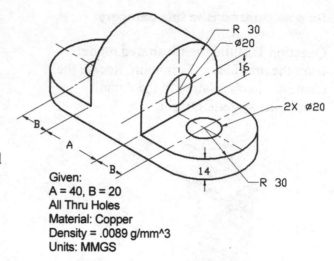

Note the Depth/Deep ⊤ symbol with a 1.50 dimension associated with the hole. The hole Ø.562 has a three decimal place precision. Hint: Insert three features to build this model: Extruded Base and two Extruded Cuts. Insert a 3D sketch for the first Extruded Cut feature. You are required to have knowledge in 3D sketching for the exam.

Given:
A = 40, B = 20
All Thru Holes
Material: Copper
Density = .0089 g/mm^3
Units: MMGS

All models for this chapter are located in the Chapter 10 CSWA Models folder.

Question 2: Build the illustrated model from the provided information. Locate the Center of mass of the part.

- A: X = 0.00 millimeters, Y = 19.79 millimeters, Z = 0.00 millimeters

- B: X = 0.00 inches, Y = 19.79 inches, Z = 0.04 inches

- C: X = 19.79 millimeters, Y = 0.00 millimeters, Z = 0.00 millimeters

- D: X = 0.00 millimeters, Y = 19.49 millimeters, Z = 0.00 millimeters

- The correct answer is A.

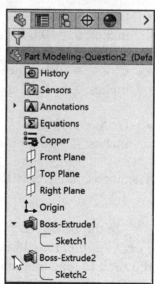

Question 3: Build the illustrated model
from the provided information. Locate the
Center of mass of the part.

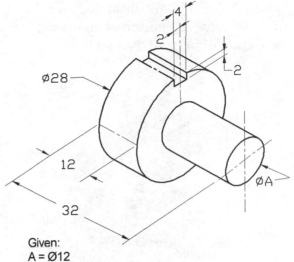

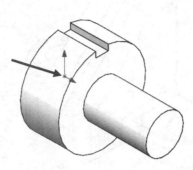

Given:
A = Ø12
Material: Cast Alloy Steel
Density = .0073 g/mm^3
Units: MMGS

- A: X = 10.00 millimeters, Y = -79.79 millimeters,
  Z: = 0.00 millimeters

- B: X = 9.79 millimeters, Y = -0.13 millimeters,
  Z = 0.00 millimeters

- C: X = 9.77 millimeters, Y = -0.10 millimeters,
  Z = -0.02 millimeters

- D: X = 10.00 millimeters, Y = 19.49 millimeters,
  Z = 0.00 millimeters

- The correct answer is B.

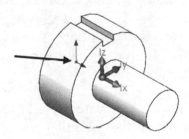

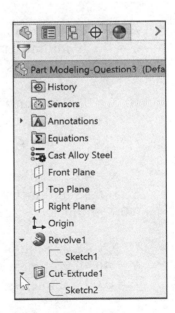

Question 4: Build the illustrated model from the provided information. Locate the Center of mass of the part.

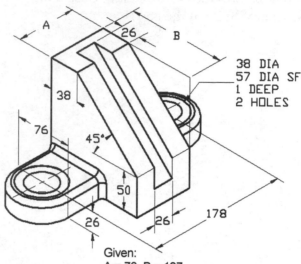

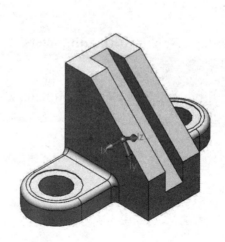

38 DIA
57 DIA SF
1 DEEP
2 HOLES

Given:
A = 76, B = 127
Material: 2014 Alloy
Density: .0028 g/mm^3
Units: MMGS
ALL ROUNDS EQUAL 6MM

There are numerous ways to build this model. Think about the various features that create the model. Hint: Insert seven features to build this model: Extruded Base, Extruded Cut, Extruded Boss, Fillet, Extruded Cut, Mirror and a second Fillet. Apply symmetry.

In the exam, create the left half of the model first, and then apply the Mirror feature. This is a timed exam.

- A: X = 49.00 millimeters,
  Y = 45.79 millimeters,
  Z = 0.00 millimeters

- B: X = 0.00 millimeters,
  Y = 19.79 millimeters,
  Z = 0.04 millimeters

- C: X = 49.21 millimeters,
  Y = 46.88 millimeters, Z =
  0.00 millimeters

- D: X = 48.00 millimeters,
  Y = 46.49 millimeters,
  Z = 0.00 millimeters

The correct answer is C.

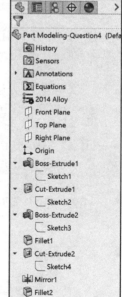

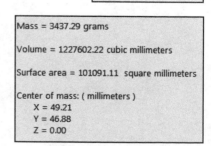

Mass = 3437.29 grams

Volume = 1227602.22 cubic millimeters

Surface area = 101091.11 square millimeters

Center of mass: (millimeters)
    X = 49.21
    Y = 46.88
    Z = 0.00

Question 5: Build the illustrated model from the provided information. Locate the Center of mass of the part. All Thru Holes.

💡 Think about the various features that create this model. Hint: Insert five features to build this part: Extruded Base, two Extruded Bosses, Extruded Cut, and Rib. Insert a Reference plane to create the Extruded Boss feature.

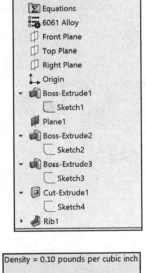

Given:
A = Ø3.00, B = 1.00
Material: 6061 Alloy
Density: .097 lb/in^3
Units: IPS
Decimal places = 2

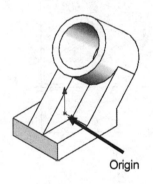

Origin

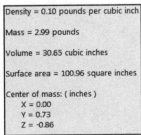

Density = 0.10 pounds per cubic inch

Mass = 2.99 pounds

Volume = 30.65 cubic inches

Surface area = 100.96 square inches

Center of mass: (inches)
  X = 0.00
  Y = 0.73
  Z = -0.86

- A: X = 49.00 inches, Y = 45.79 inches, Z = 0.00 inches

- B: X = 0.00 inches, Y = 19.79 inches, Z = 0.04 inches

- C: X = 49.21 inches, Y = 46.88 inches, Z = 0.00 inches

- D: X = 0.00 inches, Y = 0.73 inches, Z = -0.86 inches

The correct answer is D.

All models for this chapter are located in the Chapter 10 CSWA Models folder.

Question 6: Build the illustrated model from the provided information.

Calculate the overall mass and volume of the part with the provided information.

- Precision for linear dimensions = **2**.

- Material: **AISI 304**.

- Units: **MMGS**.

- All Holes ⊽ **25**mm.

- All Rounds **5**mm.

- All Holes Ø**4**mm.

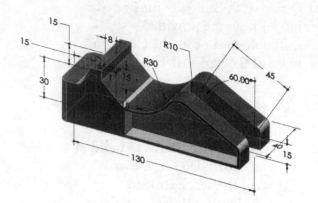

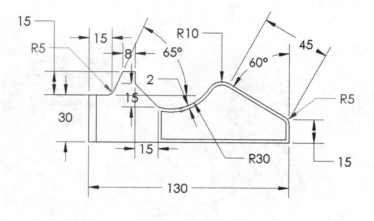

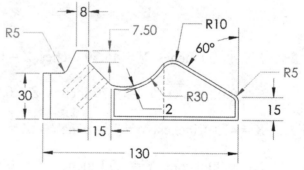

Front views

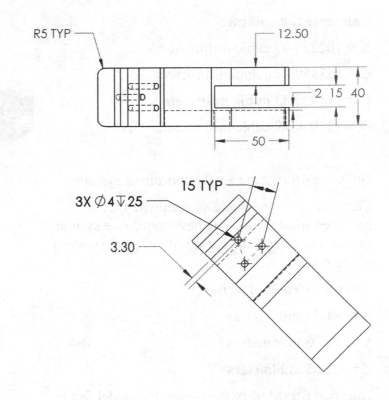

Top and Auxiliary

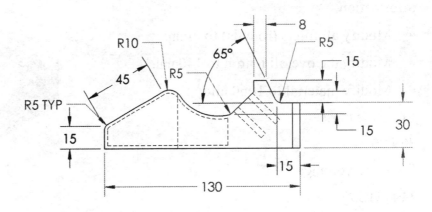

Calculate the mass:

A = 888.48grams

B = 990.50grams

C = 788.48grams

D = 820.57grams

💡 If you don't find your answer (within 1%) in the multiple choice single answer format section, recheck your solid model for precision and accuracy. It could be as simple as missing a few fillets.

Calculate the volume:

A = 102259.43 cubic millimeters

B = 133359.47 cubic millimeters

C = 111059.43 cubic millimeters

D = 125059.49 cubic millimeters

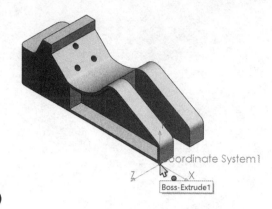

Question 6A: Create a new coordinate system.

Center a new coordinate system with the provided illustration. The new coordinate system location is at the front right bottom point (vertex) of the model.

Enter the Center of Mass:

X = -80.39 millimeters

Y = -15.93 millimeters

Z = -22.65 millimeters

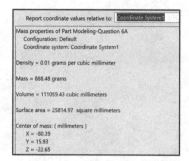

Question 6B: Modify the illustrated model from the provided information. Calculate the overall mass and volume of the part with the provided information.

- Modify all fillets (rounds) to 7mm.
- Modify the overall length to 140mm.
- Modify material to 1060 alloy.

Enter the mass:

309.75

Enter the volume:

114721.22

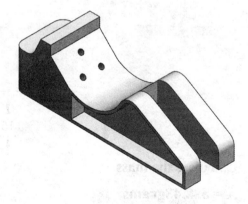

🔆 If you don't find your answer (within 1%) in the multiple choice single answer format section, recheck your solid model for precision and accuracy. It could be as simple as missing a few fillets.

🔆 All models for this chapter are located in the Chapter 10 CSWA Models folder.

Examples: Advanced Part Creation and Modification

Advanced Part Creation and Modification is one of the five categories on the CSWA exam. The main difference between the *Advanced Part Creation and Modification* and the *Basic Part Creation and Modification* category and the *Intermediate Part Creation and Modification* is the complexity of the sketches and the number of dimensions and geometric relations along with an increased number of features.

There are three questions - one multiple choice/two single answers - 15 points each. The question is either on the location of the Center of mass relative to the default part Origin or to a new created coordinate system and all of the mass properties located in the Mass Properties dialog box: total overall mass, volume, etc.

Sample Questions in the Category

In the *Advanced Part Creation and Modification* category, an exam question could read:

Question 1: Build the illustrated model from the provided information. Locate the Center of mass of the part.

| | Advanced Part (Bracket) - Step 1 |
Build this part in SolidWorks.
(Save part after each question in a different file in case it must be reviewed)

Unit system: MMGS (millimeter, gram, second)
Decimal places: 2
Part origin: Arbitrary
All holes through all unless shown otherwise.
Material: AISI 1020 Steel
Density = 0.0079 g/mm^3

A = 64.00
B = 20.00
C = 26.50

What is the overall mass of the part (grams)?

Screen shots from the exam

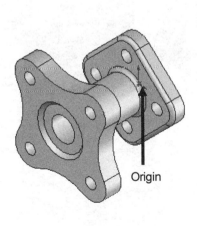

Origin

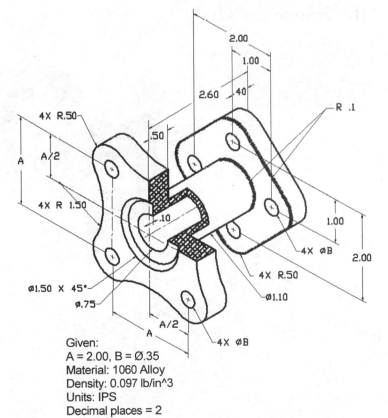

4X R.50
A/2
A
4X R 1.50
.10
Ø1.50 X 45°
Ø.75
A/2
A

2.00
1.00
2.60   .40
.50
R .1
1.00
4X ØB
2.00
4X R.50
Ø1.10
4X ØB

Given:
A = 2.00, B = Ø.35
Material: 1060 Alloy
Density: 0.097 lb/in^3
Units: IPS
Decimal places = 2

Think about the steps that you would take to build the illustrated part. Identify the location of the part Origin.

Start with the back base flange. Review the provided dimensions and annotations in the part illustration.

💡 The key difference between the *Advanced Part Creation and Modification* and the *Basic Part Creation and Modification* category and the *Intermediate Part Creation and Modification* is the complexity of the sketches and the number of features, dimensions, and geometric relations. You may also need to locate the Center of mass relative to a created coordinate system location.

- A: X = 1.00 inches, Y = 0.79 inches, Z = 0.00 inches

- B: X = 0.00 inches, Y = 0.00 inches, Z = 1.04 inches

- C: X = 0.00 inches, Y = 1.18 inches, Z = 0.00 inches

- D: X = 0.00 inches, Y = 0.00 inches, Z = 1.51 inches

The correct answer is D.

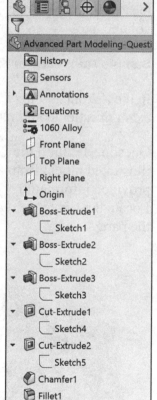

Mass = 0.59 pounds

Volume = 6.01 cubic inches

Surface area = 46.61 square inches

Center of mass: (inches)
X = 0.00
Y = 0.00
Z = 1.51

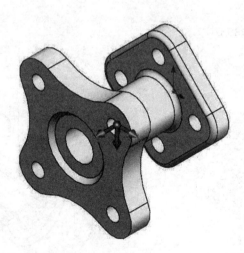

Question 2: Build the illustrated model from the provided information. Locate the Center of mass of the part.

Hint: Create the part with eleven features and a Reference plane: Extruded Base, Plane1, two Extruded Bosses, two Extruded Cuts, Extruded Boss, Extruded Cut, Extruded-Thin, Mirror, Extruded Cut, and Extruded Boss.

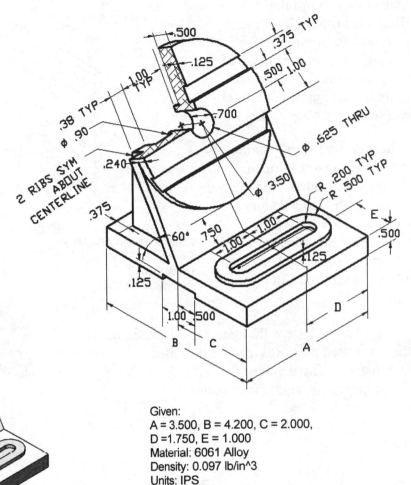

Given:
A = 3.500, B = 4.200, C = 2.000,
D =1.750, E = 1.000
Material: 6061 Alloy
Density: 0.097 lb/in^3
Units: IPS
Decimal places = 3

Think about the steps that you would take to build the illustrated part. Create the rectangular Base feature. Create Sketch2 for Plane1. Insert Plane1 to create the Extruded Boss feature: Extrude2. Plane1 is the Sketch plane for Sketch3. Sketch3 is the sketch profile for Extrude2.

- A: X = 1.59 inches, Y = 1.19 inches, Z = 0.00 inches

- B: X = -1.59 inches, Y = 1.19 inches, Z = 0.04 inches

- C: X = 1.00 inches, Y = 1.18 inches, Z = 0.10 inches

- D: X = 0.00 inches, Y = 0.00 inches, Z = 1.61 inches

The correct answer is A.

Density = 0.10 pounds per cubic inch

Mass = 1.37 pounds

Volume = 14.05 cubic inches

Surface area = 79.45 square inches

Center of mass: (inches)
  X = 1.59
  Y = 1.19
  Z = 0.00

Question 3: Build the illustrated model from the provided information. Locate the Center of mass of the part. Note the coordinate system location of the model as illustrated.

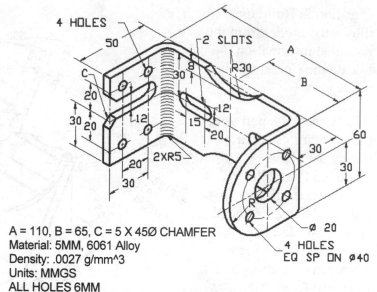

A = 110, B = 65, C = 5 X 45Ø CHAMFER
Material: 5MM, 6061 Alloy
Density: .0027 g/mm^3
Units: MMGS
ALL HOLES 6MM

Where do you start? Build the model. Insert thirteen features: Extruded-Thin1, Fillet, two Extruded Cuts, Circular Pattern, two Extruded Cuts, Mirror, Chamfer, Extruded Cut, Mirror, Extruded Cut and Mirror.

Think about the steps that you would take to build the illustrated part. Review the provided information. The depth of the left side is 50mm. The depth of the right side is 60mm.

Create Coordinate System1 to locate the Center of mass.

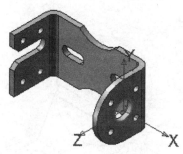

Coordinate system: +X, +Y. +Z

☀ The SOLIDWORKS software displays positive values for (X, Y, Z) coordinates for a reference coordinate system. The CSWA exam displays either a positive or negative sign in front of the (X, Y, Z) coordinates to indicate direction as illustrated (-X, +Y, -Z).

- A: X = -53.30 millimeters, Y = -0.27 millimeters, Z = -15.54 millimeters

- B: X = 53.30 millimeters, Y = 0.27 millimeters, Z = 15.54 millimeters

- C: X = 49.21 millimeters, Y = 46.88 millimeters, Z = 0.00 millimeters

- D: X = 45.00 millimeters, Y = -46.49 millimeters, Z = 10.00 millimeters

The correct answer is A.

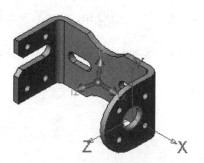

Question 4: Build the illustrated model from the provided information. Locate the Center of mass of the part.

Hint: Insert twelve features and a Reference plane: Extruded-Thin1, two Extruded Bosses, Extruded Cut, Extruded Boss, Extruded Cut, Plane1, Mirror, and five Extruded Cuts.

Think about the steps that you would take to build the illustrated part. Create an Extrude-Thin1 feature as the Base feature.

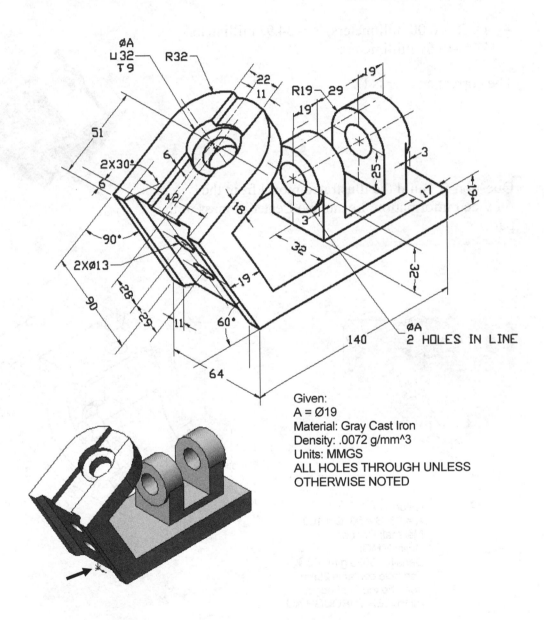

Given:
A = Ø19
Material: Gray Cast Iron
Density: .0072 g/mm^3
Units: MMGS
ALL HOLES THROUGH UNLESS
OTHERWISE NOTED

- A: X = -53.30 millimeters, Y = -0.27 millimeters, Z = -15.54 millimeters

- B: X = 53.30 millimeters, Y = 1.27 millimeters, Z = -15.54 millimeters

- C: X = 0.00 millimeters, Y = 34.97 millimeters, Z = 46.67 millimeters

- D: X = 0.00 millimeters, Y = 34.97 millimeters, Z = -46.67 millimeters

The correct answer is D.

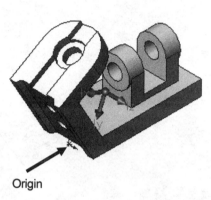

| Density = 0.01 grams per cubic millimeter |
| Mass = 2536.59 grams |
| Volume = 352304.50 cubic millimeters |
| Surface area = 61252.90 square millimeters |
| Center of mass: (millimeters)
X = 0.00
Y = 34.97
Z = -46.67 |

Question 5: Build the illustrated model from the provided information. Locate the Center of mass of the part.

Origin

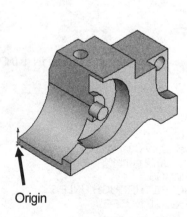

Origin

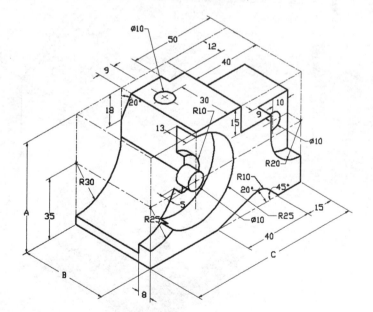

Given:
A = 63, B = 50, C = 100
Material: Copper
Units: MMGS
Density: .0089 g/mm^3
Top hole center is 20mm
from the top front edge.
All HOLES THROUGH ALL

The center point of the top hole is located 30mm from the top right edge.

Think about the steps that you would take to build the illustrated part.

- A: X = 26.81 millimeters, Y = 25.80 millimeters, Z = -56.06 millimeters

- B: X = 43.30 millimeters, Y = 25.27 millimeters, Z = -15.54 millimeters

- C: X = 26.81 millimeters, Y = -25.75 millimeters, Z = 0.00 millimeters

- D: X = 46.00 millimeters, Y = -46.49 millimeters, Z = 10.00 millimeters

The correct answer is A.

This model has thirteen features and twelve sketches.

☼ There are numerous ways to create the models in this chapter.

Density = 0.01 grams per cubic millimeter

Mass = 1280.33 grams

Volume = 143857.58 cubic millimeters

Surface area = 26112.48 square millimeters

Center of mass: (millimeters)
  X = 26.81
  Y = 25.80
  Z = -56.06

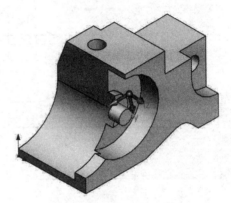

Question 6: Build the illustrated model from the provided information. Calculate the overall mass and volume of the part with the provided information.

- Precision for linear dimensions = **2**.

- Material: **Plain Carbon Steel**.

- Units: **MMGS**.

- The part is **symmetrical** about the Front Plane.

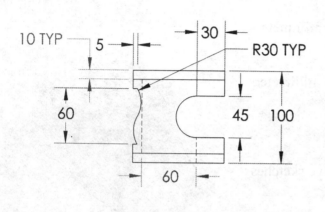

Top view   Front view

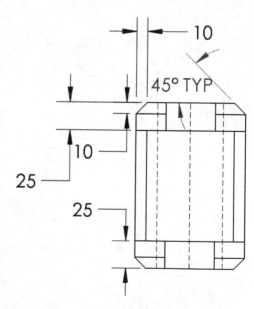

Right view

Calculate the mass:

A = 4411.5 grams

B = 4079.32 grams

C = 4234.30 grams

D = 5322.00 grams

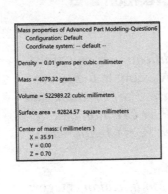

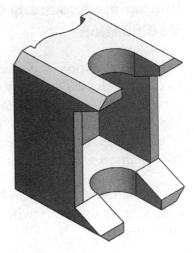

Calculate the volume:

A = 522989.22 cubic millimeters

B = 555655.11 cubic millimeters

C = 511233.34 cubic millimeters

D = 655444.00 cubic millimeters

Question 6A: Create a new coordinate system.

Center a new coordinate system with the provided illustration. The new coordinate system location is at the back right bottom point (vertex) of the model.

Enter the Center of Mass:

X = -64.09

Y = 75.00

Z = 40.70

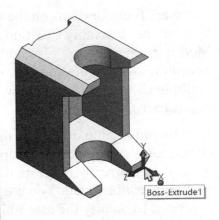

All models for this chapter are located in the Chapter 10 CSWA Models folder.

Examples: Assembly Creation and Modification

Assembly Creation and Modification is one of the five categories on the CSWA exam. In the last two sections of this chapter, a *Basic Part Creation and Modification, Intermediate Part Creation and Modification,* or an *Advanced Part Creation and Modification* was the focus.

The *Assembly Creation and Modification* category addresses an assembly with numerous sub-components.

Knowledge to insert Standard mates is required in this category.

There are four questions on the CSWA exam in the Assembly Creation and Modification category: (Two different assemblies - four questions - two multiple choice/two single answers - 30 points each).

You are required to download the needed components from a provided zip file and insert them correctly to create the assembly as illustrated. You are then requested to modify the assembly and answer fill in the blank format questions.

💡 Components for the assembly are supplied in the exam.

💡 Do not use feature recognition when you open the downloaded components for the assembly in the CSWA exam. This is a timed exam. Manage your time. You do not need this information.

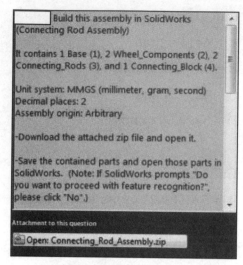

Screen shots from the exam

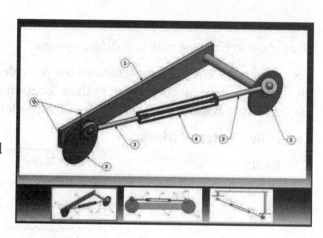

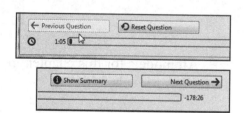

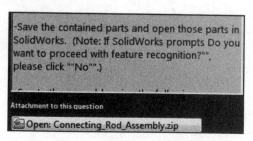

Screen shots from the exam

Sample Questions in this Category

In the *Assembly Creation and Modification* Assembly Modeling category, an exam question could read:

Build this assembly in SOLIDWORKS (Chain Link Assembly). It contains 2 long_pins (1), 3 short_pins (2), and 4 chain_links (3).

- Unit system: MMGS (millimeter, gram, second).

- Decimal places: 2.

- Assembly origin: Arbitrary.

IMPORTANT: Create the Assembly with respect to the Origin as shown in the Isometric view. This is important for calculating the proper Center of Mass. Create the assembly using the following conditions:

1. Pins are mated concentric to chain link holes (no clearance).

2. Pin end faces are coincident to chain link side faces.

A = 25 degrees, B = 125 degrees, C = 130 degrees

What is the center of mass of the assembly (millimeters)?

Hint: If you don't find an option within 1% of your answer please re-check your assembly.

A) X = 348.66, Y = -88.48, Z = -91.40

B) X = 308.53, Y = -109.89, Z = -61.40

C) X = 298.66, Y = -17.48, Z = -89.22

D) X = 448.66, Y = -208.48, Z = -34.64

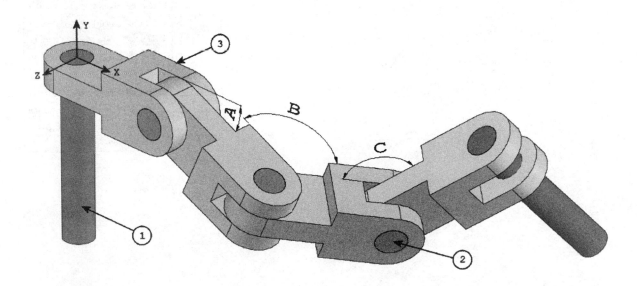

There are no step by step procedures in this section.

Download the needed components from the Chapter 6 CSWA Models folder.

The correct answer is:

A) X = 348.66, Y = -88.48, Z = -91.40

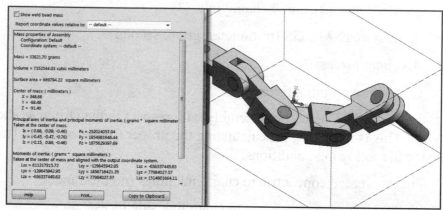

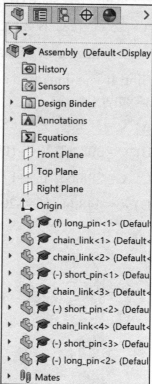

Appendix

SOLIDWORKS Keyboard Shortcuts

Below are some of the pre-defined keyboard shortcuts in SOLIDWORKS:

Action:	Key Combination:
Model Views	
Rotate the model horizontally or vertically	**Arrow** keys
Rotate the model horizontally or vertically 90 degrees	**Shift + Arrow** keys
Rotate the model clockwise or counterclockwise	**Alt** + left of right **Arrow** keys
Pan the model	**Ctrl + Arrow** keys
Magnifying glass	**g**
Zoom in	**Shift + z**
Zoom out	**z**
Zoom to fit	**f**
Previous view	**Ctrl + Shift + z**
View Orientation	
View Orientation menu	**Spacebar**
Front view	**Ctrl + 1**
Back view	**Ctrl + 2**
Left view	**Ctrl + 3**
Right view	**Ctrl + 4**
Top view	**Ctrl + 5**
Bottom view	**Ctrl + 6**
Isometric view	**Ctrl + 7**
NormalTo view	**Ctrl + 8**
Selection Filters	
Filter edges	**e**
Filter vertices	**v**
Filter faces	**x**
Toggle Selection Filter toolbar	**F5**
Toggle selection filters on/off	**F6**
File menu items	
New SOLIDWORKS document	**Ctrl + n**
Open document	**Ctrl + o**
Open From Web Folder	**Ctrl + w**
Make Drawing from Part	**Ctrl + d**
Make Assembly from Part	**Ctrl + a**
Save	**Ctrl +s**
Print	**Ctrl + p**
Additional items	
Access online help inside of PropertyManager or dialog box	**F1**
Rename an item in the FeatureManager design tree	**F2**

Action:	Key Combination:
Rebuild the model	**Ctrl + b**
Force rebuild - Rebuild the model and all its features	**Ctrl + q**
Redraw the screen	**Ctrl + r**
Cycle between open SOLIDWORKS document	**Ctrl + Tab**
Line to arc/arc to line in the Sketch	**a**
Undo	**Ctrl + z**
Redo	**Ctrl + y**
Cut	**Ctrl + x**
Copy	**Ctrl + c**
Paste	**Ctrl + v**
Delete	**Delete**
Next window	**Ctrl + F6**
Close window	**Ctrl + F4**
View previous tools	**s**
Selects all text inside an Annotations text box	**Ctrl + a**

In a sketch, the **Esc** key un-selects geometry items currently selected in the Properties box and Add Relations box.

In the model, the **Esc** key closes the PropertyManager and cancels the selections.

Use the **g** key to activate the Magnifying glass tool. Use the Magnifying glass tool to inspect a model and make selections without changing the overall view.

Use the **s** key to view/access previous command tools in the Graphics window.

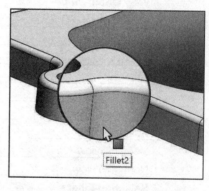

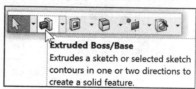

Extruded Boss/Base
Extrudes a sketch or selected sketch contours in one or two directions to create a solid feature.

Modeling - Best Practices

Best practices are simply ways of bringing about better results in easier, more reliable ways. The Modeling - Best Practice list is a set of rules helpful for new users and users who are trying to experiment with the limits of the software.

These rules are not inflexible, but conservative starting places; they are concepts that you can default to, but that can be broken if you have good reason. The following is a list of suggested best practices:

- Create a folder structure (parts, drawings, assemblies, simulations, etc.). Organize into project or file folders.

- Construct sound document templates. The document template provides the foundation that all models are built on. This is especially important if working with other SOLIDWORKS users on the same project; it will ensure consistency across the project.

- Generate unique part filenames. SOLIDWORKS assemblies and drawings may pick up incorrect references if you use parts with identical names.

- Apply Custom Properties. Custom Properties is a great way to enter text-based information into the SOLIDWORKS parts. Users can view this information from outside the file by using applications such as Windows Explorer, SOLIDWORKS Explorer, and Product Data Management (PDM) applications.

- Understand part orientation. When you create a new part or assembly, the three default Planes (Front, Right and Top) are aligned with specific views. The plane you select for the Base sketch determines the orientation.

- Learn to sketch using automatic relations.

- Limit your usage of the Fixed constraint.

- Add geometric relations, then dimensions in a 2D sketch. This keeps the part from having too many unnecessary dimensions. This also helps to show the design intent of the model. Dimension what geometry you intend to modify or adjust.

- Fully define all sketches in the model. However, there are times when this is not practical, generally when using the Spline tool to create a freeform shape.

- When possible, make relations to sketches or stable reference geometry, such as the Origin or standard planes, instead of edges or faces. Sketches are far more stable than faces, edges, or model vertices, which change their internal ID at the slightest change and may disappear entirely with fillets, chamfers, split lines, and so on.

- Do not dimension to edges created by fillets or other cosmetic or temporary features.

- Apply names to sketches, features, dimensions, and mates that help to make their function clear.

- When possible, use feature fillets and feature patterns rather than sketch fillets and sketch patterns.

- Apply the Shell feature before the Fillet feature, and the inside corners remain perpendicular.

- Apply cosmetic fillets and chamfers last in the modeling procedure.

- Combine fillets into as few fillet features as possible. This enables you to control fillets that need to be controlled separately, such as fillets to be removed and simplified configurations.

- Create a simplified configuration when building very complex parts or working with large assemblies.

- Use symmetry during the modeling process. Utilize feature patterns and mirroring when possible. Think End Conditions.

- Use global variables and equations to control commonly applied dimensions (design intent).

- Add comments to equations to document your design intent. Place a single quote (') at the end of the equation, then enter the comment. Anything after the single quote is ignored when the equation is evaluated.

- Avoid redundant mates. Although SOLIDWORKS allows some redundant mates (all except distance and angle), these mates take longer to solve and make the mating scheme harder to understand and diagnose if problems occur.

- Fix modeling errors in the part or assembly when they occur. Errors cause rebuild time to increase, and if you wait until additional errors exist, troubleshooting will be more difficult.

- Create a Library of Standardized notes and parts.

- Utilize the Rollback bar. Troubleshoot feature and sketch errors from the top of the design tree.

- Determine the static and dynamic behavior of mates in each sub-assembly before creating the top level assembly.

- Plan the assembly and sub-assemblies in an assembly layout diagram. Group components together to form smaller sub-assemblies.

- When you create an assembly document, the base component should be fixed, fully defined or mated to an axis about the assembly origin.

- In an assembly, group fasteners into a folder at the bottom of the FeatureManager. Suppress fasteners and their assembly patterns to save rebuild time and file size.

- When comparing mass, volume and other properties with assembly visualization, utilize similar units.

- Use limit mates sparingly because they take longer to solve and whenever possible, mate all components to one or two fixed components or references. Long chains of components take longer to solve and are more prone to mate errors.

Helpful On-line Information

The SOLIDWORKS URL: http://www.SOLIDWORKS.com contains information on Local Resellers, Solution Partners, Certifications, SOLIDWORKS users groups and more.

Access 3D ContentCentral using the Task Pane to obtain engineering electronic catalog model and part information.

Use the SOLIDWORKS Resources tab in the Task Pane to obtain access to Customer Portals, Discussion Forums, User Groups, Manufacturers, Solution Partners, Labs and more.

Helpful on-line SOLIDWORKS information is available from the following URLs:

- http://www.swugn.org/

List of all SOLIDWORKS User groups.

- https://www.solidworks.com/sw/education/certification-programs-cad-students.htm

The SOLIDWORKS Academic Certification Programs.

- http://www.solidworks.com/sw/industries/education/engineering-education-software.htm

The SOLIDWORKS Education Program.

- https://solidworks.virtualtester.com/#home_button

The SOLIDWORKS Certification Center - Virtual tester site.

*On-line tutorials are for educational purposes only. Tutorials are copyrighted by their respective owners.

SOLIDWORKS Document Types

SOLIDWORKS has three main document file types: Part, Assembly and Drawing, but there are many additional supporting types that you may want to know. Below is a brief list of these supporting file types:

Design Documents	Description
.sldprt	SOLIDWORKS Part document
.slddrw	SOLIDWORKS Drawing document
.sldasm	SOLIDWORKS Assembly document

Templates and Formats	Description
.asmdot	Assembly Template
.asmprp	Assembly Template Custom Properties tab
.drwdot	Drawing Template
.drwprp	Drawing Template Custom Properties tab
.prtdot	Part Template
.prtprp	Part Template Custom Properties tab
.sldtbt	General Table Template
.slddrt	Drawing Sheet Template
.sldbombt	Bill of Materials Template (Table-based)
.sldholtbt	Hole Table Template
.sldrevbt	Revision Table Template
.sldwldbt	Weldment Cutlist Template
.xls	Bill of Materials Template (Excel-based)

Library Files	Description
.sldlfp	Library Part file
.sldblk	Blocks

Other	Description
.sldstd	Drafting standard
.sldmat	Material Database
.sldclr	Color Palette File
.xls	Sheet metal gauge table

Project 8: Answer key

1. What is the Modulus of Elasticity?

- The slope of the Deflection-Stress curve
- **The slope of the Stress-Strain curve in its linear section**
- The slope of the Force-Deflection curve in its linear section
- The first inflection point of a Strain curve

2. What is Stress?

- A measure of power
- A measure of strain
- A measure of material strength
- **A measure of the average amount of force exerted per unit area**

3. Which of the following assumptions are true for a static analysis in SOLIDWORKS Simulation with small displacements?

- **Inertia effects are negligible and loads are applied slowly**
- The model is not fully elastic. If loads are removed, the model will not return to its original position
- **Results are proportional to loads**
- **All the displacements are small relative to the model geometry**

4. What is Yield Stress?

- **The stress level beyond which the material becomes plastic**
- The stress level beyond which the material breaks
- The strain level above the stress level which the material breaks
- The stress level beyond the melting point of the material

5. A high quality Shell element has _____ nodes.

- 4
- 5
- **6**
- 8

6. Stress σ is proportional to _____ in a Linear Elastic Material.

- **Strain**
- Stress
- Force
- Pressure

7. The Elastic Modulus (Young's Modulus) is the slope defined as _____ divided by
_____.

- Strain, Stress
- **Stress, Strain**
- Stress, Force
- Force, Area

8. Linear static analysis assumes that the relationship between loads and the induced
response is _____.

- Flat
- **Linear**
- Doubles per area
- Translational

9. In SOLIDWORKS Simulation, the Factor of Safety (FOS) calculations are based on
one of the following failure criteria.

- **Maximum von Mises Stress**
- **Maximum shear stress (Tresca)**
- **Mohr-Coulomb stress**
- **Maximum Normal stress**

10. The Yield point is the point where the material begins to deform at a faster rate than at the elastic limit. The material behaves _____ in the Plastic Range.

- Flatly
- Linearly
- **Non-Linearly**
- Like a liquid

11. What are the Degrees of Freedom (DOFs) restrained for a Solid?

- None
- **3 Translations**
- 3 Translations and 3 Rotations
- 3 Rotations

12. What are the Degrees of Freedom (DOFs) restrained for Truss joints?

- None
- **3 Translations**
- 3 Translations and 3 Rotations
- 3 Rotations

13. What are the Degrees of Freedom (DOFs) restrained for Shells and Beams?

- None
- 3 Translations
- **3 Translations and 3 Rotations**
- 3 Rotations

14. Which statements are true for Material Properties using SOLIDWORKS Simulation?

- **For solid assemblies, each component can have a different material**
- For shell models, each shell cannot have a different material and thickness
- **For shell models, the material of the part is used for all shells**
- For beam models, each beam cannot have a different material

15. A Beam element has _____ nodes (one at each end) with _____ degrees of freedom per node plus_____ node to define the orientation of the beam cross section.

- 6, 3, 1
- 3, 3, 1
- **3, 6, 1**
- None of the above

16. A Truss element has _____ nodes with _____ translational degrees of freedom per node.

- **2, 3**
- 3, 3
- 6, 6
- 2, 2

17. In general, the finer the mesh the better the accuracy of the results.

- **True**
- False

18. How does SOLIDWORKS Simulation automatically treat a Sheet metal part with uniform thickness?

- **Shell**
- Solid
- Beam
- Mixed Mesh

19. Use the mesh and displacement plots to calculate the distance between two _____ using SOLIDWORKS Simulation.

- **Nodes**
- Elements
- Bodies
- Surfaces

20. Surface models can only be meshed with _____ elements.

- **Shell**
- Beam
- Mixed Mesh
- Solid

21. The shell mesh is generated on the surface (located at the mid-surface of the shell).

- **True**
- False

22. In general, use Thin shells when the thickness-to-span ratio is less than _____.

- **0.05**
- .5
- 1
- 2

23. The model (a rectangular plate) has a length to thickness ratio of less than 5. You extracted its mid-surface to use it in SOLIDWORKS Simulation. You should use a _____.

- Thin Shell element formulation
- **Thick Shell element formulation**
- Thick or Thin Shell element formulation, it does not matter
- Beam Shell element formulation

24. The model, a rectangular sheet metal part, uses SOLIDWORKS Simulation. You should use a:

- **Thin Shell element formulation**
- Thick Shell element formulation
- Thick or Thin Shell element formulation, it does not matter
- Beam Shell element formulation

25. The Global element size parameter provides the ability to set the global average element size. SOLIDWORKS Simulation suggests a default value based on the model volume and _____ area. This option is only available for a standard mesh.

- Force

- Pressure

- **Surface**

- None of the above

26. A remote load applied on a face with a Force component and no Moment can result in: Note: Remember (DOFs restrain).

- **A Force and Moment of the face**

- A Force on the face only

- A Moment on the face only

- A Pressure and Force on the face

27. There are _____ DOFs restrain for a Solid element.

- **3**

- 1

- 6

- None

28. There are _____ DOFs restrain for a Beam element.

- 3

- 1

- **6**

- None

29. What best describes the difference(s) between a Fixed and Immovable (No translation) boundary condition in SOLIDWORKS Simulation?

- There are no differences

- There are no difference(s) for Shells but it is different for Solids

- **There is no difference(s) for Solids but it is different for Shells and Beams**

- There are only differences(s) for a Static Study

30. Can a non-uniform pressure of force be applied on a face using SOLIDWORKS Simulation?

- No

- Yes, but the variation must be along a single direction only

- **Yes. The non-uniform pressure distribution is defined by a reference coordinate system and the associated coefficients of a second order polynomial**

- Yes, but the variation must be linear

31. You are performing an analysis on your model. You select five faces, 3 edges and 2 vertices and apply a force of 20lb. What is the total force applied to the model using SOLIDWORKS Simulation?

- 100lb

- 1600lb

- 180lb

- **200lb**

32. Yield strength is typically determined at _____ strain.
- 0.1%

- **0.2%**

- 0.02%

- 0.002%

33. There are four key assumptions made in Linear Static Analysis: 1: Effects of inertia and damping are neglected, 2. The response of the system is directly proportional to the applied loads, 3: Loads are applied slowly and gradually, and_____ .

- **Displacements are very small. The highest stress is in the linear range of the stress-strain curve**

- There are no loads

- Material is not elastic

- Loads are applied quickly

34. How many degrees of freedom does a physical structure have?

- Zero

- Three - Rotations only

- Three - Translations only

- **Six - Three translations and three rotational**

35. Brittle materials have little tendency to deform (or strain) before fracture and do not have a specific yield point. It is not recommended to apply the yield strength analysis as a failure criterion on brittle material. Which of the following failure theories is appropriate for brittle materials?

- **Mohr-Columb stress criterion**

- Maximum shear stress criterion

- Maximum von Mises stress criterion

- Minimum shear stress criterion

36. You are performing an analysis on your model. You select three faces and apply a force of 40lb. What is the total force applied to the model using SOLIDWORKS Simulation?

- 40lb

- 20lb

- **120lb**

- Additional information is required

37. A material is orthotropic if its mechanical or thermal properties are not unique and independent in three mutually perpendicular directions.

- True

- **False**

38. An increase in the number of elements in a mesh for a part will:

- Decrease calculation accuracy and time

- **Increase calculation accuracy and time**

- Have no effect on the calculation

- Change the FOS below 1

39. SOLIDWORKS Simulation uses the von Mises Yield Criterion to calculate the Factor of Safety of many ductile materials. According to the criterion:

- **Material yields when the von Mises stress in the model equals the yield strength of the material**

- Material yields when the von Mises stress in the model is 5 times greater than the minimum tensile strength of the material

- Material yields when the von Mises stress in the model is 3 times greater than the FOS of the material

- None of the above

40. SOLIDWORKS Simulation calculates structural failure on:

- Buckling

- Fatigue

- Creep

- **Material yield**

41. Apply a uniform total force of 200lb on two faces of a model. The two faces have different areas. How do you apply the load using SOLIDWORKS Simulation for a Linear Static Study?

- Select the two faces and input a normal to direction force of 200lb on each face

- Select the two faces and a reference plane. Apply 100lb on each face

- **Apply equal force to the two faces. The force on each face is the total force divided by the total area of the two faces**

- None of the above

42. Maximum and Minimum value indicators are displayed on Stress and Displacement plots in SOLIDWORKS Simulation for a Linear Static Study.

- **True**

- False

43. What SOLIDWORKS Simulation tool should you use to determine the result values at specific locations (nodes) in a model using SOLIDWORKS Simulation?

- Section tool
- **Probe tool**
- Clipping tool
- Surface tool

44. What criteria are best suited to check the failure of ductile materials in SOLIDWORKS Simulation?

- Maximum von Mises Strain and Maximum Shear Strain criterion
- **Maximum von Misses Stress and Maximum Shear Stress criterion**
- Maximum Mohr-Coulomb Stress and Maximum Mohr-Coulomb Shear Strain criterion
- Mohr-Coulomb Stress and Maximum Normal Stress criterion

45. Set the scale factor for plots_____ to avoid any misinterpretation of the results, after performing a Static analysis with gap/contact elements.

- Equal to 0
- **Equal to 1**
- Less than 1
- To the Maximum displacement value for the model

46. It is possible to mesh _____ with a combination of Solids, Shells and Beam elements in SOLIDWORKS Simulation.

- **Parts and Assemblies**
- Only Parts
- Only Assemblies
- None of the above

47. SOLIDWORKS Simulation supports multi-body parts. Which of the following is a true statement?

- You can employ different mesh controls to each Solid body
- You can classify Contact conditions between multiple Solid bodies
- You can classify a different material for each Solid body
- **All of the above are correct**

48. Which statement best describes a Compatible mesh?

- A mesh where only one type of element is used
- **A mesh where elements on touching bodies have overlaying nodes**
- A mesh where only a Shell or Solid element is used
- A mesh where only a single Solid element is used

49. The Ratio value in Mesh Control provides the geometric growth ratio from one layer of elements to the next.

- **True**
- False

50. The structures displayed in the following illustration are best analyzed using:

- Shell elements
- Solid elements
- **Beam elements**
- A mixture of Beam and Shell elements

51. The structure displayed in the following illustration is best analyzed using:

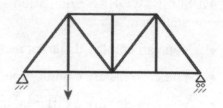

- Shell elements
- Solid elements
- **Beam elements**
- A mixture of Beam and Shell elements

52. The structure displayed in the following illustration is best analyzed using:

- **Shell elements**
- Solid elements
- Beam elements
- A mixture of Beam and Shell elements

Sheet metal model

53. The structure displayed in the following illustration is best analyzed using:

- Shell elements
- **Solid elements**
- Beam elements
- A mixture of Beam and Shell elements

54. Surface models can only be meshed with _____ elements.

- **Shell elements**
- Solid elements
- Beam elements
- A mixture of Beam and Shell elements

55. Use the _____ and _____ plots to calculate the distance between two nodes using SOLIDWORKS Simulation.

- **Mesh and Displacement**
- Displacement and FOS
- Resultant Displacement and FOS
- None of the above

56. You can simplify a large assembly in a Static Study by using the _____ or _____ options in your study.

- **Make Rigid, Fix**
- Shell element, Solid element
- Shell element, Compound element
- Make Rigid, Load element

57. A force "F" applied in a static analysis produces a resultant displacement URES. If the force is now 2x F and the mesh is not changed, then URES will:

- Double if there are no contacts specified and there are large displacements in the structure
- Be divided by 2 if contacts are specified
- The analysis must be run again to find out
- **Double if there is not source of nonlinearity in the study (like contacts or large displacement options)**

58. To compute thermal stresses on a model with a uniform temperature distribution, what type/types of study/studies are required?

- **Static only**
- Thermal only
- Both Static and Thermal
- None of these answers is correct

59. In an h-adaptive method, use smaller elements in mesh regions with high errors to improve the accuracy of results.

- **True**
- False

60. In a p-adaptive method, use elements with a higher order polynomial in mesh regions with high errors to improve the accuracy of results.

- **True**
- False

61. Where will the maximum stress be in the illustration?

- A
- B
- **C**
- D

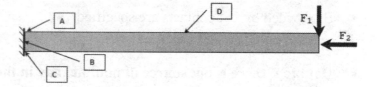

GLOSSARY

Alphabet of Lines: Each line on a technical drawing has a definite meaning and is drawn in a certain way. The line conventions recommended by the American National Standards Institute (ANSI) are presented in this text.

Alternate Position View: A drawing view superimposed in phantom lines on the original view. Utilized to show range of motion of an assembly.

Anchor Point: The origin of the Bill of Material in a sheet format.

Annotation: An annotation is a text note or a symbol that adds specific information and design intent to a part, assembly, or drawing. Annotations in a drawing include specific note, hole callout, surface finish symbol, datum feature symbol, datum target, geometric tolerance symbol, weld symbol, balloon and stacked balloon, center mark, centerline marks, area hatch and block.

ANSI: American National Standards Institute.

Area Hatch: Apply a crosshatch pattern or solid fill to a model face, to a closed sketch profile, or to a region bounded by a combination of model edges and sketch entities. Area hatch can be applied only in drawings.

ASME: American Society of Mechanical Engineering, publisher of ASME Y14 Engineering Drawing and Documentation Practices that controls drawing, dimensioning and tolerancing.

Assembly: An assembly is a document in which parts, features and other assemblies (sub-assemblies) are put together. A part in an assembly is called a component. Adding a component to an assembly creates a link between the assembly and the component. When SOLIDWORKS opens the assembly, it finds the component file to show it in the assembly. Changes in the component are automatically reflected in the assembly. The filename extension for a SOLIDWORKS assembly file name is *.sldasm.

Attachment Point: An attachment point is the end of a leader that attaches to an edge, vertex, or face in a drawing sheet.

AutoDimension: The Autodimension tool provides the ability to insert reference dimensions into drawing views such as baseline, chain, and ordinate dimensions.

Auxiliary View: An Auxiliary View is similar to a Projected View, but it is unfolded normal to a reference edge in an existing view.

AWS: American Welding Society, publisher of AWS A2.4, Standard Location of Elements of a Welding Symbol.

Axonometric Projection: A type of parallel projection, more specifically a type of orthographic projection, used to create a pictorial drawing of an object, where the object is rotated along one or more of its axes relative to the plane of projection.

Balloon: A balloon labels the parts in the assembly and relates them to item numbers on the bill of materials (BOM) added in the drawing. The balloon item number corresponds to the order in the Feature Tree. The order controls the initial BOM Item Number.

Baseline Dimensions: Dimensions referenced from the same edge or vertex in a drawing view.

Bill of Materials: A table inserted into a drawing to keep a record of the parts and materials used in an assembly.

Block: A symbol in the drawing that combines geometry into a single entity.

BOM: Abbreviation for Bill of Materials.

Broken-out Section: A broken-out section exposes inner details of a drawing view by removing material from a closed profile. In an assembly, the Broken-out Section displays multiple components.

CAD: The use of computer technology for the design of objects, real or virtual. CAD often involves more than just shapes.

Cartesian Coordinate System: Specifies each point uniquely in a plane by a pair of numerical coordinates, which are the signed distances from the point to two fixed perpendicular directed lines, measured in the same unit of length. Each reference line is called a coordinate axis or just axis of the system, and the point where they meet is its origin.

Cell: Area to enter a value in an EXCEL spreadsheet, identified by a Row and Column.

Center Mark: A cross that marks the center of a circle or arc.

Centerline: An axis of symmetry in a sketch or drawing displayed in a phantom font.

CommandManager: The CommandManager is a Context-sensitive toolbar that dynamically updates based on the toolbar you want to access. By default, it has toolbars embedded in it based on the document type. When you click a tab below the Command Manager, it updates to display that toolbar. For example, if you click the Sketch tab, the Sketch toolbar is displayed.

Component: A part or sub-assembly within an assembly.

ConfigurationManager: The ConfigurationManager is located on the left side of the SOLIDWORKS window and provides the means to create, select and view multiple configurations of parts and assemblies in an active document. You can split the

ConfigurationManager and either display two ConfigurationManager instances, or combine the ConfigurationManager with the FeatureManager design tree, PropertyManager or third party applications that use the panel.

Configurations: Variations of a part or assembly that control dimensions, display and state of a model.

Coordinate System: SOLIDWORKS uses a coordinate system with origins. A part document contains an original origin. Whenever you select a plane or face and open a sketch, an origin is created in alignment with the plane or face. An origin can be used as an anchor for the sketch entities, and it helps orient perspective of the axes. A three-dimensional reference triad orients you to the X, Y, and Z directions in part and assembly documents.

Copy and Paste: Utilize copy/paste to copy views from one sheet to another sheet in a drawing or between different drawings.

Cosmetic Thread: An annotation that represents threads.

Crosshatch: A pattern (or fill) applied to drawing views such as section views and broken-out sections.

Cursor Feedback: The system feedback symbol indicates what you are selecting or what the system is expecting you to select. As you move the mouse pointer across your model, system feedback is provided.

Datum Feature: An annotation that represents the primary, secondary and other reference planes of a model utilized in manufacturing.

Depth: The horizontal (front to back) distance between two features in frontal planes. Depth is often identified in the shop as the thickness of a part or feature.

Design Table: An Excel spreadsheet that is used to create multiple configurations in a part or assembly document.

Detail View: A portion of a larger view, usually at a larger scale than the original view. Create a detail view in a drawing to display a portion of a view, usually at an enlarged scale. This detail may be of an orthographic view, a non-planar (isometric) view, a section view, a crop view, an exploded assembly view or another detail view.

Detailing: Detailing refers to the SOLIDWORKS module used to insert, add and modify dimensions and notes in an engineering drawing.

Dimension Line: A line that references dimension text to extension lines indicating the feature being measured.

Dimension Tolerance: Controls the dimension tolerance values and the display of non-integer dimensions. The tolerance types are *None, Basic, Bilateral, Limit, Symmetric, MIN, MAX, Fit, Fit with tolerance* or *Fit (tolerance only)*.

Dimension: A value indicating the size of the 2D sketch entity or 3D feature. Dimensions in a SOLIDWORKS drawing are associated with the model, and changes in the model are reflected in the drawing, if you DO NOT USE DimXpert.

Dimensioning Standard - Metric: - ASME standards for the use of metric dimensioning require all the dimensions to be expressed in millimeters (mm). The (mm) is not needed on each dimension, but it is used when a dimension is used in a notation. No trailing zeros are used. The Metric or International System of Units (S.I.) unit system in drafting is also known as the Millimeter, Gram Second (MMGS) unit system.

Dimensioning Standard - U.S: - ASME standard for U.S. dimensioning uses the decimal inch value. When the decimal inch system is used, a zero is not used to the left of the decimal point for values less than one inch, and trailing zeros are used. The U.S. unit system is also known as the Inch, Pound, Second (IPS) unit system.

DimXpert for Parts: A set of tools that applies dimensions and tolerances to parts according to the requirements of the ASME Y.14.41-2009 standard.

DimXpertManager: The DimXpertManager lists the tolerance features defined by DimXpert for a part. It also displays DimXpert tools that you use to insert dimensions and tolerances into a part. You can import these dimensions and tolerances into drawings. DimXpert is not associative.

Document: In SOLIDWORKS, each part, assembly, and drawing is referred to as a document, and each document is displayed in a separate window.

Drawing Sheet: A page in a drawing document.

Drawing Template: A document that is the foundation of a new drawing. The drawing template contains document properties and user-defined parameters such as sheet format. The extension for the drawing template filename is .DRWDOT.

Drawing: A 2D representation of a 3D part or assembly. The extension for a SOLIDWORKS drawing file name is .SLDDRW. Drawing refers to the SOLIDWORKS module used to insert, add, and modify views in an engineering drawing.

Edit Sheet Format: The drawing sheet contains two modes. Utilize the Edit Sheet Format command to add or modify notes and Title block information. Edit in the Edit Sheet Format mode.

Edit Sheet: The drawing sheet contains two modes. Utilize the Edit Sheet command to insert views and dimensions.

eDrawing: A compressed document that does not require the referenced part or assembly. eDrawings are animated to display multiple views in a drawing.

Empty View: An Empty View creates a blank view not tied to a part or assembly document.

Engineering Graphics: Translates ideas from design layouts, specifications, rough sketches, and calculations of engineers & architects into working drawings, maps, plans and illustrations which are used in making products.

Equation: Creates a mathematical relation between sketch dimensions, using dimension names as variables, or between feature parameters, such as the depth of an extruded feature or the instance count in a pattern.

Exploded view: A configuration in an assembly that displays its components separated from one another.

Export: The process to save a SOLIDWORKS document in another format for use in other CAD/CAM, rapid prototyping, web or graphics software applications.

Extension Line: The line extending from the profile line indicating the point from which a dimension is measured.

Extruded Cut Feature: Projects a sketch perpendicular to a Sketch plane to remove material from a part.

Face: A selectable area (planar or otherwise) of a model or surface with boundaries that help define the shape of the model or surface. For example, a rectangular solid has six faces.

Family Cell: A named empty cell in a Design Table that indicates the start of the evaluated parameters and configuration names. Locate Comments in a Design Table to the left or above the Family Cell.

Fasteners: Includes Bolts and nuts (threaded), Set screws (threaded), Washers, Keys, and Pins to name a few. Fasteners are not a permanent means of assembly such as welding or adhesives.

Feature: Features are geometry building blocks. Features add or remove material. Features are created from 2D or 3D sketched profiles or from edges and faces of existing geometry.

FeatureManager: The FeatureManager design tree located on the left side of the SOLIDWORKS window provides an outline view of the active part, assembly, or drawing. This makes it easy to see how the model or assembly was constructed or to examine the various sheets and views in a drawing. The FeatureManager and the Graphics window are dynamically linked. You can select features, sketches, drawing views and construction geometry in either pane.

First Angle Projection: In First Angle Projection the Top view is looking at the bottom of the part. First Angle Projection is used in Europe and most of the world. However, America and Australia use a method known as Third Angle Projection.

Fully defined: A sketch where all lines and curves in the sketch, and their positions, are described by dimensions or relations, or both, and cannot be moved. Fully defined sketch entities are shown in black.

Foreshortened radius: Helpful when the centerpoint of a radius is outside of the drawing or interferes with another drawing view: Broken Leader.

Foreshortening: The way things appear to get smaller in both height and depth as they recede into the distance.

French curve: A template made out of plastic, metal or wood composed of many different curves. It is used in manual drafting to draw smooth curves of varying radii.

Fully Defined: A sketch where all lines and curves in the sketch, and their positions, are described by dimensions or relations, or both, and cannot be moved. Fully defined sketch entities are displayed in black.

Geometric Tolerance: A set of standard symbols that specify the geometric characteristics and dimensional requirements of a feature.

Glass Box method: A traditional method of placing an object in an *imaginary glass box* to view the six principle views.

Global Coordinate System: Directional input refers by default to the Global coordinate system (X-, Y- and Z-), which is based on Plane1 with its origin located at the origin of the part or assembly.

Graphics Window: The area in the SOLIDWORKS window where the part, assembly, or drawing is displayed.

Grid: A system of fixed horizontal and vertical divisions.

Handle: An arrow, square or circle that you drag to adjust the size or position of an entity such as a view or dimension.

Heads-up View Toolbar: A transparent toolbar located at the top of the Graphic window.

Height: The vertical distance between two or more lines or surfaces (features) which are in horizontal planes.

Hidden Lines Removed (HLR): A view mode. All edges of the model that are not visible from the current view angle are removed from the display.

Hidden Lines Visible (HLV): A view mode. All edges of the model that are not visible from the current view angle are shown gray or dashed.

Hole Callouts: Hole callouts are available in drawings. If you modify a hole dimension in the model, the callout updates automatically in the drawing if you did not use DimXpert.

Hole Table: A table in a drawing document that displays the positions of selected holes from a specified origin datum. The tool labels each hole with a tag. The tag corresponds to a row in the table.

Import: The ability to open files from other software applications into a SOLIDWORKS document. The A-size sheet format was created as an AutoCAD file and imported into SOLIDWORKS.

Isometric Projection: A form of graphical projection, more specifically, a form of axonometric projection. It is a method of visually representing three-dimensional objects in two dimensions, in which the three coordinate axes appear equally foreshortened and the angles between any two of them are 120°.

Layers: Simplifies a drawing by combining dimensions, annotations, geometry and components. Properties such as display, line style and thickness are assigned to a named layer.

Leader: A solid line created from an annotation to the referenced feature.

Line Format: A series of tools that control Line Thickness, Line Style, Color, Layer and other properties.

Local (Reference) Coordinate System: Coordinate system other than the Global coordinate system. You can specify restraints and loads in any desired direction.

Lock Sheet Focus: Adds sketch entities and annotations to the selected sheet. Double-click the sheet to activate Lock Sheet Focus. To unlock a sheet, right-click and select Unlock Sheet Focus or double click inside the sheet boundary.

Lock View Position: Secures the view at its current position in the sheet. Right-click in the drawing view to Lock View Position. To unlock a view position, right-click and select Unlock View Position.

Mass Properties: The physical properties of a model based upon geometry and material.

Menus: Menus provide access to the commands that the SOLIDWORKS software offers. Menus are Context-sensitive and can be customized through a dialog box.

Model Item: Provides the ability to insert dimensions, annotations, and reference geometry from a model document (part or assembly) into a drawing.

Model View: A specific view of a part or assembly. Standard named views are listed in the view orientation dialog box such as isometric or front. Named views can have a user-defined name for a specific view.

Model: 3D solid geometry in a part or assembly document. If a part or assembly document contains multiple configurations, each configuration is a separate model.

Motion Studies: Graphical simulations of motion and visual properties with assembly models. Analogous to a configuration, they do not actually change the original assembly model or its properties. They display the model as it changes based on simulation elements you add.

Mouse Buttons: The left, middle, and right mouse buttons have distinct meanings in SOLIDWORKS. Use the middle mouse button to rotate and Zoom in/out on the part or assembly document.

Oblique Projection: A simple type of graphical projection used for producing pictorial, two-dimensional images of three-dimensional objects.

OLE (Object Linking and Embedding): A Windows file format. A company logo or EXCEL spreadsheet placed inside a SOLIDWORKS document are examples of OLE files.

Ordinate Dimensions: Chain of dimensions referenced from a zero ordinate in a drawing or sketch.

Origin: The model origin is displayed in blue and represents the (0,0,0) coordinate of the model. When a sketch is active, a sketch origin is displayed in red and represents the (0,0,0) coordinate of the sketch. Dimensions and relations can be added to the model origin but not to a sketch origin.

Orthographic Projection: A means of representing a three-dimensional object in two dimensions. It is a form of parallel projection, where the view direction is orthogonal to the projection plane, resulting in every plane of the scene appearing in affine transformation on the viewing surface.

Parametric Note: A Note annotation that links text to a feature dimension or property value.

Parent View: A Parent view is an existing view on which other views are dependent.

Part Dimension: Used in creating a part, they are sometimes called construction dimensions.

Part: A 3D object that consist of one or more features. A part inserted into an assembly is called a component. Insert part views, feature dimensions and annotations into 2D drawing. The extension for a SOLIDWORKS part filename is .SLDPRT.

Perspective Projection: The two most characteristic features of perspective are that objects are drawn smaller as their distance from the observer increases and foreshortened: the size of an object's dimensions along the line of sight are relatively shorter than dimensions across the line of sight.

Plane: To create a sketch, choose a plane. Planes are flat and infinite. Planes are represented on the screen with visible edges.

Precedence of Line Types: When obtaining orthographic views, it is common for one type of line to overlap another type. When this occurs, drawing conventions have established an order of precedence.

Precision: Controls the number of decimal places displayed in a dimension.

Projected View: Projected views are created for Orthogonal views using one of the following tools: Standard 3 View, Model View or the Projected View tool from the View Layout toolbar.

Properties: Variables shared between documents through linked notes.

PropertyManager: Most sketch, feature, and drawing tools in SOLIDWORKS open a PropertyManager located on the left side of the SOLIDWORKS window. The PropertyManager displays the properties of the entity or feature so you specify the properties without a dialog box covering the Graphics window.

RealView: Provides a simplified way to display models in a photo-realistic setting using a library of appearances and scenes. RealView requires graphics card support and is memory intensive.

Rebuild: A tool that updates (or regenerates) the document with any changes made since the last time the model was rebuilt. Rebuild is typically used after changing a model dimension.

Reference Dimension: Dimensions added to a drawing document are called Reference dimensions, and are driven; you cannot edit the value of reference dimensions to modify the model. However, the values of reference dimensions change when the model dimensions change.

Relation: A relation is a geometric constraint between sketch entities or between a sketch entity and a plane, axis, edge or vertex.

Relative view: The Relative View defines an Orthographic view based on two orthogonal faces or places in the model.

Revision Table: The Revision Table lists the Engineering Change Orders (ECO), in a table form, issued over the life of the model and the drawing. The current Revision letter or number is placed in the Title block of the Drawing.

Right-Hand Rule: Is a common mnemonic for understanding notation conventions for vectors in 3 dimensions.

Rollback: Suppresses all items below the rollback bar.

Scale: A relative term meaning "size" in relationship to some system of measurement.

Section Line: A line or centerline sketched in a drawing view to create a section view.

Section Scope: Specifies the components to be left uncut when you create an assembly drawing section view.

Section View: You create a section view in a drawing by cutting the parent view with a cutting, or section line. The section view can be a straight cut section or an offset section defined by a stepped section line. The section line can also include concentric arcs. Create a Section View in a drawing by cutting the Parent view with a section line.

Sheet Format: A document that contains the following: page size and orientation, standard text, borders, logos, and Title block information. Customize the Sheet format to save time. The extension for the Sheet format filename is .SLDDRT.

Sheet Properties: Sheet Properties display properties of the selected sheet. Sheet Properties define the following: Name of the Sheet, Sheet Scale, Type of Projection (First angle or Third angle), Sheet Format, Sheet Size, View label, and Datum label.

Sheet: A page in a drawing document.

Silhouette Edge: A curve representing the extent of a cylindrical or curved face when viewed from the side.

Sketch: The name to describe a 2D profile is called a sketch. 2D sketches are created on flat faces and planes within the model. Typical geometry types are lines, arcs, corner rectangles, circles, polygons, and ellipses.

Spline: A sketched 2D or 3D curve defined by a set of control points.

Stacked Balloon: A group of balloons with only one leader. The balloons can be stacked vertically (up or down) or horizontally (left or right).

Standard views: The three orthographic projection views, Front, Top and Right, positioned on the drawing according to First angle or Third angle projection.

Suppress: Removes an entity from the display and from any calculations in which it is involved. You can suppress features, assembly components, and so on. Suppressing an entity does not delete the entity; you can unsuppress the entity to restore it.

Surface Finish: An annotation that represents the texture of a part.

System Feedback: Feedback is provided by a symbol attached to the cursor arrow indicating your selection. As the cursor floats across the model, feedback is provided in the form of symbols riding next to the cursor.

System Options: System Options are stored in the registry of the computer. System Options are not part of the document. Changes to the System Options affect all current and future documents. There are hundreds of Systems Options.

Tangent Edge: The transition edge between rounded or filleted faces in hidden lines visible or hidden lines removed modes in drawings.

Task Pane: The Task Pane is displayed when you open the SOLIDWORKS software. It contains the following tabs: SOLIDWORKS Resources, Design Library, File Explorer, Search, View Palette, Document Recovery and RealView/PhotoWorks.

Templates: Templates are part, drawing and assembly documents that include user-defined parameters and are the basis for new documents.

Third Angle Projection: In Third angle projection the Top View is looking at the Top of the part. First Angle Projection is used in Europe and most of the world. America and Australia use the Third Angle Projection method.

Thread Class or Fit: Classes of fit are tolerance standards; they set a plus or minus figure that is applied to the pitch diameter of bolts or nuts. The classes of fit used with almost all bolts sized in inches are specified by the ANSI/ASME Unified Screw Thread standards (which differ from the previous American National standards).

Thread Lead: The distance advanced parallel to the axis when the screw is turned one revolution. For a single thread, lead is equal to the pitch; for a double thread, lead is twice the pitch.

Tolerance: The permissible range of variation in a dimension of an object. Tolerance may be specified as a factor or percentage of the nominal value, a maximum deviation from a nominal value, an explicit range of allowed values, be specified by a note or published standard with this information, or be implied by the numeric accuracy of the nominal value.

Toolbars: The toolbar menus provide shortcuts enabling you to access the most frequently used commands. Toolbars are Context-sensitive and can be customized through a dialog box.

T-Square: A technical drawing instrument, primarily a guide for drawing horizontal lines on a drafting table. It is used to guide the triangle that draws vertical lines. Its name comes from the general shape of the instrument where the horizontal member of the T slides on the side of the drafting table. Common lengths are 18", 24", 30", 36" and 42".

Under-defined: A sketch is under defined when there are not enough dimensions and relations to prevent entities from moving or changing size.

Units: Used in the measurement of physical quantities. Decimal inch dimensioning and Millimeter dimensioning are the two types of common units specified for engineering parts and drawings.

Vertex: A point at which two or more lines or edges intersect. Vertices can be selected for sketching, dimensioning, and many other operations.

View Palette: Use the View Palette, located in the Task Pane, to insert drawing views. It contains images of standard views, annotation views, section views, and flat patterns (sheet metal parts) of the selected model. You can drag views onto the drawing sheet to create a drawing view.

Weld Bead: An assembly feature that represents a weld between multiple parts.

Weld Finish: A weld symbol representing the parameters you specify.

Weld Symbol: An annotation in the part or drawing that represents the parameters of the weld.

Width: The horizontal distance between surfaces in profile planes. In the machine shop, the terms length and width are used interchangeably.

Zebra Stripes: Simulate the reflection of long strips of light on a very shiny surface. They allow you to see small changes in a surface that may be hard to see with a standard display.

Index